Lecture Notes in Computer Scie

Edited by G. Goos, J. Hartmanis, and J. van L

T0237716

Springer
Berlin
Heidelberg
New York
Barcelona
Hong Kong
London
Milan
Paris
Tokyo

Wil van der Aalst   Eike Best (Eds.)

Applications and Theory of Petri Nets 2003

24th International Conference, ICATPN 2003
Eindhoven, The Netherlands, June 23-27, 2003
Proceedings

 Springer

Series Editors

Gerhard Goos, Karlsruhe University, Germany
Juris Hartmanis, Cornell University, NY, USA
Jan van Leeuwen, Utrecht University, The Netherlands

Volume Editors

Wil M.P. van der Aalst
Eindhoven University of Technology
Faculty of Technology and Management
P.O. Box 513, 5600 MB Eindhoven, The Netherlands
E-mail: w.m.p.v.d.aalst@tm.tue.nl

Eike Best
University of Oldenburg
Computer Science Department
Parallel Systems Group
26111 Oldenburg, Germany
E-mail: eike.best@informatik.uni-oldenburg.de

Cataloging-in-Publication Data applied for

A catalog record for this book is available from the Library of Congress.

Bibliographic information published by Die Deutsche Bibliothek
Die Deutsche Bibliothek lists this publication in the Deutsche Nationalbibliografie;
detailed bibliographic data is available in the Internet at <http://dnb.ddb.de>.

CR Subject Classification (1998): F.1-3, C.1-2, G.2.2, D.2, D.4, J.4

ISSN 0302-9743
ISBN 3-540-40334-5 Springer-Verlag Berlin Heidelberg New York

Springer-Verlag Berlin Heidelberg New York
a member of BertelsmannSpringer Science+Business Media GmbH

http://www.springer.de

© Springer-Verlag Berlin Heidelberg 2003
Printed in Germany

Typesetting: Camera-ready by author, data conversion by PTP-Berlin GmbH
Printed on acid-free paper   SPIN: 10927571   06/3142   5 4 3 2 1 0

Preface

This volume contains the proceedings of the 24th International Conference on Application and Theory of Petri Nets (ICATPN 2003). The aim of the Petri net conferences is to create a forum for discussing progress in the application and theory of Petri nets.

Typically, the conferences have 100-150 participants – one third of these coming from industry while the rest are from universities and research institutions. The conferences always take place in the last week of June.

The conference and a number of other activities are coordinated by a steering committee with the following members: Jonathan Billington (Australia), Giorgio De Michelis (Italy), Susanna Donatelli (Italy), Serge Haddad (France), Kurt Jensen (Denmark), Maciej Koutney (UK), Sadatoshi Kumagai (Japan), Tadao Murata (USA), Carl Adam Petri (Germany; honorary member), Wolfgang Reisig (Germany), Grzegorz Rozenberg (The Netherlands; chairman), and Manuel Silva (Spain).

The 2003 conference was organized by the Information Systems (IS) and Information & Technology (I&T) research groups of the Technische Universiteit Eindhoven (TU/e), Eindhoven, The Netherlands. We would like to thank the members of the program committee and the reviewers (see next page) for their enormous efforts in selecting 24 papers from the 74 papers submitted. We received contributions from 30 countries distributed over three categories: theory papers (35 submitted, 12 accepted), application papers (31 submitted, 8 accepted), and tool presentation papers (8 submitted, 4 accepted). This volume comprises the papers that were accepted for presentation. Invited lectures were given by M. Ajmone Marsan, E. Brinksma, J.M. Colom, C. Ellis, K. Jensen, and S. Miyano (whose papers are included in this volume). Moreover, a paper on the standardization of a Petri net exchange format is included (as suggested by the steering committee).

The conference was held in conjunction with the International Conference on Business Process Management: On the Application of Formal Methods to "Process-Aware" Information Systems (BPM 2003). BPM 2003 and ICATPN 2003 were supported by the following institutions and organizations: KNAW, NWO, Deloitte & Touche, Pallas Athena, BETA, Philips, Gemeente Eindhoven, Sodexho, OCE, FileNet, TU/e, SIGPAM, EMISA, and Atos Origin.

We gratefully acknowledge the considerable (technical) support of Eric Verbeek and Harro Wimmel, and the help of the people in the IS and I&T research groups involved in the organization of ICATPN 2003 and BPM 2003. Finally, we would like to mention the excellent cooperation with Springer-Verlag during the preparation of this volume.

April 2003   Wil van der Aalst and Eike Best

Yann Etesse
Giuliana Franceschinis
Rossano Gaeta
Roberto Gorrieri
Marco Gribaudo
Luuk Groenewegen
Xudong He
Keijo Heljanko
Jane Hillston
Kunihiko Hiraishi
Lawrence Holloway
Hendrik Jan Hoogeboom
András Horváth
Zhaoxia Hu
Jarle Hulaas
David Hurzeler
Le Van Huu
Jean Michel Ilié
John Jeffrey
Jens Bæk Jørgensen
Gabriel Juhás
Jorge Julvez
Victor Khomenko
Fabrice Kordon
Walter Kosters
Maciej Koutny
Lars Kristensen
Narayan Kumar
Timo Latvala
Kamal Lodaya
Robert Lorenz
Thomas Mailund
Marko Mäkelä
Daniele Manini
Dan Marinescu
José Meseguer
Vesna Milijic
Andrew Miner
Toshiyuki Miyamoto
Patrice Moreaux
Rémi Morin
Isabelle Mounier
Christian Neumair
Mogens Nielsen
Atsushi Ohta

Andrea Omicini
Paritosh Pandya
Emmanuel Paviot-Adet
Elisabeth Pelz
Marco Peña
Laure Petrucci
Claudia Picardi
Michele Pinna
Denis Poitrenaud
Agata Pólrola
Lucia Pomello
Franck Pommereau
Luigi Portinale
Jean-Francois Pradat-Peyre
Ram Ramanujam
Laura Recalde
Jose Rolim
Claus Schröter
Matteo Sereno
Radu Siminiceanu
Carla Simone
Pawel Sobocinski
Jeremy Sproston
Alin Stefanescu
Jason Steggles
Ichiro Suzuki
Maciej Szreter
Shigemasa Takai
Enrique Teruel
Dorothea Tippe
Shengru Tu
Toshimitsu Ushio
Kimmo Varpaaniemi
Jose Luis Villarroel Salcedo
Todd Wareham
Lisa Wells
Józef Winkowski
Bozena Wozna
Haiping Xu
Shingo Yamaguchi
Hideki Yamasaki
Yi Zhou
Gianluigi Zavattaro
Du Zhang

Table of Contents

Invited Papers

Full Papers

Tool Papers

PNML Paper

Author Index

Coloured Petri Nets: Status and Outlook

Kurt Jensen

CPN Group
Department of Computer Science
University of Aarhus, Denmark
kjensen@daimi.au.dk
www.daimi.au.dk/~kjensen/

Abstract. This talk presents my personal view of the current status of high-level Petri Nets, in particular Coloured Petri Nets. What have we achieved over the last 25 years and what are the main challenges that need to be addressed in order to make high-level Petri nets more useful, e.g., to the software industry? What kind of language extensions and tool support do we need to make Petri net modelling, simulation and analysis accessible to the knowledgeable engineer? I will also briefly discuss the main areas which I believe are the most obvious candidates for further theoretical development.

References

More information on Coloured Petri Nets can be found in the following web-pages, papers and books.

1. Home-page of CPN Group, University of Aarhus: www.daimi.au.dk/CPnets/.
2. L.M. Kristensen, S. Christensen and K. Jensen: *The Practitioner's Guide to Coloured Petri Nets.* In K. Jensen (ed.): *Special Section on Coloured Petri Nets.* Int. Journal on Software Tools for Technology Transfer, 2/2 (1998), Springer Verlag, 98–132. Also available from: www.daimi.au.dk/~kjensen/.
3. K. Jensen: *Coloured Petri Nets. Basic Concepts, Analysis Methods and Practical Use. Volume 1, Basic Concepts.* Monographs in Theoretical Computer Science, Springer-Verlag, 1992. ISBN: 3-540-60943-1.
4. K. Jensen: *Coloured Petri Nets. Basic Concepts, Analysis Methods and Practical Use. Volume 2, Analysis Methods.* Monographs in Theoretical Computer Science, Springer-Verlag, 1994. ISBN: 3-540-58276-2.
5. K. Jensen: *Coloured Petri Nets. Basic Concepts, Analysis Methods and Practical Use. Volume 3, Practical Use.* Monographs in Theoretical Computer Science, Springer-Verlag, 1997. ISBN: 3-540-62867-3.
6. K. Jensen: *A Brief Introduction to Coloured Petri Nets.* In: E. Brinksma (ed.): Tools and Algorithms for the Construction and Analysis of Systems. Proceeding of the TACAS'97 Workshop, Enschede, The Netherlands 1997, Lecture Notes in Computer Science Vol. 1217, Springer-Verlag 1997, 201–208. Also available from: www.daimi.au.dk/~kjensen/.
7. K. Jensen: *An Introduction to the Theoretical Aspects of Coloured Petri Nets.* In: J.W. de Bakker, W.-P. de Roever, G. Rozenberg (eds.): A Decade of Concurrency, Lecture Notes in Computer Science Vol. 803, Springer-Verlag 1994, 230–272. Also available from: www.daimi.au.dk/~kjensen/.

W.M.P. van der Aalst and E. Best (Eds.): ICATPN 2003, LNCS 2679, pp. 1–2, 2003.

8. K. Jensen: *An Introduction to the Practical Use of Coloured Petri Nets*. In: W. Reisig and G. Rozenberg (eds.): Lectures on Petri Nets II: Applications, Lecture Notes in Compute Science Vol. 1492, Springer-Verlag 1998, 237–292. Also available from: ww.daimi.au.dk/~kjensen/.
9. K. Jensen (ed.): *Special Section on Coloured Petri Nets*. Int. Journal on Software Tools for Technology Transfer, 2/2 (1998), Springer Verlag, 95–191.
10. K. Jensen (ed.): *Special Section on Practical Use of High-level Petri Nets*. Int. Journal on Software Tools for Technology Transfer,3/4 (2001), Springer Verlag, 369–430.
11. K. Jensen: *Condensed State Spaces for Symmetrical Coloured Petri Nets*. Formal Methods in System Design 9 (1996), Kluwer Academic Publishers, 7–40.
12. S. Christensen, K. Jensen, and T. Mailund: *State Space Methods for Timed Petri Nets*. In: H. Weber, H. Ehrig and W. Reisig: Proceedings of 2nd International Colloquium on Petri Net Technologies for Modelling Communication Based Systems, Berlin, Germany, September 14–15, 2001, 33–42.
13. K. Jensen (ed.): *Workshop on Practical Use of Coloured Petri Nets and Design/CPN*. Proceedings, Department of Computer Science, University of Aarhus, Denmark, PB-532, 1998. Also available from: www.daimi.au.dk/CPnets/.
14. K. Jensen (ed.): *Second Workshop on Practical Use of Coloured Petri Nets and Design/CPN*. Proceedings, Department of Computer Science, University of Aarhus, Denmark, PB-541, 1999. Also available from: www.daimi.au.dk/CPnets/.
15. K. Jensen (ed.): *Practical Use of High-level Petri Net*. Proceedings, Department of Computer Science, University of Aarhus, Denmark, PB-547, 2000. Also availble from: www.daimi.au.dk/CPnets.
16. K. Jensen (ed.): *Third Workshop and Tutorial on Practical Use of Coloured Petri Nets and the CPN Tools*. Proceedings, Department of Computer Science, University of Aarhus, Denmark, PB-554, 2001. Also available from: www.daimi.au.dk/CPnets/.
17. K. Jensen (ed.): *Fourth Workshop and Tutorial on Practical Use of Coloured Petri Nets and the CPN Tools*. Proceedings, Department of Computer Science, University of Aarhus, Denmark, PB-560, 2002. Also available from: www.daimi.au.dk/CPnets/.

Towards Biopathway Modeling and Simulation

Hiroshi Matsuno[1], Sachie Fujita[2], Atsushi Doi[2],
Masao Nagasaki[3], and Satoru Miyano[3]

[1] Faculty of Science, Yamaguchi University,
Yoshida, Yamaguchi 753-8512, Japan
`matsuno@sci.yamaguchi-u.ac.jp`
[2] Graduate School of Science and Engineering, Yamaguchi University
Yoshida, Yamaguchi 753-8512, Japan
`{fujita, atsushi}@ib.sci.yamaguchi-u.ac.jp`
[3] Human Genome Center, Institute of Medical Science, University of Tokyo
4-6-1 Shirokanedai, Minato-ku, Tokyo 108-8639, Japan
`{masao, miyano}@ims.u-tokyo.ac.jp`

Abstract. Petri net has been employed for modeling metabolic pathways as well as signal transduction pathways and gene regulatory networks. The purpose of this paper is to introduce an extension of Petri net called hybrid functional Petri net (HFPN) which allows us to model biopathways naturally and effectively. The method for creating biopathways with HFPN is demonstrated through a well-known biopathway example "*lac* operon gene regulatory mechanism and glycolytic pathway." In order to evaluate this biopathway model, simulations of five mutants of the *lac* operon are carried out by Genomic Object Net which is a biopathway simulator based on the HFPN architecture. The software Genomic Object Net and the HFPN files presented in this paper can be downloaded from `http://www.GenomicObject.Net/`.

1 Introduction

The notion of Petri net enhanced with its graphical representation creates a high affinity for the understanding of biological systems. It describes a system as a network consisting of places, transitions, and arcs that connect them. Then places contain tokens as their contents, and transitions are active components which can control the contents of places. By assigning biological molecules or stimuli to places and by defining biological reactions as transitions, Petri net can turn to be a model of a biological system. In particular, most biopathways can be abstracted with this notion.

In the literature, Petri net has been employed for modeling metabolic pathways, and qualitative properties of these pathways were discussed [20], where places represent biochemical compounds (metabolites), transitions represent biochemical reactions between metabolites which are usually catalyzed by a certain enzyme, and tokens indicate the presence of compounds. Quantitative simulations by using self-modified Petri net [23] was also made successfully [9], where the number of tokens in a place is used to represent the level of concentrations of the corresponding compound.

W.M.P. van der Aalst and E. Best (Eds.): ICATPN 2003, LNCS 2679, pp. 3–22, 2003.
© Springer-Verlag Berlin Heidelberg 2003

These Petri nets are discrete in the sense that places can contain nonnegative integers as tokens. On the other hand, it is sometimes more reasonable to regard a biopathway as a hybrid system of discrete and continuous events. We have shown that the gene regulatory network of λ phage can be more naturally modeled as a hybrid system of "discrete" and "continuous" dynamics [17] by employing hybrid Petri net (HPN) [2] and its modification [4]. It is also observed [6] that biopathways can be handled as hybrid systems, e.g. protein concentration dynamics behaves continuously being coupled with discrete switches; protein production is switched on or off depending on the expression of other genes, i.e. presence or absence of other proteins in sufficient concentration.

This paper introduces an extension of hybrid Petri net called hybrid functional Petri net (HFPN) with which we can model biopathways more smoothly. Then we demonstrate how biopathway modeling can be effectively conducted with HFPN by taking, as an example, the *lac* operon gene regulatory mechanism and glycolytic pathway. For this modeling and simulation, we used a software Genomic Object Net (GON) [11,19] that we have developed by realizing the notion of HFPN together with the GUI specially designed for biopathway modeling. HFPN is defined by slightly but inevitably extending HPN so that metabolic pathways, gene regulatory networks and signal transduction pathways may be intuitively modeled.

These extensions and usages are not enough to cover all features in biological systems, but the notion of Petri net would a very promising architecture for biopathway modeling and simulation, if extensions/modifications could be made so that they reflect the way of thinking of biologists and their intuitions and insights are appreciated in the process of modeling and simulation.

Some practical evidences have been be provided for the usefulness of this notion of HFPN together with the usage of GON. First, in addition to the biopathway modeled in this paper, we have modeled with GON (a) the signal transduction pathway for apoptosis induced by the protein Fas [11], (b) the boundary formation by Notch signaling pathway in *Drosophila* multicellular systems [18], (c) the gene regulatory network of the *Drosophila* circadian rhythms [11], (d) the gene regulatory network of λ phage, and (e) KEGG pathway models are converted to GON XML files for re-modeling and simulation. All these models are available from GON web site [11]. Second, we can give a graphical HFPN description of a system of ordinary differential equations (ODEs) which has been widely used as a representation method in biological simulation systems such as BioDrive [13], Gepasi [15], and E-Cell [22]. In particular, E-Cell can be realized as a subset of GON and we developed a program to convert rule and reactor files of E-Cell to GON XML files except for the C++ programs in E-Cell (this should be done by hand conversion only if necessary) [16]. Third, since HFPN has a graphical representation, our intuitive understanding of the whole system will be enchanced even if the system constitutes a large network of cascades. Generally speaking, in the modeling method based on ODEs, it is difficult to observe the whole system intuitively. On the other hand, GON provides an editor for HFPN which can manipulate biopathway models like pictures. As is emphasized in the design concept of the GUI of BioSPICE [10], the understandability of biopathways is important for modeling and the approach of HFPN with GON realizes this idea in a similar way. The difference between BioSPICE and GON is that GON equips the engine for simulations. The gene regulatory network of λ phage is also modeled [7] with QSIM simulation environment [12], where LISP is used to describe the model. This is a mathematically beautiful representation but the understandability from the biological viewpoint would remain as another challenge.

In the following section, we define the notion of HFPN in an informal and intuitive way. Then we show how to create the HFPN model of the *lac* operon gene regulatory mechanism and glycolytic pathway in details. This would help readers to model another biopathways of their interests. The model is evaluated with GON by simulation. We deal with various mutants in the *lac* operon for comparison with the wild type.

2   Modeling Methods

2.1   Hybrid Functional Petri Net

This section introduces the notion of *hybrid functional Petri net* (HFPN) which consititues the basic architecture of Genomic Object Net (GON)[1].

Petri net is a network consisting of *place*, *transition*, *arc*, and *token*. A place can hold tokens as its content. A transition has arcs coming from places and arcs going out from the transition to some places. A transition with these arcs defines a *firing rule* in terms of the contents of the places where the arcs are attached.

Hybrid Petri net (HPN) [2] has *discrete place* and *continuous place* as places and *discrete transition* and *continuous transition* as transitions. A discrete place and a discrete transition are the same notions as used in the traditional *discrete Petri net* [21]. A continuous place can hold a nonnegative real number as its content. A continuous transition fires continuously in an HPN and its firing speed is given as a function of values in the places. The graphical notations of a discrete transition, a discrete place, a continuous transition, and a continuous place are shown in Fig. 1, together with three types of arcs. A specific value is assigned to each arc as a weight. When a *normal arc* with weight w is attached to a discrete transition, w tokens are transferred through the normal arc. On the other hand, when a normal arc is attached to a continuous transition, the amount of token that flows is determined by the firing speed of the continuous transition. An *inhibitory arc* with weight w enables the transition to fire only if the content of the place at the source of the arc is less than or equal to w. For example, an inhibitory arc can be used to represent repressive activity in gene regulation. A *test arc* does not consume any content of the place at the source of the arc by firing. For example, a test arc can be used to represent enzyme activity, since the enzyme itself is not consumed.

The discrete Petri net has been extended so that the content of a place can be used as a parameter for the formula describing the weight on the arc from the place that represents the threshold and consumption of tokens for firing [23, 9]. With this extention, biochemical processes are well modeled [9,5,8].

---

[1] GON is realized with *hybrid functional Petri net with extension* which is enhanced with more types for places (integer, real, boolean, string, vector).

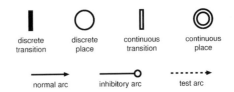

Fig. 1. Basic elements of hybrid (functional) Petri net.

By the definitions of the hybrid Petri net [2] and its modification [4], the firing speed of a continuous transition must be the same as the consuming speed through each arc from its its source place and the contents of all source places are consumed with the same speed. This speed also must be the same as the production speed through each arc from the transition. This is a severe drawback of HPN for biopathway modeling. For example, consider a reaction in which a dimer is cleaved to two monomers.

From these considerations, we define a new kind of transition called a *functional continuous transition* so that any functions can be assigned to arc/transition controling the speed/condition of consumption/production/firing. Then we define HFPN by following the restrictions arising from the hybridity of discrete and continuous features in HPN. The formal definition of HFPN requires a long tedius description and we omit it here. Further necessary details of HFPN will be explained along with the modeling process.

Another feature of HFPN which is not used in the modeling in this paper is hierarchization. The hierarchical representation has been introduced in the hybrid object net [4]. HFPN also inherits this hierarchization capability which is realized in GON. An example of hierarchization can be found in [17].

2.2 *lac* Operon Gene Regulatory Mechanism and Glycolytic Pathway

This section demonstrates how we can model the *lac* operon gene regulatory mechanism and glycolytic pathway in *E. coli* with HFPN by using the biological knowledge in the literature [1,14,24]. The HFPN modeling will start with "transcription control switch" (Fig. 3), then it will be expanded by adding "positive regulation" (Fig. 4), "negative regulation" (Fig. 5), and "hydrolyzing lactose to glucose and galactose" (Fig. 7).

All the parameters in the transitions of the HFPN model are summarized in Tables 1 and 2. We shall show a rough guideline for the usage of transitions in this modeling.

Continuous transition

The firing speed of a continuous transition is given as a simple arithmetic formula such as mX/a, $mX \times mY$, $(mX + mY)/(mX + a \times mY)$, etc., where mX, mY are variables representing the contents of the input places going into the transition and a is a constant parameter that will be manually tuned. With this speed, the contents of input places will be consumed and simultaneously each output place will receive the amount through the arc.

To each continuous place representing the concentration of some substance, a continuous transition is attached to model the degradation of the substance (e.g. $T1$). The degradation rate is set to be $mX/10$ for high speed degradation (e.g. $T7$) and $mX/10000$ for low speed degradation (e.g. $T8$).

The default weight of the arc going into a continuous transition is set to be "0". If necessary, an appropriate weight is assigned to the arc according to the underlying biological knowledge (e.g. $T67$, $T80$). From the definition, no weight is assigned to the arc outgoing from a continuous transition.

Discrete transition

The default delay time of a discrete transition is set to be "1" and the default weight of an arc from/to a discrete transition is "1". If necessary, the delay time and the weight are appropriately chosen according to the underlying biological knowledge (e.g. $T63$).

Functional continuous transition

A functional continuous transition is used for $T59$ which describes a reaction converting four monomers to one tetramer (see Fig. 6 (c)).

Initial values of places

The places which have initial values greater than zero are listed in Table 3.

2.3   Transcription Control Switch

We will sketch how places, transitions and arcs are determined in the process of modeling.

Fig. 2 shows the structure of the *lac* operon.

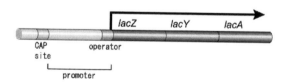

Fig. 2. The *lac* operon: The enzyme β-galactosidase, the product of the *lacZ* gene, hydrolyzes lactose to glucose and galactose. *lacY* encodes the permease that brings lactose into the cell, and *lacA* encodes an acetylase that is believed to detoxify thioglactosides, which, along with lactose, are transported into cells by *lacY*. The "operator" lies within the "promoter", and the "CAP site" lies just upstream of the promoter.

In the absence of lactose, a repressor is bound to the operator. The repressor prevents RNA polymerase from starting RNA synthesis at the promoter. On the other hand, in the presence of lactose and the absence of glucose, the catabolite activator protein (CAP) is bound to the CAP site. Since the CAP helps RNA polymerase to bind the promoter, the transcription of the *lac* operon begins in this situation.

This regulation mechanism can be expressed by the Petri net of Fig. 3 consisting of only discrete elements.

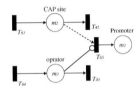

Fig. 3. HFPN representation of the control mechanism of the *lac* operon transcription switch. Only discrete elements are used for representing the switching mechanism.

The place "promoter" ($m1$) represents the status of the transcription of the *lac* operon. That is, if this place contains token(s), the *lac* operon is being transcribed, but if this place has no token, transcription of the operon does not begin. The rates of releasing CAP and repressor from the DNA are assigned to the transitions $T42$ and $T43$ as the delay time, respectively. The production rates of CAP and repressor are also assigned to the transitions $T63$ and $T64$, respectively, as the delay time.

Each time the transition $T65$ fires, the place "promoter" gets one token. This transition fires when both of the following conditions hold; (1) The place "CAP site" ($m2$) contains tokens (this is the case that the protein CAP is bound to the CAP site). (2) The place "operator" ($m3$) has no token (this is the case that the repressor is not bound to the operator).

2.4   Positive Regulation

The DNA binding of CAP is regulated by glucose to ensure that the transcription of the *lac* operon begins only in the absence of glucose. In fact, glucose starvation induces an increase in the intracellular levels of cyclic AMP (cAMP). Transcription of the *lac* operon is activated with the help of CAP to which cAMP binds. When glucose is plentiful, the cAMP level gets down; cAMP therefore dissociates from CAP, which reverts to an inactive form that can no longer bind DNA. This regulatory mechanism by CAP and cAMP is called *positive regulation* of the *lac* operon. Fig. 4 shows an HFPN model which represents positive regulation of the *lac* operon.

Continuous places are used for representing the concentrations of the substances CAP, cAMP, AMP, ADP, and glucose. Tokens in the places "CAP" ($m4$) and "cAMP" ($m5$) should not be consumed by the firing of the transition $T63$,

Table 1. Transitions in Figure 7. All transitions in this figure are listed in "Name" column. The symbol D or C in "Type" column represents the type of transition, discrete transition or continuous transition, respectively. In "Delay/Speed" column, the firing speed of continuous transition or the delay time of discrete transition is described according to the type of transition. The column "From", which represents incoming arc(s) to a transition, is divided into three sub-columns, "variable" (variable name(s) of the place(s) attached to the incoming arc(s)), "weight" (weight of the incoming arc(s)), and "type" (N, T, and I represent normal, test, and inhibitory arcs, respectively). The column "To", which represents outgoing arc(s) from a transition, is divided into two sub-columns, "variable" (variable name(s) of the place(s) attached to the outgoing arc(s)) and "weight" (weight of the outgoing arc(s)).

Name	Type	Delay / Speed	From variable	weight	type	To variable	weight	Comment
T1	C	$m4/10$	m4	0	N	—	—	degradation rate of CAP
T2	C	$m14/10$	m14	0	N	—	—	degradation rate of mRNA repressor
T3	C	$m15/10$	m15	0	N	—	—	degradation rate of repressor
T4	C	$m17/10$	m17	0	N	—	—	degradation rate of repressor binding to DNA
T5	C	$m18/10$	m18	0	N	—	—	degradation rate of repressor not binding to DNA
T6	C	$m7/10$	m7	0	N	—	—	degradation rate of repressor binding to operator region
T7	C	$m19/10$	m19	0	N	—	—	degradation rate of lacZ mRNA
T8	C	$m20/10000$	m20	0	N	—	—	degradation rate of LacZ
T9	C	$m23/10$	m23	0	N	—	—	degradation rate of lacY mRNA
T10	C	$m24/10000$	m24	0	N	—	—	degradation rate of LacY
T11	C	$m27/10$	m27	0	N	—	—	degradation rate of lacA mRNA
T12	C	$m28/10000$	m28	0	N	—	—	degradation rate of LacA
T13	C	$m29/10000$	m29	0	N	—	—	degradation rate of lactose outside of a cell
T14	C	$m9/10000$	m9	0	N	—	—	degradation rate of lactose
T15	C	$m8/2$	m8	0	N	—	—	degradation rate of arolactose
T16	C	$m30/10000$	m30	0	N	—	—	degradation rate of galactose
T17	C	$m17/10000$	m17	0	N	—	—	degradation rate of glucose
T18	C	$m10/10000$	m10	0	N	—	—	degradation rate of complex
T19	C	$m5/10000$	m5	0	N	—	—	degradation rate of cAMP
T20	C	$m11/10000$	m11	0	N	—	—	degradation rate of AMP
T21	C	$m12/10000$	m12	0	N	—	—	degradation rate of ADP
T42	D	1	m2	1	N	—	—	CAP releasing rate
T43	D	1	m3	1	N	—	—	repressor releasing rate
T45	C	1	—	—	N	m4	—	CAP production rate
T46	D	1	—	—		m13	1	activation of repressor gene
T57	D	1.082	m13	1	N	m14	1	transcription rate of repressor
T58	C	$m14$	m14	0	N	m15	—	translation rate of repressor mRNA
T59	C	—	m15		C†	m16	—	conformation rate of repressor
T60	C	$96 \times m16/100$	m16	0	N	m7	—	repressor binding rate to operator
T61	C	$399 \times m16/10000$	m16	0	N	m17	—	repressor binding rate to the DNA other than repressor site
T62	C	$m16/10000$	m16	0	N	m18	—	rate of repressor do not bind any DNA
T63	D	1	m4 m5	1 100	T T	m2	1	binding rate of CAP to the CAP site
T64	D	1	m7 m8	1 4	T I	m3	1	binding rate of repressor to the operon

† During $m15 > 0$, the amount flow into the transition $T59$ in the speed $(4 \times m15)/5$ from the place "repressor" and the amount flow from the transition $T59$ to the place "4 repressor" in the speed $m15/5$.

Table 2. Transitions in Figure 7 (continued)

Name	Type	Delay / Speed	From			To		Comment
			variable	weight	type	variable	weight	
T65	D	1	$m2$	1	T	$m1$	1	logical operation of the places
			$m3$	1	I			"CAP site" and "operator"
T66	D	3.075	$m1$	1	T	$m19$	1	transription rate of $lacZ$
						$m21$	1	
T67	C	$m19$	$m19$	1	T	$m20$	—	translation rate of $lacZ$
T68	D	0.051	$m21$	1	N	$m22$	1	moving rate of RNA polymerase
T69	D	1.254	$m22$	1	N	$m23$	1	transription rate of $lacY$
						$m25$	1	
T70	C	$m23/2$	$m23$	1	T	$m24$	—	translation rate of $lacY$
T71	D	0.065	$m25$	1	N	$m26$	1	moving rate of RNA polymerase
T72	D	0.682	$m26$	1	N	$m27$	1	transription rate of $lacA$
T73	C	$m27/5$	$m27$	1	T	$m28$	—	translation rate of $lacA$
T74	C	$\frac{m24 \times m29}{m29 + m24 \times 10}$	$m29$	0	N	$m9$	—	transforming rate of lactose
			$m24$	2.5	T			into a cell
T75	C	$\frac{m20 \times m9}{m9 + m20 \times 10}$	$m9$	0	N	$m30$	—	decomposing rate of lactose
			$m20$	5	T	$m6$		to galactose and glucose
T76	C	$m9/5$	$m9$	1	T	$m8$	—	producing rate of arolactose from lactose inside of a cell
T77	D	0.5	$m3$	1	N	$m10$	1	conformation rate of
			$m8$	1	N			repressor and arolactose
T78	C	$m8 \times m16$	$m8$	4	N	$m10$	1	conformation rate of
			$m16$	1	N			repressor and arolactose
T79	C	$m5/10$	$m5$	0	N	$m11$	—	reaction rate: cAMP to AMP
T80	C	$m11/10$	$m11$	0	N	$m5$	—	reaction rate: AMP to cAMP
			$m6$	5	I			
T81	C	$m11/10$	$m11$	0	N	$m12$	—	reaction rate: AMP to ADP
T82	C	$m12/10$	$m12$	0	N	$m11$	—	reaction rate: ADP to AMP
T94	C	$m29/10$	$m29$	0	T	$m8$	—	producing rate of arolactose from lactose outside of a cell

Table 3. The places having non-zero initial values in Figure 7.

Name	variable	initial value	Comment
CAP	$m4$	5	concentration of CAP
cAMP	$m5$	100	concentration of cAMP
glucose	$m6$	50	concentration of glucose
AMP	$m11$	200	concentration of AMP
ADP	$m12$	200	concentration of ADP
LacZ	$m20$	5	concentration of LacX
LacY	$m24$	2.5	concentration of LacY
LacA	$m28$	1	concentration of LacA
lactose outside of a cell	$m29$	50	lactose outside of a cell

since both of CAP and cAMP are not lost by forming a complex of these two substances. Then, two test arcs are used from the places "CAP" and "cAMP" to the transition $T63$. The weight of the arc from the place cAMP to the transition $T63$ is set to 100, while the weight of the arc from the place "CAP" to that transition is 1, which was determined by manual tuning and referring to the simulation results. After both of the concentrations of CAP and cAMP exceed the thresholds which are given to these two arcs as weights, the transition $T63$ can fire, transferring a token from the transition $T63$ to the place "CAP site."

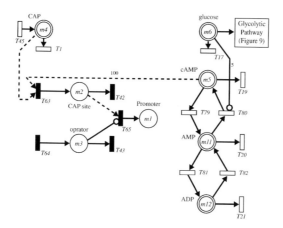

Fig. 4. Positive regulation mechanism: This figure properly contains the HFPN model of Fig. 3.

In general, reactions among cAMP, AMP, and ADP are reversible. The transition $T80$ (the transition $T82$) between the places "cAMP" ($m5$) and "AMP" ($m11$) (the places "AMP" and "ADP" ($m12$)) represents the reversible reaction together with the transition $T79$ (the transition $T81$). To the places "cAMP", "AMP", and "ADP", 1, 200, and 200 are assigned as initial values, respectively.

Recall that when glucose is plentiful, the cAMP level gets down. This phenomenon is represented by the inhibitory arc from the place "glucose" ($m6$) to the transition $T80$. When the concentration in the place "glucose" exceeds the threshold given at this inhibitory arc, the transition $T80$ stops its firing.

In this model, since we suppose that the CAP is produced continuously, this production mechanism is modeled with the place "CAP" and the transitions $T45$ and $T1$. The place "CAP" contains five tokens as an initial value, since CAP is produced by the production mechanism which is independent of the mechanism described here.

Finally, the transitions $T1$, $T17$, $T19$, $T20$, and $T21$ represent the natural degradation of the corresponding substances. Since these transitions are used for representing degradation and do not produce any, no arcs are going out from these transitions.

2.5   Negative Regulation

In the presence of lactose, a small sugar molecule called allolactose is formed in a cell. Allolactose binds to the repressor protein, and when it reaches a high enough concentration, transcription is turned on by decreasing affinity of the repressor protein for the operator site. The repressor protein is the product of the *lacI* gene which is located upstream of the *lac* operon. Actually, after forming a tetramer, the repressor protein can bind the operator site.

By adding this negative regulation mechanism to Fig. 4, we obtain Fig. 5. Since it is known from the literature [14] that repressor should be produced sufficiently prior to the production of other substances, parameters relating to this negative regulation are set faster than other parameters.

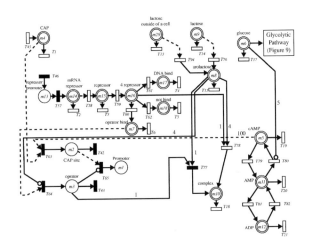

Fig. 5. Negative regulation mechanism: This figure properly contains the HFPN model of Fig. 4.

In our model, a discrete place is used for representing the promoter site of a gene. When the discrete place "repressor promoter"($m13$) gets a token, the transcription of the *lacI* gene begins. We can determine the transcription frequency by the delay rate of the transition $T46$. Transcription and translation mechanisms are modeled by the places "mRNA repressor" ($m14$) and "repressor" ($m15$) and the transitions $T57$, $T2$, $T58$, and $T3$.

The reaction composing a tetramer from monomers (Fig. 6 (a)) can be represented by HPN as is shown in Fig. 6 (b). By comparing this representation with the representation of Fig. 6 (c), we can recognize that HFPN allows us to represent such reaction naturally and intuitively. Tetramer formation is represented in the HFPN by places "repressor" and "4 repressor" ($m16$) and three transitions $T3$, $T59$, and $T60$. Actually, the function $\frac{4 \times m15}{5}$ $\left(\frac{m15}{5}\right)$ is assigned to the input (output) arc to (from) the transition $T59$ as the flow speed. Note that the speed of the input arc is four times faster than the speed of the output arc.

For the repressor forming tetramer, we determined that about 96% of them bind to operator site, about 3.99% of them bind to the other DNA sites, and about 0.01% of them do not bind to DNA. These percentages are determined based on the literature [14]. The places "operator bind" ($m7$), "DNA bind" ($m17$), and "not bind" ($m18$) represent the amount of these repressors. Accord-

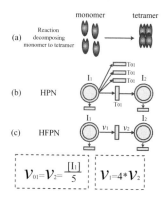

Fig. 6. HPN and HFPN representations of the reaction composing monomers to a tetramer. The speed of v_1 is four times faster than that of v_2. v_{01} is the firing speed of the transition T_{01}. $[I_1]$ ($[I_2]$) represents the content of the place I_1 (I_2).

ing to these binding rates, the firing speeds of the transitions $T60$, $T61$, and $T62$ were given by $\frac{96 \times m16}{100}$, $\frac{399 \times m16}{10000}$, and $\frac{m16}{10000}$, respectively.

We separate the concentration of lactose to two places "lactose outside of a cell" ($m29$) and "lactose" ($m9$) for the convenience of describing the function of the $lacY$ gene in the next subsection. The concentration of the arolactose is represented by the place "arolactose" ($m8$) whose accumulation rate is given by the transitions $T94$ and $T76$. It is known [14] that arolactose is produced from the lactose existing outside of a cell as well as the lactose inside a cell. Since it is natural to consider that a production speed of arolactose is faster than a passing rate of arolactose through the cell membrane, the speed of transition $T76$ ($m9/5$) is set to be faster than the speed of transition $T94$ ($m29/10$).

The negative regulation in Fig. 5 is realized in the following way: The place "operator" gets tokens if

- the concentration of the place "operator bind" exceeds the threshold 1 given at the test arc to the transition $T64$ as a weight, and
- the concentration of the place "arolactose" does not exceed the threshold 4 given at the inhibitory arc to the transition $T64$.

Note that the threshold values 1 and 4 are determined according to the observation that four arolactose molecules are required for one tetramer of repressor proteins to bind at the operator site.

The transition $T78$ gives the complex forming late of the arolactose and the tetramer of repressor proteins. The place "complex" ($m10$) represents the concentration of the complex. Arolactose can also release the tetramer of repressor protein from the operator site by forming a complex with it. The transition $T77$ and the arcs from/to the transition realize this mechanism. Discrete transitions are used for the transitions $T77$, since only a discrete transition is available for the arc from the discrete place (continuous amount can not be removed from

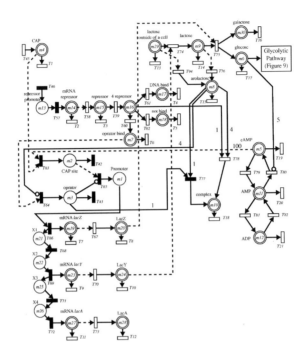

Fig. 7. HFPN modeling of the *lac* operon gene regulatory mechanism. This figure properly contains the HFPN model in Fig. 5.

a discrete place). In order to realize smooth removal of arolactose, small delay time (0.5) is assigned to the transition $T77$.

The transitions $T4$, $T5$, $T6$, $T13$, $T14$, $T15$ and $T18$ represent the natural degradation of the corresponding substances.

2.6   Hydrolyzing Lactose to Glucose and Galactose

The *lac* operon transcription and translation mechanisms are described in Fig. 7 together with the effects of two products of the genes *lacZ* and *lacY* on hydrolyzing lactose to glucose and galactose. The effect of the gene *lacA* is not included in this figure, since the gene *lacA* does not work in both of the *lac* operon gene regulatory mechanism and glycolytic pathway.

The places "mRNA lacZ" ($m19$), "mRNA lacY" ($m23$), and "mRNA lacA" ($m27$) represent the concentrations of mRNAs transcribed from the genes *lacZ*, *lacY*, and *lacA*, respectively. The transcription rates are given at the discrete transitions $T66$, $T69$, and $T72$ as the delay time.

The places "LacZ" ($m20$), "LacY" ($m24$), and "LacA" ($m28$) represent the concentrations of proteins translated from the *lacZ*, *lacY*, and *lacA* mRNAs, and the translation rates are given at the continuous transitions $T67$, $T70$, and $T73$. As is shown in Table 3, we set the initial values of proteins LacZ, LacY, and LacA as 5, 2.5, 1, respectively. These values are chosen according to the production ratios of LacZ, LacY, and LacA proteins [14]. Actually, the formulas $m19$, $\frac{m23}{2}$, and $\frac{m27}{5}$ are assigned to the transitions $T67$, $T70$, and $T73$ as their speeds, according to the fact that the proteins of $lacZ$, $lacY$, and $lacA$ are produced in the ratio $1 : \frac{1}{2} : \frac{1}{5}$.

Degradation rates of mRNAs (proteins) are assigned to the transitions $T7$, $T9$, and $T11$ ($T8$, $T10$, and $T12$).

In this model, to represent the lac operon DNA, only discrete elements, discrete transitions $T66$, $T68$, $T69$, $T71$, and $T72$, and discrete places X1 ($m21$), X2 ($m22$), X3 ($m25$), and X4 ($m26$), are used. The discrete places X1 (X3) represents the Boolean status of the transcription of $lacZ$ gene ($lacY$ gene). That is, each time transcription of $lacZ$ ($lacY$) is finished, the place X1 (X3) gets a token. At the discrete transition $T68$ ($T71$), the delay time 0.051 (0.065), which is required for RNA polymerase moving from the end of $lacZ$ gene to the beginning of $lacY$ gene (the end of $lacY$ gene to the beginning of $lacA$ gene), is assigned. The delay times 3.075 and 1.254 are assigned to the transitions $T66$ and $T69$, respectively, according to the fact that the length of $lacZ$ gene ($lacY$ gene) is 3075bp (1254bp). The delay time 0.682 at the transition $T72$ represents the length of $lacA$ gene (the length of $lacA$ gene is 682bp). The lengths of the genes are obtained from the website [3]. Note that, we can know the transcription status of the gene $lacY$ (the gene $lacA$) by observing whether the discrete places X2 (X4) contains token(s) or not.

Recall that the product of the gene $lacZ$ is an enzyme which hydrolyzes lactose to glucose and galactose. This reaction is modeled by using the places "lactose", "galactose" ($m30$), and "glucose", and the transitions $T75$, $T16$, and $T17$. In our model, 20 is assigned to each of the places "lactose" and "lactose outside of a cell" as an initial value. Test arc is used from the place "lacZ" to the transition $T75$, since the enzyme is not consumed. We consider that the production rates of glucose and lactose depend on both of the concentration of lactose and the concentration of product of $lacZ$ gene. The formula $\frac{m24 \times m29}{m29 + m24 \times 10}$ representing the speed of the transition $T74$ reflects this idea.

We mentioned that the gene $lacY$ encodes the permease that brings lactose into the cell. In Fig. 7, this function is realized with the places "LacY", "lactose outside of a cell", and "lactose", and the transitions $T74$, $T13$, and $T14$. Since the product of $lacY$ gene is an enzyme, test arc is used from the place "LacY" to the transition $T74$, and the speed of this transition is given by the formula $\frac{m20 \times m9}{m9 + m20 \times 10}$ according to the same idea above.

The weight 2.5 (5) of the arc from the place LacY ($m24$) (LacZ ($m20$)) to the transition $T77$ ($T75$) corresponds to the basal concentration of LacY (LacZ) presented in Table 3.

2.7   Glycolytic Pathway

In the glycolytic pathway, a glucose is converted into two molecules of pyruvate, where the enzymatic reactions shown in Fig. 8 and Table 4 are involved. Fig. 9 shows the HFPN model of the glycolytic pathway.

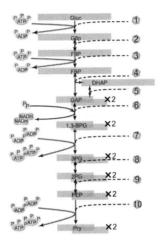

Fig. 8. A part of the glycolytic pathway participating glucose (Gluc), glucose 6-phosphate (G6P), fructose 6-phosphate (F6P), fructose 1,6- diphosphate (FBP), dihydroxyacetone phosphate (DHAP), glyceraldehyde 3-phosphate (GAP), 1,3-diphosphoglycerate (1,3-BPG), 3-phosphoglycerate (3PG), 2-phosphoglycerate (2PG), phosphoenolpyruvate (PEP), and pyruvate (Pry).

Table 4. Mapping between numbers in Fig. 8 and metabolites in reactions

Index	Enzyme / Reaction	Index	Enzyme / Reaction
1	hexokinase	2	phosphoglucose isomerase
3	phosphofructokinase	4	aldolase
5	triosephosphate isomerase	6	glyceraldehyde-3-phosphate dehydrogenase
7	phosphoglycerate kinase	8	phosphoglycerate mutase
9	enolase	10	pyruvate kinase
11	lactate dehydrogenase		

Main pathway from glucose to pyruvate acid

First we create continuous places corresponding to glucose ($m6$), intermediates ($m31$–$m39$), and pyruvate ($m40$). Default continuous places are introduced at this step though these places are meant to represent the concentrations of the corresponding substrates. Then, by following the pathway in Fig. 8, we put continuous transitions ($T83$–$T93$) together with normal arcs between two consecutive places in the pathway. These transitions and arcs shall represent the reactions, but default transitions and arcs are initially introduced without any parameter tuning. By considering the natural degradation of substrates, we put continuous transitions $T17$ for glucose and $T23$–$T31$ for intermediates with normal arcs. By taking into account the fact that natural degradation is very slow in glycolysis, each of the firing speed of these transitions is given by the formula $mX/10000$ for $X = 17, 23, 24, \ldots, 31$.

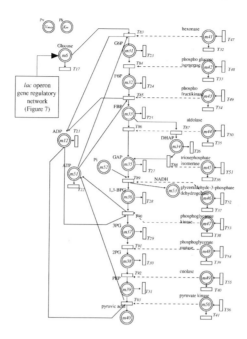

Fig. 9. HFPN model of glycolytic pathway. Michaelis-Menten's equation is applied to the reactions in the main pathway from glucose to pyruvic acid. Test arcs are used to represent enzyme reactions.

Production of ATP from ADP and NADH from phosphoric acid

Next we consider ADP, ATP, Pi (phosphoric acid) and NADH. In the pathway shown in Fig. 8, two ADP molecules and two Pi's are invested to produce four ATP molecules and two NADH molecules. Continuous places "ADP" ($m12$), "ATP" ($m51$), "Pi" ($m52$, initial value=200), and "NADH" ($m53$) are created to represent ADP, ATP, Pi and NADH. We attach continuous transitions $T21$ and $T22$ representing the natural degradation of ADP and ATP. Their firing speeds are set to be very slow ($T21 : m21/10000$, $T22 : m22/10000$) as is set for intermediates in the above. For the process of ATP→ADP in the reaction (1), the normal arc from the place "ATP" to the transition $T83$ and the normal arc from the transition $T83$ to the place "ADP" are introduced. In the same way, normal arcs connected to $T85$ are introduced for the process of ATP→ADP in the reaction (3). Similarly, to represent reactions (7) and (10), transitions $T90$ and $T93$ are used, respectively. Reaction (6) (Pi→NADH) is realized with the transition $T89$. Since ATP and ADP degrade slowly, the formulas $m22/10000$ and $m21/10000$ are assigned to the transitions $T22$ and $T21$ as the degradation speed, respectively.

Installing enzyme reactions

Places having variables $m41$, $m42$,...,$m50$ represent enzyme concentrations. Initial values of these variables are set to 5. The production rates of these enzymes are assigned to the transitions $T47$,$T48$,...,$T56$ whose speeds are set to 1. The transitions $T32$, $T33$,...,$T41$ have the firing speeds $mX/10$'s ($X = 41, 42, ..., 50$), which represent the speed of natural degradation of these enzymes.

Test arcs are used for enzyme reactions, since an enzyme itself is not consumed in the reaction. To each of these test arcs, the value 3 is chosen as a weight of the arc.

Reaction speeds in the main pathway

To represent the reaction speeds in the main pathway, we adopt the Michaelis-Menten equation such as

$$\frac{V_{max}[S]}{K_m + [S]},$$

where $[S]$ is the substance concentration, V_{max} is the maximum reaction speed, and K_m is a Michaelis constant. In our model, we let $V_{max} = 1$ and $K_m = \frac{1}{2}$. The two independent places "Pv" and "Pk" in Fig. 9 represent these variables V_{max} and K_m. The values of V_{max} and K_m can be easily manipulated by changing the contents of the places Pv and Pk, respectively. The Michaelis-Menten equation is used for representing each of the firing speeds of the transitions $T83$, $T84$,..., and $T93$. For example, for the transition $T84$, the formula $\frac{V_{max}m31}{K_m+m31}$ is used.

ADP molecules and two Pi's are invested to produce four ATP molecules and two NADH molecules.

For each of the following two reactions, an HFPN model is created in the following way by assigning different functions to the arcs entering and leaving the transition which represents the reaction:

1. One fructose 1,6-diphosphate (FBP) molecule is invested to produce two glyceraldehyde 3-phosphate (GAP) molecules;
   - "FBP" → $T86$: $\frac{V_{max}m33}{K_m+m33}$ and
   - $T86$ → "GAP" : $\frac{2 \times V_{max}m33}{K_m+m33}$.
2. One dihydroxyacetone phosphate (DHAP) is invested to produce two GAP molecules;
   - "DHAP" → $T88$: $\frac{V_{max}m34}{K_m+m34}$ and
   - $T88$ → "GAP" : $\frac{2 \times V_{max}m34}{K_m+m34}$.

This demonstrates the effectiveness of representing biopathways with HFPN.

3   Evaluation

3.1   Simulation Results of Mutants by GON

The five mutants taken into consideration in this simulation are listed below together with the realization methods of the mutants in the HFPN model of Fig. 8.

lacZ⁻ a mutant which can not produce β-galactosidase,
 – delete the transition *T*69,

lacY⁻ a mutant which can not produce β-galactoside permease,
 – delete the transition *T*70,

lacI⁻ a mutant in which 4 repressor monomers can not constitute one active repressor tetramer,
 – delete the transition *T*59,

lacIˢ a mutant to which arolactose can not bind,
 – delete the transitions *T*77 and *T*78 together with the inhibitory arc from the place "arolactose" to the transition *T*64,

lacI⁻ᵈ a mutant which can not bind to the DNA,
 – delete the transitions *T*60 and *T*61.

Behavior of the wild type

In both of Fig. 10 and Fig. 11, the concentration behaviors of lactose (outside of a cell), lactose, glucose, LacZ (β-galactosidase), and LacY (β-galactoside permease) are shown. From the beginning, glucose is degraded, since it is consumed in the glycolytic pathway. At time point 55, the glucose was consumed, the transcription of the *lac* operon begins, producing the LacZ protein (time=60), and successively, LacY protein begins to be produced. By comparing the concentration behavior of lactose and lactose (outside of a cell), we can recognize that the LacY protein works well. At time point 65, the concentration of LacZ exceeds 10, the decomposition of lactose to glucose and galactose starts, and the concentration of glucose increases again. Just after the glucose is once again completely consumed, the transcription of the *lac* operon is stopped, keeping the concentration of LacZ and LacY proteins at some levels (from the assumption that the degradation speed of these proteins is very show) [1,14].

Behavior of the lac repressor mutant

Fig. 10 shows the simulation results of the mutants *lacI⁻*, *lacIˢ*, and *lacI⁻ᵈ* obtained from GON. In the *lacI⁻* and *lacI⁻ᵈ* mutants, LacZ protein and LacY protein are produced, while these proteins are not produced in the *lacIˢ* mutant. Furthermore, in the *lacI⁻* and *lacI⁻ᵈ* mutants, the concentrations of LacZ and LacY proteins keep growing (except the period of glucose re-production), even after stopping the decomposition of lactose. Note that these simulation results support the experimental observation [1,14].

Behavior of the lac operon mutant

Fig. 11 shows the simulation results of the mutants *lacZ⁻* and *lacY⁻* obtained from GON. From this figure, we can observe that, in the *lacZ⁻* mutant, once glucose is completely consumed, it is never produced again. On the other hand, in the *lacY⁻* mutant, the concentration of lactose (inside of a cell) never grows, since a lactose can not pass a cell membrane in this mutant. Note that these observations from the simulation results also support the experimental observation [1,14].

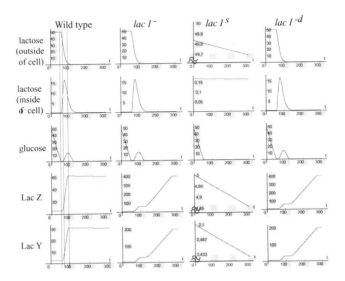

Fig. 10. Simulation results of lac repressor mutant. Concentrations of proteins LacZ and LacY keep growing, since the mutants $lacI^-$ and $lacI^{-d}$ lose abilities to bind at the operator site. In the mutant $lacI^s$, the transcription of the *lac* operon does not begin, since the repressor can not be removed from the operator site.

4   Conclusion

In this paper, we demonstrated how to build up HFPNs in modeling biopathways as the *lac* operon gene regulatory mechanism and glycolytic pathway as an example. This example was modeled and simulated with GON and the results of five mutants of the *lac* operon gene were shown, which correspond well to the facts described in the literature. Although this result is a well known biological fact, with GON, we have succeeded in discovering one unknown biological phenomenon in multicellular systems [18].

We should emphasize that, essentially, any differential equations can be modeled with HFPN. This means that GON has the potential to simulate biopathway models for other biosimulation tools such as E-Cell and Gepashi. With GON, we have succeeded in modeling some kinds of biopathways [11]. However, at the same time, we have also recognized that the current notion of HFPN is still insufficient to model more sophisticated biopathways including more complex information such as localization, cell interaction, etc. This is one of the reasons to motivate us to develop a new software "Genomic Object Net ver.1.0" (GON 1.0) [19]. GON 1.0 employs the notion of *hybrid functional Petri net with extension* (HFPNe) which allows more "types" for places (integer, real, boolean, string, vector) with which complex information such as localization, etc., can be handled. Furthermore, HFPNe can define a hybrid system of continuous and discrete events together with hierarchization of objects for intuitive creation of complex

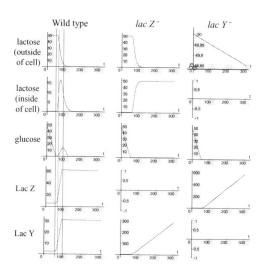

Fig. 11. Simulation results of $lacZ^-$ and $lacY^-$ mutants. We can see that the glucose is not produced in the $lacZ^-$ mutant and the lactose can not enter a cell in the $lacY^-$ mutant.

objects. The news of GON 1.0 including the release date will be announced in the webpage at http://www.GenomicObject.Net/.

References

1. Alberts, B., Bray, D., Lewis, J., Raff, M., Roberts, K., Watson, J.: The Molecular Biology of the Cell, Third Edition. Garland Publishing, Inc. (1994)
2. Alla, H., David, R.: Continuous and hybrid petri nets. J. Circuits, Systems, and Computers **8** (1998) 159–188
3. Colibri (*Escherichia coli* genolist browser)
   http://genolist.pasteur.fr/Colibri/genome.cgi.
4. Drath, R.: Hybrid object nets: An object oriented concept for modeling complex hybrid systems. Proc. Hybrid Dynamical Systems, 3rd International Conference on Automation of Mixed Processes, ADPM'98 (1998) 437–442
5. Genrich, H., Küffner, R., Voss, K.: Executable petri net models for the analysis of metabolic pathways. International Journal on Software Tools for Technology Transfer **3** (2001) 394–404
6. Ghosh, R., Tomlin, C.J.: Lateral inhibition through Delta-Notch signaling: a piecewise affine hybrid model. Proc. 4th International Workshop on Hybrid Systems: Computation and Control, Lecture Notes in Computer Science **2034** (2001) 232–246
7. Heidtke, K.R., Schulze-Kremer, S.: Design and implementation of qualitative simulation model of λ phage infection. Bioinformatics **14** (1998) 81–91
8. Heiner, M., Koch, I., Voss, K.: Analysis and simulation of steady states in metabolic pathways with petri nets. Third Workshop and Tutorial on Practical Use of colored Petri Nets and CPN Tools, Universität Aarhus, DAIMI PB–554 (2001) 15–34

 9. Hofestädt, R., Thelen, S.: Quantitative modeling of biochemical networks. In Silico Biology **1** (1998) 39-53
10. http://www.biospice.org.
11. http://www.GenomicObject.Net/.
12. Kuipers, B.: Qualitative reasoning. Modeling and simulation with incomplete knowledge. MIT Press, Cambridge, MA (1994)
13. Kyoda, K., Muraki, M., Kitano, H.: Construction of a generalized simulator for multi-cellular organisms and its application to smad signal transduction. Pacific Symposium on Biocomputing **5** (2000) 317-329
14. Lewin, B.: Genes VI. Oxford University Press and Cell Press. (1997)
15. Mendes, P.: GEPASI: a software for modeling the dynamics, steady states and control of biochemical and other systems. Comput. Appl. Biosci. **9** (1993) 563-571
16. Matsui, M., Doi, A., Matsuno, H., Hirata, Y., Miyano, S.: Biopathway model conversion from E-CELL to Genomic Object Net. Genome Informatics **12** (2001) 290-291
17. Matsuno, H., Doi, A., Nagasaki, M., Miyano, S.: Hybrid petri net representation of gene regulatory network. Pacific Symposium on Biocomputing **5** (2000) 338-349
18. Matsuno, H., Murakami, R., Yamane, R., Yamasaki, N., Fujita, S., Yoshimori, H., Miyano, S.: Boundary formation by notch signaling in *Drosophila* multicellular systems: Experimental observations and a gene network modeling by Genomic Object Net. Pacific Symposium on Biocomputing **8** (2003) 152-163
19. Nagasaki, M., Doi, A., Matsuno, H., Miyano, S.: Genomic Object Net: a platform for modeling and simulating biopathways. submitted for publication.
20. Reddy, V., Mavrovouniotis, M., Liebman, M.: Petri net representations in metabolic pathways. Proc. First ISMB (1993) 328-336
21. Reisig, W.: Petri Nets. Springer-Verlag (1985)
22. Tomita, M., Hashimoto, K., Takahashi, K., Shimizu, T., Matsuzaki, Y., Miyoshi, F., Saito, K., Tanida, S., Yugi, K., Venter, J.C., Hutchison, C.: E-CELL: software environment for whole cell simulation. Bioinformatics **15** (1999) 72-84
23. Valk, R.: Self-modifying nets, a natural extension of petri nets. Lecture Notes in Computer Science **62** (ICALP '78) (1978) 464-476
24. Watson, J.D., Hopkins, N.H., Roberts, J.W., Steitz, J.A., Weiner, A.M.: Molecular Biology of the Gene, Fourth Edition. The Benjamin/Cummings Publishing Company Inc (1987)

The Resource Allocation Problem in Flexible Manufacturing Systems*

Extended Abstract

J.M. Colom

Instituto de Investigación en Ingeniería de Aragón
Universidad de Zaragoza
María de Luna, N. 3, 50.018 Zaragoza, España
jm@posta.unizar.es

Abstract. The analysis of resource allocation related aspects is a precondition for the design and control of Flexible Manufacturing Systems. The formulation of this application-driven problem in terms of Petri nets leads to a class of models, with a specific structure-based characterization, which we explore in this presentation. We will concentrate our efforts on the characterization of the liveness of such models. We will also discuss the structural causes of the non-liveness (deadlock of some manufacturing processes) that will allow to state the foundations to introduce control elements which eliminate all the bad states.

1   Introduction

Flexible Manufacturing Systems (FMS) are a kind of production systems where a variety of products with relatively low production volumes must be manufactured. They are designed to conciliate the need of efficiency and flexibility in product scenarios that must change in order to react to market changes. An FMS is an automatically controlled system involving automated flexible machines, material handling, transport and storage systems.

The behaviour of FMSs is often extremely complex, and subtle or even paradoxical phenomena can appear. Among many diverse problems concerning FMSs we consider here some issues in the design from a systems theory perspective [18] (i.e. we disregard technological aspects that depend on the nature of the production process, and concentrate on the interactions of the subsystems of the FMS).

The adopted perspective in this presentation is at the global coordination level in the typical hierarchical architecture of the control of an FMS. This perspective perceives an FMS as a set of processes (parts, jobs, etc), following predefined production plans, requesting, in a competitive way, different quantities of a finite number of shared conservative resources (machines, tools, buffer space,

---

* This work has been partially supported by the Spanish research project CICYT-FEDER 2001-1819.

W.M.P. van der Aalst and E. Best (Eds.): ICATPN 2003, LNCS 2679, pp. 23–35, 2003.
© Springer-Verlag Berlin Heidelberg 2003

etc). The attention is focussed on the study of the problems arising when the shared resources must be granted to a set of concurrent processes. This view of an FMS corresponds to a class of concurrent systems called *Resource Allocation Systems* (RAS) [15,16].

The use of Petri Nets (PN) in RASs is an active research field devoted to define and exploit different subclasses of Petri Nets allowing to model the widest set of RAS. The definition of these subclasses is based on the net structure, allowing to obtain structure-based characterizations of the partial/total deadlocks. The goal of these characterizations is to synthesize controllers preventing/avoiding deadlocks arising from the resource allocation.

Most of the approaches to construct the PN model of a RAS assume a modular methodology in three steps:

- *Characterization of the Production Plans.* Each part or part component that enters the system is a process. Processes are modelled as tokens that move through a PN, $\mathcal{N}$, representing the production plan for a type of product. The places (partial states) of $\mathcal{N}$ are related to the different operations (either transformations, handling or assembly/disassembly operations) to be carried out over the parts contained in these places. The transitions of $\mathcal{N}$ allow to progress a part towards its final state. A production plan has distinguished input points of raw materials and output points of terminated products. The execution of a production plan is achieved by the execution of a production path, and several of them can exist in the same production plan. A production path is a sequence of transitions firable in $\mathcal{N}$, whose occurrence represents the production of a finished product.
- *Incorporation of Resources to each Production Plan.* A physical element composing the FMS (a machine, a buffer, a robot, a tool, etc.) is a resource with a given capacity (the number of parts that, at a given time, the resource is able "to store/to be used by"). Each state of a production plan has associated the (multi-)set of resources needed for the corresponding processing step (including the buffering capacity to hold the part itself). A resource type is represented by means of a place whose initial marking represents either the number of available copies of the resource or its capacity. A resource place has input (output) arcs to (from) those transitions of a production plan that moves a process to (from) a state that requires (was using) a number of copies of this resource type. In all cases, the considered resources can neither be created nor destroyed.
- *Construction of the global model by composition of the Production Plans with resources.* In a FMS there is a set of production plans: one type per each type of product. In order to obtain the global model of the RAS we must compose the production plans with the needed resources. This composition is based on the fusion of the resource places representing the same resource type in the different production plans. The initial marking of the resources, after the fusion, normally is computed as the maximum of the initial markings of the instances that have been merged.

This common methodology has been applied to the modelling of restricted classes of RASs by means of PNs that are defined imposing restrictions either on the class of production plans to be considered or on the way that system resources can be used by a production plan at a given state.

A first group of RASs are those where a raw material only suffers succesive transformations until its final state. In this group, a production plan is represented (or modeled) by means of a state machine. Additional restrictions on the production plans refer to the availability of different routings in the system; another important question is whether a part can choose different paths once it is in the system. The first feature is offered in some models, but many of them do not allow on–line decisions, and the path is fixed once the part selects one of the available routes [1,24,6,9,8,26]. Studies allowing on–line decisions for part routing can be found in [2,4,22,5].

Restrictions related to resources, in this first group, refer to the number and type of resources that can be used by a process at a given state. In most previous work only one resource of a unique type was allowed at each state of each process (the "Single–Unit RAS", as named in [16]). This restriction was relaxed in [22] and solved for the more general case in [21,20,2,5]. This last set of works solve the "Disjunctive–Conjunctive (OR/AND) RAS" problem [16], which corresponds to the more general case when conservative resources are considered: alternative routings are allowed in processes and no restriction is imposed on the resources that can be used at a given state of a process.

A second group of RAS presents a concurrent processing nature. This is usually due to some assembly/disassembly operations that introduce the possibility of independent processing steps of different part components. In this case, more complicated models are needed to represent the production plans of the involved components. As a matter of fact, while a lot of work related to the first group of RAS can be found in the literature, it is much more difficult to find solutions for this second group (see for example [3,25]).

The restrictions imposed in the definition of these subclasses of RAS are mainly oriented to obtain a non–liveness characterization for the class of nets. Most of these characterizations are based on the existence of a siphon insufficiently marked at a reachable marking. The important property of insufficiently marked siphons is that some of its output transitions are dead for ever. This kind of characterization was firstly presented in [4] and later developed for other subclasses in [17,3,25,12,14,2,22,20,7].

The interest of these siphon based characterizations is in their use for the synthesis of controllers preventing the the system from deadlock states. The strategy is based on the computation of some places from the detected bad siphons that, when added to the net, prevent that the siphon can become insufficiently marked. Different methods using this approach can be found in [2,4,22, 24,13].

This work is an overview of a design methodology for RAS that uses Petri nets as the underlying model for specification, deadlock analysis and controller synthesis. This extended abstract is organized as follows: the class of problems, RASs

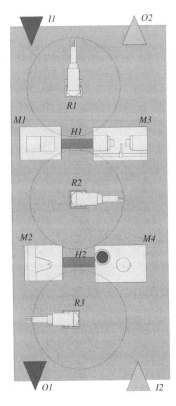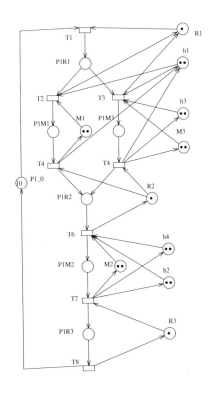

Fig. 1. Layout of a manufacturing cell and the Petri net model of the production plan for parts of Type WP1 in the cell with the needed resources.

and models we are considering are presented in Section 2. Section 3 presents a set of liveness characterizations for the proposed net models and their use to prevent deadlock states. Some conclusions and suggestions for further research are provided in the last section.

2   Models for Resource Allocation Systems

This section defines a class of PNs to model RASs, following the methodology outlined in the introduction. Before introducing the class of nets in a formal way, let us introduce a simple example. In these nets, the production plans are modelled by means of strongly connected state machines and they share a set of non–consumable and reusable resources. This simple class of nets allows to identify the set of requirements defining the RASs that traditionally have been studied using PNs. We will see that some of them are expressed in PN terms in a very strict way, reducing the number of RASs that can be considered.

Figure 1 sketches a production cell, where two types of parts have to be processed. The cell is composed of four machines, *M1*, *M2*, *M3*, *M4*, whose

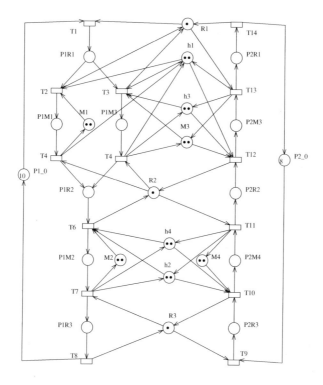

Fig. 2. The Petri net modelling the RAS in the cell of the Figure 1.

processing capacity is of two. They are able to carry out different operations aided with the tools available in two tool stores, $H1$ and $H2$; the first one contains two classes of tools, $h1$ and $h3$, which have to be shared by $M1$ and $M3$. There are two copies of each one of these tools. Machine $M3$ uses one copy of each tool for the processing of a part, while machine $M1$ uses one copy of $h1$. $H2$ contains two copies of $h2$ tool and two copies of $h4$ tool. Machines $M2$ and $M4$ use one $h2$ tool and one $h4$ tool. In order to transport the parts along the cell there are three robots, $R1,R2,R3$. $R1$ can load machines $M1$ and $M3$ from point $I1$, and can unload machine $M3$ towards point $O2$. $R2$ loads and unloads the four machines. Finally, the cell also contains a robot $R3$, which can load parts into machine $M4$ from point $I2$ and unload parts from $M2$ to $O1$. In this cell two different types of parts have to be processed. Parts of type $WP1$ are taken from a conveyor at point $I1$, processed in machine $M1$ or $M3$, then in machine $M2$ and finally unloaded on a conveyor at point $O1$. Parts of type $WP2$ are first loaded into the system from a conveyor at point $I2$, then processed in machine $M4$, then in machine $M3$ and finally unloaded to another conveyor at point $O2$.

The PN of Figure 1 models the production plan for parts of type $WP1$ with the needed resources. Places whose name begins with P model states that can be reached by a part (except $P1\_0$, which models parts out of the system). For example, a token in $P1M2$ corresponds to a part being processed in $M2$. Each part in one of these states needs some number of system resources; this fact is modelled by the places whose names start with R, M, and h. The initial marking of each one of them represents the availability of the corresponding resource. For example, a token in the place $R1$ means that the robot $R1$ is free, while no tokens in the place means that $R1$ is busy. At the state modelled by place $P1M2$, the part is "using" machine $M2$ and a copy of tools $h2$ and $h4$. Arcs joining state places model the flow of parts in the system. Arcs joining resource places with state places model the use/release of system resources. For instance, the arc joining $P1R2$ and $P1M2$ models the fact of the part being loaded from robot $R2$ into machine $M2$, while the arc joining $M1$ and $P1M1$ models the fact that, to reach this state, one unit of machine $M1$ capacity needs to be available. Then, when there is a token in $P1M2$, to move it out of the system the use of robot $R3$ is needed (this is modelled by means of the arc from $R3$ to $T7$). This is modelled by transition $T7$, which will be enabled when the robot is available (that is, when there is a token in the place $R3$). When this action takes place, the robot $R3$ is busy loading the part (and the corresponding place loses its token), the machine $M2$ recovers the availability for a new part, and one copy of each $h2$ and $h4$ tools is released (a token is put back in $M2$, $h2$, and $h4$ places, respectively).

Following similar guidelines, the PN of the production plan for parts of type $WP2$ with the needed resources can be obtained. To construct the RAS model, we compose these two nets via fusion of the resource places of the two nets representing the same type of resource. The Figure 2 shows the PN obtained after the composition of the two PNs. This model belongs to the class called S^4PR [19]. This class is similar to some classes presented in previous works, named WS^3PR in [23], S^4R in [2] and S^3PGR^2 in [13]. The following is the formal definition of S^4PR class.

Definition 1 (The class of S^4PR nets). Let $I_N = \{1, 2, ..., m\}$ be a finite set of indices. An S^4PR net is a connected generalised self–loop free Petri net $\mathcal{N} = \langle P, T, \mathbf{C} \rangle$ where:

1. $P = P_0 \cup P_S \cup P_R$ is a partition such that:
    a) $P_S = \bigcup_{i \in I_N} P_{S_i}$, $P_{S_i} \neq \emptyset$ and $P_{S_i} \cap P_{S_j} = \emptyset$, for all $i \neq j$.
    b) $P_0 = \bigcup_{i \in I_N} \{p_{0_i}\}$.
    c) $P_R = \{r_1, r_2, \ldots, r_n\}$, $n > 0$.
2. $T = \bigcup_{i \in I_N} T_i$, $T_i \neq \emptyset$, $T_i \cap T_j = \emptyset$, for all $i \neq j$
3. For all $i \in I_N$, the subnet $\mathcal{N}_i$ generated by $P_{S_i} \cup \{p_{0_i}\} \cup T_i$ is a strongly connected state machine, such that every cycle contains p_{0_i}.
4. For each $r \in P_R$ there exists a minimal P–Semiflow, $\mathbf{y}_r \in \mathbb{N}^{|P|}$, such that $\{r\} = \|\mathbf{y}_r\| \cap P_R$, $\mathbf{y}_r[r] = 1$, $P_0 \cap \|\mathbf{y}_r\| = \emptyset$, and $P_S \cap \|\mathbf{y}_r\| \neq \emptyset$.
5. $P_S = \bigcup_{r \in P_R} (\|\mathbf{y}_r\| \setminus \{r\})$.

$\square$

Places of P_S are called *process places*. Each place p_{0_i} is called *idle place*, and represents the state in which the corresponding processes (or parts) are idle. Each strongly connected state machine in Definition 1.3 models a production plan to produce a part type. Places of P_R are called *resource places*. This definition must be completed with the definition of the *acceptable initial markings* [19]: initial markings representing no activity in the system and allowing the processing of each part in isolation.

Definition 2. *Let* $\mathcal{N} = \langle P_0 \cup P_S \cup P_R, T, \mathbf{C} \rangle$ *be a* S^4PR *net. An initial marking* $\mathbf{m_0}$ *is acceptable for* $\mathcal{N}$ *if and only if: (1)* $\forall i \in I_{\mathcal{N}}$, $\mathbf{m_0}[p_{0_i}] > 0$. *(2)* $\forall p \in P_S$, $\mathbf{m_0}[p] = 0$. *(3)* $\forall r \in P_R$, $\mathbf{m_0}[r] \geq \max_{p \in ||\mathbf{y}_r|| \setminus \{r\}} \mathbf{y}_r[p]$. □

Using nets of the S^4PR class we are able to model RASs that have the following abstract properties.

1. *The production paths of a production plan are reproducible.* In S^4PR nets the production paths are the minimal t–semiflows of the state machines (the production plans). Each cycle in a state machine represents the firing sequence that completes the processing of a part from the idle place until the end. The reproducible nature of this sequence is captured in PNs by the concept of t–semiflow.

2. *A production plan is composed of production paths.* A production plan, in S^4PR nets, is a strongly connected state machine that is covered by t–semiflows: the set of production paths. All these production paths are able to produce a same type part.

3. *Each production path must be executable in isolation from the initial state (equivalence of production paths).* In the case of S^4PR nets, this property holds because the production plans have an structure of strongly connected state machine and the the initial marking of the idle places is non–empty.

4. *Given an intermediary execution state of a production plan, it is always possible to terminate recovering the initial state.* This property holds in S^4PR nets because a marked strongly connected state machine can move all tokens to any place of the net.

5. *All transitions between production operations defined in a process plan are live.* This property holds in S^4PR nets for the same reasons than the previous one.

6. *Production plans use resources in a conservative way.* In S^4PR nets, for each $r \in P_R$, Definition 1 imposes the existence of a minimal p–semiflow $\mathbf{y}_r$ such that $\mathbf{y}_r[r] = 1$. Moreover, it is easy to prove that for each $r \in P_R$, $\mathbf{y}_r$ is the only minimal p–semiflow such that $r \in ||\mathbf{y}_r||$. For a given $p \in P_S$, $\mathbf{y}_r[p] = k(\geq 0)$ means that k copies of resource r are used by each part (token) in the state modelled by means of place p. Moreover, the invariant imposed by $\mathbf{y}_r$ ($\forall \mathbf{m} \in \mathrm{RS}(\mathcal{N}, \mathbf{m_0})$, $\mathbf{y}_r \cdot \mathbf{m} = \mathbf{y}_r \cdot \mathbf{m_0} = \mathbf{m_0}[r]$) represents the fact that resources can neither be created nor destroyed.

7. *All partial states of a production plan use some resource to carry out the operation they represent.* This property is the last condition in Definition 1 of S^4PR. Therefore, the net models of this kind of RAS are conservative.

8. *A production plan with resources is well defined.* This means that initially there is no activity in system, and each possible production path has enough

resources to be executed in isolation. As proved formally in [20], the initial marking satisfying these conditions is the presented in Definition 2.

Observe that the kind of RAS that can be modelled by S^4PR nets is limited, e.g. assembly/disassembly operations cannot be represented.

3   Liveness Characterizations of S^4PR Models of RASs

The following theorem presents a liveness characterisation for S^4PR nets. This characterisation is fully behavioural. The structural causes for non-liveness are related to the existence of certain siphons and this will be presented later.

Given a marking $\mathbf{m}$ in a S^4PR net, a transition t is said to be $\mathbf{m}$-process-enabled ($\mathbf{m}$-process-disabled) iff it is (not) enabled by its input process place, and $\mathbf{m}$-resource-enabled ($\mathbf{m}$-resource-disabled) iff its input resource places have (not) enough tokens to fire it.

Theorem 1 ([19]). *An S^4PR, $\langle \mathcal{N}, \mathbf{m_0} \rangle$, is non–live if and only if there exists a marking $\mathbf{m} \in \mathrm{RS}(\mathcal{N}, \mathbf{m_0})$ such that the set of $\mathbf{m}$–process–enabled transitions is non–empty and each one of these transitions is $\mathbf{m}$–resource–disabled.*

Theorem 1 relates non–liveness to the existence of a marking where active processes are blocked. Their output transitions need resources that are not available. These needed resources cannot be generated (released by the corresponding processes) by the system (the transitions are dead) because there exist a set of circular waits between the blocked processes.

This concept of circular waits is captured in the model by the existence of a siphon (in Petri Net terms) whose resource places are the places preventing the firing of the process–enabled transitions. The following theorem characterises non–liveness in terms of siphons establishing the bridge between behavior and model structure.

Theorem 2 ([20,11]). *An S^4PR, $\langle \mathcal{N}, \mathbf{m_0} \rangle$, is non–live if and only if there exists a marking $\mathbf{m} \in \mathrm{RS}(\mathcal{N}, \mathbf{m_0})$ and a siphon D such that: i) There exists at least an $\mathbf{m}$–process–enabled transition; ii) Every $\mathbf{m}$–process–enabled transition is $\mathbf{m}$–resource–disabled by resources in D; iii) Process places in D are not marked at $\mathbf{m}$.*

Given a siphon D, and a marking $\mathbf{m} \in \mathrm{RS}(\mathcal{N}, \mathbf{m_0})$ as in Theorem 2, D is called a *bad siphon*, and $\mathbf{m}$ a *D–deadlocked marking*. As an example, consider the S^4PR in Figure 3, and its reachable marking $\mathbf{m} = 2 \cdot P1\_0 + 2 \cdot P1\_2 + R1 + R2 + 4 \cdot P2\_0$. Transition $T3$ is dead at $\mathbf{m}$, and the siphon $D = \{R2, P1\_3, P2\_1\}$ satisfies the conditions of the Theorem 2: $R2$ is preventing the firing of $T3$, which is $\mathbf{m}$–process–enabled; moreover, all process places in D have zero tokens.

The previous non–liveness characterization has been used to define deadlock prevention control policies to make live the original net system [20,11]. The first step is to compute a bad siphon and a marking $\mathbf{m}$ satisfying the conditions of the

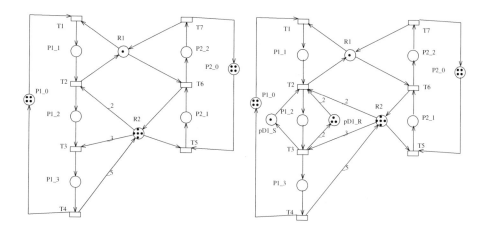

Fig. 3. A net with bad siphons and the same net with two alternative control places associated to the bad siphon $D = \{R2, P1\_3P2\_1\}$.

Theorem 2. For this, different methods have been proposed, most of them are based on the use of an integer linear system of inequalities whose solutions are the markings (solutions of the net state equation), and the siphons of Theorem 2. These methods are affected for the existence of spurious markings of the net state equation. A drawback of these methods is that unnecessary control elements can be introduced.

The second step is to compute, from the obtained marking and bad siphon, the control to be added to the net. This control is usually implemented by means of the addition of a new place to the net. This place restricts the behaviour of the net in such a way that it is not possible to reach a marking for the siphon as those described in Theorem 2 in the controlled net. To do that two different strategies can be adopted:

1. *The place is obtained as a non–negative linear combination of the places of the siphon.* The initial marking of the control place will be less than the same linear combination of the initial markings of the siphon places. This added place restricts the number of tokens that can flow out of the siphon. That is, the place forbids markings such that the number of tokens in the siphon is under a minimum that considers the situations in which the resources of the siphon are preventing the firing of process–enabled transitions. The place $pD1\_R$ in Figure 3 is an example of the application of this strategy for the bad siphon D.

2. *The place is the complementary place of a non–negative linear combination of the destination places of the tokens flowing out the siphon.* The initial marking of the control place will be less than the same linear combination of the marking of these places at the bad marking. This added place restricts

the number of tokens, in the destination places of the tokens flowing out of the siphon, to be under a maximum corresponding to the bad marking of Theorem 2. The place $pD1\_S$ in Figure 3 is an example of the application of this strategy for the bad siphon D.

From a structural point of view, the added control place can be considered as a new (virtual) resource. Then, the initial net plus this control place will have the structure of an S^4PR net with, perhaps, deadlock problems. The process of computing a bad siphon for the new net can be iterated until, as it is proved for example in [19], a final live controlled S^4PR is obtained. It is important to remark that not all the siphons of the system need to be computed, just the ones that can be problematic.

One of the problems of this approach is that the added places (they are linear constraints) can remove from the net state space more markings than necessary. This leads to define several heuristics trying to obtain more and more permissive controllers making the net live.

4   Conclusions

The analysis of resource allocation related aspects is a precondition for the design and control of Flexible Manufacturing Systems. This analysis pursues efficient characterizations of the partial/total deadlocks, arising from the resource allocation perspective, that can be used to synthesize controllers preventing/avoiding system deadlocks. The RAS perspective perceives an FMS as a set of processes (parts, jobs, etc), following predefined production plans, requesting, in a competitive way, different quantities of a finite number of shared resources (machines, tools, buffer space, etc).

Petri Nets have been used to model RASs for this purpose, and an active research field arose ten years ago, devoted to define and exploit different subclasses of Petri Nets allowing to model the widest set of RAS.

The methodologies used to construct Petri net models of RASs usually have a modular nature, where the modules are the production plans that are composed via the fusion of a set of shared resources.

In general, PN models are defined imposing structural restrictions to the nets modelling a RAS. In many cases, these constraints are too strict and are imposed to obtain a liveness characterization for the class.

Future research directions must try to extend the class of RAS that can be studied. These extensions must lie on the basic requirements of the RASs in order to characterize them in terms of Petri Net elements. In particular, with respect to the outlined methodology, some techniques to construct good process plans are needed. They must allow to obtain well behaved process plans via composition of process paths. On the other hand, some additional mechanisms of comunication and synchronization between process plans (and/or process paths) must be considered. A first step could be oriented to the addition of message passing mechanisms.

With respect to the techniques for the synthesis of controllers making live the RASs, many problems remain open from the theoretical and practical point of view. For example, it is very important to consider more complex kinds of controllers than a set of places, and also to develop techniques to synthesize them in an automatic and efficient way.

References

1. Z.A. Banaszak and B.H. Krogh. Deadlock avoidance in flexible manufacturing systems with concurrently competing process flows. *IEEE Trans. on Robotics and Automation*, 6(6):724–734, December 1990.

2. K. Barkaoui, A. Chaoui, and B. Zouari. Supervisory control of discrete event systems based on structure of Petri nets. In *Procs of the 1997 Int. Conf. on Systems, Man and Cybernetics*, pages 3750–3755, Oralndo (USA), October 1997.

3. F. Chu and X. Xie. Deadlock analysis of petri nets using siphons and mathematical programming. *IEEE Transactions on Robotics and Automation*, 13(6):793–804, December 1997.

4. J. Ezpeleta, J.M. Colom, and J. Martínez. A Petri net based deadlock prevention policy for flexible manufacturing systems. *IEEE Transactions on Robotics and Automation*, 11(2):173–184, April 1995.

5. J. Ezpeleta, F. Tricas, F. García-Vallés, and J.M. Colom. A Banker's solution for deadlock avoidance in FMS with flexible routing and multiresource states. *IEEE Transactions on Robotics and Automation*, 18(4):621–625, August 2002.

6. M.P. Fanti, B. Maione, S. Mascolo, and B. Turchiano. Event–based feedback control for deadlock avoidance in flexible production systems. *IEEE Trans. on Robotics and Automation*, 13(3):347–363, June 1997.

7. M.P. Fanti, B. Maione, and T. Turchiano. Comparing digraph and petri net approaches to deadlock avoidance in FMS modelling and performance analysis. *IEEE Transactions on System Man and Cybernetics. Part B*, 30:783–798, 2000.

8. F. Hsieh and S. Chang. Dispatching-driven deadlock avoidance controller synthesis for flexible manufacturing systems. *IEEE Trans. on Robotics and Automation*, 10(2):196–209, April 1994.

9. F. S. Hsieh and S. C. Chang. Deadlock avoidance controller synthesis for flexible manufacturing systems. In *Procs of the 3th. Int. Conf. on Computer Integrated Manufacturing*, pages 252–261, Troy, New York, May 1992. Rensselaer Polytechnic Institute.

10. T. Murata. Petri nets: Properties, analysis and applications. *Proceedings of the IEEE*, 77(4):541–580, April 1989.

11. J. Park and S. Reveliotis. A polynomial-complexity deadlock avoidance policy for sequential resource allocation systems with multiple resource acquisitions and flexible routings. In *Proceedings of the IEEE International Conference on Decision & Control*, pages 2663–2669, Australia, December 2000. IEEE.

12. J. Park and S.A. Reveliotis. Algebraic synthesis of efficient deadlock avoidance policies for sequential resource allocation systems. *IEEE Transactions on Robotics and Automation*, 16(2):190–195, April 2000.

13. J. Park and S.A. Reveliotis. Algebraic deadlock avoidance policies for conjunctive/disjunctive resource allocations systems. In *Procs of the 2001 Int. Conf. on Robotics and Automation*, Seoul, Korea, May 2001.

14. J. Park and S.A. Reveliotis. Deadlock avoidance in sequential resource allocation systems with multiple resource acquisitions and flexible routings. *IEEE Transactions on Automatic Control*, 46:1572–1583, 2001.

15. J.L. Peterson and A. Silberschatz. *Operating System Concepts*. Addison-Wesley, 1985.

16. S.A. Reveliotis. Accommodating FMS operational contingencies through routing flexibility. In *Procs of the 1998 Int. Conf. on Robotics and Automation*, pages 573–579, Leuven, Belgium, May 1998.

17. S.A. Reveliotis, M.A. Lawley, and P.M. Ferreira. Polynomial complexity deadlock avoidance policies for sequential resource allocation systems. *IEEE Transactions on Automatic Control*, 42(6):1344–1357, 1997.

18. M. Silva, E. Teruel, R. Valette, and H. Pingaud. Petri nets and production systems. In W. Reisig and G. Rozenberg, editors, *Lectures on Petri Nets II: Applications*, volume 1492 of *Lecture Notes on Computer Science*, pages 85–124. Springer-Verlag, 1998.

19. F. Tricas. *Analysis, Prevention and Avoidance of Deadlocks in Sequential Resource Allocation Systems*. PhD thesis, Zaragoza. España, Departamento de Ingeniería Eléctrica e Informática, Universidad de Zaragoza, May 2003.

20. F. Tricas, J.M. Colom, and J. Ezpeleta. A solution to the problem of deadlocks in concurrent systems using Petri nets and integer linear programming. In G. Horton, D. Moller, and U. Rude, editors, *Proc. of the 11th European Simulation Symposium*, pages 542–546, Erlangen, Germany, October 1999. The society for Computer Simulation Int.

21. F. Tricas and J. Ezpeleta. A Petri net solution to the problem of deadlocks in systems of processes with resources. In *Proc. of the 7th IEEE Int. Conf. on Emerging Technologies and Factory Automation (ETFA)*, pages 1047–1056, Barcelona, October 1999.

22. F. Tricas, F. García-Vallés, J.M. Colom, and J. Ezpeleta. A structural approach to the problem of deadlock prevention in processes with resources. In *IEE International Workshop on Discrete Event Systems. WODES 98*, pages 273–278, Cagliary (Italy), August 1998. IEE, IEE.

23. F. Tricas and J. Martínez. An extension of the liveness theory for concurrent sequential processes competing for shared resources. In *Proc. of the 1995 IEEE Int. Conf. on Systems, Man and Cybernetics.*, pages 4119–4124, Vancouver, Canada, October 1995.

24. N. Viswanadham, Y. Narahari, and T.L. Johnson. Deadlock prevention and deadlock avoidance in flexible manufacturing systems using Petri net models. *IEEE Trans. on Robotics and Automation*, 6(6):713–723, December 1990.

25. X. Xie and M. Jeng. ERCN–merged nets and their analysis using siphons. *IEEE Transactions on Robotics and Automation*, 15(4):692–703, August 1999.

26. K.Y. Xing, B.S. Hu, and H.X. Chen. Deadlock avoidance policy for Petri-net modeling of flexible manufacturig systems with shared resources. *IEEE Transactions on Automatic Control*, 41(2):289–295, February 1996.

Appendix: Petri Net Concepts and Notations

We assume the reader is familiar with basic Petri net concepts. They can be found, for instance, in [10].

A Petri net (or Place/Transition net) is a 3–tuple $\mathcal{N} = \langle P, T, W \rangle$ where P and T are two non–empty disjoint sets whose elements are called *places* and *transitions*. $W : (P \times T) \cup (T \times P) \longrightarrow \mathbb{N}$ is the *flow relation*: if $W(x, y) > 0$, then we say that there is an arc from x to y, with weight $W(x, y)$. Ordinary nets are those for which $W : (P \times T) \cup (T \times P) \to \{0, 1\}$. Given a net $\mathcal{N} = \langle P, T, W \rangle$ and a node $x \in P \cup T$, $^\bullet x = \{y \in P \cup T \mid W(y, x) > 0\}$, and $x^\bullet = \{y \in P \cup T \mid W(x, y) > 0\}$. An ordinary net will be called a *state machine* if and only if $\forall t \in T$, $\mid {}^\bullet t \mid = 1$ *and* $\mid t^\bullet \mid = 1$. A *marking* is a mapping $\mathbf{m} : P \longrightarrow \mathbb{N}$. The pair $\langle \mathcal{N}, \mathbf{m_0} \rangle$, where $\mathcal{N}$ is a net and $\mathbf{m_0}$ is an (initial) marking, is called a *net system*. A transition $t \in T$ is *enabled* for a marking $\mathbf{m}$ if and only if $\forall p \in {}^\bullet t$, $\mathbf{m}[p] \geq W(p, t)$; this fact will be denoted as $\mathbf{m} \overset{t}{\longrightarrow}$; when fired (in the usual way), this gives a new marking $\mathbf{m}'$; this will be denoted as $\mathbf{m} \overset{t}{\longrightarrow} \mathbf{m}'$. A marking $\mathbf{m}'$ is *reachable* from another marking $\mathbf{m}$ if and only if there exists a firing sequence $\sigma = t_1 t_2 ... t_n$ so that $\mathbf{m} \overset{t_1}{\longrightarrow} \mathbf{m_1} \overset{t_2}{\longrightarrow} \mathbf{m_2} ... \mathbf{m}_{n-1} \overset{t_n}{\longrightarrow} \mathbf{m}'$, and is denoted as $\mathbf{m} \overset{\sigma}{\longrightarrow} \mathbf{m}'$. The set of *reachable* markings from $\mathbf{m_0}$ in $\mathcal{N}$ is denoted as $RS(\mathcal{N}, \mathbf{m_0})$. A Petri net is self–loop free if and only if $\neg(W(x, y) > 0 \wedge W(y, x) > 0)$. A self–loop free Petri net $\mathcal{N} = \langle P, T, W \rangle$ can be alternatively represented as $\mathcal{N} = \langle P, T, \mathbf{C} \rangle$ where $\mathbf{C}$ is the net *flow matrix*: a $P \times T$ integer matrix so that $\mathbf{C} = \mathbf{Post} - \mathbf{Pre}$ where for each $t \in T$, $p \in P$, $\mathbf{Post}[p, t] = W(t, p)$ $\mathbf{Pre}[p, t] = W(p, t)$.

We consider structural analysis techniques based on linear algebra using a *linear relaxation* of the behaviour of a net system. Let $\langle \mathcal{N}, \mathbf{m_0} \rangle$ be a net system, and σ a fireable sequence of transitions from $\mathbf{m_0}$. The (integer) linear relaxation looks as follows: $\mathbf{m_0} \overset{\sigma}{\longrightarrow} \mathbf{m} \Rightarrow \mathbf{m} = \mathbf{m_0} + \mathbf{C} \cdot \overline{\sigma} \geq 0$, $\overline{\sigma} \geq 0$ where $\mathbf{m}$ is reachable from $\mathbf{m_0}$ firing σ, $\overline{\sigma}$ is the Parikh (or firing count) vector of σ and $\mathbf{C}$ the incidence matrix of the net, $\mathcal{N}$. This linear system is known as the *state equation* of the net system. Unfortunately, the reverse of the above implication is not true. More precisely, the state equation has integer solutions, $(\mathbf{m}, \sigma)$, not reachable on the net system. We call them *spurious* solutions.

A set of places D is a *siphon* if and only if $^\bullet D \subseteq D^\bullet$. A marked Petri net $\langle \mathcal{N}, \mathbf{m_0} \rangle$ is *deadlock–free* if and only if $\forall \, \mathbf{m} \in RS(\mathcal{N}, \mathbf{m_0})$, $\{t \in T \mid \mathbf{m} \overset{t}{\longrightarrow}\} \neq \emptyset$. A transition t is *live* if and only if $\forall \, \mathbf{m} \in RS(\mathcal{N}, \mathbf{m_0})$, $\exists \, \mathbf{m}' \in RS(\mathcal{N}, \mathbf{m})$ so that $\mathbf{m}' \overset{t}{\longrightarrow}$. A marked net is live if and only if every transition is live. We will say that a transition t is *dead for a reachable marking* $\mathbf{m}$ if and only if there is no reachable marking from $\mathbf{m}$ that enables t. *Semiflows* are natural annuller of matrix $\mathbf{C}$. Right and left natural annullers are called *t–semiflows* ($\mathbf{x} \in \mathbb{N}^{|T|}$ such that $\mathbf{C} \cdot \mathbf{x} = 0$) and *p–semiflows* ($\mathbf{y} \in \mathbb{N}^{|P|}$ such that $\mathbf{y} \cdot \mathbf{C} = 0$), respectively. A net is said to be *conservative* if there exists a p–semiflow $\mathbf{y} > 0$.

Colored GSPN Models for the QoS Design of Internet Subnets

M. Ajmone Marsan, M. Garetto, R. Lo Cigno, and M. Meo

Electronics Department of Politecnico di Torino, Italy

Abstract. In this paper we develop an approximate colored GSPN model to study the behavior of short-lived TCP connections sharing a common portion of the Internet for the transfer of TCP segments. The CGSPN model is then paired with a very simple approximate model of the IP network used by TCP connections, and the two models are jointly solved through an iterative procedure. The combined model allows the investigation of the Quality of Service (QoS) tradeoffs between network cost and network parameters, thus allowing a QoS-based design, dimensioning and planning of portions of the Internet. The QoS predictions generated by the model are validated against results of detailed simulation experiments in a realistic networking scenario, proving that the proposed modeling approach is very accurate.

W.M.P. van der Aalst and E. Best (Eds.): ICATPN 2003, LNCS 2679, p. 36, 2003.

Compositional Theories of Qualitative and Quantitative Behaviour

Ed Brinksma

Chair of Formal Methods and Tools
Department of Computer Science, University of Twente
PO Box 217, 7500AE Enschede, Netherlands
brinksma@cs.utwente.nl
http:/home.cs.utwente.nl/ brinksma

Extended Abstract

The integrated modelling and analysis of functional and non-functional aspects of system behaviour is one of the important challenges in the field of formal methods today. Our ever-increasing dependence upon of all sorts of critical applications of networked and/or embedded systems, often including sophisticated multi-media features, lends this intellectual challenge also great practical relevance. In this talk we will report on work in this area in the past decade or so on the use of techniques from so-called formal methods in the area of performance modelling and analysis, and in particular on the theory of stochastic process algebra (SPA) and its application.

Traditional performance models like Markov chains and queueing networks are widely accepted as simple but effective models in different areas, yet they lack the notion of hierarchical system (de)composition that has proved so useful for conquering the complexity of systems in the domain of funtional system properties. Compositional, hierarchical description and analysis of functional system behaviour is the domain *process algebra* [24,3,17]. It offers a mathematically well-elaborated framework for reasoning about the structure and behaviour of reactive and distributed systems in a compositional way, including abstraction mechanisms that allow for the treatment of system components as black boxes, encapsulating their internal structure. Process algebras are typically equipped with a formally defined structured operational semantics (SOS [26]) that maps process algebra terms onto *labelled transition systems* in a compositional manner. Such labelled transition systems consist of a set of states and a transition relation that describes how the system evolves from one state to another. These transitions are labelled with action names that represent the (inter)actions that may cause the transitions to occur. Such transition systems can be visualised by drawing states as nodes of a graph and transitions as directed edges (labelled with action names) between them.

The labelled transition model is very close to the usual representation of Markov chains as transition systems or automata. Also there system states are connected

W.M.P. van der Aalst and E. Best (Eds.): ICATPN 2003, LNCS 2679, pp. 37–42, 2003.
© Springer-Verlag Berlin Heidelberg 2003

by directed transition arcs that are labelled. In the case of discrete time Markov chains the labels are probabilities, and in the case of continuous time Markocv chains the labels are the *rates* that correspond to the (nagative) exponential distributions that represent the stochastic delays associated with the state transitions. This structural correspondence between the two models motivated the beginning of research in the early 1990's on *stochastic process algebras* [4,16,15], which sought to integrate performance modelling with Markov chains with functional analysis, and to transfer the process algebraic notion of (de)composition and hierarchy to Markov chain theory.

This marriage of process algebra with performance modelling requires careful re-examinination of the interpretation of some classical process algebraic concepts, in particular that of choice, concurrent composition, and synchronization. In standard process algebra's choice operator '+' offers a qualitative selection between alternative behaviours. Its idempotency law $B + B = B$ has a natural interpretation as a poor man's choice: identical choices are as good as no choice at all. In most performance models, however, branching behaviour implies a race condition between the alternatives that have a quantitative effect. For example, if a is an action offered with rate μ then the choice between two such actions offers a with a rate of 2μ. This gives rise to additive laws such as $(a, \mu).B + (a, \lambda).B = (a, \mu + \lambda).B$ (see [15]), or $(\mu).B + (\lambda).B = (\mu + \lambda).B$ (see [22], where (μ) denotes an exponentially distributed delay with rate μ). The standard interleaving interpretation of concurrent composition in process algebra matches well with the memorylessness of exponential (or geometric) distributions in continuous (discrete) Markov chains, allowing for so-called expansion laws to remove explicit parallelism from system descriptions. This becomes more problematic if general distributions are allowed, such as, e.g., in semi-Markov chains. An elegant solution here is to treat stochastic delays in a way similar to clocks in timed automata, with separate operators to set them and test for their expiration [18]. Synchronization, finally, in the setting of stochastically delayed actions, poses the question of what is the (distribution of the) delay of the synchronizations between them. Here, different strategies have been proposed, ranging from (normalized) products of rates [15], synchronizations only between passive (no rates) and active action occurences (determining the synchronized rate) [4], to apparent rates [16]. Elegant is the solution to have rates associated only to pure delay actions that do not synchronize, but interleave, as in IMC. This induces a delay of synchronized actions with a distribution of the maximum of the delays preceeding the synchronizing actions, which is an intuitively appealing choice. A more complete overview of Markovian stochastic algebras can be found in [10]; an account of the non-Markovian case is given in [18].

The fruitfulness of the process algebraic approach to the specification and generation of Markov chains has been demonstrated by a number results. In the stochastic setting, *bisimulation* equivalence [25], a central notion of equivalence for comparing labelled transition systems, has been shown to coincide with *lumpability*, a key concept for the aggregation of Markov chains [16]. Moreover, as bisimulation can be shown to be preserved under system composition operators (algebraically: bisimulation is a *congruence*), Markov chain aggregation can be carried out compositionally, i.e. component-wise. Case studies have shown the practicality of this compositional approach, and important progress has been made in exploiting the syntactic structure of specifications for performance analysis purposes.

A second area where approaches from formal methods are being put to use successfully is the evaluation of Markov chain models. Once a continuous-time Markov chain (or CTMC) has been generated, the next step is to evaluate the measure(s) of interest such as time to failure, system throughput or utilisation, with the required accuracy. Whereas various techniques have been developed for the specification of CTMCs, such as stochastic process algebras and Petri nets, the specification of measures of interest has remained fairly cumbersome and is typically done in a rather informal, ad-hoc manner. In particular, usually only simple state-based performance measures – such as long-run and transient probabilities – can be defined and analysed with relative ease.

In contrast, in the area of formal methods powerful means have been developed to express temporal properties of systems, e.g., based on temporal logics. Logics such as CTL (Computation Tree Logic) [14] allow one to express state-based properties as well as properties over paths, i.e., state sequences through transition systems. One thus may express, for instance, that along all (or some) paths a certain set of goal states can eventually be reached while visiting only states of a particular kind before reaching one of these goal states. The validity of CTL-formulas over finite automata can be established by automated techniques such as *model checking* [13]. These techniques are based on a systematic, usually exhaustive, state-space exploration to check whether a property is satisfied in each state, thereby using effective methods to combat the state-space explosion problem. Model checking is supported by software tools such as SMV [12] and SPIN [21] and has been successfully used in various industrial case studies.

Model-checking of CTL formulas is usually done by a recursive descent over the construction of the logical property to be checked, exploiting compositionality on the level of the logical operators. CTL has recently been extended with ample means to specify state- as well as path-based performance and dependability measures for CTMCs in a compact and unambiguous way. Besides the standard steady-state and transient measures, the logic CSL (Continuous Stochastic Logic) [2,8] allows for the specification of (constraints over) probabilistic measures over paths through CTMCs. For instance, it can be expressed what the probability is, that starting from a particular state, within t time units a set of goal-states is reached, thereby avoiding or deliberately visiting particular intermediate states before. This is a useful feature for dependability analysis that goes beyond the standard measures. Other types of non-standard, but practically interesting measures that can conveniently be expressed are, for example, response times that are conditioned on the equilibrium state of a CTMC, properties that are typically analysed using dedicated and rather involved techniques. An indication of the adequacy of CSL is that logical equivalence coincides with the lumpability equivalence over Markov Chains mentioned above. Lumping-equivalent CTMCs thus satisfy the same formulas, and there is no formula that can distinguish between lumping-equivalent CTMCs.

Given a finite CTMC and a performance measure specified in CSL, an automated procedure can be applied – á la model checking – to establish the validity of the (constraint over the) measure [7]. To that purpose, the traditional model-checking algorithms are extended with numerical methods such as matrix-vector multiplication, techniques for solving systems of linear equations, and uniformization (or solvers for Volterra integral equation systems). For path-formulas, measure-driven transformations are employed: for a given CTMC M and state s in M, the probability for s to satisfy path-formula φ is calculated by means of a transient analysis of another (smaller) CTMC M', which can easily be derived from M using φ. The time and space complexity of the model-checking algorithms is polynomial in the size of the model and linear in the length of the formula. Tool-implementations such as $E \vdash MC^2$ [20], PRISM [23] and the APNNTOOLBOX [9] are available.

Outlook. The work on stochastic process algebras has, apart from the technical achievements, brought the formal methods and performance analysis communities closer together, with now at least a qualified group of people being active in both communities. Markov chains, and CTMCs in particular, have received scant attention in concurrency theory for a long time: whenever probabilities have been considered, they were mostly of a purely discrete nature. Verification of discrete-time Markov chains, for instance, dates back to the early nineties [19]. The scientific interest in verifying CTMCs is steadily increasing. Its main advantages are that it offers a flexible and precise means to succinctly specify standard and complex performance and dependability measures, complemented by an automated technique to compute these measures in a uniform way. Specialized algorithms are thus hidden from the performance engineer. Main current research topics are: extensions to Markov reward models, and development of techniques to deal with very large state spaces, e.g., symbolic techniques, Kronecker algebra, and abstraction techniques.

Stochastic process algebraic specifications mostly do not yield Markov chain models directly, but mixed transition systems that involve both system actions and stochastic distributions. Abstracting the actions away is generally insufficient to obtain the embedded Markov models, as the resulting transition system, known as a Markov decision process, may still contain non-stochastic elements in the form of nondeterministic transitions. To obtain a Markov chain such non-determinism must first be resolved using, for example, reduction or scheduling techniques using adversaries. Another option is to study the properties of the Markov decison processes directly. Some verification theory for discrete Markov decision processes is available [6,5], as well as for the stationary behaviour of discrete semi-Markov decision processes [1]. A theory for the continuous-time case is yet to be formulated, however. Such a theory and effective model-checking algorithms for continuous-time Markov decision processes remain formidable open problems to be challenged.

A third direction of work is the further elaboration of the stochastic process algebraic theory for the non-Markovian case, which is closely linked to the performance model of generalized semi-Markov processes (GSMP [27]. They are usually analysed by discrete-event simulation techniques. But so far little has been done to categorize interesting subcases for which more powerful analytic techniques exist, and which could be exploited in the form of interesting equational laws on the process algebraic level.

Acknowledgements. For more than a decade the author has had the privilege to discuss and work on stochastic process algebraic topics with a number of great colleagues from whom he has learned a lot. In particular, he wished to acknowledge his interactions with the enthousiastic members of the Twente team Pedro D'Argenio, Holger Hermanns, Joost-Pieter Katoen and Rom Langerak.

References

1. L. de Alfaro. How to specify and verify the long-run average behavior of probabilistic systems. In *IEEE 13th Symp. on Logic in Comp. Sc.*, pp. 174–183, IEEE CS Press, 1998.

2. A. Aziz, K. Sanwal, V. Singhal and R. Brayton. Model checking continuous time Markov chains. *ACM Transactions on Computational Logic*, **1**(1): 162–170, 2000.

3. J.A. Bergstra, A. Ponse, and S.A. Smolka, editors. *Handbook of Process Algebra.* Elsevier Science Publishers, 2001.

4. M. Bernardo and R. Gorrieri. Extended Markovian Process Algebra. In Ugo Montanari and Vladimiro Sassone, editors, *CONCUR '96: Concurrency Theory (7th International Conference, Pisa, Italy, August 1996)*, volume 1119 of *Lecture Notes in Computer Science.* Springer, 1996.

5. A. Bianco and L. de Alfaro. Model checking of probabilistic and nondeterministic systems. In *Found. of Softw. Technology and Th. Comp. Sc.*, LNCS 1026: 499–513, Springer, 1995.

6. C. Baier and M.Z. Kwiatkowska. Model checking for a probabilistic branching time logic with fairness. *Distr. Comp.*, **11**: 125–155, 1998.

7. C. Baier, B.R. Haverkort, H. Hermanns and J.-P. Katoen. Model checking continuous-time Markov chains by transient analysis. In: E.A. Emerson and A.P. Sistla, *Computer-Aided Verification,*LNCS 1855: 358–372, 2000.

8. C. Baier, J.-P. Katoen and H. Hermanns. Approximate symbolic model checking of continuous-time Markov chains. In J.C.M. Baeten and S. Mauw, *Concurrency Theory*, LNCS 1664: 146–162, 1999.

9. P. Buchholz, J.-P. Katoen, P. Kemper and C. Tepper. Model-checking large structured Markov chains. *Journal of Logic and Algebraic Programming*, 2003 (to appear).

10. E. Brinksma and H. Hermanns. Process Algebra and Markov Chains. In [11], LNCS 2090: 183–231, 2001.

11. E. Brinksma and H. Hermanns and J.-P. Katoen. Lectures on Formal Methods and Performance Analysis. LNCS 2090, 2001.

12. A. Cimatti, E. Clarke, F. Giunchiglia and M. Roveri. NuSMV: a new symbolic model checker. *J. on Software Tools for Technology Transfer*, **2**: 410–425, 2000.

13. E. Clarke, O. Grumberg and D. Peled. *Model Checking*. MIT Press, 1999.
14. E.M. Clarke and E.A. Emerson. Design and synthesis of synchronisation skeletons using branching time temporal logic. In *Logic of Programs*, LNCS 131: 52–71, 1981.
15. N. Götz, U. Herzog, and M. Rettelbach. Multiprocessor and Distributed System Design: The Integration of Functional Specification and Performance Analysis using Stochastic Process Algebras. In *Performance'93*, 1993.
16. J. Hillston. *A Compositional Approach to Performance Modelling*. PhD thesis, University of Edinburgh, 1994.
17. C.A.R. Hoare. *Communicating Sequential Processes*. Prentice-Hall, Englewood Cliffs, NJ, 1985.
18. J.-P. Katoen and P.R. D'Argenio. General Distributions in Process Algebra. In [11], LNCS 2090: 375–430, 2001.
19. H. Hansson and B. Jonsson. A logic for reasoning about time and reliability. *Formal Aspects of Computing* **6**: 512–535, 1994.
20. H. Hermanns, J.-P. Katoen, J. Meyer-Kayser and M. Siegle. A Markov chain model checker. In S. Graf and M.I. Schwartzbach, *Tools and Algorithms for the Construction and Analysis of Systems*, LNCS 1785: 347–362, 2000.
21. G.J. Holzmann. The model checker SPIN. *IEEE Tr. on Softw. Eng.*, **23**(5): 279–295, 1997.
22. H. Hermanns. Interactive Markov Chains. LNCS 2428, 2002.
23. M.Z. Kwiatkowska, G. Norman, and D. Parker. Probabilistic symbolic model checking with prism: A hybrid approach. In J-P. Katoen and P. Stevens (eds), *Tools and Algorithms for the Construction and Analysis of Algorithms*, LNCS 2280: 52–66, 2002
24. R. Milner. Calculi for Synchrony and Asynchrony. *Theoretical Computer Science*, 25:269–310, 1983.
25. R. Milner. *Communication and Concurrency*. Prentice Hall, London, 1989.
26. G.D. Plotkin. A Structured Approach to Operational Semantics. Technical Report DAIMI FM-19, Computer Science Department, Aarhus University, 1981.
27. G.S. Shedler. *Regenerative Stochastic Simulation*. Academic Press, 1993.

Net Models Supporting Human and Humane Behaviors

C.A. Ellis

Department of Computer Science, University of Colorado, CO, USA

Abstract. Numerous computer systems are guilty of decreasing the interaction and informal communication within groups. For example, some severe and well acknowledged problems of Workflow Management Systems stem from their rigorous and formal nature. Implementations of workflow tend to be coersive, isolationistic, and inflexible; whereas the natural interaction of people frequently incorporates flexibility, opportunistic behaviors, social awareness, and compromise. To combat this problem, there has been some fledgling work on workware (socially and organizationally aware groupware systems). In this talk we will summarize the state of this work, and propose a theoretical model that can help lay the foundations for further progress in this domain.

W.M.P. van der Aalst and E. Best (Eds.): ICATPN 2003, LNCS 2679, p. 43, 2003.

Deciding Life-Cycle Inheritance on Petri Nets

H.M.W. Verbeek and T. Basten

Eindhoven University of Technology
P.O. Box 513, NL-5600 MB Eindhoven, The Netherlands
h.m.w.verbeek@tm.tue.nl

Abstract. One of the key issues of object-oriented modeling is inheritance. It allows for the definition of a subclass that inherits features from some superclass. When considering the dynamic behavior of objects, as captured by their life cycles, there is no general agreement on the meaning of inheritance. Basten and Van der Aalst introduced the notion of life-cycle inheritance for this purpose. Unfortunately, the search tree needed for deciding life-cycle inheritance is in general prohibitively large. This paper presents a backtracking algorithm to decide life-cycle inheritance on Petri nets. The algorithm uses structural properties of both the base life cycle and the potential sub life cycle to prune the search tree. Test cases show that the results are promising.

Keywords. Object-orientation, workflow, life-cycle inheritance, branching bisimilarity, backtracking, Petri nets, structural properties, T-invariants.

1 Introduction

Inheritance of behavior. One of the main goals of object-oriented design is the reuse of system components. A key concept to achieve this goal is the concept of inheritance. The inheritance mechanism allows the designer to specify a class, the subclass, that inherits features of some other class, its superclass. Thus, it is possible to specify that the subclass has the same features as the superclass, but that in addition it may have some other features.

The Unified Modeling Language (UML) [19, 10, 17] has been accepted throughout the software industry as the standard object-oriented framework for specifying, constructing, visualizing, and documenting software-intensive systems. The development of UML began in late 1994, when Booch and Rumbaugh of Rational Software Corporation began their work on unifying the OOD [9] and OMT [18] methods. In the fall of 1995, Jacobson and his Objectory company joined Rational, incorporating the OOSE method [16] in the unification effort.

The informal definition of inheritance in UML states the following: "The mechanism by which more specific elements incorporate structure and behavior defined by more general elements." [19]. However, only the class diagrams, describing purely *structural* aspects of a class, are equipped with a concrete notion of inheritance. It is implicitly assumed that the *behavior* of the objects of a subclass, as defined by the object life cycle (OLC), is an extension of the behavior of the objects of its superclass.

In the literature, several formalizations of what it means for an OLC to extend the behavior of another OLC have been studied; see [8] for an overview. Combining the usual definition of inheritance of methods and attributes with a definition of inherit-

W.M.P. van der Aalst and E. Best (Eds.): ICATPN 2003, LNCS 2679, pp. 44–63, 2003.

ance of behavior yields a complete formal definition of inheritance, thus, stimulating the reuse of life-cycle specifications during the design process. One possible formalization of behavioral inheritance is called life-cycle inheritance (LCI) [8]:

> An OLC is a subclass of another OLC under LCI if and only if it is not possible to distinguish the external behavior of both when the new methods, that is, the methods only present in the potential subclass, are either blocked or hidden.

The notion of LCI has been shown to be a sound and widely applicable concept. In [8], it has been shown that it captures extensions of life cycles through common constructs such as parallelism, choices, sequencing and iteration. In [5], it is shown how LCI can be used to analyze the differences and the commonalities in sets of OLCs. Furthermore, in [3], the notion of LCI has been successfully lifted to the various behavioral diagram techniques of UML. Finally, LCI is successfully applied to the workflow-management domain. There is a close correspondence between OLCs and workflow processes. Behavioral inheritance can be used to tackle problems related to dynamic change of workflow processes [6]; furthermore, it has proven to be useful in producing correct interorganizational workflows [4].

Exponential-size search space. The basis of deciding LCI is an equivalence check, namely a branching bisimilarity (BB) check on the state spaces of both OLCs (the base OLC and the potential sub OLC). Particularly in the workflow domain, these state spaces can be large (up to millions of states each). Therefore, such a check might be time-consuming despite the fact that efficient algorithms exist to check BB on state spaces [15]. An exhaustive search algorithm (ESA) for deciding LCI might require many equivalence checks on these state spaces: One check for every possible partitioning of hiding and blocking the new methods in the potential sub OLC. The number of possible partitionings is exponential in the number of new methods. The combination of the large state spaces and the exponential factor results in an algorithm that is prohibitively expensive in many cases.

Reducing the size of the search space. This paper introduces a backtracking algorithm (BA) that is based on efficient pruning of the possible partitionings. Our main goal is to reduce the number of BB checks. To be able to do so, we assume that Petri nets are used to model the OLCs.

We develop the concept of constraints indicating that certain methods must be hidden or blocked in order to allow a successful BB check. These constraints are efficiently generated using structural analysis techniques for Petri nets. Our first experiments show that the BA using these constraints does indeed efficiently and effectively reduce the search space.

Overview. The remainder of this paper is organized as follows. Section 2 gives a formal definition of LCI. Section 3 presents the BA. Section 4 compares the BA with the ESA for several test cases. Section 5 concludes the paper.

2   Life-Cycle Inheritance

This section formalizes the concepts of object life cycles (OLCs) and life-cycle inheritance (LCI). After discussing some general notations, we define the subclass of Petri nets that we assume are used to model the OLCs. On these Petri nets, we define a branching bisimilarity (BB) relation [14]. This BB relation forms the basis of the LCI relation. Using these definitions, we lay down the concepts of OLCs and LCI.

2.1   General

Let U be some universe of identifiers, and let L be some set of action labels.

A bag over some alphabet A is a function from A to $\mathbb{N}$ that assigns only a finite number of elements from A a positive value. For a bag b over alphabet A and $a \in A$, $b(a)$ denotes the number of occurrences of a in b, often called the cardinality of a in b. Note that a finite set of elements from A is also a bag over A, namely the function yielding 1 for every element in the set and 0 otherwise. The set of all bags over A is denoted μA. We use brackets to explicitly enumerate a bag and superscripts to denote cardinalities. For example, $[a^2, b^3, c]$ is the bag with two a's, three b's and one c; the bag $[a^2|P(a)]$, where P is a predicate on A, contains two elements a for every a such that $P(a)$ holds. The sum of two bags b_1 and b_2, denoted $b_1 + b_2$, is defined as $[a^n|a \in A \wedge n = b_1(a) + b_2(a)]$. The difference of b_1 and b_2, denoted $b_1 - b_2$, is defined as $[a^n|a \in A \wedge n = (b_1(a) - b_2(a)) \max 0]$. Bag b_1 is a subbag of b_2, denoted $b_1 \leq b_2$, if and only if for all $a \in A$: $b_1(a) \leq b_2(a)$. Bag b_1 is a strict subbag of b_2, denoted $b_1 < b_2$, if and only if $b_1 \leq b_2$ and $b_1 \neq b_2$. Bag b_1 is minimal in a set of bags if and only if there is no bag b_2 in that set such that $b_2 < b_1$. The set of all minimal bags in a set of bags S is denoted $\lfloor S \rfloor$.

The range of a function $f \in D \to R$, denoted $\text{rng}(f)$, is defined as the set of $r \in R$ for which a $d \in D$ exists such that $f(d) = r$.

2.2   Labeled P/T Systems and Branching Bisimilarity

The class of Petri nets used as the basis for specifying OLCs is the class of labeled P/T nets. Labels are an abstract representation of object methods.

Definition 1. (Labeled P/T net) A labeled P/T net is a tuple (P, T, F, l) where

  (i) $P \subseteq U$ is a finite set of places,
  (ii) $T \subseteq U$ is a finite set of transitions such that $P \cap T = \varnothing$,
  (iii) $F \subseteq (P \times T) \cup (T \times P)$ is a set of directed arcs, called the flow relation, and
  (iv) $l \in T \to L$ is a labeling function.

The labeling function connects transitions to methods. The use of labels allows multiple occurrences of a method in an OLC specification.

Let $N = (P, T, F, l)$ be a labeled P/T net. Elements of $P \cup T$ are referred to as nodes. A node $n_1 \in P \cup T$ is an input node of another node $n_2 \in P \cup T$ if and only if there exists a directed arc from n_1 to n_2, that is, if and only if $n_1 F n_2$. Node n_1 is an output node of n_2 if and only if there exists a directed arc from n_2 to n_1. If n_1 is a place, it is called an input place or an output place; if it is a transition, it is called an input transition or an output transition. The set of all input nodes of some node n is called the preset and is denoted $\bullet n$; its set of output nodes is called the postset and is denoted $n\bullet$. To avoid confusion, a subscript x of a net N_x is also used as subscript for the various elements (P_x, $\bullet_x$, ...).

The current state of a labeled P/T net $N = (P, T, F, l)$ is given by a bag (or marking) $M \in \mu P$. A transition $t \in T$ is enabled at a marking M, denoted $M[t\rangle$, if and only if each of its input places is marked, that is, if and only if $\bullet t \leq M$. (Note that $\bullet t$ is interpreted as a bag.) If $M[t\rangle$, transition t can fire, resulting in a new marking $M_1 = M - \bullet t + t\bullet$. This is denoted $M[t\rangle M_1$. If M_1 can be reached from M by firing a sequence of transitions $s = t_1 \cdot t_2 \cdot t_3 \cdot ...$, this is denoted $M[s\rangle M_1$. The set of markings that are reachable from a marking M by firing transitions is denoted $[M\rangle$.

The constraint generation for our backtracking algorithm (BA) uses the well-known structural technique of T-invariants. Usually, T-invariants are defined to be vectors over T, that is, mappings from T to the integer numbers $\mathbb{Z}$. For the BA, we are only interested in *semi-positive* T-invariants, that is, mappings from T to the natural numbers $\mathbb{N}$. This allows us to define semi-positive T-invariants as bags.

Definition 2. (Semi-positive T-invariant) Let $N = (P, T, F, l)$ be a labeled P/T net. A bag $b \in \mu T$ is a semi-positive T-invariant (STI) of N if and only if for all $p \in P$:

$$\sum_{t \in \bullet p} b(t) = \sum_{t \in p\bullet} b(t)$$

The set of all STIs in N is denoted σN .

For an arbitrary place, the transitions in a semi-positive T-invariant produce (left-hand side of the equation) as many tokens as they consume (righthand side).

In this paper, we are particularly interested in elementary (or minimal) STIs.

Definition 3. (Elementary STI) Let $N = (P, T, F, l)$ be a labeled P/T net and let $b \in \mu T$ be an STI of N, that is, $b \in \sigma N$. STI b is elementary if and only if it is minimal in σN. The set of all elementary STIs in N is denoted $\lfloor \sigma N \rfloor$.

A labeled P/T net extended with an initial marking and a notion of successful termination defines a labeled P/T system.

Definition 4. (Labeled P/T system) A labeled P/T system is a tuple $S = (N, I, O)$ where

(i) $N = (P, T, F, l)$ is a labeled P/T net,

(ii) $I \in \mu P$ is the initial marking, and

(iii) $O \in \mu P$ is the marking indicating successful termination.

The notion of successful termination is not standard for P/T nets. However, when modeling OLCs, it is crucial to distinguish successful termination from unsuccessful termination (or deadlock). Hence, we include the notion of successful termination already in the definition of a labeled P/T system.

Definition 5. (Cycle) Let $S = (N, I, O)$ be a labeled P/T system, and let $s = t_1 \cdot t_2 \cdot t_3 \cdot ...$ be a sequence of transitions. Sequence s is a cycle of S if and only if there exists an $M \in [I\rangle$ such that $M[s\rangle M$. The set of cycles of a labeled P/T system S is denoted σS .

As with STIs, we are particularly interested in elementary cycles, that is, in cycles with minimal transition support.

Definition 6. (Elementary cycle) Let S be a labeled P/T system and let $c = t_1 \cdot t_2 \cdot t_3 \cdot \ldots$ be a cycle of S, that is, $c \in \sigma S$. Cycle c is elementary if and only if its transition bag, that is, $[t_1] + [t_2] + [t_3] + \ldots$, is minimal in the set of transition bags of all cycles of S. The set of all elementary cycles in S is denoted $\lfloor \sigma S \rfloor$.

It is straightforward to check that every cycle is a combination of elementary cycles, that is, from the set of elementary cycles we can generate the set of cycles. As a result, the set of elementary cycles uniquely defines the set of cycles. Note that when all transitions in a semi-positive T-invariant would be fired (assuming sufficient tokens are available for consumption) in some sequence, this sequence would be a cycle.

The complete behavior of a labeled P/T system is captured by its reachability graph.

Definition 7. (Reachability Graph) Let $S = ((P, T, F, l), I, O)$ be a labeled P/T system. Let $G = (V, E)$ be a graph which satisfies the following requirements:

  (i) $V = [I \rangle$, and
  (ii) $E = \{(M_1, l(t), M_2) | M_1, M_2 \in V \wedge t \in T \wedge M_1[t \rangle M_2\}$.

Graph G is called the reachability (or occurrence) graph of S.

Let $G = (V, E)$ be a reachability graph, $v_1, v_2 \in V$, and $a \in L$. We write $v_1[a \rangle v_2$ if node v_2 can be reached from node v_1 by following an a-labeled edge, that is, if $(v_1, a, v_2) \in E$.

All labels $a \in L$ are externally observable, except for the designated label τ. Transitions that are labeled τ are silent and not visible to the environment. The notion of silent actions forms the basis for the hiding operation. We write $v_1[\tau^* \rangle v_2$ if v_2 can be reached from v_1 by following any number of τ-labeled edges, and we write $v_1[(a) \rangle v_2$ if either (i) $a = \tau$ and $v_1 = v_2$, or (ii) $v_1[a \rangle v_2$. Thus, $v_1[(\tau) \rangle v_2$ means that v_2 can be reached from v_1 by following zero (i) or one (ii) τ-labeled edges; for any other $a \in L$, $v_1[(a) \rangle v_2$ is equivalent to $v_1[a \rangle v_2$ because (i) can never be satisfied.

The next definition introduces branching bisimilarity (BB) [14]. BB is a behavioral equivalence that equates systems with the same externally observable behavior but possibly different internal behavior. Although BB is not the only equivalence suitable for this purpose, we build on this well-known equivalence.

Definition 8. (Branching bisimilarity) Let $G_1 = (V_1, E_1)$ be the reachability graph of a labeled P/T system $S_1 = (N_1, I_1, O_1)$, let $G_2 = (V_2, E_2)$ be the reachability graph of a labeled P/T system $S_2 = (N_2, I_2, O_2)$, and let $R \subseteq V_1 \times V_2$ be a binary relation on the nodes of both graphs. The relation R is called a branching bisimulation relation on G_1 and G_2 if and only if,

  (i)   if $v_1 R v_2$ and $v_1[a \rangle v_1'$, then there exist $v_2', v_2'' \in V_2$ such that $v_2[\tau^* \rangle v_2''$, $v_2''[(a) \rangle v_2'$, $v_1 R v_2''$, and $v_1' R v_2'$,
  (ii) if $v_1 R v_2$ and $v_2[a \rangle v_2'$, then there exist $v_1', v_1'' \in V_1$ such that $v_1[\tau^* \rangle v_1''$, $v_1''[(a) \rangle v_1'$, $v_1'' R v_2$, and $v_1' R v_2'$,

(iii) if $v_1 R v_2$ then $(v_1 = O_1) \Rightarrow (v_2 [\tau^* \rangle O_2)$ and $(v_2 = O_2) \Rightarrow (v_1 [\tau^* \rangle O_1)$.

Systems S_1 and S_2 are branching bisimilar, denoted $S_1 \cong S_2$, if and only if a branching bisimulation R exists between G_1 and G_2 such that $I_1 R I_2$.

2.3 Object Life Cycles and Life-Cycle Inheritance

An OLC specifies the order in which the methods of an object may be executed. When modeling a life cycle with a Petri net, a transition firing corresponds to the execution of a method. The emphasis of an OLC is on the execution order of methods and not on their implementation details. Therefore, the uncolored formalism of P/T nets as introduced before is well suited as the basic framework for modeling life cycles and studying LCI. As mentioned, transition labels correspond to method identifiers.

Definition 9. (Object Life Cycle) Let $N = (P, T, F, l)$ be a labeled P/T net such that:

(i) there is exactly one $p \in P$ such that $\bullet p = \varnothing$; this place denotes the state that an OLC has just been created and is denoted i;

(ii) there is exactly one $p \in P$ such that $p \bullet = \varnothing$; this place corresponds to successful termination of an OLC and is denoted o;

(iii) for all $n \in P \cup T$: $i F^* n$ and $n F^* o$, where F^* is the reflexive and transitive closure of F; this requirement means that every node in an OLC must lie on a path from creation to termination.

The labeled P/T system $S = (N, [i], [o])$ is an object life cycle (OLC) if and only if:

(iv) for all $M \in [[i] \rangle$, there exists an $M_1 \in [M \rangle$ such that $[o] \le M_1$ (an OLC always has the option to terminate);

(v) for all $M \in [[i] \rangle$, if $[o] \le M$ then $[o] = M$ (termination of an OLC is always successful);

(vi) for all $t \in T$, there exists an $M \in [[i] \rangle$ such that $M[t \rangle$ (no dead transitions; any transition in an OLC has a meaningful contribution to some execution of the OLC).

Figure 1 shows examples of OLCs modeling simple production units (rcmd : receive command; pmat : process material; omat : output material; repp : repeat processing; cerr : correct error; ssps : send start-processing signal; ppmat : preprocess material).

OLCs coincide with the class of sound WF-nets as described in [1]. Therefore, from Theorem 11 in [1], we may conclude that for any OLC $S = ((P, T, F, l), I, O)$ the labeled P/T system $((P, T \cup \{t\}, F \cup \{(t, i), (o, t)\}, l \cup \{(t, \tau)\}), I, O)$, where $t \notin P \cup T$, is live and bounded. This labeled P/T system is called the short-circuited system of S, denoted φS, because it adds a short-circuiting transition from the output place o to the input place i. Likewise, the short-circuited net underlying an OLC $S = (N, I, O)$ is denoted φN. As a result of the boundedness of short-circuited OLCs, the reachability graph of any OLC is finite.

Definition 10. (Encapsulation) Let $S = ((P, T, F, l), I, O)$ be an OLC and let $\Delta \subseteq L \setminus \{\tau\}$ be a set of labels. The encapsulation operator ∂_Δ removes from a given OLC all transitions with a label in Δ. Formally, $\partial_\Delta(S) = ((P, T_1, F_1, l_1), I, O)$ such

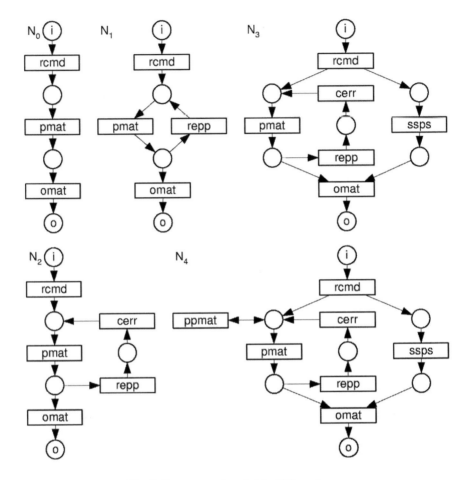

Fig. 1. Some examples of object life cycles

that $T_1 = \{t \mid t \in T \wedge l(t) \notin \Delta\}$, $F_1 = F \cap ((P \times T_1) \cup (T_1 \times P))$,   and
$l_1 = l \cap (T_1 \times L)$.

Encapsulation can be used to block methods in an OLC. Note that the reachability graph of an encapsulated P/T system $\partial_\Delta(S)$ is a subgraph of the reachability graph of the OLC S. As a result, it is also finite. Also note that encapsulating methods in an OLC may result in a labeled P/T system that is no longer an OLC.

Definition 11. (Abstraction) Let $S = ((P, T, F, l), I, O)$ be an OLC and let $\Delta \subseteq L \backslash \{\tau\}$ be a set of labels. The abstraction operator τ_Δ renames all transitions with a label in Δ to the silent action τ. Formally, $\tau_\Delta(S) = ((P, T, F, l_1), I, O)$ such that, for any $t \in T$, $l(t) \in \Delta$ implies $l_1(t) = \tau$ and $l(t) \notin \Delta$ implies $l_1(t) = l(t)$.

Abstraction provides a means to hide methods in an OLC.

Definition 12. (Life-cycle inheritance [7, 8]) Let S_1 and S_2 be OLCs with reachability graphs G_1 and G_2. OLC S_2 is a subclass of OLC S_1 under life-cycle inheritance, denoted $S_2 \leq S_1$, if and only if $\Delta_H, \Delta_B \subseteq L \backslash \{\tau\}$ exist such that $\Delta_H \cap \Delta_B = \emptyset$ and $S_1 \cong \tau_{\Delta_H} \circ \partial_{\Delta_B}(S_2)$.

Consider again the example OLCs in Figure 1. It is not difficult to verify that $N_i \leq N_j$ if and only if $i \geq j$. For example, hiding cerr and ssps and blocking ppmat in N_4 yields a P/T net branching bisimilar to N_1, thus showing that $N_4 \leq N_1$.

3   Backtracking Algorithm

First, we explain the basic backtracking algorithm (BA). Then, we discuss two techniques to compute constraints.

3.1   General Structure

The BA needs to decide whether some OLC is a subclass of another OLC under LCI. For sake of convenience, we refer to the first OLC as the potential sub OLC and to the second as the base OLC.

An exhaustive algorithm for deciding LCI has to test all partitionings of the methods new in the potential sub OLC into methods to be hidden and methods to be blocked. The BA prunes the search tree of partitionings when possible. For this pruning, we introduce a fast check on necessary conditions that have to hold when the base OLC and the potential sub OLC (after hiding and/or blocking the new labels in the potential sub OLC) are branching bisimilar. If the fast check fails, the branching bisimilarity (BB) check will also fail. The fast check can be applied to a partial partitioning of new labels, whereas the BB check can only be applied to a full partitioning. So, when the fast check fails on some partial partitioning, a BB check on any extension of that partial partitioning will fail. As a result, we can prune an entire subtree.

We investigate some properties of the set of new labels that have to hold when we assume BB. These properties are called constraints. When such a constraint is violated, that is, when such a property does not hold, we know that we cannot achieve BB any more. For the remainder of this section, we use the following notations:

(i) $S_b = ((P_b, T_b, F_b, l_b), I_b, O_b)$ is the base OLC.

(ii) $S_s = ((P_s, T_s, F_s, l_s), I_s, O_s)$ is the potential sub OLC.

(iii) Δ is the set of new labels; that is, $\Delta = \mathrm{rng}(l_s) \backslash \mathrm{rng}(l_b)$.

(iv) Δ_H and Δ_B partition Δ, where Δ_H is the set of new labels that are hidden and Δ_B is the set of new labels that are blocked.

(v) S_Δ is a shorthand for $\tau_{\Delta_H} \circ \partial_{\Delta_B}(S_s)$.

Using these notations, we define the concept of constraints. A constraint is simply a pair of sets of new labels. To allow a successful BB check *either* some labels must be hidden from the first set *or* some labels must be blocked from the second set. This notion of constraints provides a very general and flexible way of expressing constraints on partitionings of new labels. If the first set in a constraint is empty, then the

```
1    compute constraints
2    if (constraint (∅, ∅) exists) {
3        return without solution
4    }
5    set Π_H to ∅, Π_B to ∅, backtrack to false, and n to |Δ|
6    while (true) {
7        if (some constraint is violated by Π_H and Π_B) {
8            set backtrack to true
9        } else {
10           if (n equals 0) {
11               set Δ_H to Π_H and Δ_B to Π_B
12               if (branching bisimilar) {
13                   return with solution
14               }
15               set backtrack to true
16           } else { /* backtrack is false */
17               decrement n
18               add Δ[n] to Π_B
19       } }
20       if (backtrack) {
21           if (Π_B is empty) {
22               return without solution
23           }
24           while (Δ[n] in Π_H) {
25               remove Δ[n] from Π_H
26               increment n
27           }
28           /* Π_B not empty */
29           move Δ[n] from Π_B to Π_H
30           set backtrack to false
31   } }
```

Fig. 2. The basic backtracking algorithm

second set is called a block constraint; if the second set is empty, the first set is called a hide constraint.

Definition 13. (Constraint) Let $h, b \subseteq \Delta$. The pair (h, b) is a constraint if and only if for any Δ_H and Δ_B: $(S_b \cong S_\Delta) \Rightarrow \left(h \not\subseteq \Delta_B \vee b \not\subseteq \Delta_H \right)$ If $h = \emptyset$, then b is called a block constraint; if $b = \emptyset$, then h is called a hide constraint.

Note that for the BA to work properly, we are allowed to ignore a valid constraint (less pruning than possible), but are not allowed to take into account an invalid constraint (too much pruning, that is, possible pruning of valid solutions).

At this point, we can explain the basic BA. This BA is shown in Figure 2. First, we compute constraints (line 1). We return to the computation of constraints in later sub-

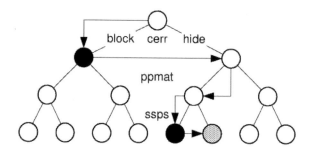

Fig. 3. The algorithm in action

sections. If the unsatisfiable constraint $(\varnothing, \varnothing)$ exists (line 2), we return without a solution (line 3). Otherwise, we initialize some necessary variables (line 5). Π_H holds the new labels that are hidden, Π_B holds the new labels that are blocked, backtrack indicates that we detected a dead end, and n equals the number of new labels still to assign. So, when n equals 0, Π_H and Π_B partition Δ and are therefore valid candidates for Δ_H and Δ_B. Note that the partial partitioning of labels to Π_H and Π_B may already violate some constraint (h, b), namely if h is fully contained in Π_B and if b is fully contained in Π_H. It is clear that it is of no use to extend a partial partitioning that violates some constraint: A violated constraint will remain violated. Repeatedly, we check the current partitioning, Π_H and Π_B (lines 6–31). If the current partitioning violates some constraint, we decide to backtrack (line 8). If the current partitioning does not violate some constraint, we check whether the partitioning is full (that is, is not partial, line 10). If so, we set Δ_H and Δ_B to Π_H and Π_B and check BB between S_b and S_Δ (line 11). If branching bisimilar, we return with the current partitioning as a solution (line 13), otherwise we decide to backtrack (line 15). If the current partitioning was only partial, we block the next unassigned new label (lines 16–18), generating a new partial partitioning for the next iteration. Note that the BA assumes some kind of total ordering on the new labels in Δ. Also note that the BA prefers to block new labels (because blocking reduces the size of the state space where hiding has no effect on this size). Thus, the last partitioning to check is the one that assigns all new labels in Δ to Π_H, and, if a (partial) partitioning assigning only labels to Π_H violates the constraints, no solution is possible anymore. When it was decided to backtrack, we first check whether there are partitionings left to check (line 21). If not, we return without a solution (line 22). Otherwise, we repeatedly unassign the last assigned new label until we find a blocked label (lines 24–27), hide this blocked label (line 29), and reset the backtrack variable (line 30).

Figure 3 shows how the BA performs when we take N_1 of Figure 1 as base OLC and N_4 as potential sub OLC. An exhaustive search algorithm (ESA) would traverse this search tree from left to right, testing every leaf node for BB. These leaf nodes correspond to partitionings of the new labels cerr, ppmat, and ssps into labels to be blocked and to be hidden. From the structure of both OLCs, we can deduce that both {cerr} and {ssps} are hide constraints, see below. The arrows in Figure 3 indicate the actual tree traversal of the BA, using both hide constraints. A black node indicates

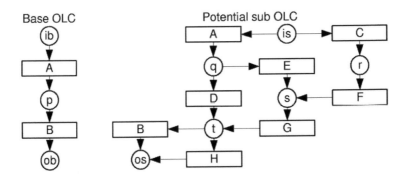

Fig. 4. OLCs used to explain constraints

that some constraint is violated, a grey node indicates that a (successful) BB check is performed. Note that the BA needs only one BB check to show that $N_4 \leq N_1$, whereas an ESA that prefers blocking over hiding would need six BB checks.

Now, we only have to find suitable constraints. In the following subsections, we derive (i) constraints from the fact that a base OLC cannot complete unsuccessfully, and (ii) hide and block constraints from the comparison of cycles in both (short-circuited) systems.

3.2 Successful Termination

The labels that the BA operates on are, by definition, only present in the potential sub OLC. Therefore, the base OLC will not be affected by blocking or hiding them. Because the base OLC is an OLC, it can only terminate successfully. BB distinguishes successful termination from unsuccessful termination. As a result, the potential sub OLC can only be branching bisimilar to the base OLC if it too can only terminate successfully. That is, we cannot allow the potential subclass to terminate unsuccessfully.

Consider the base and potential sub OLC in Figure 4. Recall that successful termination for the base OLC and the potential sub OLC coincides with the markings O_b and O_s, that is, with [ob] and [os]. Because of what we mentioned above, we cannot allow a token to get stuck in places is , q , r , s , or t of the potential sub OLC. If we block labels D and E , a token in place q will be stuck; if we block label F , a token in place r will be stuck (unless also C is blocked); if we block label G , a token in place s will be stuck (unless also E, and C or F are blocked). Note that we cannot block labels A or B ; therefore, a token cannot get stuck in places is or t . In general, if the potential sub OLC contains a place $p \in P_s \backslash \{o_s\}$ from which every output transition is labeled with a new label, we should not block all labels on these transitions. Thus, such a set of labels indicates a possible hide constraint. In Figure 4, the label sets {D, E}, {F}, and {G} are example candidates. However, the attentive reader has noticed that we made provisions. In the potential sub OLC of Figure 4, places r and s might not be markable. If we block label C , place r will be unmarkable. If we block label E and either label C or F , place s will be unmarkable. If a place is unmarkable, we *are* allowed to block all outgoing labels. Thus, {F} is not a hide constraint because of the partitionings where C is blocked; {G} is not a hide constraint because of the partitionings where either (i) C and E or (ii) E and F are blocked. Although {F} and {G} are not hide constraints, the pairs ({F}, {C}) and ({G}, {C, E, F}) are both constraints: If we do not hide some labels of the first set, we have to block at least one of the labels of the second set.

In the remainder of this subsection, we formalize this idea of deriving constraints from the successful termination requirement of OLCs. All places in the potential sub OLC of which all output transitions are labeled with new labels in Δ induce a constraint (h, b), where h is the set of all labels of output transitions, and where b is a set of labels whose encapsulation may render the place unmarkable. The first main result shows that if we block all output transitions of some place p and it is still possible to prove an LCI relationship, then it must be the case that p is unmarkable.

Theorem 1. Let Δ_H and Δ_B partition Δ such that $p \in P_s \backslash \{o_s\}$ is a place with for all $t \in p\bullet_s : l_s(t) \in \Delta_B$. Then $S_b \cong S_\Delta$ implies that p is unmarkable.

Proof. Suppose $S_b \cong S_\Delta$. A crucial property of a branching bisimulation is that, given two related markings, it relates any marking reachable from one of these to a marking reachable from the other (see Lemma 2.2.21 of [7]). Suppose p is markable in S_Δ. Then a marking containing p can be reached in S_Δ. However, because we cannot remove the token in p, from this marking we cannot reach O_s. Thus, property (iii) of Definition 8 and the fact that O_b is reachable in S_b from any reachable marking in S_b (Requirements (iv) and (v) of Definition 9) imply that no branching bisimulation can relate any marking with a token in p to a marking reachable in S_b, which leads to a contradiction with the assumption that $S_b \cong S_\Delta$. Therefore, p is unmarkable. □

The properties of an OLC (Requirements (iii) and (vi) of Definition 9) guarantee that all places are markable from the initial marking. Thus, only blocking transitions in an OLC can make places unmarkable. We use a structural property to find the new labels that, when blocked in the potential sub OLC, *might* render a place with an induced constraint unmarkable. When one of these new labels gets blocked, the associated constraint is satisfied. However, it is possible that after blocking some of these new labels the place is markable after all. If so, this can only result in a suboptimally pruned search space and is thus safe. It is for our purposes more important that the structural property is efficiently computable.

The basis for the second set b of our constraints is the following: If the base OLC and the potential sub OLC (after hiding and blocking new labels) are BB, and if a place in the potential sub OLC is unmarkable, there must be a blocked transition with (i) a directed path of arcs leading to that place, (ii) all its input places markable, and (iii) all its input places having at least one other output transition (to remove tokens possibly put into them). Thus, a place inducing a constraint (h, b) *may* become unmarkable if any transition satisfying the above three conditions is blocked. The set b is the set of labels of all transitions satisfying these conditions.

In the potential sub OLC of Figure 4, the pair ({F}, {C}) is a constraint, the pair ({G}, {C, E}) is a constraint, and the pair ({D, E}, ∅) is a constraint (thus, {D, E} is a hide constraint). Note that, because of requirement (iii) above, the constraint ({G}, {C, E, F}) mentioned earlier can be strengthened to ({G}, {C, E}): if F is blocked, it would only shift the problem (a token that gets stuck) from place s to place r. Also note that G needs to be hidden s long as not both C and E are blocked, whereas the constraint $(\{G\}, \{C, E\})$ is satisfied as soon as either C or E is blocked. Although this is safe, it could lead to suboptimal pruning of the search tree.

Theorem 2. Let Δ_H and Δ_B partition Δ such that $S_b \cong S_\Delta$ and $p \in P_s$ is an unmarkable place. Then there exists a $t \in T_s$ such that $l_s(t) \in \Delta_B$, (i) tF_s^*p, and for all $q \in \bullet_s t$: (ii) q is markable and (iii) $|q\bullet_s| > 1$.

Proof. S_s is an OLC, but p is unmarkable in S_Δ. Requirements (iii) and (vi) of Definition 9 imply that any place in an OLC is markable. Therefore, there has to exist a $t \in T_s$ such that tF_s^*p, and $l_s(t) \in \Delta_B$ (p has become unmarkable because of blocking transition t), and for all $q \in \bullet_s t$: q is markable (blocking transition t can only have effects, like rendering a place unmarkable, if it can be enabled). Because $S_b \cong S_\Delta$, a token in $q \in \bullet_s t$ needs to be removable (see the proof of Theorem 1). Because $t \in q\bullet_s$ is blocked, $|q\bullet_s| > 1$ should hold (there has to be another transition that removes the token).   □

Theorem 1 implies that, if some place of which all output transitions have new labels is markable, then from a specified set of labels h at least one label must be hidden for a successful BB check. Theorem 2 states that from a specified set of new labels b at least one label must be blocked if such a place is unmarkable. Thus, the pair (h, b) is a constraint. However, one of the conditions in Theorem 2 is non-structural, non-efficiently computable, namely the one concerning the markability of the input places of the mentioned transition t. As explained before, weakening a constraint is safe, thus we can replace this condition with the obvious requirement that t is not an output transition of the unmarkable place p. Therefore, we obtain the following theorem.

Theorem 3. Suppose $p \in P_s \setminus \{o_s\}$ is a place such that for all $t \in p\bullet_s$: $l_s(t) \in \Delta$. Then the pair (h, b) is a constraint, where

   (i) $h = \{l_s(t) \mid t \in p\bullet_s\}$ and

   (ii) $b = \{l_s(t) \mid t \in T_s \setminus \{p\bullet_s\} \wedge l_s(t) \in \Delta \wedge tF_s^*p \wedge \forall q \in \bullet_s t : (|q\bullet_s| > 1)\}$.

Proof. Theorem 1 and Theorem 2.   □

3.3   Visible Label Supports

3.3.1 Cycles

Recall that LCI is based on BB after hiding and blocking new labels. A fundamental property of BB is that if some system can generate a certain sequence of visible labels, any other branching bisimilar system can generate the same sequence of visible labels. Cycles can generate infinite sequences of labels. Because OLCs are bounded, cycles are the only way to generate infinite sequences of labels. If some bounded system contains a cycle, then any other branching bisimilar bounded system should contain a cycle that can generate the same sequences of visible labels. As a result, the visible label supports (VLSs) of these two cycles, that is, the sets of visible labels in the cycles, should be identical. Note that the cycles themselves need not be identical:

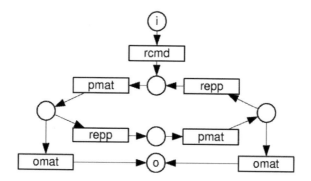

Fig. 5. OLC N_1 (see Figure 1) with the cycle unfolded once

Figure 5 shows an OLC that is branching bisimilar to OLC N_1 of Figure 1, although the (labeled) cycles pmat · repp and pmat · repp · pmat · repp are not identical. However, if two bounded systems have different visible label *supports* for their cycles, they cannot be branching bisimilar.

Definition 14. (Visible label support (VLS) of cycles) Let $S = ((P, T, F, l), I, O)$ be a labeled P/T system, and let $c = t_1 \cdot t_2 \cdot t_3 \cdot \ldots$ be a cycle of S, that is, $c \in \sigma S$. The VLS of c, denoted $\langle c \rangle$ is the set of visible labels occurring in c, that is, $\{l(t_1), l(t_2), l(t_3), \ldots\} \setminus \{\tau\}$. If C is a set of cycles, we use $\langle C \rangle$ to denote the set of VLSs of all cycles in C.

Theorem 4. Let S_1 and S_2 be two bounded labeled P/T systems. Then $(S_1 \cong S_2) \Rightarrow (\langle \sigma S_1 \rangle \not\models \langle \sigma S_2 \rangle)$.

Proof. Assume $S_1 \cong S_2$ and $\langle \sigma S_1 \not\models \langle \sigma S_2 \rangle$ Consider the set V of VLSs in $\langle \sigma S_1 \cup \langle \sigma S_2 \rangle$but not in $\langle \sigma S_1 \cap \langle \sigma S_2 \rangle$ Take a $v \in V$. Assume without loss of generality that $v \in \langle \sigma S_1 \rangle$and $v \notin \langle \sigma S_2 \rangle$Apparently, S_2 cannot generate an infinite sequence containing *exactly* the labels from v, whereas S_1 can. This contradicts the assumption of branching bisimilarity. ☐

As mentioned before, for two bounded systems to be branching bisimilar, the VLSs of their cycles should be equivalent. However, some parts of the systems, OLCs in our case, need not be covered by cycles. We use the fact that short-circuited OLCs usually have many more cycles than the original OLCs and are just like OLCs bounded.

Theorem 5. $(S_b \cong S_\Delta) \Leftrightarrow (\varphi S_b \cong \varphi S_\Delta)$, where $\varphi S_\Delta = \tau_{\Delta_H} \circ \partial_{\Delta_B}(\varphi S_s)$.

Proof. Let R be a branching bisimulation between S_b and S_Δ (φS_b and φS_Δ). It follows immediately from Definitions 8 and 9, and the definition of short-circuited OLCs that R is also a branching bisimulation between φS_b and φS_Δ (S_b and S_Δ). ☐

Corollary 1. A potential sub OLC can only be a subclass under life-cycle inheritance of a base OLC if, after short-circuiting the systems and hiding and blocking all new labels, the VLSs of their cycles are equivalent.

We can derive both hide and block constraints from this corollary. Consider for example the OLCs in Figure 4. If we short-circuit both OLCs, the base OLC will have the label set $\{A, B\}$ as the VLS of the only cycle. If we block all new labels, the short-circuited potential sub OLC will have no cycles at all. Therefore, blocking all labels is not an option. Apparently, we have blocked too many new labels. We should make sure that we do not block all possible cycles that have $\{A, B\}$ as VLS. This will lead to hide constraints. If, on the other hand, we hide all new labels, the short-circuited potential sub OLC will have cycles with four possible VLSs: $\varnothing$, $\{A\}$, $\{B\}$, and $\{A, B\}$. Because of the first three VLSs, hiding all labels is also not an option. Apparently, we hid too many new labels. We should make sure that we do block all possible cycles that have $\varnothing$, $\{A\}$, or $\{B\}$ as VLS. This will lead to block constraints.

3.3.2 Block Constraints

Assume that we hide all new labels in the short-circuited potential sub OLC of Figure 4. The resulting system has cycles with VLSs $\varnothing$, $\{A\}$, or $\{B\}$ that the short-circuited base OLC does not have. Clearly, if we want a successful BB check, we should prevent the cycles with VLSs $\varnothing$, $\{A\}$, or $\{B\}$ from occurring. If no new labels are present in such a cycle, then we cannot prevent it from occurring; that is, we have the empty set as a block constraint, which cannot be satisfied: There is no partitioning of new labels yielding BB. This fact is used in line 2 of the BA of Figure 2. Otherwise, we can possibly achieve branching bisimilarity by blocking at least one of the new labels present in every cycle with VLSs $\varnothing$, $\{A\}$, or $\{B\}$. To block cycles with VLS $\varnothing$, we should block one or more of the labels C, F, G, or H; to block cycles with VLS $\{A\}$, we should block one or more of the labels D or H, and one or more of the labels E, G, or H; to block cycles with VLS $\{B\}$, we should block one or more of the labels C, F or G. As a result, we conclude that the label sets $\{C, F, G, H\}$, $\{D, H\}$, $\{E, G, H\}$, and $\{C, F, G\}$ are block constraints. The following is a direct corollary of Corollary 1.

Corollary 2. Let $v \in \langle \sigma \varphi S_s \rangle$ be a VLS such that $v \backslash \Delta \notin \langle \sigma \varphi S_b \rangle$ The set $v \cap \Delta$ is a block constraint.

3.3.3 Hide Constraints

Assume that we block all new labels in the short-circuited potential sub OLC of Figure 4. The resulting system has no cycles, while the short-circuited base OLC has a cycle with VLS $\{A, B\}$. Clearly, if we want both systems to be branching bisimilar, we should hide some of the new labels in such a way that a cycle appears in the short-circuited potential sub OLC with the VLS $\{A, B\}$. This can be done because such cycles exist when we would hide all new labels. If no such cycle would exist, then both systems cannot become branching bisimilar for any partitioning of the new labels. When they exist, we should not block them all. As a result, we should hide all new labels of one or more of these cycles. Considering the example of Figure 4, we should (i) hide D or (ii) hide E and G. This implies that the label set $\{D, E, G\}$ is a hide constraint. Note that we do not use the fact that we need to hide both label G and E if we block label D. Because this information does not fit our notion of constraints, we cannot use it at this point. However, as mentioned earlier, we prefer simplicity wherever possible (as long as it leads to good results). The following is a second direct corollary of Corollary 1.

Corollary 3. Let $v_b \in \langle \sigma\varphi S_b \rangle$ be a VLS such that $v_b \notin \langle \sigma\varphi S_s \rangle$ The set $\{l | l \in v_s \cap \Delta \wedge v_s \in \langle \sigma\varphi S_s \rangle \wedge v_s \backslash \Delta = v_b\}$ is a hide constraint.

3.3.4 Structural Properties

At this point, we deduced from the VLSs of all cycles in the two (short-circuited) OLCs under consideration a set of hide constraints and a set of block constraints. However, as mentioned earlier, we prefer to use structural properties rather than behavioral properties. Therefore, we prefer to use elementary STIs instead of cycles. First, we argue that we only need to take minimal VLSs into account. Second, we argue that when taking only minimal VLSs into account, it suffices to take only elementary cycles into account. Third, we argue that, with some odd exceptions perhaps, we can use elementary STIs as an alternative for elementary cycles. We also discuss what to do when the systems at hand happen to be one of these odd exceptions.

Theorem 6. Let S_1 and S_2 be two bounded labeled P/T systems. Then $(S_1 \cong S_2) \Rightarrow (\lfloor \langle \sigma S_1 \rfloor = \lfloor \langle \sigma S_2 \rfloor)$.

Proof. This follows immediately from Theorem 4 and the fact that two sets can only be identical if their sets of minimal elements are identical. □

As a result of Theorem 6, we only need to take the sets of minimal VLSs into account. Thus, in Corollaries 2 and 3 we can replace the occurrences of $\langle \sigma\varphi S_b \rangle$ and $\langle \sigma\varphi S_s \rangle$ by $\lfloor \langle \sigma\varphi S_b \rfloor$ and $\lfloor \langle \sigma\varphi S_s \rfloor$.

Theorem 7. Let S be a labeled P/T system and let $v \in \lfloor \langle \sigma S \rfloor$. There exists an elementary cycle $c \in \lfloor \sigma S \rfloor$ such that $\langle c \rangle \models v$.

Proof. From $v \in \lfloor \langle \sigma S \rfloor$, it follows that $C_1 = \{c_1 \in \sigma S | \langle c_1 \rangle \models v\}$ is not empty. Let $c_2 \in C_1$. Assume $c_2 \notin \lfloor \sigma S \rfloor$. Because $c_2 \notin \lfloor \sigma S \rfloor$, there has to exist a $c_3 \in \lfloor \sigma S \rfloor$ such that the transition bag of c_3 is a strict subbag of the transition bag of c_2, and thus, $\langle c_3 \rangle \sqsubseteq \langle c_2 \rangle$ Recall that $\langle c_2 \rangle \models v$ and that v is minimal. From this, it follows that $\langle c_3 \rangle \models v$. □

As a result of Theorem 7, we only need to take elementary cycles into account in Corollaries 2 and 3. Thus, we can replace the occurrences of $\langle \sigma\varphi S_b \rangle$ and $\langle \sigma\varphi S_s \rangle$ by $\lfloor \langle \lfloor \sigma\varphi S_b \rfloor \rfloor$ and $\lfloor \langle \lfloor \sigma\varphi S_s \rfloor \rfloor$.

Recall that σN denotes the set of STIs of a labeled P/T net N.

Definition 15. (VLSs of STIs) Let $N = (P, T, F, l)$ be a labeled P/T net and let $b \in \sigma N$. Then the VLS of b, denoted $\langle b \rangle$ is the set of visible labels occurring in b: $\{l(t) | t \in T \wedge b(t) > 0\} \backslash \{\tau\}$. If B is a set of STIs, we use $\langle B \rangle$ to denote the set of VLSs of all STIs in B.

It is easy to check that, by definition, every cycle induces an STI. For free choice labeled P/T nets, the reverse is also true [12]. Unfortunately, for non-free-choice labeled P/T nets, this is not true: Figure 6 shows an OLC that, when short-circuited,

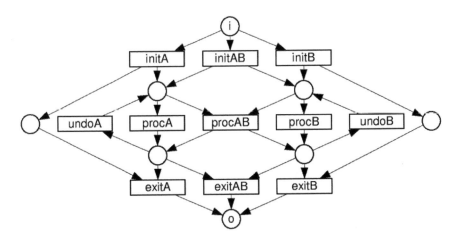

Fig. 6. Semi-positive T-invariants do not always induce cycles

contains STIs that do not induce a cycle: Their VLSs are {initA, procAB, undoB, exitA} and {initB, procAB, undoA, exitB}. (The induced cycles would all block on the transition labeled procAB.) As a result, the set of VLSs of cycles might be a strict subset of the set of VLSs of STIs: $\langle \sigma S \not\sqsubseteq \langle \sigma N \rangle$ where S is an arbitrary system of the labeled P/T net N. However, in the context of OLCs, we expect such a situation (an STI that does not induce a cycle) to be *extremely* rare. Therefore, we *assume* that every STI *does* induce a cycle. Under this assumption, $\lfloor \langle \lfloor \sigma S \rfloor \rangle = \lfloor \langle \lfloor \sigma N \rfloor \rangle$, and we can replace in Corollaries 2 and 3 the occurrences of $\langle \sigma \varphi S_b \rangle$ and $\langle \sigma \varphi S_s \rangle$ by $\lfloor \langle \lfloor \sigma \varphi N_b \rfloor \rangle$ and $\lfloor \langle \lfloor \sigma \varphi N_s \rfloor \rangle$.

For computing elementary STIs, a reasonably efficient algorithm exists [11]. However, if the assumption is not correct, the BA might prune incorrectly possible solutions from the search tree. Therefore, if the BA does not find a solution, we need to check whether the assumption holds, that is, we need to check whether every elementary STI induces an elementary cycle. First, we check whether the base OLC and the potential sub OLC are free choice. If so, the assumption holds. Otherwise, we need to check whether every elementary STI induces an elementary cycle using the reachability graphs of both OLCs. These graphs are usually already computed for the BB checks. If the assumption holds, there is indeed no solution, otherwise, we need to update the constraints or simply give way to an exhaustive search of the partitionings that have not yet been checked to find a solution, if any.

4   Test Samples

To test the effectivity of the backtracking algorithm (BA), we asked, without offering any explanations or limitations, an assistant professor to take a number of OLCs made by workflow students (base OLCs) and to add new functionality to these OLCs, yielding the same number of potential sub OLCs. We took these pairs of OLCs as samples to compare our BA to an exhaustive search algorithm (ESA, that is, the BA but without

Table 1. Results on the test samples

base OLC	A_1	B_1	C_1	D_1	E_1	F_1	G_1
Places	36	22	19	18	20	19	30
Transitions	32	24	19	16	25	21	37
Reachable states	469	24	30	22	21	20	30
potential sub OLC	A_2	B_2	C_2	D_2	E_2	F_2	G_2
Places	37	25	19	22	20	20	34
Transitions	35	30	22	18	29	24	42
Reachable states	477	29	30	30	21	23	34
Partitionings	8	16	8	4	8	4	8
— satisyfing constraints	1	0	1	1	1	2	2
Solutions	1	0	1	1	1	2	0
BB checks: BA	1	0	1	1	1	1	2
— ESA	3	16	1	4	1	1	8
Time: BA (in s/100)	536.80	0.31	0.52	0.29	1.28	0.37	3.83
— ESA (in s/100)	1654.21	2.19	0.36	0.81	0.12	0.13	2.82

computing constraints). Table 1 shows the test results. The top rows of the table show basic numbers concerning the complexities of the OLCs, the middle rows show qualitative results of both algorithms, and the bottom rows quantitative results.

Recall that our main goal is to reduce the number of BB checks. For six out of seven samples, the number of BB checks needed by the BA to find a solution is reduced to an absolute minimum: one for A, C, D, E, and F, and none for B. For G, we need to check BB twice before concluding that there exists no solution. Note that although this is not optimal, it is far better than the number of BB checks needed by the ESA (eight). We conclude that, for the samples, the BA is very effective.

For four out of seven samples, the ESA outperforms the BA when considering computation time. In our opinion, this is acceptable, because none of the samples is very complex. In the test samples, the overhead for computing the constraints make a comparison of run-times not very meaningful. Also, in three cases, the first BB check of the ESA is successful, meaning that the BA cannot be faster. Recall that the problem we try to tackle with the BA is the combination of large state spaces (that is, up to millions of states for workflows) and the exponential factor in the number of possible partitionings. From all samples, sample A is, by far, the most typical. It is satisfactory to see that for this sample, we reduce the computation time from over sixteen seconds to less than six. Of course, we would like to test the BA on more, and more complex, samples before concluding that it is efficient in general. Unfortunately, since behavioral equivalence is not a concept in common use, it is not straightforward to obtain good testing material.

5   Conclusions and Future Research

In this paper, we presented a backtracking algorithm (BA) to decide whether an object life cycle (OLC) is a subclass of another OLC under life-cycle inheritance (LCI) in a Petri-net setting. Based on structural properties of both OLCs, the BA computes a set of constraints that have to hold when the first OLC is a subclass of the second under LCI. These constraints are used to prune as many of the branching bisimilarity (BB) checks underlying LCI as possible. Our first experiments show that this reduction is very effective. However, future experiments, preferably in industrial settings, are needed. Particularly the workflow domain appears to be an interesting application area for our BA because of the complexity of today's industrial workflows [2].

The BA can be further improved, if future experiments show that this is necessary. In the current setup, we kept constraints as simple as possible: all are based on structural properties. However, if necessary, the approach can be extended with more accurate constraints, as long as they do not become too expensive to compute. And even if constraints may be too expensive to compute for certain OLCs, it is always safe to omit such constraints in those cases, and resort to constraints that are efficiently computable for the OLCs.

Another topic for further work is the translation of our approach to non-Petri-net-based formalisms that are being used for the specification of workflows and OLCs, such as the statechart-based techniques of UML [13]. The current techniques for computing constraints use Petri-net-specific techniques. However, the ideas behind these constraints are more general and can be translated to other settings.

Acknowledgements. The authors would like to thank the workflow students and Jacques Bouman for providing us with a set of test samples.

Abbreviations. Throughout the paper, the following abbreviations are used:
   BA: Backtracking Algorithm
   BB: Branching Bisimilarity
  ESA: Exhaustive Search Algorithm
  LCI: Life-Cycle Inheritance
  OLC: Object Life Cycle
  STI: Semi-positive Transition Invariant
  VLS: Visible Label Support

References

[1]   W.M.P. van der Aalst. Verification of Workflow Nets. In P. Azéma and G. Balbo, editors, *Application and Theory of Petri Nets 1997*, volume 1248 of *Lecture Notes in Computer Science*, pages 407–426, Toulouse, France, June 1997. Springer, Berlin, Germany, 1997.

[2]   W.M.P. van der Aalst. Chapter 10: Three good reasons for using a Petri-net-based workflow management system. In T. Wakayama, S. Kannapan, C.M. Khoong, S. Navathe, and J. Yates, editors, *Information and Process Integration in Enterprises: Rethinking Documents*, volume 428 of *The Kluwer International Series in Engineering and Computer Science*, pages 161–182. Kluwer Academic Publishers, Boston, Massachusetts, 1998.

[3]   W.M.P. van der Aalst. Inheritance of Dynamic Behaviour in UML. In D. Moldt, editor, *MOCA'02, Second Workshop on Modelling of Objects, Components, and Agents*, pages 105–120, Aarhus, Denmark, August 2002. University of Aarhus, Report DAIMI PB - 561, 2002. http://www.daimi.au.dk/CPnets/workshop02/moca/papers/.

[4]   W.M.P. van der Aalst. Inheritance of Interorganizational Workflows to Enable Business-to-Business E-commerce. *Electronic Commerce Research*, 2(3):195–231, 2002.

[5]   W.M.P. van der Aalst and T. Basten. Identifying Commonalities and Differences in Object Life Cycles using Behavioral Inheritance. In J.-M. Colom and M. Koutny, editors, *Applications and Theory of Petri Nets 2001*, volume 2075 of *Lecture Notes in Computer Science*, pages 32–52, Newcastle, UK, 2001. Springer, Berlin, Germany, 2001.

[6]   W.M.P. van der Aalst and T. Basten. Inheritance of Workflows: An Approach to Tackling Problems Related to Change. *Theoretical Computer Science*, 270(1-2):125–203, 2002.

[7]   T. Basten. *In Terms of Nets: System Design with Petri nets and Process Algebra*. PhD thesis, Eindhoven University of Technology, Eindhoven, The Netherlands, December 1998.

[8]   T. Basten and W.M.P. van der Aalst. Inheritance of Behavior. *Journal of Logic and Algebraic Programming*, 47(2):47–145, 2001.

[9]   G. Booch. *Object-Oriented Analysis and Design: With Applications*. Benjamin/Cunnings, Redwood City, California, USA, 1994.

[10]   G. Booch, J. Rumbaugh, and I. Jacobson. *The Unified Modeling Language User Guide*. Addison-Wesley, Reading, Massachusetts, USA, 1998.

[11]   J.-M. Colom and M. Silva. Convex Geometry and Semiflows in P/T nets: A Comparative Study of Algorithms for Computation of Minimal P-semiflows. In G. Rozenberg, editor, *Advances in Petri Nets 1990*, volume 483 of *Lecture Notes in Computer Science*, pages 79–112. Springer, Berlin, Germany, 1990.

[12]   J. Desel and J. Esparza. *Free Choice Petri Nets*, volume 40 of *Cambridge Tracts in Theoretical Computer Science*. Cambridge University Press, Cambridge, UK, 1995.

[13]   H.-E. Eriksson and M. Penker. *Business Modeling with UML: Business Patterns at Work*. John Wiley and Sons, New York, USA, January 2000.

[14]   R.J. van Glabbeek and W.P. Weijland. Branching Time and Abstraction in Bisimulation Semantics. *Journal of the ACM*, 43(3):555–600, 1996.

[15]   J.F. Groote and F.W. Vaandrager. An Efficient Algorithm for Branching Bisimulation and Stuttering Equivalence. In M.S. Paterson, editor, *Automata, Languages and Programming*, volume 443 of *Lecture Notes in Computer Science*, pages 626–638, Warwick University, England, July 1990. Springer, Berlin, Germany, 1990.

[16]   I. Jacobson, M. Ericsson, and A. Jacobson. *The Object Advantage: Business Process Reengineering with Object Technology*. Addison-Wesley, Reading, Massachusetts, USA, 1991.

[17]   Object Management Group. OMG Unified Modeling Language. http://www.omg.com/uml/.

[18]   J. Rumbaugh, M. Blaha, W. Premerlani, F. Eddy, and W. Lorensen. *Object-Oriented Modeling and Design*. Prentice-Hall, Englewoord Cliffs, New Jersey, USA, 1991.

[19]   J. Rumbaugh, I. Jacobson, and G. Booch. *The Unified Modeling Language Reference Manual*. Addison-Wesley, Reading, Massachusetts, USA, 1998.

Nets Enriched over Closed Monoidal Structures

Eric Badouel[1] and Jules Chenou[2]

[1] Ecole Nationale Supérieure Polytechnique, B.P. 8390, Yaoundé, Cameroon
ebadouel@polytech.uninet.cm
[2] Faculté des Sciences, Université de Douala, B.P. 24157 Douala, Cameroon
jchenou@caramail.com

Abstract. We show how the firing rule of Petri nets relies on a residuation operation for the commutative monoid of natural numbers. On that basis we introduce closed monoidal structures which are residuated monoids. We identify a class of closed monoidal structures (associated with a family of idempotent group dioids) for which one can mimic the token game of Petri nets to define the behaviour of these generalized Petri nets whose flow relations and place contents are valued in the closed monoidal structure.

1   Introduction

This paper reports on an ongoing research whose intent is to provide a uniform presentation of various families of Petri nets by recasting them as nets enriched over some algebraic structures, thus following the line of research best illustrated in [7] , a special issue of *Advances in Petri nets* dedicated to this theme. We aim at a general definition of nets parametric in algebraic structures corresponding to the kind of processes being modelled. We shall consider as a guideline the similar approaches that have been followed in the field of automata theory. Moreover the similarities between the approach undertaken here and what have been considered with automata seems a prequisite to our long term objective of achieving an enriched theory of regions [3] following the categorical approach based on schizophrenic objects ([2], [1]).

In Section2 we compare two different algebraic approaches to generalized automata and justify the choice that we have made: a skeletal and non-commutative variant of Lawvere's generalized logics [8]. These structures that we term *Closed Monoidal Structures* in order to stress their connection to enriched category theory [9] are also known as *residuated monoids* [11] ; they are introduced in Section 3. In Section 4 we illustrate how Petri nets appear as nets enriched over the closed monoidal structure on the commutative monoid of integers. Based on this example we try to circumvent the class of properties that a closed monoidal structure should satisfy so that it could be associated with a family of Generalized Petri nets. This part of our work is reported in Section 5.

W.M.P. van der Aalst and E. Best (Eds.): ICATPN 2003, LNCS 2679, pp. 64–81, 2003.

2   Algebraic Approaches to Generalized Automata

The theory of Kleene algebras, by providing an axiomatization of regular expressions, has paved the way to an algebraic theory of automata. It gives a modelisation of the three fundamental operations of choice $(+)$, sequencing $(\bullet)$ and iteration $(\star)$ under the form of an idempotent semiring $(K, +, \bullet, 0, 1)$ together with specific axioms for iteration. When the semiring is complete, iteration can be obtained as a derived operation : $x^\star = \sum_{n \in N} x^n$. Any Kleene algebra gives rise to a special kind of automata. The crucial observation is that the set of square matrices of dimension n with entries in a Kleene algebra is also a Kleene algebra whose choice and sequencing operations are given respectively by $(M + N)_{i,j} = M_{i,j} + N_{i,j}$, and $(M \bullet N)_{i,j} = \sum_{1 \le k \le n} M_{i,k} \bullet N_{k,j}$. Iteration is somewhat more complicated to define however. It is then possible to define an automaton with n states

over a Kleene algebra K as a triple (λ, A, γ) where $\lambda \in K^{1 \times n}$ is the vector of initial states, $\gamma \in K^{n \times 1}$, the vector of final states, and $A \in K^{n \times n}$ is the transition matrix. This automaton then recognizes $\lambda \bullet A^\star \bullet \gamma \in K$, and the entry $A^\star(i, j) \in K$ can be interpreted as the "language" leading from state s_i to state s_j. For instance if we let $K = \wp(X^\star)$ be the set of languages over an alphabet X, with $A + B = A \cup B$, and $A \bullet B = \{uv \mid u \in A \ \& \ v \in B\}$ we obtain a complete Kleene algebra $(S, +, \bullet, 0, 1)$ with $0 = \emptyset$, the empty set, $1 = \{\varepsilon\}$, the language reduced to the empty word, and $A^\star = \cup_{n \in N} A^n$ the usual iteration on languages. Automata over this Kleene algebra are the usual finite automata. But one may also consider $K = \wp(X^2)$ with $A + B = A \cup B$, the union of relations, $A \bullet B = \{(x, y) \mid \exists z \in X \ . \ (x, y) \in A \ \& \ (y, z) \in B\}$, the composition of relations $0 = \emptyset$, the empty relation, and $1 = \Delta = \{(x, x) \mid x \in X\}$, the diagonal or identity relation. We then obtain another complete Kleene algebra with $A^\star$ the reflexive and transitive closure of relation A. Automata over this Kleene algebra can be interpreted as finite relational automata in which $(x, y) \in A^\star(i, j)$ if and only if there exists some path from state s_i to state s_j such that $(x, y) \in R$ where R is the relation obtained by composition of the various relations encountered along this path. Thus the relation recognized by the automaton can be interpreted as the input/output relation of the program whose control graph is given by the automaton. Many more (complete) Kleene algebras exist such for instance $(R_+ \cup \{\infty\}, \min, +, \infty, 1)$ for which $A^\star(i, j)$ gives the minimal cost of a path leading from state s_i to state s_j where the cost of a path is given by the sum of the values associated with each transition ; and $(R_+ \cup \{\infty\}, \max, \min, 0, \infty)$ for which $A^\star(i, j)$ gives the maximal flow of some path from state s_i to state s_j where the flow along some path is given by the minimum of the (maximal) flow enabled on each transition. Since $A^{\star\star} = A^\star$, the automaton $(\lambda, A^\star, \gamma)$ is equivalent to (i.e. recognizes the same element as) automaton (λ, A, γ). An automaton (λ, A, γ) such that $A^\star = A$ is said to be *saturated*.

A second approach to an algebraic description of generalized automata is to consider automata over an alphabet X and values in a semiring K, as introduced by Schützenberger [12] (see also [4]). We let $K \langle\langle X \rangle\rangle$ denote the set of formal power series with coefficients in K and set of variables X. Such a series is a map $s : X^\star \longrightarrow K$ that can be interpreted as a "generalized set" of words in which the degree of membership of a word is measured by an element of K (its coefficient). Indeed if K is the boolean semiring $K = \{0, 1\}$ one has $K \langle\langle X \rangle\rangle = \wp(X^\star)$. The set of formal power series $K \langle\langle X \rangle\rangle$ is a semiring whose operations are given by : $(s + t)(w) = s(w) + t(w)$, $(s \bullet t)(w) = \sum_{w=uv} s(u) \bullet t(v)$. The same holds, by restriction, for its subset of polynomials $K \langle X \rangle = \{s \in K \langle\langle X \rangle\rangle \mid \exists^{<\infty} w \in X^\star \; s(w) \neq 0\}$ (formal power series with a finite domain). Usually $K \langle\langle X \rangle\rangle$ is not a Kleene algebra, however a partially defined iteration operation exists. Indeed let say that a family of formal power series $\{s_i \in K \langle\langle X \rangle\rangle \mid i \in I\}$ is *locally finite* when $\forall w \in X^\star \; \exists^{<\infty} i \in I \; s_i(w) \neq 0$ then the sum of such a family can be defined : $(\sum s_i)(w) = \sum (s_i(w))$. For any proper

formal power series $(s(\varepsilon) = 0)$, the family $\{s^n \in K \langle\langle X \rangle\rangle \mid n \in N\}$ is locally finite, and if we let $s^\star = \sum_{n \in N} s^n$ then for any t the unique solution of the equation $x = t + sx$ (respectively of the equation $x = t + xs$) is $x = s^\star t$ (resp. $x = ts^\star$). Now an automaton over the semiring K consists of a finite alphabet X, an integer n (the dimension of the automaton), a morphism of monoids $\mu : X^\star \longrightarrow K^{n \times n}$, a vector $\lambda \in K^{1 \times n}$ of initial states, and a vector $\gamma \in K^{n \times 1}$ of final states. This automaton recognizes the formal power series s such that $s(w) = \lambda \bullet \mu(w) \bullet \gamma$. The triple (λ, μ, γ) is called an n-dimensional linear representation of s, and s is said to be *recognizable*. Again, in the particular case where $K \langle\langle X \rangle\rangle = \wp(X^\star)$, it corresponds to the usual definitions known for finite automata. We say that a formal power series is *rational* if it belongs to the rational closure (i.e. closure by sum, product and star operation) of the semiring of polynomials. Then the theorem of Schützenberger stating that a formal power series is rational if and only if it is recognizable provides a generalization of Kleene theorem for finite automata. Probabilistic automata can also be associated similarly to the semiring $(R_+, +, \times, 0, 1)$, however extra conditions should be added constraining the definition of automata : an automaton consists of an initial distribution of probability λ (hence assumed to satisfy $\sum \lambda_i = 1$), $\mu_{i,j}(w)$ gives the probability to reach state s_j from state s_i when reading word $w \in X^\star$, and γ_i is the probability that state s_i be a final state, then if $M_{i,j} = \sum_{x \in X} \mu_{i,j}(x)$ represents the probability to reach state s_j from state s_i in one step, we further assume that (the normalisation property) : $\forall i \; \left(\sum_{1 \leq j \leq n} M_{i,j} + \gamma_i = 1 \right)$, i.e. either s_i is an accepting state or one can reach some other state in one step. Then the recognized formal power series is such that $p(w) = \lambda \bullet \mu(w) \bullet \gamma$ is the probability of recognizing word $w \in X^\star$.

A *dioid*[1] is a semiring for which the relation $x \leq y \Leftrightarrow \exists z \; : y = x + z$ is antisymmetric, i.e. is an order relation. This situation happens if the sum is idempotent, but also if the two following conditions are satisfied : the sum is cancellative $(x + y = x + z \Rightarrow y = z)$ and the neutral element cannot be decomposed as a sum $(x + y = 0 \Rightarrow x = y = 0)$.

---

[1] Some authors however use the term dioids for the restricted subclass of idempotent semirings.

Claim. In any dioid $x \vee y \leq x + y$, and a dioid is idempotent if and only if $x \vee y = x + y$

Proof. By definition of the order relation $x + y \geq x$ and $x + y \geq y$ and thus $x \vee y \leq x + y$. If $x \vee y = x + y$ then the sum is clearly idempotent. Conversely let us assume the idempotency of sum, observe that in that case the order relation can equivalently be defined as $x \leq y \Leftrightarrow y = x + y$. Indeed if $y = x + z$ then by idempotency : $y = x + x + z = x + y$ and the converse implication is immediate. Let z such that $x \leq z$ and $y \leq z$ then by the preceding remark $z = x + z = y + z$ and then $z = z + z = (x + z) + (y + z) = (x + y) + z$, i.e. $x + y \leq x \vee y$.

A dioid $(K, +, \times, 0, 1)$ is said to be complete if it is a complete lattice w.r.t. the induced order and for all $x \in K$ and $\{y_i \mid i \in I\} \subset K$, the following infinite distributive laws are satisfied :

$$x \bullet \left(\bigvee_{i \in I} y_i \right) = \bigvee_{i \in I} (x \bullet y_i) \quad \text{and} \quad \left(\bigvee_{i \in I} y_i \right) \bullet x = \bigvee_{i \in I} (y_i \bullet x)$$

A quantale (with unit) is a complete idempotent dioid. Since, by the above claim, the sum then coincides with the join, a quantale is usually presented as a structure $(K, \vee, \bullet, 1)$ consisting of a complete lattice with an infinitary joint operator $\vee$, and a monoid $(K, \bullet, 1)$ such that both infinite distributive laws are satisfied.

We saw that probabilistic automata are the automata over the semiring $(R_+, +, \times, 0, 1)$ satisfying certain normalisation properties. This semiring is a complete dioid but it is not idempotent and thus can certainly not be extended into a Kleene algebra structure. For a semiring to be part of a Kleene algebra it is necessary to be idempotent and sufficient to be idempotent and complete, i.e. a quantale. How do both notions of automata then compare ? It is readily verified that for any semiring K, the semiring $K \langle\langle X \rangle\rangle$ is idempotent (respectively is a quantale) whenever K is idempotent (respectively is a quantale). An automaton (λ, A, γ) over the quantale $K \langle\langle X \rangle\rangle$ viewed as a (complete) Kleene algebra is saturated if and only if $I + AA \leq A$ which reads as :

$$1 \leq A_{i,i}(\varepsilon) \quad \text{and} \quad \bigvee_{j} \bigvee_{uv=w} A_{i,j}(u) \bullet A_{j,k}(v) \leq A_{i,k}(w)$$

Since 1 is the greatest element the first condition reads as $A_{i,i}(\varepsilon) = 1$. Now matrix A is equivalent to the data $\mu : X^\star \longrightarrow K^{n \times n}$ where $\mu(w)(i, j) = A_{i,j}(w)$, and the preceding conditions rewrite as

$$\mu(\varepsilon)(i, i) = 1 \quad \text{and} \quad \bigvee_{uv=w} \bigvee_{j} \mu(u)(i, j) \bullet \mu(v)(j, k) \leq \mu(w)(i, k)$$

i.e. $\mu(\varepsilon) = I$ and $\mu(u) \bullet \mu(v) \leq \mu(w)$ whenever $uv = w$. Now $\mu : X^\star \longrightarrow K^{n \times n}$ is morphism of monoid, i.e. (λ, μ, γ) is an automaton over the semiring K with variables in X, if and only if

$$\bigvee_j A_{i,j}(u) \bullet A_{j,k}(v) = A_{i,k}(w) \quad \text{whenever} \quad uv = w$$

If K is the tropical semiring $(R \cup \{\infty\}, \min, +)$ the above identity says

$$\min_j (A_{i,j}(u) + A_{j,k}(v)) = A_{i,k}(w) \quad \text{whenever} \quad uv = w$$

i.e. the "distance" from state s_i to state s_k (associated with some word) is the minimal length of a path from s_i to s_k (labelled with the same word). In analogy with the corresponding notion borrowed from the theory of metric spaces, we thus term *geodesic* any saturated automaton (λ, A, γ) such that $\bigvee_j A_{i,j}(u) \bullet A_{j,k}(v) = A_{i,k}(w)$ whenever $uv = w$. We thus have established the following result :

Proposition 1. *If K is a quantale, a geodesic automaton (λ, A, γ) over the quantale $K \langle\langle X \rangle\rangle$ viewed as a (complete) Kleene algebra is the same thing as an automaton (λ, μ, γ) over the semiring K with variables in X, with the correspondence given by $\mu(w)(i,j) = A_{i,j}(w)$.*

Let us again consider the case of probabilistic automata. The semiring $K_1 = (R_+, +, \times, 0, 1)$ is a complete dioid hence it induces a quantale $Q = (R_+, \bigvee, \times, 1)$ however it should not be confused with that quantale. Probabilistic automata do constitute a class of automata over $Q \langle\langle X \rangle\rangle$ viewed as a (complete) Kleene algebra (because $x \vee y \leq x + y$) however, by forgetting the sum operation, we have lost all possibility of identifying this subclass of automata. In particular the class of geodesic automata over $Q \langle\langle X \rangle\rangle$ corresponds to the automata with variables X over the idempotent semiring $K_2 = (R_+, \vee, \times, 1)$ where $\vee$ is the least upper bound operation and these automata have no relation with probabilistic automata !

The above proposition compare the two different approaches in the particular case where they both apply (i.e. when K is a quantale). Since we would like to be able to deal with the largest possible variety of generalized automata, we are rather searching for an algebraic structure that would allow to encompass these two approaches. The closed monoidal structures introduced in the following section will realize this goal at least for large subclasses of dioids including all complete dioids and all group dioids.

3   Closed Monoidal Structures

A simple approach to generalized automata compatible with those described in the previous section is to identify such an automaton to a set of states S together with a map $A : S \times S \to K$ so that $A(s, s')$ measures the (possibly structured) set of trajectories from state s to state s' into some closed monoidal structure. This approach was first proposed by Lawvere [8] who termed them *generalized metric spaces*. As compared to this original definition, the approach taken here is in some respect more specific (because we use skeletal monoidal categories, here simply called monoidal structures) and in some other respect more general (because the tensor may not be symmetric in our case). An extension of Lawvere's approach to the non symmetric case was already proposed by Kasanghian, Kelly and Rossi (also in order to define generalized automata over closed monoidal categories). The presentation below can be seen as the simplification of their approach to the skeletical case.

Definition 1. *A **monoidal structure** $\vartheta = (K, \leq, \otimes, 1)$ is a set K equipped with an order relation $\leq$ and a structure of monoid $(K, \otimes, 1)$ where the composition $\otimes$, called tensor, is monotonic in both arguments. The structure is **closed** (or is a **residuated monoid**) when there exists two binary operators $/$ and $\backslash$, called the right and left residual operations, verifying the universal property that for any elements x, y, and z of K one has*

$$x \leq z/y \Leftrightarrow x \otimes y \leq z \Leftrightarrow y \leq x\backslash z$$

Monotonicity of tensor follows from the existence of residuals (see [11]). The following identities are immediate consequences of the definition

$$
\begin{aligned}
&x \leq x/1 \quad \text{and} \quad x \leq 1\backslash x \\
&1 \leq x/x \quad \text{and} \quad 1 \leq x\backslash x \\
&y/x \otimes x \leq y \quad \text{and} \quad x \otimes x\backslash y \leq y \\
&(x' \leq x \quad \text{and} \quad y \leq y') \Rightarrow (y/x \leq y'/x' \quad \text{and} \quad x\backslash y \leq x'\backslash y') \\
&z/(x \otimes y) = (z/y)/x \quad \text{and} \quad (x \otimes y)\backslash z = y\backslash(x\backslash z) \\
&x\backslash(z/y) = (x\backslash z)/y
\end{aligned}
$$

When the tensor is commutative, the monoidal structure is termed *commutative*, we then usually adopt additive notations: $\oplus$ in place of $\otimes$, and 0 in place of 1. In that case both residual operations coincide and we denote it $\ominus$. By analogy to semigroups we say that a closed monoidal structure is *complete* if it is a complete lattice and the tensor is continuous in both arguments :

$$x \otimes \left(\bigvee_{i \in I} y_i \right) = \bigvee_{i \in I} (x \otimes y_i) \quad \text{and} \quad \left(\bigvee_{i \in I} y_i \right) \otimes x = \bigvee_{i \in I} (y_i \otimes x)$$

In that case the residuals are given by the formulas :

$$z/y = \bigvee \{x \mid x \otimes y \leq z\} \quad \text{and} \quad x\backslash z = \bigvee \{y \mid x \otimes y \leq z\}$$

hence there are derived operators and complete closed monoidal structures may be identified with continuous ordered monoids (this situation is similar to the identification of complete Kleene algebras with quantales). Let us enumerate some families of closed monoidal structures.

3.1   Complete Closed Monoidal Structures

As for Kleene algebras most closed monoidal structures will be complete. Nevertheless, as we shall see in the next sections, the residual operations are the basic operations needed for representing Petri net-like computations and, in order to remain as general as possible, we don't want to enforce completeness as long as this assumption is not strictly necessary. However let us mention some families of complete closed monoidal structures.

Complete Dioids. For instance, the complete dioid $(R_+ \cup \{\infty\}, +, \times, 0, 1)$ induces the commutative closed monoidal structure $(R_+ \cup \{\infty\}, \leq, \times, 1)$. whose residual is given by : $x \ominus y = x/y$ when $x, y \in R_+$, $\infty \ominus \infty = 0$, $\infty \ominus x = \infty$, and $x \ominus \infty = 0$, for any $x \in R_+$

Quantales. Quantales are just complete monoids whose sum is idempotent (equivalently coincides with join).

1. The most typical example (already mentioned) is the set $K = \wp(X^\star)$ of languages on an alphabet X with set-theoretic inclusion as order relation, concatenation of languages as tensor product

$$L \otimes M = \{u.v \in X^* \mid u \in L \wedge v \in M\}$$

$1 = \{\varepsilon\}$, the language reduced to the empty word, as unit, and the usual residual operations on languages :

$$N/L = \{v \in X^* \mid \forall u \in L \ v.u \in N\} \text{ and } L\backslash N = \{v \in X^* \mid \forall u \in L \ u.v \in N\}$$

This monoidal structure stems from the quantale $(\wp(X^\star), \cup, \otimes, \emptyset, \{\varepsilon\})$. Similarly one can obtain a closed monoidal structure $\wp(\mathcal{M})$ by replacing the free monoid $X^\star$ by an arbitrary monoid $\mathcal{M}$.

2. Similarly the set $K = rel(U)$ of binary relations over a set U (for universe) is also a closed monoidal structure with set-theoretic inclusion as order relation, composition of relations as tensor product

$$R \otimes S = \{(x, z) \in U \times U \mid \exists x \in U \ (x, y) \in R \wedge (y, z) \in S\}$$

the diagonal $\Delta = \{(x, x) \in U \times U \mid x \in U\}$ as unit, and the residuals operations on relations :

$$R/S = \{(x, y) \in U \times U \mid \forall z \in U \ (y, z) \in S \Rightarrow (x, z) \in R\}$$

$$S\backslash R = \{(x, y) \in U \times U \mid \forall z \in U \ (z, x) \in S \Rightarrow (z, y) \in R\}$$

This monoidal structure stems from the quantale $(rel(U), \cup, \otimes, \emptyset, \Delta)$. The set of binary relations is also equipped with two unary operators giving respectively the inverse $R^{-1} = \{(x, y) \in R \mid (y, x) \in R\}$ and the complement $R^c = \{(x, y) \in U \times U \mid (y, x) \notin R\}$ of a relation R. One may then check that residuals can be expressed using these operators and composition as :

$R/S = \left(R^c \otimes S^{-1}\right)^c$ and $S\backslash R = \left(S^{-1} \otimes R^c\right)^c$. The calculus of binary relations has a long history involving works of famous logicians like De Morgan, Russel and Tarski (see the survey by Pratt [10]), and there is a recent increasing interest in computer science community due to its similarities with the Lambek Calculus (a logical calculus used for the treatment of natural languages) and linear logic [11].

3. Our third example corresponds to Petri nets: the set of non negative integers with addition, and the opposite of the usual order relation is a commutative closed monoidal structure whose residuals are given by truncated difference : $x \ominus y = x - y$ if $x \geq y$ and $x \ominus y = 0$ otherwise:

$$x + y \geq z \Leftrightarrow y \geq z \ominus x$$

If we add an element ∞ such that $x \leq \infty$, and $\infty \ominus x = \infty$ for every $x \in N$, and $x + \infty = \infty + x = \infty$, and $x \ominus \infty = 0$ for every $x \in N \cup \{\infty\}$ we obtain a complete commutative closed monoidal structure which stems from the quantale $(N \cup \{\infty\}, \min, +, \infty, 0)$.

Complete Heyting Algebras. A complete algebra is a quantale whose tensor coincides with the meet of the lattice. It then induces a commutative closed monoidal structure whose residual $y \ominus x$ is the relative complement $x \Rightarrow y = \bigvee \{z \mid x \wedge z \leq y\}$. The universal property of residuals: $x \wedge y \leq z \Leftrightarrow y \leq x \Rightarrow z$ is modus ponens which explain why Lawvere coined the term *generalized logics*. A boolean ring $\Re$ of sets ordered by inverse inclusion, with set union as tensor is a commutative closed monoidal structure whose residuals are given by set-theoretic difference :

$$X \cup Y \supseteq Z \Leftrightarrow Y \supseteq Z\backslash X$$

This monoidal structure stems from the complete Heyting algebra $(\Re, \supseteq)$.

3.2   Group-Like Monoidal Structures

In many respects closed monoidal structures appear as weak forms of groups, namely any group $\left(G, \otimes, 1, (.)^{-1}\right)$ is a closed monoidal structure whose order relation is equality and whose residuals are given by : $x/y = x \otimes y^{-1}$ and $y\backslash x = y^{-1} \otimes x$. A group $\left(G, \bullet, 1, (.)^{-1}\right)$ is said to be partially ordered if there exists a partial order $\leq$ with respect to which the product is monotonic in both arguments. A non trivial group can be ordered by letting $x < y \Leftrightarrow x \in y \bullet N$ if and only there exists a non-empty subset $N \subset G$ such that $N \cap N^{-1} = \emptyset$, $N \bullet N \subseteq N$, and $x \bullet N = N \bullet x$ for all $x \in G$ and in that case $N = \{x \in G \mid x < 1\}$. We then obtained a closed monoidal structure $(K, \leq, \otimes, 1)$, where $K = N \cup \{\bot, \top\}$ where $\bot$ and $\top$ new elements (i.e. not elements of G). The order relation is given by $x \leq y \Leftrightarrow x < y \ \vee \ x = y$ for $x, y \in N$ (order induced from the order on G) and $\bot \leq x \leq \top$ for all $x \in N$. The tensor is given by $x \otimes y = x \bullet y$ if

$x, y \in N$, $x \otimes \bot = \bot \otimes x = \bot$, and $x \otimes \top = \top \otimes x = x$ for all $x \in K$ (i.e. the least element is absorbing and the greatest element is neutral). The residuals are given as follows where a, b range over N, x ranges over K, and y ranges over $K \setminus \{\bot\}$: $a/b = a \bullet b^{-1}$ and $b\backslash a = b^{-1} \bullet a$ if $a \leq b$ and else $a/b = b\backslash a = \top$; $x/\top = \top\backslash x = x$; $x/\bot = \bot\backslash x = \top$; $\top/x = x\backslash\top = \top$; $\bot/y = y\backslash\bot = \bot$. The details of the verification are easy and left to the reader. It can also be verified that with the prescribed order on K, the above is the unique possible closed monoidal structure on K such that for all a, b in N :

$$a/b = a \bullet b^{-1} \quad \text{and} \quad b\backslash a = b^{-1} \bullet a \quad \text{if} \quad a \leq b \quad \text{else} \quad a/b = b\backslash a = \top$$

Notice that the element $1 \in G$ could have play the role of $\top$ since it satisfies the required properties w.r.t. to the elements in N : it is a neutral element for $\otimes$, and a greatest element for the order. The group of integers $(Z, +, 0)$ with its usual ordering induces in this way a closed monoidal structure isomorphic to the one associated with the quantale $(N \cup \{\infty\}, \min, +, \infty, 0)$. This suggests to look at the particular case where the order is induced from a sum.

A dioid $(K, \oplus, \otimes, \varepsilon, e)$ is a *group dioid* if every element of $K \setminus \{\varepsilon\}$ has an inverse for $\otimes$. Hence $(K \setminus \{\varepsilon\}, \otimes, e)$ is a group ordered by the relation induced by the sum : $x \leq y \Leftrightarrow \exists z : y = x \oplus z$. Now since ε is neutral for $\oplus$, it is a least element for this order and we can verify that it is an absorbing element for the tensor. Hence it can play the role of $\bot$ in the above construction, i.e. one has

Proposition 2. *Any group dioid $(K, \oplus, \otimes, \varepsilon, e)$ induces a closed monoidal structure $(M, \leq, \otimes, e)$ where M is the interval $]\varepsilon; e] = \{x \in K \mid \varepsilon < x \leq e\}$ and $\leq$ and $\otimes$ are respectively the order and the tensor of the dioid induced on this interval. The residuals are given by $a/b = a \otimes b^{-1}$ and $b\backslash a = b^{-1} \otimes a$ if $a \leq b$ and else $a/b = b\backslash a = e$; and $x/e = e\backslash x = x$, and $e/x = x\backslash e = e$ for any $a, b \in]\varepsilon; e[$, and $x \in]\varepsilon; e]$. If the dioid is idempotent (sum is join : $x \oplus y = x \vee y$), then it is a lattice whose meet is given by $x \wedge y = (x^{-1} \vee y^{-1})^{-1}$ and then residuations are given by : $a/b = e \wedge (a \otimes b^{-1})$ and $b\backslash a = e \wedge (b^{-1} \otimes a)$.*

It induces also a *complete* closed monoidal structure $(M', \leq, \otimes, e)$ where M' is the interval $[\varepsilon; e] = \{x \in K \mid \varepsilon \leq x \leq e\}$. The residuals further satisfy : $x/\varepsilon = \varepsilon\backslash x = e$ and $\varepsilon/y = y\backslash\varepsilon = \varepsilon$ for $x \in [\varepsilon; e]$, and $y \in]\varepsilon; e]$.

4   Nets over a Closed Monoidal Structure

Petri nets are associated with the monoidal structure $(N, \leq, +, 0)$ where $\leq$ is the opposite of the usual order $\sqsubseteq$ on N. As we saw it is a commutative closed monoidal structure whose residual is given by the truncated difference : $x \ominus y = x - y$ if $x \geq y$ else $x \ominus y = 0$. Indeed a Petri net is a structure $(P, E, Pre, Post)$ where P is a finite set of places, E a finite set of events (disjoint from P), and $Pre, Post : E \to N^P$ are called the flow relations. We can inductively extend these maps : $Pre, Post : E^* \to N^P$ by letting $Pre(\varepsilon)(p) = Post(\varepsilon)(p) = 0$ and

$$Pre(ue) = [Pre(e) \ominus Post(u)] + Pre(u)$$
$$Post(ue) = [Post(u) \ominus Pre(e)] + Post(a)$$

with a componentwise definition of the closed monoidal strcuture on N^P. A marking is a map from P into N. We readily verify that

$$M\,[u\rangle\,M' \iff M \sqsupseteq Pre(u) \ \wedge \ M' = [M \ominus Pre(u)] + Post(u)$$

where $M\,[u\rangle\,M'$ is the usual firing relation for Petri nets and where all operators are defined componentwise. Moreover reversibilty can be expressed by the fact that

$$M\,[u\rangle\,M' \iff M' \sqsupseteq Post(u) \ \wedge \ M = [M' \ominus Post(u)] + Pre(u)$$

Another equivalent formulation is

$$M\,[u\rangle\,M' \iff M \sqsupseteq Pre(u) \ \wedge \ M' \sqsupseteq Post(u) \ \wedge \ M \ominus Pre(u) = M' \ominus Post(u)$$

As in the above equivalences, when speaking of the firing rule of nets we shall allow ourselves to use the *notation* $\sqsupseteq$ in place of $\leq$, the order relation of the closed monoidal structure, since the order relation $\sqsubseteq = (\leq)^{-1}$ better reflects the intuition when dealing with nets. One can argue that we could have use this relation in the first place by using the equivalences

$$x \sqsupseteq z/y \Leftrightarrow x \otimes y \sqsupseteq z \Leftrightarrow y \sqsupseteq x \backslash z$$

for the definition of closed monoidal structures. We have indeed hesitated between these two possibilities for a long time and have finally taken the choice that allows a better comparison with semirings and dioids. This reversing of order relations reflects the duality between automata and nets that we intend to investigate in the future.

A sequence $u \in E^*$ of events can always be fired in some marking : indeed it can be fired in any marking M such that $M \sqsupseteq Pre(u)$. We assume $Pre(u)$ to give the "minimal amount of resources" in places so that the sequence $u \in E^*$ is firable. For some classes of nets however, like Elementary Net Systems, there exists sequences of events that are firable in no markings. A value $x \sqsupseteq Pre(u)$ where u is such an unfirable sequence should not be a legitimate place value. We can indeed handle Elementary Net Systems in this way. But making a distinction between the set of place contents and the set of flow relation values raises new issues that we can treat in this particular case and in similar cases, but for which we have not yet a global satisfactory answer. We propose then in the present paper to stick to base case where place contents and flow relation values belongs to the same set.

Definition 2. *A net over a closed monoidal structure* $\vartheta = (K, \sqsupseteq, \otimes, 1)$ *is a structure* $(P, E, Pre, Post)$ *where* P *is a finite set of places,* E *a finite set of events (disjoint from* P*), and* $Pre, Post : E \to K^P$ *are called the flow relations.*

A marking is a map $M : P \to K$. An event $e \in E$ is said to be firable in marking M and leads then to marking M', in notation $M \,[e\rangle\, M'$, if

$$M \sqsupseteq Pre(e) \ \wedge \ M' = M/Pre(e) \otimes Post(e)$$

however in order to meet intuition it should be the case that the first condition of this conjunct states that $e \in E$ is firable in marking M, and the second part gives the computation of the resulting marking M'. The interpretation is that when $M \sqsupseteq Pre(e)$, then marking M should decompose as $M = M/Pre(e) \otimes Pre(e)$; hence assuming

$$a \sqsupseteq b \Rightarrow a = (a/b) \otimes b$$

In order to get marking M', we then replace $Pre(e)$ by $Post(e)$ in the above decomposition of marking M. Due to the curryfying law of closed monoidal structures (namely $z/(x \otimes y) = (z/y)/x$) one can interpret z/u as "popping u from z" ; hence marking M' is obtained by popping $Pre(e)$ from marking M and then pushing $Post(e) : z \otimes (x \otimes y) = (z \otimes x) \otimes y$. In order to keep reversibility we need to ensure that both operations of popping and pushing are invertible, which amounts to the condition : $(a \otimes b)/b = a$. Then the firing rule can equivalently be stated as

$$M \,[e\rangle\, M' \iff M' \sqsupseteq Post(e) \ \wedge \ M = M'/Post(e) \otimes Pre(e)$$

by reversing the roles of $Post(e)$ and $Pre(e)$; or still as

$$M \,[e\rangle\, M' \iff M \sqsupseteq Pre(e) \ \wedge M' \sqsupseteq Post(e) \ \wedge \ M/Pre(e) = M'/Post(e)$$

Of course we could also chose a fifo structure (first in first out) rather than a lifo (last in first out) structure for places in which case we need the following equivalences to hold true :

$$M \,[e\rangle\, M' \iff M \sqsupseteq Pre(e) \ \wedge \ M' = Post(e) \otimes M/Pre(e)$$
$$\iff M' \sqsupseteq Post(e) \ \wedge \ M = Post(e) \backslash M' \otimes Pre(e)$$
$$\iff M \sqsupseteq Pre(e) \ \wedge \ M' \sqsupseteq Post(e) \ \wedge \ M/Pre(e) = Post(e) \backslash M'$$

this further requires:

$$a \sqsupseteq b \Rightarrow a = b \otimes (b \backslash a)$$
$$b \backslash (b \otimes a) = a$$

Now comes the problem of extending inductively the definitions of $Pre(u)$ and $Post(u)$ for words $u \in E^*$ so that

$$M \,[u\rangle\, M' \iff M \sqsupseteq Pre(u) \ \wedge \ M' = M/Pre(u) \otimes Post(u)$$
$$\iff M' \sqsupseteq Post(u) \ \wedge \ M = M/Post(u) \otimes Pre(u)$$
$$\iff M \sqsupseteq Pre(u) \ \wedge \ M' \sqsupseteq Post(u) \ \wedge \ M/Pre(u) = M/Post(u)$$

(and similarly for the fifo case) where $M \sqsupseteq Pre(u)$ is equivalent to $M \lfloor u\rangle$ (the enabling of sequence u in marking M : $\exists M'\ M \lfloor u\rangle M'$) and $M' \sqsupseteq Post(u)$ is dually equivalent to $\lfloor u\rangle M'$ (the co-enabling of sequence u in marking M' : $\exists M\ M \lfloor u\rangle M'$). For that purpose we proceed to some computations :

$$M \lfloor uv\rangle \iff M \sqsupseteq Pre(u) \ \wedge \ M/Pre(u) \otimes Post(u) \sqsupseteq Pre(v)$$
$$\iff M \sqsupseteq Pre(u) \ \wedge \ M/Pre(u) \sqsupseteq Pre(v)/Post(u)$$
$$\implies M \sqsupseteq (Pre(v)/Post(u)) \otimes Pre(u)$$

The first equivalence is how we would like $M \lfloor uv\rangle$ to be defined (this will prove to be independent of the decomposition of $w = uv$ by associativity of the binary relation to be defined below). The second equivalence is just an application of residuation, and the last implication comes from the equivalence

$$M \sqsupseteq Pre(u) \iff M = M/Pre(u) \otimes Pre(u)$$

This suggests the definition $Pre(uv) = (Pre(v)/Post(u)) \otimes Pre(u)$ and similarly $Post(uv) = (Post(u)/Pre(v)) \otimes Post(v)$. Now this requires that the above implication is an equivalence, i.e.

$$a \sqsupseteq b \otimes c \iff [a \sqsupseteq c \ \wedge \ a/c \sqsupseteq b]$$

for every a, b, and c in K. In the fifo case we should also add the requirement that

$$a \sqsupseteq b \otimes c \iff [a \sqsupseteq b \ \wedge \ b\backslash a \sqsupseteq c]$$

and let $Pre(uv) = (Post(u)\backslash Pre(v)) \otimes Pre(u)$ and $Post(uv) = Post(v) \otimes (Post(u)/Pre(v))$ since in that case :

$$M \lfloor uv\rangle \iff M \sqsupseteq Pre(u) \ \wedge \ Post(u) \otimes M/Pre(u) \sqsupseteq Pre(v)$$
$$\iff M \sqsupseteq Pre(u) \ \wedge \ M/Pre(u) \sqsupseteq Post(u)\backslash Pre(v)$$
$$\iff M \sqsupseteq (Post(u)\backslash Pre(v)) \otimes Pre(u)$$

and

$$\lfloor uv\rangle M' \iff M' \sqsupseteq Post(v) \ \wedge \ Post(v)\backslash M' \otimes Pre(v) \sqsupseteq Post(u)$$
$$\iff M' \sqsupseteq Post(v) \ \wedge \ Post(v)\backslash M' \sqsupseteq Post(u)/Pre(v)$$
$$\iff M' \sqsupseteq Post(v) \otimes (Post(u)/Pre(v))$$

We can then define a binary operation on $K \times K$ as follows. We denote Pre and $Post$ the two canonical projections from $K \times K$ to K, so that an element $\alpha \in K \times K$ be written $\alpha = (Pre(\alpha), Post(\alpha))$ and the binary relation $\boxtimes$ is defined by the two above identities (for the lifo case)

$$Pre(\alpha \boxtimes \beta) = (Pre(\beta)/Post(\alpha)) \otimes Pre(\alpha)$$
$$Post(\alpha \boxtimes \beta) = (Post(\alpha)/Pre(\beta)) \otimes Post(\beta)$$

K should contain a least element $\perp$ (for $\sqsubseteq$) corresponding to no constraint ($M(p) \sqsupseteq \perp$ is always satisfied) ; and $(\perp, \perp)$ should be the neutral element of this composition, so that we can let $Pre(\varepsilon)(p) = Post(\varepsilon)(p) = \perp$. Then we can inductively define the maps $Pre, Post : E^* \to K^P$ by letting

$(Pre(u)(p), Post(u)(p)) = \varphi(u)(p)$ where $\varphi : E^* \to (K^2)^P$ is the unique morphism of monoids such that the images $\varphi(e)(-) = (Pre(e)(-), Post(e)(-))$ of the generators $e \in E$ are given by the flow relations of the net. We then obtain

$$Pre(\varepsilon)(p) = Post(\varepsilon)(p) = \perp$$
$$Pre(u \otimes v)(p) = (Pre(v)(p) / Post(u)(p)) \otimes Pre(u)(p)$$
$$Post(u \otimes v)(p) = (Post(u)(p) / Pre(v)(p)) \otimes Post(v)(p)$$

5   Petri Monoidal Structures

The following definition provides a set of conditions sufficient to ensure the different requirements mentioned in the previous section. We omit most of the proofs of the various following propositions. Hint of most of these proofs and their main arguments were already sketched in the informal discussion of the previous section. An important point however that was not touched upon in that discussion and that we shall consider here is how associativity of operation $\boxtimes$ is ensured.

Definition 3. *A Petri monoidal structure is a closed monoidal structure $\vartheta = (K, \sqsupseteq, \otimes, 1)$ such that :*

1. *1 is the least element of $(K, \sqsubseteq)$*
2. *$(a \sqsupseteq b \otimes c) \Longleftrightarrow (a \sqsupseteq c \wedge a/c \sqsupseteq b) \Longleftrightarrow (a \sqsupseteq b \wedge b \backslash a \sqsupseteq c)$*
3. *$(a \otimes b)/b = a$ and $b \backslash (b \otimes a) = a$*

Proposition 3. *A Petri monoidal structure satisfies the following:*

1. *There are no divisor of the unit : $[a \otimes b = 1 \Rightarrow a = b = 1]$, and $1/a = a \backslash 1 = 1$; $a/1 = 1 \backslash a = a$; $a/a = a \backslash a = 1$; and $a \sqsupseteq b \Leftrightarrow b/a = 1 \Leftrightarrow a \backslash b = 1$.*
2. *It has a join given by $a \vee b = a/b \otimes b = b \otimes b \backslash a = b/a \otimes a = a \otimes a \backslash b$.*
3. *It has a meet given by $a \wedge b = a/(b \backslash a) = (a/b) \backslash a = b/(a \backslash b) = (b/a) \backslash b$.*
4. *$a \sqsupseteq b \Longrightarrow (a = a/b \otimes b = b \otimes b \backslash a)$.*
5. *$a \otimes b \sqsupseteq a$; $a \otimes b \sqsupseteq b$.*

Definition 4. *A Petri monoidal structure is said to be* lifo *when*

$$(b \otimes c)/a = b/(a/c) \otimes c/a \quad and \quad a \backslash (b \otimes c) = a \backslash b \otimes (b \backslash a) \backslash c$$

It is said to be fifo *when :*

$$(b \otimes c)/a = b/a \otimes c/(b \backslash a) \quad and \quad a \backslash (b \otimes c) = (a/c) \backslash b \otimes a \backslash c$$

If the tensor is commutative the lifo and fifo conditions are equivalent and can be expressed (with the additive notation) as :

$$(b \oplus c) \ominus a = [b \ominus (a \ominus c)] \oplus (c \ominus a)$$

but of course the terminology "lifo" and "fifo" makes no much sense in that case.

The respective conditions in the above definition are both "internalisations" of condition (2) in Def.3, this fact is expressed by the following two propositions.

Proposition 4. *A lifo Petri monoidal structure is a closed monoidal structure $\vartheta = (K, \sqsupseteq, \otimes, 1)$ such that :*

1. *1 is the least element of $(K, \sqsupseteq)$.*
2. *There are no divisor of the unit : $[a \otimes b = 1 \Rightarrow a = b = 1]$.*
3. *$(a \otimes b)/b = a$ and $b \backslash (b \otimes a) = a$.*
4. *$(b \otimes c)/a = b/(a/c) \otimes c/a$ and $a \backslash (b \otimes c) = a \backslash b \otimes (b \backslash a) \backslash c$*

Proof. By the residuation property $(a \sqsupseteq b \otimes c) \Longleftrightarrow [(b \otimes c)/a = 1] \Longleftrightarrow [b/(a/c) \otimes c/a = 1]$. Since $[a \otimes b = 1 \Rightarrow a = b = 1]$, this latter condition is equivalent to $[b/(a/c) = 1] \wedge [c/a = 1]$; i.e. $(a \sqsupseteq c \wedge a/c \sqsupseteq b)$. The equivalence $(a \sqsupseteq b \otimes c) \Longleftrightarrow (a \sqsupseteq b \wedge b \backslash a \sqsupseteq c)$ is proved similarly

In the same manner we obtain the analogous proposition :

Proposition 5. *A fifo Petri monoidal structure is a closed monoidal structure $\vartheta = (K, \sqsupseteq, \otimes, 1)$ such that :*

1. *1 is the least element of $(K, \sqsupseteq)$.*
2. *There are no divisor of the unit : $[a \otimes b = 1 \Rightarrow a = b = 1]$.*
3. *$(a \otimes b)/b = a$ and $b \backslash (b \otimes a) = a$.*
4. *$(b \otimes c)/a = b/a \otimes c/(b \backslash a)$ and $a \backslash (b \otimes c) = (a/c) \backslash b \otimes a \backslash c$.*

Proposition 6. *A Petri monoidal structure $\vartheta = (K, \sqsupseteq, \otimes, 1)$ with a total ordering is lifo.*

Corollary 1. *The closed monoidal structure induced by an idempotent group dioid with a total ordering is a lifo Petri monoidal structure.*

This corollary provides a reasonable class of group-like closed monoidal structures that are lifo Petri monoidal structures. Unfortunately we don't have an analogue of Proposition6 (and hence of its corollary) in the fifo case. Condition 4 in Prop. 5 is much harder to establish. And this is easy to understand : as long as we pop and push "on the same side", things go quite nicely with group-like closed monoidal structures (in which the right residuals looks like ab^{-1} with the negative part "consuming the data" coming first into the lifo due to the curryfication law : $u/(ab^{-1}) = (u/b^{-1})/a$. When on the contrary consumption and production of data are made at the opposite sides of a fifo, we need to express how a data just entered can have an effect on the enabling or on the contrary on the inhibition of some event. This information need to "flow" through the entire fifo and take into account some parts of the value of the place that comes from its past history. This flow of information is obtained by mixing both residual operations as described by the Condition 4 in Prop. 5. And it is very hard to find concrete models that realize this computation.

Proposition 7. *Let $\vartheta = (K, \sqsupseteq, \otimes, 1)$ be a fifo Petri monoidal structure . We have a monoid $(K^2, \boxtimes, e)$ with unit $e = (1, 1)$ and whose composition law is given by :*

$$(u, u') \boxtimes (v, v') = (v/u' \otimes u, u'/v \otimes v')$$

Proof. It is immediate that $e = (1, 1)$ is a left end right unit of the given product. Let us prove the associativity of the operation.

$$\begin{aligned}
[(u, u') \boxtimes (v, v')] \boxtimes (w, w') &= (v/u' \otimes u, u'/v \otimes v') \boxtimes (w, w') \\
&= (w/ [u'/v \otimes v'] \otimes v/u' \otimes u, [u'/v \otimes v'] /w \otimes w')
\end{aligned}$$

$$\begin{aligned}
(u, u') \boxtimes [(v, v') \boxtimes (w, w')] &= (u, u') \boxtimes (w/v' \otimes v, v'/w \otimes w') \\
&= ([w/v' \otimes v] /u' \otimes u, u'/ [w/v' \otimes v] \otimes v'/w \otimes w')
\end{aligned}$$

For the left-hand side:

$$\begin{aligned}
[w/v' \otimes v] /u' \otimes u &= (w/v') / (u'/v) \otimes v/u' \otimes u :: (b \otimes c) /a = b/ (a/c) \otimes c/a \\
&= w/ [u'/v \otimes v'] \otimes v/u' \otimes u \quad :: (a/b) /c = a/ (c \otimes b)
\end{aligned}$$

For the right-hand side:

$$\begin{aligned}
[u'/v \otimes v'] /w \otimes w' &= (u'/v) / (w/v') \otimes v'/w \otimes w' :: (b \otimes c) /a = b/ (a/c) \otimes c/a \\
&= u'/ [w/v' \otimes v] \otimes v'/w \otimes w' \quad :: (a/b) /c = a/ (c \otimes b)
\end{aligned}$$

We have the same result in case of the fifo semantics. In that case the composition is given by

$$(u, u') \boxtimes (v, v') = (u'\backslash v \otimes u, v' \otimes u'/v)$$

and then

$$\begin{aligned}
[(u, u') \boxtimes (v, v')] \boxtimes (w, w') &= (u'\backslash v \otimes u, v' \otimes u'/v) \boxtimes (w, w') \\
&= ([v' \otimes u'/v] \backslash w \otimes u'\backslash v \otimes u, w' \otimes [v' \otimes u'/v] /w)
\end{aligned}$$

$$(u, u') \boxtimes [(v, v') \boxtimes (w, w')] = (u, u') \boxtimes (v'\backslash w \otimes v, w' \otimes v'/w)$$
$$= (u'\backslash [v'\backslash w \otimes v] \otimes u, w' \otimes v'/w \otimes u'/[v'\backslash w \otimes v])$$

For the left-hand side:

$$u'\backslash [v'\backslash w \otimes v] \otimes u = (u'/v)\backslash (v'\backslash w) \otimes u'\backslash v \otimes u :: a\backslash (b \otimes c) = (a/c)\backslash b \otimes a\backslash c$$
$$= [v' \otimes u'/v]\backslash w \otimes u'\backslash v \otimes u :: (c \otimes b)\backslash a = b\backslash (c\backslash a)$$

For the right-hand side:

$$w' \otimes [v' \otimes u'/v]/w = w \otimes v'/w \otimes (u'/v)/(v'\backslash w) :: (b \otimes c)/a = b/a \otimes c/(b\backslash a)$$
$$= w' \otimes v'/w \otimes u'/[v'\backslash w \otimes v] :: a/(b \otimes c) = (a/c)/b$$

Let $(P, E, Pre, Post)$ be a net over a lifo Petri monoidal structure $\vartheta = (K, \sqsupseteq, \otimes, 1)$. We then can define $Pre, Post : E^* \to K^P$ by letting $(Pre(u)(p), Post(u)(p)) = \varphi(u)(p)$ where $\varphi : E^* \to (K^2)^P$ is the unique morphism of monoids such that the images $\varphi(e)(-) = (Pre(e)(-), Post(e)(-))$ of the generators $e \in E$ are given by the flow relations of the net. We then obtain

$$Pre(\varepsilon)(p) = Post(\varepsilon)(p) = 1$$
$$Pre(u \otimes v)(p) = (Pre(v)(p)/Post(u)(p)) \otimes Pre(u)(p)$$
$$Post(u \otimes v)(p) = (Post(u)(p)/Pre(v)(p)) \otimes Post(v)(p)$$

Proposition 8. *Let $(P, E, Pre, Post)$ be a net over a lifo Petri monoidal structure $\vartheta = (K, \sqsupseteq, \otimes, 1)$. The following three statements are equivalent*

1. $M \sqsupseteq Pre(u) \ \wedge \ M' = M/Pre(u) \otimes Post(u)$
2. $M' \sqsupseteq Post(u) \ \wedge \ M = M/Post(u) \otimes Pre(u)$
3. $M \sqsupseteq Pre(u) \ \wedge \ M' \sqsupseteq Post(u) \ \wedge \ M/Pre(u) = M/Post(u)$

Proof.

$$\left(M' \sqsupseteq M/Pre(u) \otimes Post(u)\right) \Leftrightarrow \left(M' \sqsupseteq Post(u) \ \wedge \ M/Post(u) \sqsupseteq M/Pre(u)\right)$$

together with

$$(M/Pre(u) \otimes Post(u) \sqsupseteq M') \Leftrightarrow (M/Post(u) \sqsupseteq M/Pre(u))$$

establishes the equivalence (1) $\Leftrightarrow$ (3). Equivalence (2) $\Leftrightarrow$ (3) follows in the same manner.

The case of lifo nets is treated similarly.

We let $M[u\rangle M'$ when one of the three equivalent statements of the previous proposition is met and we say that the sequence $u \in E^*$ is enabled in marking M and leads to marking M'. We let $M[u\rangle \Leftrightarrow (\exists M' \ M[u\rangle M')$ and $[u\rangle M' \Leftrightarrow (\exists M \ M[u\rangle M')$

Proposition 9. $M[u\rangle \Leftrightarrow (M \sqsupseteq Pre(u))$ and $[u\rangle M' \Leftrightarrow (M' \sqsupseteq Post(u))$

Proof. By definition $M[\varepsilon\rangle M$ always holds. The firing relation of an event $e \in E$ is given by

$$M[e\rangle M' \iff M \sqsupseteq Pre(e) \ \land \ M' = M/Pre(e) \otimes Post(e)$$

from which the equivalence $M[e\rangle \Leftrightarrow (M \sqsupseteq Pre(e))$ immediately follows. We then proceed by induction by showing that $M[uv\rangle \Leftrightarrow \exists M' \ M[u\rangle M' \ \land \ M'[v\rangle$:

$M[uv\rangle \Leftrightarrow [M \sqsupseteq Pre(uv) \quad = (Pre(v)/Post(u)) \otimes Pre(u)] \Leftrightarrow$
$[M \sqsupseteq Pre(u) \ \land \ M/Pre(u) \sqsupseteq Pre(v)/Post(u)] \Leftrightarrow$
$M[u\rangle M' \ \land \ M' = M/Pre(u) \otimes Post(u) \sqsupseteq Pre(v)/Post(u) \otimes Post(u) \sqsupseteq Pre(v)$
$\Leftrightarrow [M[u\rangle M' \ \land \ M'[v\rangle]$

The second equivalence relation is proved similarly.

6   Conclusion

We have suggested in this paper a definition of generalized Petri nets as nets enriched over certain closed monoidal structures. In the case of non commutative structures two semantics have been considered. The purpose of this paper was to identify algebraic structures that allow us to mimic the token game of Petri nets. However in order to be able to represent meaningful classes of nets, like for instance continuous Petri nets ([5],[6]), this work should be extended, and this can be done in several directions. First one should not necessarily consider that place contents and flow relations take their values in the same domain : one might consider that flow relations act on the place contents by making these values range in some module of which events act through flow relations. In the same manner one might also consider more complex trajectories than simply those indexed by words. Finally it would be necessary to be able to internalize the set of net computations in order to derive a duality between this enriched nets and corresponding enriched automata.

References

1. E. Badouel, M. Bednarczyk, and Ph. Darondeau. Generalized automata and their net representations. In H. Ehrig, J. Padberg, and G. Rozenberg, editors, *Uniform approaches to Petri nets*, Advances in Petri nets, Lectures Notes in Computer Sciences, pages 304–345. Springer Verlag, 2001.
2. E. Badouel and Ph. Darondeau. Dualities between nets and automata induced by schizophrenic objects. In *6th International Conference on Category Theory and Computer Science*, volume 953 of *Lecture Notes in Computer Science*, pages 24–43, 1995.
3. E. Badouel and Ph. Darondeau. Theory of regions. In *Third Advance Course on Petri Nets, Dagstuhl Castle*, volume 1491 of *Lecture Notes in Computer Science*, pages 529–586. Springer-Verlag, 1998.

4. J. Berstel and Ch. Reutenauer. *Les séries rationnelles et leurs langages*. Masson, Paris, 1984.
5. R. David and H. Alla. Petri nets for modeling of dynamic systems - a survey. *Automatica*, 30:175–202, 1994.
6. M. Droste and R.M. Shortt. Continuous petri nets and transition systems. In H. Ehrig, J. Padberg, and G. Rozenberg, editors, *Uniform approaches to Petri nets*, Advances in Petri nets, Lectures Notes in Computer Sciences. Springer Verlag, 2001.
7. H. Ehrig, J. Padberg, and G. Rozenberg, editors. *Uniform approaches to Petri nets*, Advances in Petri nets, Lectures Notes in Computer Sciences. Springer Verlag, 2001.
8. F.W. Lawvere. Metric spaces, generalized logics, and closed categories. *Rendiconti del seminario Matematico e Fisico di Milano*, XLIII:135–166, 1973.
9. J. Meseguer and U. Montanari. Petri nets are monoids. *Information and Computation*, 88(2):105–155, 1990.
10. V.R. Pratt. Origins of the calculus of binary relations. In *7th Annual IEEE Symp. on Logic in Computer Science, Santa Cruz, CA*, pages 248–254, 1992.
11. Ch. Retoré. The logic of categorical grammars. In *Lecture Notes of the European Summer School in Logic, Language and Information, ESSLLI'2000*, Birmingham, 2000.
12. M.P. Schutzenberger. On the definition of a family of automata. *Information and Control*, 4:245–270, 1961.

Automatic Symmetry Detection in Well-Formed Nets

Yann Thierry-Mieg, Claude Dutheillet, and Isabelle Mounier

(LIP6-SRC, France)
Laboratoire d'Informatique de Paris 6, France
Yann.Thierry-Mieg,Claude.Dutheillet,Isabelle.Mounier@lip6.fr

Abstract. Formal verification of complex systems using high-level Petri Nets faces the so-called state-space explosion problem. In the context of Petri nets generated from a higher level specification, this problem is particularly acute due to the inherent size of the considered models. A solution is to perform a symbolic analysis of the reachability graph, which exploits the symmetry of a model.

Well-Formed Nets (*WN*) are a class of high-level Petri nets, developed specifically to allow automatic construction of a symbolic reachability graph (SRG), that represents equivalence classes of states. This relies on the definition *by the modeler* of the symmetries of the model, through the definition of "static sub-classes". Since a model is self-contained, these (a)symmetries are actually defined *by the model itself*.

This paper presents an algorithm capable of automatically extracting the symmetries inherent to a model, thus allowing its symbolic study by translating it to *WN*. The computation starts from the assumption that the model is entirely symmetric, then examines each component of a net to deduce the symmetry break it induces. This translation is transparent to the end-user, and is implemented as a service for the AMI-Net package. It is particularly adapted to models containing large value domains, yielding combinatorial gain in the size of the reachability graph.

Keywords: Well-Formed Petri nets, symmetry detection, symbolic model-checking, partial symmetry

1 Introduction

Formal verification of complex systems using high-level Petri nets (HLPN) faces the state-space size explosion problem. This is mainly due to interleaving between sequences of possible execution, and to the increase in the size of the variable domains. Lastly, and to a lesser extent, we have the impact of the number of tokens regardless of their value.

Well Formed nets (*WN*) [2] are a class of HLPN that allows automatic generation of a Symbolic Reachability Graph (SRG) [3]. An SRG allows compact representation of the state-space by defining classes of equivalent concrete accessible states. This is made possible by the correct definition *by the modeler* of the classes of structurally equivalent objects (tokens). In this paper, we will show that such a definition is not truly necessary, as it can be deduced from the model itself. The gains of the symbolic approach are tremendous, as it reduces the variable domain size to the lesser token cardinality problem. Furthermore it leaves room to apply structural reductions to fight the interleaving problem [11,4].

W.M.P. van der Aalst and E. Best (Eds.): ICATPN 2003, LNCS 2679, pp. 82–101, 2003.
© Springer-Verlag Berlin Heidelberg 2003

We propose here a syntactic sugar for the definition of *WN*, which does not require prior definition of the symmetries but allows to calculate them automatically. This is not an extension of the *WN* formalism, as we retain exactly the same expression power. Calculating the symmetry allowed by a given model automatically has several advantages:

- The fact that *symbolic* model-checking is being performed is transparent to the user. Therefore non-expert users may use the advantages of symbolic study more easily.
- Even expert users have a difficult time defining symmetries, and the syntax of *WN* is quite constraining in many respects. It forces to define all the elements of the net in terms of these equivalence classes. By allowing use of concrete predicates, we lighten the burden of the modeler.
- Although there is no risk of under-specification when defining a *WN*, there is a risk of over-specification. In other words, though one cannot put in the same equivalence class objects that have different structural behaviors, there is a risk of distinguishing between objects that are actually equivalent with respect to the model, or of defining operations like ordering that constrain the symmetry when they are not needed.
- Modifying the model is easier, for the same reasons.
- Moreover, if the net is being generated from a higher level specification language, such as **L***f***P**[12,6], we require a process capable of detecting such symmetries without human intervention.

This work has been implemented as a tool in the CPN-AMI [1] software package, and relies on GreatSPN [8] of the university of Torino for the construction of the SRG. Beyond the context of this tool, the extraction of symmetries from the elements of a model has other applications being developed.

- It gives us insights on how to increase the symmetry of a model, by allowing to deduce which elements of the net are the most constraining on its symmetry.
- It allows, as an extension the ESRG technique [10], to take these constraints into account only when they are truly required. This allows symbolic study of only partially symmetric models.

The problem addressed here is also encountered in a variety of other situations where our technique might be successfully applied. It could be used to determine relevant limit values for automatic symbolic test-case generation such as in [13,9]. It also might be used with other classes of models, such as the extensions to murphy [5] that take into account the same type of symmetries as *WN*.

This paper is organized as follows: section 2 introduces the *WN* formalism; section 3 defines the symmetries that are considered and their impact on the construction of a symbolic reachability graph; section 4 presents the syntactic extensions that are made to *WN* syntax, and their effect on the symmetry of the system; section 5 develops the rules that allow automatic symmetry extraction from a guard predicate; finally section 6 reports the performances of the tool implementing this technique on a realistic example.

2 Context

We recall here the main features of the WN model, and the way it can be used to automatically build a quotient graph, namely the Symbolic Reachability Graph. The construction of the SRG relies on the exploitation of symmetries that are explicitly contained in the model definition.

2.1   Well-Formed Nets: Definition

Well-formed Nets (WN) are a high-level Petri net model in which color domains and color functions must respect a simple, rigorous syntax, which automatically takes into account the symmetries of the system. WN tokens carry a composite information expressed as a tuple of colors (called objects), taken from possibly ordered *basic color classes*. Each color class represents system components of a given kind (e.g. the process class, the processor class,...).

   If all the objects in a class do not behave the same way, the class must be partitioned into static subclasses. Objects in the same static subclass represent entities that always behave in a symmetric (homogeneous) way, while objects belonging to different static subclasses may have different behaviors. This constraint is reflected by the syntax of the color functions of the model, which makes it impossible to distinguish objects that belong to the same static subclass.

   The general color functions of the WN model are built from three types of basic functions: the *projection* function, the *successor* function and the *diffusion/synchronization* function. The syntax used for the projection function is x, where x is one of the transition variables (i.e., one of the parameters that appears on an arc adjacent to the transition). It is called projection because it selects one element from the tuple of parameter values defining the transition color instance). The syntax used for the successor function is $!x$ where x is again one of the transition variables, it applies only to ordered classes and returns the successor of the color assigned to x in the transition color instance. Finally, the syntax for the diffusion/synchronization function is $C_i.all$ (or $C_i^j.all$): it is a constant function that returns the whole set of colors of class C_i (of static subclass $C_i^j \subset C_i$). It is called synchronization when used on a transition input arc because it implements a synchronization among a set of colored tokens contained into a place, while it is called diffusion when used on a transition output arc because it puts one token of each color of the (sub)class it applies to into a place.

   Transitions can be associated with a predicate which restricts the set of their possible instantiations.

Definition 1. *Standard Predicates. A standard predicate (or guard) associated with a transition t is a boolean expression of* basic predicates. *The allowed basic predicates are:* $x = y$, $x =!y$, $d(x) = C_i^j$, $d(x) = d(y)$, *where* $x, y \in Var_i(t)$ *are parameters of t of the same type, $!y$ denotes the successor of y (assuming that the type of y is an ordered class), and $d(x)$ denotes the domain of x, which is the static subclass x belongs to.*

   General color (arc) functions are defined as weighted sums of tuples. The elements composing the tuples are in turn weighted sums of *basic functions*, defined on basic color classes and returning multi-sets of colors in the same class.

Definition 2. *An arc function $\mathcal{F}$ associated with an arc connecting place p and transition t has the form:* $\mathcal{F} = \sum_k \alpha_k.F_k$ *where α_k is a positive integer and F_k is a function which maps an element of the color domain of transition t ($cd(t)$) onto a multi-set of colors of $cd(p)$, color domain of p.*
   Formally, $F_k : cd(t) \rightarrow Bag(cd(p))$ is a function of the form

$$F = \bigotimes_{C_i \in \mathcal{C}} \bigotimes_{j=1,\dots,e_i} f_i^j = \langle f_1^1, \dots, f_1^{e_1}, \dots, f_n^1, \dots, f_n^{e_n} \rangle$$

with e_i representing the number of occurrences of class C_i in color domain of place p, and $\otimes$ denoting the Cartesian product, i.e.,
$$cd(p) = \bigotimes_{C_i \in \mathcal{C}} \bigotimes_{j=1,\ldots,e_i} C_i.$$
Each function f_i^j in turn is defined as:

$$f_i = \sum_{q=1}^{ns_i} \beta_{i,q}.C_i^q.all + \sum_{x \in Var_i(t)} (\gamma_x.x + \delta_x.!x)$$

where $C_i^q.all$, x and $!x$ are the diffusion, projection and successor functions, $\beta_{i,q}$, γ_x and δ_x are integer numbers taken such that the number of selected objects in a class is never negative.

The syntax of a WN initial marking is quite similar to the syntax of color functions, except it does not use variables. The f_i are thus limited to the first term of the above equation when defining an initial marking. Although the syntax of color functions makes it possible to write complex expressions, the functions practically used in models are rather simple.

Example 1. Let a be an arc linking a place p of color domain $cd(p) = C_1 \times C_2$ with a transition t such that $cd(t) = C_1 \times C_1 \times C_2$, a possible color function of a is: $\langle C_1.all - x, y \rangle$, where x and y are variables instantiated in C_1 and C_2 respectively. This function takes a set of tokens in p, composed of every element of C_1 except for one (the instance of x), and all having an identical associated value (the instantion of y). Let a' be another arc linking place p' to t such that $cd(p') = C_1 \times C_1$. A possible color function of a' is $\langle x \rangle + 2.\langle z \rangle$, where $x, z \in C_1$ are variables. This function requires three tokens in p' of which two must be identical.

The key consequence of these definitions is that the syntax of Well-formed Nets makes it *impossible* for objects belonging to the same static subclass to have different behaviors. Hence, the partition of color classes into static subclasses directly corresponds to the definition of the symmetries of the system. The complete definition of the model is summarized in the following paragraph.

Definition 3. *Well-formed Nets*
A Well-formed Net is a seven-tuple:

$$\mathcal{N} = \langle P, T, \mathbf{Pre}, \mathbf{Post}, \mathcal{C}, cd, m_0 \rangle$$

where:

1. P and T are disjoint finite non empty sets (the places and transitions of $\mathcal{N}$),
2. $\mathcal{C} = \{C_1, \ldots, C_n\}$ is the finite set of finite basic color classes. Some of the color classes may be ordered.
3. cd is a function defining the color domain of each place and transition; for places it is expressed as cartesian product of classes of $\mathcal{C}$ (repetitions of the same class are allowed), for transitions it is expressed as a pair $\langle$ variable types, guard $\rangle$ defining the possible values that can be assigned to transition variables in a transition instance; guards must be expressed in the form of standard predicates,
4. $\mathbf{Pre}[p, t], \mathbf{Post}[p, t] : cd(t) \to \mathrm{Bag}(cd(p))$ are the pre- and post- incidence matrices, expressed in the form of arc functions,
5. $m_0 : m_0(p) \in \mathrm{Bag}(cd(p))$ is the initial marking of place p.

2.2   SRG

The syntax of Well-formed Nets makes it possible to directly build a quotient graph, the so-called Symbolic Reachability Graph (SRG). Classes group markings that have the same distribution of tokens in places but differ only by the identity of tokens. Hence, a class can be built by forgetting the colors of tokens and replace them by variables. The markings represented by the class correspond to the possible instantiations of the variables.

Of course, a variable is associated with a Cartesian product of object classes and can be instantiated only by colors belonging to that product. However, not any instantiation is allowed within a class. If we want the SRG to preserve the firing sequences of the original RG, all the markings belonging to the same class must correspond to states in which the system behaves similarly.

In the WN formalism, the similarity of behavior is taken into account by the definition of static subclasses. If we consider a marking m obtained by some instantiation, only the markings obtained from m by a permutation of objects that respects static subclasses belong to the same class as m.

As a consequence, the cardinality of the classes of markings, hence the efficiency of the approach, is directly related to the cardinality of static subclasses. This is all the more true if there are ordered classes in the model. Actually, only rotations are allowed among objects of an ordered class thus reducing the possibility of grouping markings. The worst case appears when there are static subclasses inside an ordered class. In this case, an object of the class can never be substituted another object of the same class in an attempt to find another marking belonging to the same class. It is thus crucial that static subclasses be defined in a maximal way, i.e., that objects with similar behaviors do not belong to different static subclasses.

3   Symmetry of a Model and Representation

The goal of this work is to show that the partition into static sub-classes can be automatically deduced from the model itself. A partition of color domains into static sub-classes is such that an unfolding of the colored net into a black and white one will produce the same motif for any two elements of a given sub-class. This means that static sub-classes capture the **structural** symmetries of the model. As each element of a colored net is liable to introduce structural asymmetry, our goal is to independently analyze all the components of the model, to deduce what structural asymmetry -if any- it induces.

We start our exploration of the model with the basic assumption that all classes are symmetric and unordered. Of course for a net without any places or transitions this assumption is true. As each element is analyzed, the asymmetry it induces is added (by refinement) to the partitioning of the classes corresponding to the global structural symmetry.

Definition 4. Partition refinement *Let C be a color class and Π_1, Π_2 be two partitions of C. We recursively define the refinement $\Pi = \Pi_1 \sqcap \Pi_2$, as a partition of C obtained by intersecting every set S_1 in Π_1 with every set S_2 in Π_2:*

– if $S1 = S2$ a set $S = S1$ is added to Π, the construction proceeds with the next $S1$ from Π_1

- if $S1 \cap S2 = \emptyset$ the construction proceeds, no sets are added to Π.
- if $S1 \cap S2 \neq \emptyset$ the set $S1 \cap S2$ is added to Π, and the construction proceeds with $S_1 \setminus S_2$ in lieu of S_1.

Example 2. For instance, let $C = \{1, 2, ..6\}$, $\Pi_1 = \{1, 2, 3\} \uplus \{4, 5, 6\}$ and $\Pi_2 = \{1, 2\} \uplus \{3, 4\} \uplus \{5, 6\}$, $\Pi = \Pi_1 \sqcap \Pi_2 = \{1, 2\} \uplus \{3\} \uplus \{4\} \uplus \{5, 6\}$.

A partition of a color domain is a way to represent allowable symmetries: it represents all permutations that preserve the partition, i.e. any bijective function $f : C \to C$ such that $\forall e \in C_i \subset \Pi, f(e) \in C_i$.

Thus we start by considering that all elements of a color class C are interchangeable by defining a partition composed of a single static sub-class containing all its elements. As we find elements of the net that break the symmetry, we extract the partition corresponding to the symmetry they allow, and obtain the new globally admissible symmetries by common refinement of these two partitions. This process is iterated until either all the elements of the considered color class have been separated into an individual subclass, or all the components of the net have been analyzed.

A partition of a class of n elements into n sub-classes is a worst-case scenario, in which symbolic study will yield the same results as the classic approach. It corresponds to a totally asymmetric color domain.

An important aspect of the symmetry of a color class is whether it is ordered or not. In *WN* an ordered class is necessarily circular, and allows use of the successor operator. On ordered classes the only permitted permutations are rotations, thus the circularity helps to increase the symmetry of a problem.

For instance in the classic philosophers problem, we wish to distinguish the relative positions of the philosophers around the table (i.e. $!P$ is left of P), but not set a "first" philosopher. In this manner, we can study what happens when **a** philosopher has started eating, and his left neighbor wishes to eat, instead of studying independently $Phi0$ is eating and $Phi1$ wishes to eat, $Phi1$ is eating and $Phi2$ wishes to eat ... etc.

An exploration of the components of the net is performed to find any use of the successor operator on a given class, thus giving it the status "ordered". A problem sometimes encountered is when we have an ordered class from which we wish to distinguish an element or group of elements. Since the element thus distinguished is structurally different from the others of the class, its predecessor has to be distinguished, since the successor operation when applied to it yields a structurally different token. Recursively, we obtain that we have to distinguish all the elements from each other, thus falling in our worst-case scenario described above.

4   Components of Net and Induced (a)Symmetry

By taking advantage of symmetries, *WN* offer the possibility of efficient model-checking, but the syntax is a bit constraining. It requires careful study to maximize the potential for symbolic analysis of the reachability graph.

The model being self-contained, the actual (a)symmetries are defined by the model itself. The goal of our work is to identify and exploit the information that induces asymmetry in the model *automatically*, thus allowing to use *WN* with a less constrained syntax. This section successively studies the effect of place markings, arc color functions and transition guard predicates on the model symmetries.

4.1 Places

First let us consider a place P of domain $cd(P)$ composed of a single class $C = cd(P)$: a marking in a *WN* is a function from $\mathcal{M} : P \rightarrow \text{Bag} (cd(P))$, and is noted $\mathcal{M}(P) = \sum \alpha_i C_i . ALL$, $\alpha_i \in \mathbb{N}$, where C_i are static sub-classes of C. Let $C = \{e_1, \dots, e_n\}$ be its elements. We allow markings to directly reference these elements, therefore markings have the form $\mathcal{M}' : P \rightarrow \text{Bag} (\{e_1, \dots, e_n\})$. In other words we allow markings to directly reference the elements of a color domain. This form can easily be rewritten in the former using the following technique: let $\Pi = \Pi_0 \cup \dots \cup \Pi_k$ be a partition of C such that the elements of Π_i are present exactly i times in in place P. This partition is trivially the minimal partition into static sub-classes that allows to describe $\mathcal{M}(P)$, and describes the asymmetry generated by the marking of this place. The marking of P is therefore expressed as $\mathcal{M}(P) = \sum_i i \cdot \Pi_i . ALL$.

Example 3. For example, let C be a class composed of $\{e1..e6\}$ and the marking of a place P be
$$\mathcal{M}(P) = \langle e1 \rangle + \langle e1 \rangle + \langle e2 \rangle + 2 * \langle e3 \rangle + \langle e4 \rangle,$$
We rewrite this under the form:
$$\mathcal{M}(P) = 0 * (\langle e5 \rangle + \langle e6 \rangle) + 1 * (\langle e2 \rangle + \langle e4 \rangle) + 2 * (\langle e1 \rangle + \langle e3 \rangle),$$
In this form we obtain:
$$\Pi_0 = \{e5, e6\}; \Pi_1 = \{e2, e4\}; \Pi_2 = \{e1, e3\}.$$
This partition of C defines the asymmetry induced by the place P. We represent such a marking under the form:
$$\mathcal{M}(P) = \langle [e2, e4] \rangle + 2 \langle [e1, e3] \rangle.$$

However in the case of a composite color domain the calculation is more difficult. We use an intermediate notation to obtain a homogeneous form for individual marks: in *WN* syntax, a marking is of the form $\mathcal{M}(P) = \sum \alpha_i L_i, \alpha_i \in \mathbb{N}$, and L_i are tuples of the form $\langle C_{i1,j1} . ALL, .., C_{i_n,j_n} . ALL \rangle$ where a $C_{i,j}$ represents the jth static subclass of color class C_i.

Example 4. For example $\mathcal{M} = 2 \langle C_0, D_1 \rangle + \langle C_1, D_0 \rangle$ is a possible marking $\mathcal{M}$ for a place of color domain $C \times D$, if we do not represent the $.ALL$ to simplify the notation. Let us now assume that $C = C_0 \uplus C_1, C_0 = \{a, b\}, C_1 = \{c\}$ and $D = D_0 \uplus D_1, D_0 = \{x\}, D_1 = \{y, z\}$, this marking represents the following concrete marking: $\mathcal{M} = 2 * (\langle a, y \rangle + \langle a, z \rangle + \langle b, y \rangle + \langle b, z \rangle) + \langle c, x \rangle$. Since the subclasses are not yet defined when we begin our analysis, we use a notation explicitly enumerating the elements of a subclass. For example we write:
$$\mathcal{M} = 2 \langle [a, b], [y, z] \rangle + \langle [c], [x] \rangle = 2(\langle [a], [y, z] \rangle + \langle [b], [y, z] \rangle) + \langle [c], [x] \rangle$$
$$\mathcal{M} = 2 * (\langle [a], [y] \rangle + \langle [a], [z] \rangle + \langle [b], [y] \rangle + \langle [b], [z] \rangle) + \langle [c], [x] \rangle$$

We turn up here upon the beginnings of some algebraic laws that might allow us to factorize a given concrete marking into a reduced symbolic one. Let a vector of elements L_i represent the tuple of a token, and a multiplicity α_i assigned to a token represent a *mark* M, and a list or sum of marks represent a marking $\mathcal{M}$. There are generally different equivalent *representations* or formulae F of a given marking $\mathcal{M}$, as shown in example 4 that exhibits three representations of a marking $\mathcal{M}$. Let us define more precisely the process of factorization:

Definition 5. Marks:

A mark is of the form $M = \alpha_i L_i$ *,* $\alpha_i \in \mathbb{N}, L_i = \langle v_1, .., v_n \rangle$ *with* v_i *a vector of elements of a color domain.* α_i *is the multiplicity of the mark in the considered marking and* L_i *is its associated tuple. Let* $M = \alpha L, L = \langle v_1, .., v_n \rangle$ *, and* $M' = \alpha' L', L' = \langle v'_1, .., v'_n \rangle$ *. The following rules hold true:*

Marking Factorization:

Let F_1 *and* F_2 *be representations of a marking* $\mathcal{M} = F_1 = F_2$ *.* F_1 *is more factorized than* F_2 *if less marks are used to represent it. Since for any non-empty marking at least one mark is necessary, we define the factorized form of a marking as any form that uses a minimal number of marks to represent it.*

Multiplicity distributivity rule:

$\forall \alpha$ *and* $\alpha', \alpha L + \alpha' L = (\alpha + \alpha')L$

Merge rule:

At equal multiplicity, if L *and* L' *have a single mismatch* $m \neq m'$ *, such that* $L = \langle v_1, ..v_{k-1}, m, v_{k+1}, ..v_n \rangle$ *and* $L' = \langle v_1, ..v_{k-1}, m', v_{k+1}, ..v_n \rangle$ *,* $M + M'$ *can be merged into a single symbolic marking, by defining an additive operation on vectors of elements as the union operator on sets:*

$\alpha = \alpha', L = \langle v_1, ..v_{k-1}, m, v_{k+1}, ..v_n \rangle$ *and* $L' = \langle v_1, ..v_{k-1}, m', v_{k+1}, ..v_n \rangle$
$\Leftrightarrow M + M' = \alpha \langle v_1, ..v_{k-1}, m \cup m' \setminus m \cap m', v_{k+1}, ..v_n \rangle + 2\alpha \langle v_1, ..v_{k-1}, m \cap m', v_{k+1}, ..v_n \rangle.$

To simplify our calculations we will depend upon the underlying data structure to ensure that $m \cap m' = \emptyset$. This property is easily obtained if the algorithm starts from a completely developed form (i.e $\forall v_i, |v_i| = 1$), and the multiplicity distributivity rule is systematically applied before any merging occurs. Example 5 exhibits an application of these rules.

Lemma 1. *Even at equal multiplicity, if* L *and* L' *have more than one mismatch, no factorization of* $M + M'$ *is possible.*

Proof. The number of concrete marks represented by a symbolic mark is obtained by multiplying the cardinality of the v_i. We initiate a recursion by considering a tuple composed of two color domains. Let $M'' = M + M' \Leftrightarrow \langle v''_1, v''_2 \rangle = \langle v_1, v_2 \rangle + \langle v'_1, v'_2 \rangle$, we obtain $|M''| = |v''_1| * |v''_2| = |M| + |M'| = |v_1| * |v_2| + |v'_1| * |v'_2|$. Since all the individual elements mentioned in v_1 and v'_1 [resp. v_2 and v'_2] must be mentioned in v''_1 [resp. v''_2], we obtain that $v''_1 = v_1 \uplus v'_1$ and $v''_2 = v_2 = v'_2$ (or vice versa). Because a cardinality is a strictly positive integer, $|M''| = |M| + |M'|$ is true if and only if $|v''_1| = |v_1| + |v'_1|$ and $|v''_2| = |v_2| = |v'_2|$ or vice versa. In other words there is a single mismatch between M and M' or M'' cannot be defined. This result is trivially extensible to tuples of arbitrary length.

Corollary 1. *Any form for a marking* $\mathcal{M}$ *, in which any two marks have two or more mismatch is in factorized form.*

An algorithm computing such a form tries to match off pairs of marks, if their tuples L are equal it applies the multiplicity rule, if they have one mismatch it applies the merge rule, and if they have more than a single mismatch it tries the next pair. Because when a new mark is produced by applying the merge rule, it has to be compared (again) to all the other marks we obtain a combinatorial effect. The total complexity of the algorithm is in $O(setcard * tuplesize * nbterm!)$, where $setcard$ is the cardinality of a mentioned set, $tuplesize$ is the number of elements that constitute a tuple, and $nbterm$ is the number of terms encountered in the marking expression. This complexity is therefore strongly dominated by the factorial effect of the number of terms in a given formula, and the size of the domains studied only feebly impacts the complexity through $setcard$. This complexity is manageable, as most marking functions contain a limited number of terms. The total asymmetry generated by a given marking is thus calculated by intersecting the sets mentioned in each mark of the factorized marking. Once this partition has been found, we trivially rewrite the marks in term of this partition, because for any v_i of a mark of the factorized marking, $v_i = C_{j_1} \uplus .. \uplus C_{j_n}$ where $C = C_1 \uplus .. \uplus C_n$ is the partition obtained for color domain C that v_i is part of. In other words the partition into static subclasses is "finer grain" than that encountered in any set mentioned.

Example 5. Let us study the asymmetry generated by the marking:
$$\mathcal{M} = 2 * (\langle [a], [y] \rangle + \langle [a], [z] \rangle + \langle [b], [y] \rangle) + \langle [b], [z] \rangle + \langle [b], [z] \rangle + \langle [c], [z] \rangle$$
By applying distributivity we obtain:
$$\mathcal{M} = 2 * (\langle [a], [y] \rangle + \langle [a], [z] \rangle + \langle [b], [y] \rangle + \langle [b], [z] \rangle) + \langle [c], [z] \rangle$$
Then at equal multiplicity, $\langle [a], [y] \rangle + \langle [a], [z] \rangle \Rightarrow$ Single mismatch $\Rightarrow \langle [a], [y, z] \rangle$
...

$\mathcal{M} = 2 \langle [a, b], [y, z] \rangle + \langle [c], [z] \rangle$ which in turn gives us $C = (C_0 = [a, b]) \uplus (C_1 = [c])$ and $D = (D_0 = [y]) \uplus (D_1 = [z]) \uplus (D_2 = [x])$ since the $\langle c, z \rangle$ term forces to isolate z from y.

Thus we obtain $\mathcal{M} = 2(\langle [a, b], [y] \rangle + \langle [a, b], [z] \rangle) + \langle c, z \rangle$, and $\mathcal{M} = 2(\langle C_0, D_0 \rangle + \langle C_0, D_1 \rangle) + \langle C_1, D_1 \rangle$ where $.ALL$ is implicit. Since all the marks mentioned now have at least two mismatches, this symbolic marking is the factorized minimal representation of the initial concrete marking in *WN* syntax.

4.2   Arcs

WN basic arc functions are limited to diffusion-synchronization over a static sub-class, identity and successor functions.

- Identity is a totally symmetric function and thus does not introduce asymmetry in the model
- The successor function implies an ordering of the associated color domain. The first occurrence of a successor function on a variable X of a color domain C will add the tag ordered to this class. Any subsequent partitioning of C into sub-classes will give rise to a worst-case scenario in which each element of color C will be isolated in a static sub-class containing only that element. A class thus partitioned will be ignored in further analysis, as no further gain for symbolic exploration is possible.
- The diffusion-synchronization function separates the sub-classes it mentions from the rest of the elements of the class. Thus the static sub-classes mentioned are part of the minimal partition expressing the asymmetry generated by the considered arc, and will be used to further refine the partition corresponding to the global asymmetries.

In addition to these "classic *WN*" color functions, we add the possibility to directly reference elements e_i of class C:

– The constant reference function expressed as $\sum \alpha_i e_i$ (i.e. direct reference to constant elements e_i of C with multiplicity α_i) can be rewritten in terms of the diffusion-synchronization function given an appropriate partition Π of C of the same form as that defined for markings. Thus, though this is an extension of syntax for ease of use, the semantics of *WN* are preserved.

The asymmetry generated by a color function is computed like the asymmetry generated by a marking. The only difference is that a color function may reference formal parameters instead of just sets (our v_i in the study of the marking).

Definition 6. Color function:
A color function is defined as $\mathcal{F} = \sum \alpha_i L_i$, with

– $L_i = \langle V_1, ..V_n \rangle$ *and*
  • $V_i = [e_{i_1}, ..e_{i_n}]$ *a vector of elements representing a subclass, or*
  • $V_i = [(!)var_1, .., (!)var_n]$ *a list of formal parameters with possible use of the successor operator.*

Mismatch *A vector of elements always mismatches a list of variables, and they cannot be added. A variable mismatches any other variable, including itself with a successor operator. Lists of formal parameters can be added using the union operation on sets.*

The rules defined on markings in the previous section apply with this enhancement of the comparison operation.

Example 6. Let $\mathcal{F} = \langle [!X], [a] \rangle + \langle [!X], [b] \rangle + \langle [X], [a, b] \rangle$ be a color function, where X is a formal parameter and $a, b \in C$ a color domain.
$\langle [!X], [a] \rangle + \langle [!X], [b] \rangle \Rightarrow$ Single mismatch $\Rightarrow \langle [!X], [a, b] \rangle$
$\langle [!X], [a, b] \rangle + \langle [X], [a, b] \rangle \Rightarrow$ Single mismatch $\Rightarrow \langle [X, !X], [a, b] \rangle$
This defines $C_0 = [a, b]$ as a candidate static subclass of C

Though defining the possibility of mentioning several variables in a single V_i allows for more compact representation, it does not put in evidence additional symmetry of the model. Since only the actual element sets mentioned influence the computation of the partition of C into subclasses, and that merging on a variable implies that all the V_i match, the partition obtained under any form is the same.

4.3 Transitions

The study of transitions and guard predicates is slightly more complex than that of markings or color functions, and is the focus of the next section. This is due to the fact that we wish to allow expressions that are Boolean functions (AND $\wedge$, OR $\vee$ and NOT $\neg$) of basic predicates, instead of the basic additive composition allowed in both marking and color function definition. We will present here the basic predicates that we wish to allow, and show how we can write them in a homogeneous normalized way.

Definition 7. Normalized Basic Predicate: *A normalized basic predicate is a domain restriction predicate of the form $X \in E$, where X is a variable and E is a set of elements of the color domain C of X, or is a variable comparison predicate of the form $X = Y$ or $X \neq Y$ possibly with use of the successor ! operator. The set E of a domain restriction predicate will be referred to as the acceptation set of this predicate.*

We allow the "classic" *WN* basic predicates:

- $X = Y$ **synchronization**: this predicate and its dual $X \neq Y$ is permitted as is, since they do not introduce additional asymmetry. More precisely, its effect on the model is already correctly taken into account when building the SRG.
- $X =!Y$ **successor**: this predicate implies that the class X and Y belong to is ordered. Thus it will be treated in the same way as the successor color function found on arcs, by adding the information that class C is ordered. It should be noted again that this information is extremely constraining, and modifies the way subsequent partitions of C are treated. As a consequence, a first exploration of the model will be performed looking for these successor functions before beginning any analysis, to reveal which classes should be considered ordered.
- $d(X) = d(Y)$ **domain synchronization**: this predicate introduces all the static sub-classes as initially introduced by the modeler. At most, the partition Π corresponding to the asymmetry introduced by this predicate -if by itself- is the partionning of C into its static sub-classes C_i. This type of predicate is normalized as $\bigvee_i ((X \in C_i) \wedge (Y \in C_i))$.
- $d(X) = C_i$ **domain identification**: This predicate introduces the sub-class C_i, and is written $(X \in C_i)$ in our normalized form

In addition to the "classic *WN*" predicates, we introduce the possibility of explicitly referencing elements of a color domain:

- $X = e_i$ and $X \neq e_i$ **constant reference**: allows to directly compare a variable X to an element of its color domain. This predicate will be normalized in the form $(X \in \{e_i\})$ (respectively $(X \in C \backslash e_i)$)
- $X \in E$ **explicit partition**: where E is a set explicitly naming the referenced elements (i.e.: $state \in \{interrupted, error\}$). This is the basic predicate in our normalized form.

For completeness reasons, and to provide shorthands, we will also allow use of comparison operators. It should be noted that these operators are forbidden on *WN* because if a class is unordered the meaning is unclear, and if it is ordered it is circular therefore the assertion is even more meaningless. Despite this, there is a commonly understood way of interpreting these assertions, if considering classes to be non-circular ; the ordering that is referred to is given by the definition of the color, but is independent and incompatible with the ordering as defined by the successor operator:

- $X < e_i$ **constant comparison**: This will be interpreted as a shorthand referring to the $i - 1$ first elements of the class, the ordering being given by the definition of the color domain. It is thus normalized in the form $X \in \{e_1, .., e_{i-1}\}$. We will similarly allow use of $\leq, >, \geq$.
- $X < Y$ **variable comparison**: (X and Y variables of same color domain) In the same way, this predicate can only have meaning if the class is considered non-circular. This predicate gives rise to a worse case scenario for the color C concerned, as we will have to explicit all the possible cases in the normalized form: $\bigvee_{i=1}^{n} ((X \in \{e_1, .., e_i\}) \wedge (Y \in \{e_{i+1}, .., e_n\}))$. We will similarly allow use of $\leq, >, \geq$.

However, we emphasize that use of these operator does **not** introduce an ordering on the class in the sense of a successor operation, it even forbids use of successor in the rest of the net.

5 Study of Guard Predicates

Because the study of guard predicates is more complex than that of markings or color functions it is separated in this section. The technique developped in this section relies on a canonization process, such that the canonized form we obtain directly exhibits the symmetries allowed by a guard. We rely on a binary tree representation of the Boolean formula of the guard to ensure a unicity property essential to the algorithm (5.2). It requires a separate application of the canonization process for each formal parameter mentionned in the guard (5.3). We further prove that the symmetries exhibited by this canonized form are **strictly** the symmetries allowed by the guard (5.4), and outline an algorithm to obtain it (5.5).

5.1 Foreword: Normalized Forms of Predicates

As we have seen, the syntax proposed allows complex expressions for guard. To clarify the goal of this section, we will use a small example that exhibits the problems encountered and solved by our approach.

Example 7. Let us consider two color domains $C = \{1, 2, 3, 4, 5\}$ and $D = \{a, b, c\}$, and two variables $X \in C$ and $Y \in D$. If we consider a guard G where $\vee$ denotes OR and $\wedge$ denotes AND:

$G = (X \leq 3 \wedge Y = a) \vee (X \in \{3, 4, 5\} \wedge Y > a)$

With the normalized form described above we obtain:

$G = (X \in \{1, 2, 3\} \wedge Y \in \{a\}) \vee (X \in \{3, 4, 5\} \wedge Y \in \{b, c\})$

Our algorithm will allow us to obtain the form (normalized over X):

$G = (X \in \{1, 2\} \wedge Y \in \{a\}) \vee (X \in \{3\} \wedge Y \in \{a, b, c\}) \vee (X \in \{4, 5\} \wedge Y \in \{b, c\})$

Giving the partition $C = (C_0 = \{1, 2\}) \uplus (C_1 = \{3\}) \uplus (C_2 = \{4, 5\})$. Then by normalizing over Y we obtain:

$G = (Y \in \{a\} \wedge X \in \{1, 2, 3\}) \vee (Y \in \{b, c\} \wedge X \in \{3, 4, 5\})$

Giving the partition $D = (D_0 = \{a\}) \uplus (D_1 = \{b, c\})$. The final form expressed in terms of subclasses would be:

$G = (X \in C_0 \wedge Y \in D_0) \vee (X \in C_1 \wedge (Y \in D_0 \vee Y \in D_1)) \vee (X \in C_2 \wedge Y \in D_1)$

We will explain here how these different forms are computed, and prove that the obtained expression minimally represents the asymmetry induced by a guard. The rest of this section is more involved and the casual reader should skip to the *Implementation and results* in section 6.

5.2 Predicate Data Representation

We use a canonical binary tree for a guard G's representation:

- the leaves of the tree, noted L_i, are normalized predicates of the form $(X \in P)$ as introduced in section 4.3
- the internal nodes are the operators AND $\wedge$ and OR $\vee$.

The negation is directly applied to obtain the dual form of a negated sub-expression (i.e. $\neg(X \in E)$ becomes $(X \in C\backslash E)$). Variable comparison predicates are switched as needed to obtain a lexicographical order (i.e. $X = Y$ and not $Y = X$). This makes comparison of subtrees easier.

To apply the normalization algorithms, we require an essential unicity constraint, that will ensure that any two sets mentioned in predicates L_i over the same variable have an empty intersection. Let L_1 and L_2 be two leaf predicates of a guard G:

Property 1. **Unicity:**
$\forall L_1, L_2 \in G$ such that $L_1 = (X \in P_1)$ and $L_2 = (X \in P_2)$,
$\quad P_1 \cap P_2 = \emptyset$
$\quad$ or $P_1 = P_2$ and L_1 and L_2 share a single physical representation.

Our tree structure uses the following algorithm to ensure unicity of representation by construction.

Algorithm ensuring unicity
When inserting a leaf predicate $L = X \in E$ in a guard structure G:

- Explore G to find any predicates of the form $L_i = X \in E_i$ already present in G
- Compute the intersection $E \cap E_i$ of the new predicate set E with previously constructed ones E_i, for all i:
  - $E = E_i$: interrupt the iteration by returning the physical address of the $X \in E_i$ leaf ; L and L_i *will share a single physical representation.*
  - $E \cap E_i = \emptyset$: no action is taken, iterate for the next E_{i+1} set ; $P \cap P_i = \emptyset$
  - $E \cap E_i \neq \emptyset$: Compute a partition of $E \cup E_i$ such that $E \cup E_i = P_1 \uplus P_2 \uplus P_3$, such that $P_1 = E\backslash E_i$, $P_2 = E \cap E_i$ and $P_3 = E_i\backslash E$. Then create three leaves L_1, L_2, and L_3 corresponding to these sets (i.e. $L_1 = (X \in P_1)$ etc.). Finally modify **in place** the leaf node $L_i = (X \in E_i)$ into an OR $\vee$ node of left son L_2 and right son L_3. The construction ensures that these sets (P_2 and P_3) do not intersect any other already present in G. Continue iteration but using L_1 instead of L ; P_1, P_2 *and P_3 do not overlap.*
- When this iteration is concluded, we construct a tree composed of an OR between the L_i and the resultant L that constitute the partition of the original E (i.e. the P_1 and P_2 of the iteration phase), and return the physical address of its root.

Insertion of a TRUE or FALSE statement, or of a variable comparison leaf follows a similar procedure in that only one physical instance of any of these will ever be created for a given tree G.

Inserting an internal operator node A *op* B uses the above procedure to obtain A and B (recursively) then creates an *op* node of sons A and B and returns it.

Thus this data structure, by **construction** ensures the unicity property.

5.3   Normal Developed Form w.r.t. a Formal Parameter

The goal here is to define a normal form for the guard expression G considered that allows to calculate the asymmetry induced by G. This normal form will be computed independently for each formal parameter (variable) mentioned in G, as shown in the foreword (Example 7).

Definition 8. Normal developed form
Let the developed form of a guard G with respect to a formal parameter X be:
$$G = (\textstyle\sum_i X_i \wedge M_i) \vee (\textstyle\sum_j M_j)$$
Where X_i is a predicate over variable x of the form $(x \in E_i)$, such that for all $i, k, i \neq k$, $E_i \cap E_k = \emptyset$, and M_i, M_j are Boolean guard expressions that do not mention variable x except possibly through a variable comparison, with the additional constraint that no two M_i be equal.

Property 2. **Normal Form property** The permutation of an instance e_1 of formal parameter x with an instance e_2 does not modify the evaluation of the guard **if and only if** e_1 and e_2 belong to the acceptation set of the same predicate in the developed form of the guard formula.

The proof of this property may be found at the end of the next section, as it requires some further definitions.

Thus the developed form defines a **minimal** partition of a color domain C, for a given variable x, as the partition of C into the acceptation sets of the X_i predicates of the developed form. The tokens in a given E_i are structurally equivalent with respect to the considered guard, any permutation of elements that respects this partition leaves the truth value of the guard predicate unchanged.

This is the property that defines static sub-classes of a *WN*, thus it defines the asymmetry induced by the guard over the color domain C of the formal parameter considered. To obtain the full set of asymmetry induced by a guard, we need to compute the canonized form for all formal parameters mentioned, and refine the partitions obtained if more than one parameter belongs to the same color domain.

5.4   Computation of the Normal Form

Now that we have defined the form we wish to obtain to explicit the symmetry allowed by a guard, we need an algorithm to obtain it. To this end we define the *depth* of a node in an expression tree, such that **a predicate on a variable X is under developed form if its depth is at most 1**:

Definition 9. Node depth
Let the depth of a node be recursively defined as follows:

- *We define the depth of the root's parent node as **0**.*
- *Let n be the depth of the parent of the node studied.*
    - *if $n = 0$:*
        - *If the node is a leaf or a $\vee$ OR operator its depth is 0*
        - *Else the node is a $\wedge$ AND operator and its depth is 1*
    - *else $n > 0$:*
        - *If the node is a leaf its depth is n*
        - *Else the node is an operator and its depth is $n + 1$*

Table 1. Boolean Algebra laws

Name	Expression	Dual
Identity:	$x \vee 0 = x$	$x \wedge 1 = x$
Commutativity:	$x \vee y = y \vee x$	$x \wedge y = y \wedge x$
Distributivity:	$x \wedge (y \vee z) = (x \wedge y) \vee (x \wedge z)$	$x \vee (y \wedge z) = (x \vee y) \wedge (x \vee z)$
Complementarity:	$x \vee \neg x = 1$	$x \wedge \neg x = 0$
Idempotency:	$x \vee x = x$	$x \wedge x = x$
Associativity:	$x \vee (y \vee z) = (x \vee y) \vee z = x \vee y \vee z$	$x \wedge (y \wedge z) = (x \wedge y) \wedge z = x \wedge y \wedge z$
Absorption:	$x \vee (x \wedge y) = x$	$x \wedge (x \vee y) = x$
Morgan's law:	$\neg(x \vee y) = \neg x \wedge \neg y$	$\neg(x \wedge y) = \neg x \vee \neg y$
Dominance:	$x \vee 1 = 1$	$x \wedge 0 = 0$
Involution:	$\neg(\neg x) = x$	

To reduce the depth of predicates on the target variable, we apply the classic laws of Boolean algebra (table 1), where 0 represent *FALSE* and 1 represents *TRUE*.

We add the following operation, derived from the form for of basic predicates, that is similar to our merge rule defined on markings:

$$Predicate\ fusion : (X \in E_1) \vee (X \in E_2) = (X \in E_1 \cup E_2)$$
$$Dual\ (X \in E_1) \wedge (X \in E_2) = (X \in E_1 \cap E_2)$$

We further define an operation essential to the factorization process, and that is made possible by the unicity property (1):

Definition 10. Quotient tree:
*Let X be a leaf predicate bearing on variable x of the form $(x \in E)$, and T be an expression tree, the **quotient tree** T' of T by X, noted $T' = T/X$ is obtained by replacing any leaf predicate of T of the form $(x \in F)$ bearing on variable x by $TRUE$ if $E = F$, and $FALSE$ if $E \neq F$.*

This definition ensures the following property:

Property 3. A quotient tree $T' = T/X$ does not contain any leaf predicate bearing on variable x, and $T \wedge X = T' \wedge X$.

This definition and subsequent property is the result of applying distributivity and dual predicate fusion rules.

Proof. Let $F = T \wedge X$ and $F' = T' \wedge X$,

 – If the assertion X is false, both F and F' are false (Dominance)
 – Else if X is true, the evaluation of T will evaluate any predicate on x as true if it is X and false if it is a different predicate, because the acceptation sets are distinct (unicity property). This is precisely the way T' is defined.

We can now prove the normal form property given above:

Proof. **Normal form:**(proof of property 2)

Let us remind that a normal form w.r.t. a formal parameter x is defined as:

$$G = (\sum_i X_i \wedge M_i) \vee (\sum_j M_j)$$

Where the X_i are predicates over variable x of the normalized form $(x \in E_i)$, such that the acceptation sets of any two X_i be distinct, and M_i, M_j are Boolean guard expressions that do not mention variable x except possibly through a variable comparison, with the additional constraint that no two M_i be equivalent.

Let $(F)_{x=e_i}$ be the value of formula F when evaluated with the constraint $x = e_i$, e_i being a constant value of the color domain C of variable x. The following equivalences hold:

$$(F)_{x=e_1} = (F)_{x=e_2} \Leftrightarrow ((x \in \{e_1\}) \wedge F)_{x=e_1} = ((x \in \{e_2\}) \wedge F)_{x=e_2} \text{(Identity dual)}$$
$$\text{Let } F_1' = F/(x \in \{e_1\}) \text{and } F_2' = F/(x \in \{e_2\}). \text{Thus}$$
$$\Leftrightarrow ((x \in \{e_1\}) \wedge F_1')_{x=e_1} = ((x \in \{e_2\}) \wedge F_2')_{x=e_2}$$
$$\text{Since } (x \in \{e_1\})_{x=e_1} = (x \in \{e_2\})_{x=e_2} = TRUE$$
$$\text{And since a quotient tree does not mention variable } x,$$
$$(F_1')_{x=e_1} = F_1' \text{ and } (F_2')_{x=e_2} = F_2',$$
$$(F)_{x=e_1} = (F)_{x=e_2} \Leftrightarrow F_1' = F_2'$$

F is of the form $F = (\sum_i X_i \wedge M_i) \vee (\sum_j M_j)$ Let $i1$ and $i2$ be such that $X_{i1} = (x \in E_{i1})$ and $X_{i2} = (x \in E_{i2})$ be two predicates such that $e_1 \in E_{i1}$ and $e_2 \in E_{i2}$. Thus, $F_1' = M_{i1} \vee \sum_j M_j$ and $F_2' = M_{i2} \vee \sum_j M_j$. Therefore:

$$(F)_{x=e_1} = (F)_{x=e_2} \Leftrightarrow F_1' = F_2'$$
$$\Leftrightarrow M_{i1} = M_{i2}$$
$$\Leftrightarrow i1 = i2 \text{ since the } M_k \text{ are all distinct}$$
$$\Leftrightarrow \exists i, \text{suchthat } X_i = x \in E_i \text{and that } e_1 \in E_i \text{ and } e_2 \in E_i$$
$$(F)_{x=e_1} = (F)_{x=e_2} \Leftrightarrow e_1 \text{ and } e_2 \text{ belong to the acceptation set of}$$
$$\text{the same term of } F \text{ in developped form}$$

Thus the permutation of an instance e_1 of formal parameter x with an instance e_2 does not modify the evaluation of the guard **if and only if** e_1 and e_2 belong to the acceptation set of the same predicate in the developed form of the guard formula.

5.5 Algorithm to Obtain a Developed Form

The full algorithm used to obtain a normal form is composed of four phases:

- we first *develop* the terms on the target variable x to have them all at depth 0 or 1, thus obtaining a sum of products form.
- we then *merge* the identical occurrences of predicates on x to obtain a single occurrence of each by comparing them.

- we then *canonize* the M_i terms to allow their comparison (we need to obtain $\forall i, j$ $M_i = M_j \Leftrightarrow i = j$)
- Finally we *group* X_i of same M_i.

Example 8. Let $\mathcal{G} = (M_2 \vee X_1) \wedge (M_1 \wedge X_2) \vee X_1 \wedge M_3$, where $X_1 = x \in E_1, X_2 = x \in E_2$:

- Phase 1: $\mathcal{G} = X_1 \wedge M_1 \vee M_1 \wedge M_2 \vee X_2 \wedge M_2 \vee X_1 \wedge M_3$. The $X_1 \wedge X_2$ necessarily evaluates to *false* because of the unicity property.
- Phase 2: $\mathcal{G} = X_1 \wedge (M_1 \vee M_3) \vee X_2 \wedge M_2 \vee M_1 \wedge M_2$
- Phase 3: Let us suppose that canonization has shown $M_1 \vee M_3 = M_2$.
- Phase 4: $\mathcal{G} = (X_1 \vee X_2) \wedge M_2 \vee M_1 \wedge M_2 = (x \in E_1 \cup E_2) \wedge M_2 \vee M_1 \wedge M_2$ which is the normalized form for $\mathcal{G}$ over x, and defines $E_1 \cup E_2$ as a candidate subclass.

For lack of space we will not develop the four aforementioned phases of the algorithm, but they may be obtained by contacting the authors.

6   Implementation and Results

To illustrate our work we consider a part of an industrial application, an Electrical Flight Control System [7], that manages the spoilers of an airplane during takeoff and landing. Opening angles applied to the spoilers are computed with respect to the value given by sensors (altitude, speed, ...) and the angle the pilot wants to apply.

We present the function that allows to detect if the airplane is on ground by checking the altitude, speed of the wheels and weight on the wheels. The altitude and speed sensors are monitored and a corrupt behavior leading to false information may be detected. The possible values provided by the altitude sensor are in the set composed of $\{Altitude\_Min..Altitude\_Max\} \cup \{Altitude\_Corrupt\}$. In the same way, the speed sensor may return any value in $\{Speed\_Min..Speed\_Max\} \cup \{Speed\_Corrupt\}$. We represent each sensor domain by a place that initially contains all the possible values, therefore considering all possible sensor values in our model-checking.

The main difficulty when formally studying such a application is the infinite domain of possible values produced by the continuous domain sensors. A first step towards analysis is to represent these infinite domains by discrete ones. Even after this operation we obtain domains too large for conventional state graph exploration. Therefore, it is necessary to automatically detect the limit values of speed and altitude that have a relevant impact on the behavior of the model.

The Petri net model of the studied function, with classes and variables declaration, is represented by Fig.1. Places Pi correspond to the successive steps of the computation of the function result. Place $WeightPossibleVal$ contains all the possible values for a sensor weight. Places $Weight\_Left\_Wheel$ and $Weight\_Right\_Wheel$ represent the value produced by the sensor at a given instant. The domain of these three places is $Class\_Weight$. It is the same for places $AltitudePossibleVal$, $Altitude$ and $Class\_Altitude$ domain and for places $SpeedPossibleVal$, $Speed\_Right\_Wheel$, $Speed\_Left\_Wheel$ and $Class\_Speed$ domain. To legibility reasons the domain of each place is not represented on the figure. The constant values $Altitude\_Limit$ and

Class
  Class_Weight is [on,off];
  Class_Speed is Speed_Min..Speed_Corrupt;
  -- We suppose that Speed_Corrupt = Speed_Max+1
  Class_Altitude is Altitude_Min..Altitude_Corrupt;
  -- We suppose that Altitude_Corrupt = Altitude_Max+1
  Class_Signal is [T,F];

Var
  W in Class_Weight;
  A in Class_Altitude;
  S in Class_Speed;

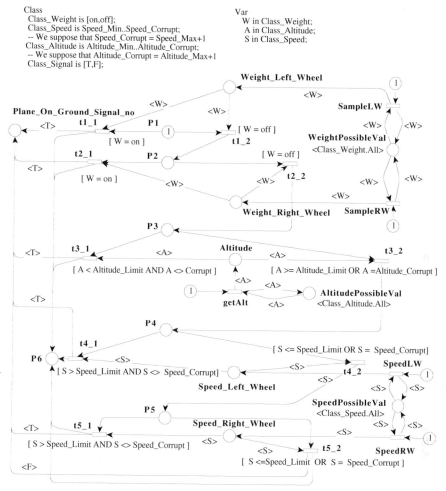

Fig. 1. Petri net model of the function

Speed_limit correspond to the values that allow to determine if the airplane is on ground, they appear on transition guards.

The application of the tool presented in this paper to this example leads to a partition of the altitude and speed domains into two classes each. The speed and altitude classes are decomposed into two classes each:

$$Class\_Altitude = \{Altitude\_Min..Altitude\_Limit - 1\}$$
$$\uplus \{\{Altitude\_Limit..Altitude\_Max\} \cup \{Altitude\_Corrupt\}\}$$
$$Class\_Speed = \{\{Speed\_Min..Speed\_Limit\} \cup \{Speed\_Corrupt\}\}$$
$$\uplus \{Speed\_Limit + 1..Speed\_Max\}$$

To compute these subclasses, we have to fix the constants values *Altitude_Min, Altitude_Max, Altitude_Limit, Altitude_Corrupt, Speed_Min, Speed_Max, Speed_Limit, Speed_Corrupt*. The results of the decomposition are independent from these values since we consider that *Speed_Corrupt = Speed_Max + 1*

and $Altitude\_Corrupt = Altitude\_Max + 1$. If we don't have this assumption each class is divided into three subclasses $\{Altitude\_Min..Altitude\_Limit - 1\} \uplus \{Altitude\_Limit..Altitude\_Max\}$
$\uplus \{Altitude\_Corrupt\}$ and $\{Speed\_Min..Speed\_Limit\} \uplus \{Speed\_Corrupt\} \uplus \{Speed\_Limit + 1..Speed\_Max\}$. The designer can modify values of the constants (in the class definition as well as on the guards) without being concerned by the subclasses of its model since they can be automatically computed if necessary.

Though this is not generally the case, the size of the SRG of this system has a constant 783 accessible symbolic states, independently of the size of the domains considered. The number of concrete states represented by this SRG is combinatorial, from 175k RG nodes for domains of cardinality 20 to about $14 \cdot 10^9$ RG nodes for the values extracted from the original specification: 2000 values of speed and 1000 of altitude. The SRG calculation time hardly increases, when the RG tool was saturated for 100 values per domain.

7   Conclusion

This paper presented a novel algorithm for automatically computing the symmetry break induced by the elements of a class of high-level Petri nets. The resulting information is exploited through translation to Well Formed nets, that may be symbolically analyzed with a symbolic reachability graph. This work has been implemented as a tool for the CPN-AMI software, and relies on GreatSPN for the construction of the SRG. This transformation is transparent for the user, and guarantees that the equivalence classes obtained are minimal with respect to the behavior of the tokens of a net. The analysis is based on the independent computation of the symmetry break induced by each element of a net, and leaves room to apply further structural reductions on the obtained net.

We are currently working at extending this mechanism to optimize the calculation of ESRG, by only taking into account the reachable elements of a net in a given state. This may be done by limiting the computation to the asymmetry induced by a current marking, the possibly fireable transitions for this marking, and their connected arcs. Such an approach will allow symbolic analysis of partially symmetric systems. This work is further used in the development of **Lf P**, a high-level specification language, that relies on the generation of Petri nets for verification of properties. Another extension in progress is to integrate in the calculation of allowed symmetries the effect of a given LTL or CTL formula, thus allowing property-based model checking. We are further interested in trying to apply a similar exploration technique to automatically deduce symmetries of other formalisms.

References

1. CPN-AMI: a Petri net based CASE environment. url: http://www-src.lip6.fr/cpn-ami.
2. G. Chiola, C. Dutheillet, G. Franceschinis, and S. Haddad. Stochastic well-formed colored nets and symmetric modeling applications. *IEEE Transactions on Computers*, 42(11):1343–1360, 1993.
3. G. Chiola, C. Dutheillet, G. Franceschinis, and S. Haddad. A symbolic reachability graph for coloured Petri nets. *Theoretical Computer Science*, 176(1–2):39–65, 1997.

4. G. Chiola and G. Franceschinis. Structural colour simplification in Well-Formed coloured nets. In *Proc. 4th Int. Workshop on Petri Nets and Performance Models*, pages 144–153, Melbourne, Australia, December 1991.
5. C.N. Ip and D.L. Dill. Better verification through symmetry. In D. Agnew, L. Claesen, and R. Camposano, editors, *Computer Hardware Description Languages and their Applications*, pages 87–100, Ottawa, Canada, 1993. Elsevier Science Publishers B.V., Amsterdam, Netherland.
6. D. Regep, Y. Thierry-Mieg, F. Gilliers, and F. Kordon. Modélisation et vérification de systèmes répartis : une approche intégrée avec **L**f**P**. In *AFADL 2003,Approches Formelles dans l'Assistance au Développement de Logiciels*. INRIA, proceedings, January 2003.
7. M. Doche, I. Vernier-Mounier, and F. Kordon. A modular approach to the specification and validation of an electrical flight control system. In *FME'01, Formal Methods for Increasing Software Productivity*, pages 590–610, Berlin, Germany, March 2001. Springer Verlag.
8. GreatSPN: GRaphical Editor, Analyzer for Timed, and Stochastic Petri Nets. url: http://www.di.unito.it/ greatspn/.
9. J-C. Fernandez, C. Jard, T. Jeron, and C. Viho. An experiment in automatic generation of test suites for protocols with verification technology. *Science of Computer Programming*, 29(1-2):123–146, 1997.
10. Serge Haddad, Jean Michel Ilie, M. Taghelit, and B. Zouari. Symbolic reachability graph and partial symmetries. In *Application and Theory of Petri Nets*, pages 238–257, 1995.
11. D. Poitrenaud and J-F. Pradat-Peyre. Pre- and post-agglomeration for ltl model-checking. *Lecture Notes in Computer Science*, 1825:387–408, 2000.
12. D. Regep and F. Kordon. **L**f**P**: a specification language for rapid prototyping of concurrent systems. In *12th IEEE International Workshop on Rapid System Prototyping*, June 2001.
13. V. Rusu, L. du Bousquet, and T. Jéron. An approach to symbolic test generation. In *2nd International Workshop on Integrated Formal Method (IFM'00)*, number 1945 in LNCS, pages 338–357, Dagstuhl, Germany, 2000. Springer-Verlag.

A Proposal for Structuring Petri Net-Based Agent Interaction Protocols

Lawrence Cabac, Daniel Moldt, and Heiko Rölke

Department of Computer Science, TGI, University of Hamburg
{6cabac, moldt, roelke}@informatik.uni-hamburg.de

Abstract. In this paper we introduce net components as means for structuring Petri net-based agent interaction protocols. We provide a tool for effortless application of net components to nets. Thus we facilitate the construction of nets and unify their appearance. Net components can be used to derive code for interaction protocols from a subset of extended AUML (Agent Unified Modeling Language) interaction protocol diagrams. This allows for a smooth integration of some traditional software development specification approaches with high-level Petri nets. By using net components we do not only unify the structure of Mulan agent protocols but also succeed to build a common language within a community of developers who share the net components.

Keywords: agents, agent interaction protocols, AUML, high-level Petri nets, Mulan, reference nets, Renew.

1 Introduction

From the beginning of computer science the readability of program code is a well known problem. There are many means of improving readability. The goals are to make code intelligible, easily accessible and clear. Generally, the problem that lies behind the readability is the complexity of code.

Methods to achieve readability are for example structured programming (see [4]) or object-oriented programming (see [19]). Also integrated development environments (IDE), syntax high-lighting and indentation of syntactical entities add to readability. While the first ones of these concepts provide orientation on an abstract level, the latter ones provide orientation on a basic cognitive level. These problems do not only pertain to text but also to other representations of code.

Visual programming languages (VPL) are subject to similar problems. The statement which is known as 'Limit of Deutsch' (see [15]) - the limitation of the number of elements on a screen to fifty graphical objects - gives a glimpse of the problem. Many advances try to hide the complexity by modularizing the views. Complex structures are represented and thus hidden behind simple elements, e.g.: icons. This leads to complex content of simple elements which increases the number of available simple elements. Nevertheless, this is a good

W.M.P. van der Aalst and E. Best (Eds.): ICATPN 2003, LNCS 2679, pp. 102–120, 2003.
© Springer-Verlag Berlin Heidelberg 2003

approach, since it clarifies the overall structure of code through abstraction. However, the arrangement of elements of code for visual programming languages like LabVIEW (see [14]) or Prograph (see [20]) is restricted by their development tools to guarantee a unified style.

Workflows are not considered to be programs but they describe processes just like programs. During modeling of workflows very similar problems of complexity occur. VAN DER AALST et al. (see [22]) offer a proposal of "Workflow-Pattern" for a unification of recurring patterns in workflows. The unification of parts of the code by modeling with patterns and the naming of those parts of the workflow are useful methods to facilitate the process of understanding the workflow.

Petri nets can also be used to describe processes. They can be simulated (executed) - just like VPL. Thus they can be regarded as programs, and nets themselves can be considered the code of the program.

Petri nets tend to grow in size with their complexity (see [8]). High-level Petri nets use various concepts to handle complex coding. They are capable of expressing complex structures in folded nets by introducing named tokens, thus adding new elements to the formalism. Another approach to handle complexity is to combine Petri nets with concepts of object orientation which leads to the object-oriented Petri nets (OOPN) (see [16]). Agent-oriented Petri nets (AOPN), such as those presented in [10], help develop multi-agent systems, especially through their inherent concurrency.

There are general recommendations for the look and feel of Petri nets. These relate usually to the simple elements of the Petri nets: Transitions, Places, Arcs and Inscriptions. JENSEN's recommendations (see [9]) - based on the work of OBERQUELLE (see [18]) - cover either the elements of the nets, or the rules are fairly general and can be interpreted quite broadly. For special nets and for the arranging of net elements - except for beautification - there exist no additional rules; consequently the appearance of Petri nets varies extremely. Naturally, the appearance depends on the programmer / modeler, the used Petri net tool and the domain. In general, it can be of advantage to have a great variety of appearances of Petri nets. In contrast, the representation of similar nets in the same domain requires conformity. This is the case when implementing application software with Petri nets.

The tool of choice for developing and modeling Petri nets at the Department of Computer Science at the University of Hamburg is *Renew* (see [12]). Not only does it allow to construct Petri nets but it also has the ability to simulate them efficiently. The nets that are processed by Renew are reference nets (see [11]), an extension to Coloured Petri Nets (CPN, see [9]). It is possible to use Java code as inscriptions of net elements which can be of advantage, for example when combining nets with a graphical user interface.

Mulan (Multi-Agent Nets) is based on Renew. It is a multi-agent system that uses the advantages of Petri nets such as concurrency. Together with Renew it provides the possibility to develop software with Petri nets by the agent-oriented paradigm. Since agents are defined as individual and independent components they offer an interesting approach to the development of software for concurrent processes. Especially for concurrent and adaptive processes strong limitations exist with conventional methods so that building a multi-agent system with inherent concurrency is of advantage. This can easily be achieved by basing the system on Petri nets.

Development of application software based on an agent-oriented software system requires the development of interaction protocols so that the behavior of the agents is well defined. The basic task for application programming with Mulan is to develop Mulan protocols, which are Petri nets. The purpose of the Mulan protocols is to define the behavior and the communication or interactions of agents.

Mulan protocols perform several basic tasks which frequently re-occur. They vary in the overall structure but perform similar tasks within this structure like sending or receiving a message or deciding on a condition. However, those nets can be rather huge so that implementation and debugging can be quite difficult and time consuming. We model agent interaction using AUML (Agent Unified Modeling Language) interaction protocol diagrams (see [7]), which is in the context of UML (Unified Modeling Language) in the version of interaction diagrams a commonly used and elaborated modeling technique.

In this paper we introduce net components as means for structuring Petri net-based agent interaction protocols. We achieve to unify the structure of Mulan protocols which increases their readability and build a common language within a community of developers who share the net components.

In the next section we will present our multi-agent system infrastructure. Then we introduce net components in general, the net components for Mulan protocols and a tool that supports the construction of Petri nets with net components. Finally we present a way to model the communication of agents with AUML interaction protocol diagrams, describe the construction of Mulan protocols with net components and show that net component-based Mulan protocols reveal their structure due to the geometrical forms of the net components.

2   Petri Net-Based Multi-agent System Infrastructure

In this section we present a short introduction to the Petri net-based multi-agent system Mulan. Mulan is implemented with reference nets and runs within Renew. Renew (Reference Net Workshop, see [12]) is a Petri Net editor and simulator. Mulan (Multi-Agent Nets, see [10]) is a reference architecture to a multi-agent system which complies with the FIPA specification for multi-agent systems. CAPA (Concurrent Agent Platform Architecture, see [5]) extents Mulan to provide FIPA-compliant communication and agent management.

2.1   Renew

With Renew it is possible to draw and simulate Petri nets and reference nets. A net that is loaded into the editor can be executed by the simulation engine. For this an instance of the net is created by the simulator. Any simulated net can instantiate another net. Hence it is possible to produce many instances of different nets. The relationship between net (also called net template) and net instance can be compared to the relationship of class and object.

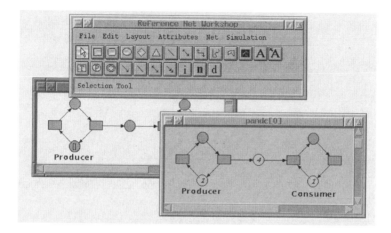

Fig. 1. Renew GUI, Petri net and net instance (producer-consumer example).

Figure 1 shows the graphical user interface (GUI) of Renew, a simple Petri net in the back and a net instance. The user interface consists of the menu bar, two palettes and a status line. The menu bar offers menus for general operations, attribute manipulations, layout adjustment and Petri net-specific operations. It also provides the possibility to handle the simulation. Of the two palettes the first one consists of usual drawing tools while the second one holds the Petri net drawing tools. The latter palette provides the tools for creating transitions, places, virtual places, arcs, test arcs, reserve arcs, inscriptions, names and description nodes. In addition to these tools the editor reacts in a context sensitive manner to facilitate the drawing of nets. One example is the dropping of arcs on the background which can create a new place if the arc starts at a transition and vice versa. Another example is the right click on inscribable elements which produces an inscription for this element with a context sensitive default value.

Nets hold the initial marking where net instances hold the current marking. In figure 1 the producer-consumer example has been started. In the net (background) one of two black tokens of the initial marking can be seen in the place labeled "Producer". While the net instance by default only shows the number of tokens in a place it is also possible to show the contents of the places by clicking on the numbers.

A special feature of Renew is that it can operate with reference nets. Renew also allows to use any kind of Java objects as tokens. It is implemented in Java and extendible through a plug-in mechanism. A plug-in for net components is presented in section 3.3. Figure 2 shows the architecture of the whole system.

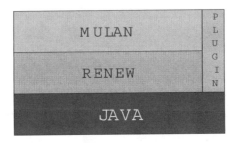

Fig. 2. The architecture of the multi-agent system infrastructure.

2.2   Reference Nets

Reference nets are special high-level Petri nets in which the tokens can be nets again. For these "nets within nets" referential semantics is assumed. Tokens in one net instance can be references to other net instances. The benefit of this feature for Renew is that it is modular and extendible. Nets can call other nets just like method calls of objects by using synchronous channels (see [11]).

A synchronous channel consists of two transitions which are called *down-link* and *up-link*. These two transitions can only fire simultaneously and only if both transitions are activated. *Down-link* and *up-link* can belong to one single net or they can belong to two different nets. In both cases any object can be transferred from either transition to the other. If two different net instances are involved it is thus possible to synchronize those two nets and to transfer objects in either direction by the synchronous channel.[1]

Mulan agents use the synchronous channels to start and stop their protocols. Also the communication between the Mulan agent and its Mulan protocols is realized with them. Messages are transferred from the agents to the protocols and back. While the agent provides the functionality to transmit the message to another agent the protocol is in charge of the processing of the message itself.

2.3   Mulan

Mulan is a multi-agent system that is based on Renew. It is implemented with Petri nets as a system of reference nets (Mulan: Multi-Agent Nets) and it complies with the open specifications of the Foundation for Intelligent Physical Agents (FIPA, see [6]) for multi-agent systems.

The system consists of numerous Petri nets. This is illustrated in figure 3. The figure shows the net within a net hierarchy of the system. Agents are nets which exist on platforms which are also nets. There can be many platforms and the agents can communicate with each other within and across platforms.

---

[1] Additional information for Renew, reference nets and synchronous channels can be found in [12].

Protocols are nets within the agents and control their behavior. The figure is taken from [10].

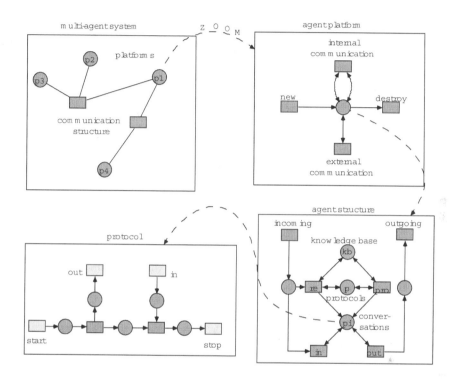

Fig. 3. The structure of Mulan (compare [10]).

Agents communicate by sending messages to each other. A set of messages sent back and forth between two or more agents is called conversation. The conversation is determined by the protocols which the agents use during this communication. So the conversation describes the agents' interactions whereas one or more protocols describe the behavior of one agent during the conversation.[2] Conversations can be compared to FIPA interaction protocols while the Mulan protocols as described here have no equivalent and are parts of the FIPA interaction protocols.

2.4   Terminology

In this section some terminology shall be clarified. Descriptions of the terms agent, protocol and net component are provided to give the reader a first notion of them while the details are postponed to a later section.

---

[2] This terminology differs slightly from the FIPA terminology.

Agent. An agent is an independent software component that follows a goal, which can be achieved alone or in combination with other agents. Agents communicate via messages and can act independently, or reactively. They can change their behavior as needed, i.e. they are adaptive. Each agent has its own knowledge base in which a part of the information of the agent's environment and also its means of reacting is stored. The adaptation of the agent's behavior is achieved by modification of the knowledge base. The interaction with other agents constitutes the social behavior of the agents. All these features are the basis for intelligent behavior. Informally agents can be regarded as software robots.

Protocol. A protocol defines a certain behavior. In the agent-oriented view a protocol determines the communicational behavior of agents. Mulan protocols are - just like Mulan agents - Petri nets. An agent can use numerous protocols and it can instantiate any number of instances of protocols.

Net Component. A net component is a set of Petri net elements that belong together in a syntactical sense. In addition, it also has a visual meaning. Both the geometrical and the directional arrangements are defined. So in addition to the syntactical unity also a visual unity is achieved. This visual character makes it easy to identify the net component.

3   Net Components

This section introduces net components and their concept. In addition, an example implementation of net components for Mulan protocols is presented. Net components are meant to be combined with each other to form a Petri net.

3.1   Notions

A net component is a set of net elements that fulfills one basic task. The task should be so general that the net component can be applied to a broad variety of nets. Furthermore, the net component can provide additional help, such as a default inscription or comments. One of the used components contains a predefined but adjustable declaration node. In a formal way net components can be seen as transition-bordered subnets. This suits the notion of net components covering tasks.

Every net component has a unique geometrical form and orientation which results from the arrangement of the net elements. A unique form is intended so that each net component can easily be distinguished from the others and identified. The geometrical figure also holds the potential to provide a defined structure for the Petri net. The structuring of Mulan protocols is achieved by the unique form of the net components and the notion that Mulan protocols can be read from left to right.

In the default implementation a state is added at the outward connecting transition (Interface Place) for convenient net component connection. Only one arc has to be drawn to connect one net component to another. This is a simple and efficient method that also emphasizes the control flow. The connection of net components is provided by this place, which at all times should only contain anonymous tokens.

Direct data exchange between net components is not desired to guarantee an easy connecting interface. Instead data is handed to the data-containing places via *virtual places*[3]. By adding an appropriate *virtual place* to the net component, data can be transferred indirectly to the transition that uses a variable. In the usual case this is done by using a test arc. Data is handled and stored in a data block, which is located above the control flow part of the protocol. Annotations of the data-containing places should be adjusted to the appropriate name as well as the annotations of the corresponding virtual place.

3.2   The Mulan Protocol Net Components

We explain a selection of the net components so that their form and application will be clear. In this section the net components for messaging and for basic flow control are presented. There exist further net components which cover sequences, sub calls and manual synchronization.[4]

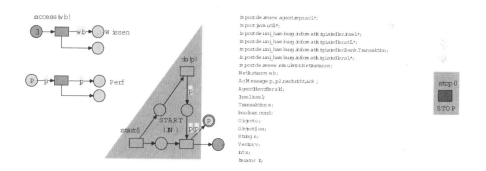

Fig. 4. Essential net components: *NC start* and *NC stop*.

Essential Net Components. Beginning (*NC start*) and Ending (*NC stop*) are needed in all protocols. There is exactly one start in every Mulan protocol, but

---

[3] *Virtual places* can be regarded as references to the original places. Another well known name for this is *fusion-place*. In Renew *virtual places* can be identified by their doubled outline.

[4] The full set of net components can be found in [3].

there may be more than one stop. The protocol is started when the transition
:start() is fired and stopped when one transition with the inscription *:stop()*
is fired. In addition the *NC start* also provides the declaration of the imports
and all variables which are used by the net components and the access to the
knowledge base (*:access(wb)*). The *NC start* always receives a message so this
functionality is also provided together with a data block, which holds the received
message. The message (performative *"p"*) is received by the transition *:in(p)* and
is handed to the data block of the net component by a virtual place (*"P"*). The
message is finally held in the place *Perf* and information from the message can
be extracted at the preceding transition and stored in additional places. The four
transitions *:start()*, *:stop()*, *:access(wb)* and *:in(p)* are the uplinks of synchronous
channels. Interfaces of the net components - i.e. the elements that can connect
to other net components - are marked with *">"*.

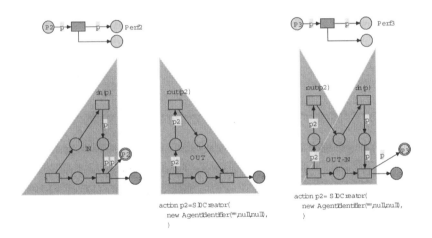

Fig. 5. The net components for message transport: *NC in, NC out, NC out-in.*

Messaging Net Components. These are the net components that provide
the means of communication. The *NC in* receives a message in the same manner
as the *NC start* (described in the preceding section). The message is handed to
the data block of the net component. Additional data containing places can be
added to the data block as desired. These places can contain elements that were
extracted from the messages, for example the name of the sender or the type
of the performative. The *NC out* provides the outgoing message task. The *NC
out-in* is a shorter implementation for the combination of both *NC out* and *NC
in* which provides a send request and wait-for-answer situation. It does not add
functionality but shortens the protocol significantly.

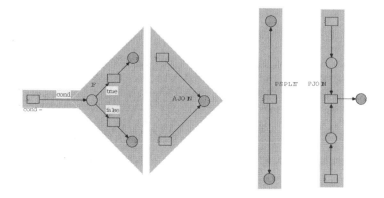

Fig. 6. Conditional and concurrent processing: *NC cond, NC ajoin, NC psplit* and *NC pjoin.*

Control Flow Net Components: Alternatives, Concurrency. The conditional can be used to add an alternative to the protocol. It provides an exclusive or (XOR) situation. To resolve the conflict the boolean variable *cond* should be adjusted as desired. The *NC psplit* (parallel split) and the *NC pjoin* are provided to enable a concurrent processing within a protocol. Note that the forms of these differ significantly from *NC cond* and *NC ajoin* to have a clear separation.

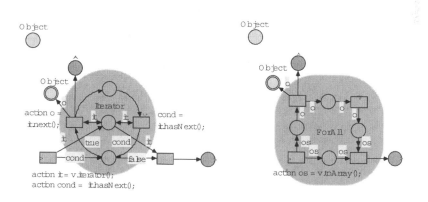

Fig. 7. Loops. *NC iterator and NC forall.*

Loops. These are the equivalent to the basic loops. The *NC iterator* provides a loop through all elements of a set described by the *Java Iterator*. It processes the core of the loop in a sequential order. The *NC forall* uses flexible arcs to provide a concurrent processing of all elements of an array. Flexible arcs allow to move multiple tokens with one single arc (see [21] and [13]). The number of tokens moved by the flexible arc may vary, thus its name. In Renew the flexible arcs are indicated by two arrowheads. A flexible arc puts all elements of an array into the output place and it removes all elements of a pre-known array from the input place. The cores of the loops are marked with ∧ (beginning) ∨ (ending).

Petri nets can be drawn with Renew in a fast and comfortable way. To be able to use net components in a similar way it is desirable to have a seamless integration of net components in Renew. This is provided by a simple palette which is the usual container for the buttons of all drawing tools for net elements.

3.3 Realization

Renew supports a highly sophisticated plug-in architecture.[5] It is appropriate to extend Renew with a plug-in, so that the usual functionality is still completely available. Once the palette is loaded into the system the net components are always available for drawing until the palette is unloaded again. Figure 8 shows the graphical user interface with the extension palette loaded.

Fig. 8. The graphical user interface of Renew with the net component extension.

All net components are realized as Renew drawings, so they can easily be adjusted to the need of the programmer by editing within Renew. The net component drawings are held in a repository, thus a general set of net components can be shared by a group of programmers. Nevertheless users can also copy and modify the repository to adjust the net components to their needs or build new net components with Renew. It is also possible to use multiple palettes of different repositories.

Net components are added to the drawing in the same way as the usual net elements. The only difference is that after the new net component is drawn all elements of it are selected automatically. This provides the possibility to adjust the position of the net component in relation to the rest of the drawing.

---

[5] Ongoing work of Jörn Schumacher.

4   Application of Net Components

This section deals with the modeling and the implementation of Mulan protocols. We show the advantages of structured net component-based development of Petri nets for Mulan protocols. Furthermore we investigate how to achieve a suitable structure. First we show how modeling agent interaction can be done with AUML interaction protocol diagrams. Then we present an example to show how to derive code from the agent interaction protocol diagram. By joining net components together we build the Mulan protocols for the agents. At last we want show that net component-based Mulan protocols reveal their structure due to the geometrical forms of the net components.

4.1   Modeling Agent Interactions

Modeling agent interaction can be done by using several means. The FIPA defines the AUML interaction protocol diagrams for modeling interactions between agents. These diagrams are an extension of the Unified Modeling Language (UML) sequence diagrams (see [1]) but they are more powerful in their expressiveness. They can fold several sequences into one diagram. Thus they can describe a set of scenarios.

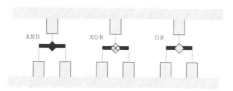

Fig. 9. New elements of interaction protocol diagrams: AND, XOR, OR

Some additional elements are added to the usual sequence diagram. Those additional elements provide alternative, concurrent and arbitrary splitting in a manner of the three gates AND, XOR and OR.

Figure 9 shows the new elements in a horizontal version which are applied to split the life line of an agent. In addition the FIPA also defines the vertical versions of those three elements to split the messages. Figure 10 shows an example protocol diagram for the contract net protocol as presented in [7]. It shows the other variant of the additional elements.

We use AUML interaction protocol diagrams to model the behavior of the Mulan agents. The models of agent interaction are then implemented as reference nets by using the net components.

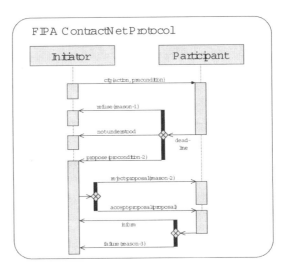

Fig. 10. The FIPA contract net protocol diagram.

4.2   Using and Applying Net Components

We restrict the interaction protocol diagrams to the usage of split figures for the life line. By doing this we can achieve a structure that can directly be transformed into a Petri net by using net components. Instead of using message split figures we favor the opposite, i.e. message join figures.

Especially for alternatively sent messages a message join figure can at times be of advantage, for example when the message is a reply. This is also intuitive since the receiver of a reply expects only one answer.

The development of a Mulan protocol using net components illustrates the procedure. For the reasons of clarity we present an agent version of the *producer-consumer* example which is adjusted to be compliant with the FIPA request protocol. Figure 11 shows the FIPA request protocol (see [7]) and an interaction protocol diagram of the FIPA-request compliant *producer-consumer* example. This version of the diagram only uses the split of the life line and the join of the message arc in addition to the usual UML diagram elements.

To demonstrate the process of transforming the diagram into a net we would like to show the development in detail in three steps. First, we divide the protocol diagram into the two parts which belong to each agent and rotate the resulting two diagrams by 90° so that they can be read from left to right.

As the second step we add the geometrical forms of the net component to the appropriate parts of the diagrams. This shows that the net components can be used for the implementation and it also shows the overall structure of the Petri nets. Figure 13 shows the same diagram augmented with the geometrical symbols of the net components.

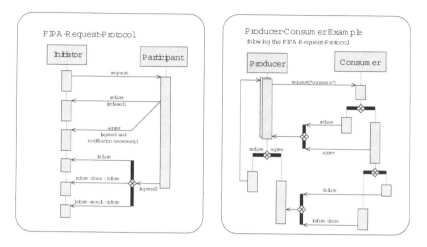

Fig. 11. Interaction protocol diagram of the FIPA Request and a FIPA Request-compliant *producer-consumer* example.

In the third and last step we use the net component extension of Renew to draw the Mulan protocols with the net components analogous to the symbols in Figure 13. Still some work has to be done regarding the actions, the adjustments of messages and other inscriptions. Both figures 14 and 15 show resulting Petri nets of the *produce* and the *consume* protocol. Note that the nets are not shown here to be read as Petri nets, although they are fully operational and can be executed in Renew and Mulan without any changes. Instead they are presented to get an impression of the structured layout, the application of the net components and the analogies to the structure of the interaction protocol diagrams.

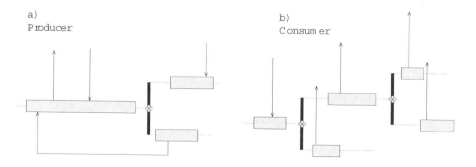

Fig. 12. Two parts of the example diagram; one for each agent.

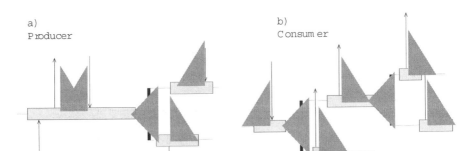

Fig. 13. Both parts augmented by the symbols for the net components.

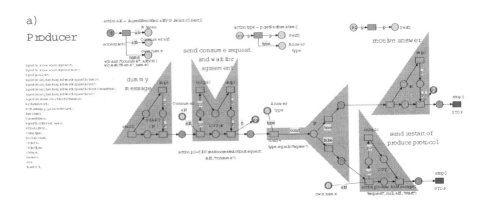

Fig. 14. The resulting *produce* Mulan protocol.

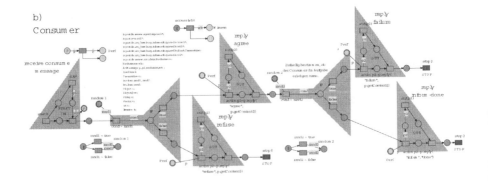

Fig. 15. The resulting *consume* Mulan protocol.

4.3   Mulan Protocols Structured by Net Components

The visual aspects of net components play a crucial role in recognition and thus in readability. Since net components have a fixed geometrical structure they can always be identified without reading any details of the net elements. The geometrical form of the net itself becomes readable to the programmer without using any modeling abstraction. Nevertheless a net constructed with net components can be transformed directly into an interaction protocol diagram. We claim that a well-structured Mulan protocol that uses net components exclusively is readable without reading any of the net elements. Figure 16 shows the registration of a player at the game control.[6] Again this net is presented here not to be read; instead the net components should be regarded. Once they are identified they can easily be mapped onto an interaction protocol diagram.

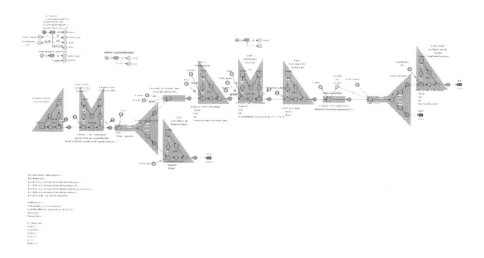

Fig. 16. Mulan protocol for the registration of a player.

Similar to the *producer-consumer* example in section 4.2 the registration protocol in figure 16 can be read like a sentence from left to right. Due to the usage of net components the basic tasks performed by this protocol and the structure of the interaction can be read without the need to interpret the details of the net itself. It can be seen that the game control is involved in several communicative acts and two decisions. The general structure of the interaction can thus be derived from the net itself. Only the participants of the conversation are not identified yet. For this we recommend to add a comment for each net component. Figure 17 shows the full conversation between the four agents as the final interaction protocol augmented with the symbols of the net components.

---

[6] For the multi-agent version of the "The Settlers of Catan" board game (see [2]).

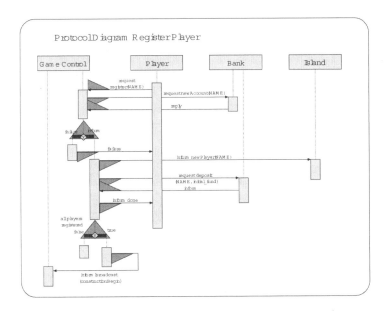

Fig. 17. Mulan conform-structured agent interaction protocol for the registration of a player.

5   Conclusion

The gap between traditional system modeling and the use of high-level Petri nets is relatively large. Both areas can contribute to the requirements of actual software development today. The latest developments in this area are AUML (see www.auml.org) where concepts from agent orientation are integrated mostly for a more compact representation. However, distributed systems require more elaborated features. These are covered by high-level Petri nets like reference nets (see [11]) in combination with an agent-oriented architecture (see [10]).

In this paper we were able to integrate all directions. For this we started with the introduction of our multi-agent system infrastructure. The advantage of using a Petri net-based system is its inherent concurrency. The implementation of a multi-agent system with Petri nets is only possible if it can be backed up with a powerful and efficient simulation tool. This was achieved by building the system on Renew.

We showed how modeling of agent interaction can be done by using FIPA interaction protocol diagrams. The notion of net components and a corresponding set for Mulan agent interaction were presented. Together with the seamless integration of the net components into Renew we now have a simple but powerful tool to support net component-based Mulan protocol implementation. AUML provides methods for modeling interactions which can be used to facilitate and structure the agent interaction. The structure of the protocols can be unified and clarified through the usage of net components. Unification is especially desired when implementing in project groups. Through the net components a common language and style can be achieved. Refactoring of protocols is facilitated because the components are loosely interconnected. So remodeling of Mulan protocols is supported.

We want to point out that by using net components we achieve a structure for a Petri net layout that improves the readability significantly. Furthermore the overall structure of the model is retained through the analogous construction of interaction protocol diagrams and Mulan protocols. Developing and debugging time for Mulan protocols can be reduced significantly and reuse of code is facilitated. Although the structuring of the Mulan protocols is not achieved automatically, the results can be seen in analogy to the structuring of program code by using indentation, syntax highlighting or conventions like capitalizing.

Net components were presented here as a part of Mulan protocols. Nevertheless, it is possible to apply the same principles to other domains. As an example we would like to mention the obvious solution for a Petri net-based workflow engine (see [17]). It is possible to realize workflow patterns for this domain using the Renew extension presented in section 3.3.

References

1. Grady Booch, James Rumbaugh, and Ivar Jacobson. *The Unified Modeling Language User Guide*. Addison-Wesley, Reading, Massachusetts, 1996.
2. Tobias Bosch, Oliver Gries, Heiko Kausch, Maxim Klenski, Kolja Lehmann, Michael Morales, Valentin Seegert, and Anatolij Vilner. *Agentenorientierte Implementierung des Spiels "Die Siedler von Catan"*. Internal report, University of Hamburg, Department of Computer Science, 2002.
3. Lawrence Cabac. *Entwicklung von geometrisch unterscheidbaren Komponenten zur Vereinheitlichung von Mulan-Protokollen*. Studienarbeit, University of Hamburg, Department of Computer Science, 2002.
4. O.-J. Dahl, E. W. Dijkstra, and C. A. R. Hoare. *Structured Programming*. Acad. Press, London, 7th edition, 1975.
5. Michael Duvigneau, Daniel Moldt, and Heiko Rölke. Concurrent architecture for a multi-agent platform. In *Proceedings of the 2002 Workshop on Agent-Oriented Software Engineering (AOSE'02)*. Springer Lecture Notes, 2002.
6. Foundation for Intelligent Physical Agents. http://www.fipa.org.
7. FIPA. FIPA Interaction Protocol Library Specification, August 2001. http://www.fipa.org/specs/fipa00025/XC00025E.pdf.
8. K. Jensen and G. Rozenberg, editors. *High-level Petri Nets – Theory and Application*. Springer-Verlag, Berlin Heidelberg, 1991.
9. Kurt Jensen. *Coloured Petri Nets*, volume 1. Springer-Verlag, Berlin, 2nd edition, 1996.
10. Michael Köhler, Daniel Moldt, and Heiko Rölke. Modeling the behaviour of Petri net agents. In *Proceedings of the 22nd Conference on Application and Theory of Petri Nets*, pages 224–241, 2001.
11. Olaf Kummer. *Referenznetze*. PhD thesis, University of Hamburg, Department of Computer Science, Logos-Verlag, Berlin, 2002. R35896-7.

12. Olaf Kummer, Frank Wienberg, and Michael Duvigneau. Renew – The Reference Net Workshop. In *Tool Demonstrations – 22nd International Conference on Application and Theory of Petri Nets*, 2001. See also http://www.renew.de.
13. Olaf Kummer, Frank Wienberg, and Michael Duvigneau. Renew – user guide. Dokumentation, University of Hamburg, Department of Computer Science, 2001. http://www.renew.de.
14. LabVIEW. Labview home, 2002. http://www.labview.com.
15. David McIntyre. Comp.lang.visual – Frequently Asked Questions List, 1998. ftp://rtfm.mit.edu/pub/usenet/comp.lang.visual/comp.lang.visual_Frequently-Asked_Questions_(FAQ).
16. Daniel Moldt. *Höhere Petrinetze als Grundlage für Systemspezifikationen*. PhD thesis, University of Hamburg, Department of Computer Science, August 1996.
17. Daniel Moldt and Heiko Rölke. Pattern based workflow design using reference nets. In W.M.P. van der Aalst, A.H.M. ter Hofstede, and M. Weske, editors, *International Conference on Business Process Management*, 2003.
18. Horst Oberquelle. *Sprachkonzepte für benutzergerechte Systeme*. Springer-Verlag, Berlin, 1987.
19. Kirsten Nygaard Ole-Johan Dahl. SIMULA: An ALGOL-based Simulation Language. Communication of the ACM, September 1966.
20. Inc. Pictorius. The home of visual object-oriented development environments., 2002. http://www.pictorius.com/home.html.
21. Wolfgang Reisig. *Elements of Distributed Algorithms: Modeling and Analysis with Petri Nets*. Springer-Verlag New York, October 1997.
22. W.M.P. van der Aalst, A.H.M. ter Hofstede, B. Kiepuszewski, and A.P. Barros. Workflow Patterns, 2000. http://tmitwww.tm.tue.nl/research/patterns/wfs-pat-2000.pdf.

Modelling Mobility and Mobile Agents Using Nets within Nets

Michael Köhler, Daniel Moldt, and Heiko Rölke

University of Hamburg, Computer Science Department
Vogt-Kölln-Str. 30, D-22527 Hamburg
{koehler,moldt,roelke}@informatik.uni-hamburg.de

Abstract. Mobility creates a new challenge for dynamic systems in all phases of the life cycle, like modelling, execution, and verification. In this work we apply the paradigm of "nets within nets" to this area since it is well suited to express the dynamics of open, mobile systems. The advantages of Petri nets – intuitive graphical representation and formal semantics – are retained and supplemented with a uniform way to model mobility and mobile (agent) systems.

First the modelling of mobility is introduced in general, the results are carried forward to model mobility in the area of agent systems. The usefulness of the approach is shown in a second step by modelling a small case study, the implementation of a household robot system.

Keywords: agent, high-level Petri nets, mobile agent system, mobility, Mulan, nets within nets, Renew

1 Introduction

The general context of this paper is mobility in open multi agent systems. The main question is how to model mobility in an elegant and intuitive manner without losing formal accuracy. The modelling language should feature a graphical representation to be used in a software engineering process. The modelling paradigm should be capable of expressing the different kinds of agent mobility. The models should build upon a formalism that has a formal semantics to support verification and execution. The (direct) execution of the models prohibits errors of manual translation from models e.g. to program code. Executable models support the validation process.

Here, we present a proposal, how the paradigm of "nets within nets" can be used to describe mobility. The paper consists of two parts: First it is shown how "mobility" in general can be expressed (environment, moved entity, different types of movement). In the second part these results are used to support the engineering of mobile agent systems.

Structure of the paper. In section 2 general statements on mobility are given: minimal preconditions for a formalism to express mobility are suggested as well as our needs for an intuitive modelling. We present a short introduction to the

W.M.P. van der Aalst and E. Best (Eds.): ICATPN 2003, LNCS 2679, pp. 121–139, 2003.

paradigm of nets within nets and show how this paradigm can be used to model different kinds of mobility. In section 3 we give a description of how these mobility models may be used in the specific context of open multi agent systems. Section 4 presents a small case study of a mobile agent system using the modelling proposals. A system overview is built upon the modelling proposals. The work presented in this paper is close-by to several research fields like mobility calculi, (graphical) modelling of agents and agent systems and other Petri net approaches. Section 5 gives some examples. The paper closes with a conclusion and an outlook to further work.

2   Nets within Nets and Mobility

Figure 1 shows the mandatory elements of a system to become a *mobile* system. The overall system is divided into separate locations. At least two different locations are necessary. Locations host entities (some of) which are able (under certain circumstances) to undertake a movement from one location to another, which means that the environment of the entity changes.[1]

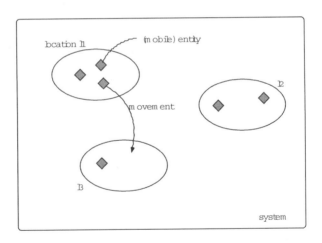

Fig. 1. Elements of mobile systems

An important point of mobility is the embeddedness of the mobile entity: each entity is embedded in a local environment (location) that assists the entity by offering some services and restricts it by declining others (or not having the potential to offer them). If all locations look the same to the mobile entity and no difference is made in local versus remote communication, movements are

---

[1] This changing may either be logical or physical. The modelling of real-world scenarios makes it necessary to cope with the additional problems of physical movements – which may be mapped to a logical changing of the environment.

transparent to the mobile entity. This is an important feature for example in load balancing systems. The systems we are considering usually show several differences between the locations and hence are more interesting to model, since these differences cause the complexity of the systems.

2.1 Nets within Nets

The paradigm of "nets within nets" due to Valk [36,37] is based on the former work on task-flow nets [35]. The paradigm formalises the aspect that tokens of a Petri net can also be nets. Taking this as a view point it is possible to model hierarchical structures in an elegant way.

We will now give a short introduction to the paradigm of Reference nets [24,25], an implementation of certain aspects of nets within nets. It is assumed throughout this text that the reader is familiar with Petri nets in general as well as coloured Petri nets. Reisig [31] gives a general introduction, Jensen [20] describes coloured Petri nets.

A net is assembled from places and transitions. Places represent resources that can be available or not, or conditions that may be fulfilled. Places are depicted in diagrams as circles or ellipses. Transitions are the active part of a net. Transitions are denoted as rectangles or squares. A transition that fires (or occurs) removes resources or conditions (for short: tokens) from places and inserts them into other places. This is determined by arcs that are directed from places to transitions and from transitions to places.

Reference nets [23] are so-called high-level Petri nets, a graphical notation that is especially well suited for the description and execution of complex, concurrent processes. As for other net formalisms there exist tools for the simulation of reference nets [26]. Reference nets offer some extensions related to "ordinary" coloured Petri nets: net instances, nets as token objects, communication via synchronous channels, and different arc types. Beside this they are quite similar to coloured Petri nets as defined by Jensen. The differences are now shortly introduced.

Net instances. Net instances are similar to objects of an object oriented programming language. They are instantiated copies of a template net like objects are instances of a class. Different instances of the same net can take different states at the same time and are independent from each other in all respects.

Nets as tokens. Reference nets implement the "nets within nets" paradigm of Valk [36]. This paper follows his nomenclature and denominates the surrounding net *system net* and the token net *object net*. Certainly hierarchies of net within net relationships are permitted, so the denominators depend on the beholders viewpoint: a (system) net containing object net tokens may itself be an object net token of another net.

Synchronous channels. A synchronous channel [6] permits a fusion of transitions (two at a time) for the duration of one occurrence. In reference nets (see [23]) a channel is identified by its name and its arguments. Channels are directed, i.e. exactly one of the two fused transitions indicates the net instance in which the counterpart of the channel is located. The other transition can correspondingly be addressed from any net instance. The flow of information via a synchronous channel can take place bi-directional and is also possible within one net instance. It is possible to synchronise more than two transitions at a time by inscribing one transition with several synchronous channels.

Arc types. In addition to the usual arc types reference nets offer *reservation arcs* and *test arcs*. Reservation arcs carry an arrow tip at both endings and reserve a token solely for one occurrence of a transition. They are a short hand notation for two opposite arcs with the same inscription connecting a place and a transition. Test arcs do not draw-off a token from a place allowing a token to be tested multiple times simultaneously, even by more than one transition (test on existence).

2.2   Modelling Mobility

The intuition of nets within nets is, that the token nets are "lying" as tokens in places just as ordinary (black or coloured) tokens. This is illustrated in the figures 2 and 3. When modelling more widespread nets a displaying as in the mentioned figures is not practical. Therefore the modelling tool Renew implements a kind of pointer concept: net tokens are references (hence the name) to nets each displayed in a window of their own.[2]

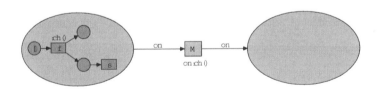

Fig. 2. Object net embedded in system net

To give an example we consider the situation where we have a two-level hierarchy. The net token is then called "object net", the surrounding net is called "system net". Figure 2 illustrates this situation: The object net in the left place of the system net can be bound to the arc inscription on. Doing so, transition M is activated with this possible binding. In addition M is inscribed

---

[2] There is another difference between the Reference nets of Renew and the intuitive modelling, namely the use of reference versus value semantics. This topic is covered by Valk [37].

with a synchronous channel (on:ch()). This inscription means that for the object net on to become an actual binding of transition M an adequate counterpart has to be found within on – an enabled transition inscribed with the channel :ch(). This precondition is fulfilled by transition f of the object net. So the synchronous firing of object and system net can take place and leads to the situation in Figure 3.

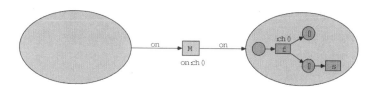

Fig. 3. Object and system net after firing

The object net is moved to the place on the right of the system net. Synchronously the marking of the object net changed so that another firing of transition f is not possible.

The example gives an idea how the interplay between object net and system net can be used to model mobile entities manoeuvring through a system net, where the system net offers or denies possibilities to move around while the mobile object net moves at the right time by activating a respective transition that is inscribed with the counterpart of the channel of the move transition of the system net.

Without the viewpoint of nets as tokens, the modeller would have to encode the agent somehow, for example as a data structure. The disadvantage of such an approach is that the inner actions of the mobile entity cannot be modelled directly, so, they have to be lifted up to the system net, which seems quite unnatural. By using nets within nets we can investigate the concurrency of the system and the agent in one model without losing the needed abstraction.

2.3   Types of Mobility

The interplay between object net and system net induces four possibilities for an object net to move or to be moved, respectively.

1. The object net is moved inside the system net, neither object nor system net controls the move. (Spontaneous Move)
2. The object net triggers the movement, the system net has no influence. (Subjective Move)
3. The system net forces the object net to move. (Transportation, Objective Move)
4. Both nets come to an agreement on the movement. (Consensual Move)

In contrast to publications that take a practitioner's view on (agent) mobility, this enumeration is complete. For example the FIPA standardisation specification [15] only distinguishes between *simple* and *full* mobility, which are variants of the subjective and the consensual move.

Other publications consider only consensual movements and discriminate between *weak* and *strong* mobility [8,3]. This discrimination is based on details of the migration of an agent:

1. Weak mobility: Only data and state of an agent is transmitted, a new agent is invoked at the target platform.
2. Strong mobility: The agent is transmitted together with its runtime environment (e.g. data, state, stack trace, and so on).[3]

Strong mobility is difficult to implement using a programming language like Java [3]. Inside a Petri net simulator it is the most natural form of migration. Enabling strong mobility between different simulators is a little bit more difficult, but no real problem – at least this holds for the Petri net simulator Renew. Weak mobility as a special kind of strong mobility is possible anyway.

Therefore we can abstract from the implementation details of the migrations and focus on the higher-level interactions between mobile entity and environment.

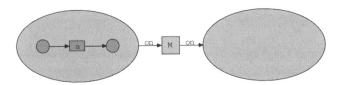

Fig. 4. Spontaneous Move

Spontaneous Move. If neither object net nor system net influence the movement it may happen *spontaneously*. This is the situation in Figure 4. There is no pre- or side condition for the movement.

Subjective Move. One may argue that the second possibility does not exist, because the system net offers the ways for object nets to move from one location to another. So the system net "decides" which movements can be carried out and which can not. But in a *given* system net it is possible for an object net to control the movement as shown in Figure 5.

In the figure the only condition for the movement to be carried out is the synchronous channel (see transition M). The synchronous channel is activated if

---

[3] Strong mobility can be extended to *transparent* mobility: The agent does not notice the migration.

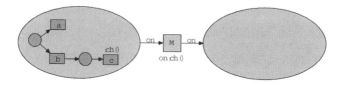

Fig. 5. Subjective Move: Object net triggers movement

its counterpart inside the object net is also activated (that means, it is inscribed to an enabled transition). So the movement depends on the (firings of the) object net only. This movement is called *subjective* because the mobile entity itself is subject of the execution.

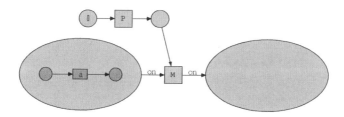

Fig. 6. Transportation: System net triggers movement

Objective Move. If Figure 4 is extended with some kind of condition for transition M, the system net may control the movement, the object net is *transported* from one location to another. Figure 6 shows only one possibility of how the system net may control the transition M, another one is for example a guard (see Figure 7).

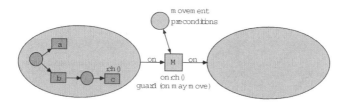

Fig. 7. Consensual Move

Consensual Move. By combining Figure 5 and Figure 6 both object and system net influence transition M. For the transition to be enabled, they have to agree upon the movement. For this reason this kind of movement is called *consensual*. An example for such a move is shown in Figure 7.

The figure shows another way of modelling a (side) condition for transition M: a combination of a place holding movement conditions and an appropriate guard. The guard monitors the movement conditions, transition M is only enabled if the object net is allowed to move.

3   Agent Systems

In the following section we lift the general insights of how mobility can be modelled to a special form of multi agent systems. The modelling of a mobile agent (object net) moving through a "world" of several locations (system net) allows for an intuitive reproduction of real-world scenarios.

First the multi agent architecture MULAN is introduced, that also profits from the use of nets within nets.

3.1   Multi Agent Architecture MULAN

The multi agent system architecture MULAN [21] is based on the "nets within nets" paradigm, which is used to describe the natural hierarchies in an agent system. MULAN is implemented in RENEW [25], the IDE (Integrated Development Environment) and simulator for reference nets. MULAN has the general structure as depicted in Figure 8: Each box describes one level of abstraction in terms of a system net. Each system net contains object nets, which structure is made visible by the ZOOM lines.[4] The figure shows a simplified version of MULAN, since for example several inscriptions and all synchronous channels are omitted. Nevertheless this is an executable model.

The net in the upper left of Figure 8 describes an agent system, which places contain agent platforms as tokens. The transitions describe communication or mobility channels, which build up the infrastructure. The multi agent system net shown in the figure is just an illustrating example, the number of places and transitions or the interconnections have no further meaning.

By zooming into the platform token on place p1, the structure of a platform becomes visible, shown in the upper right box. The central place agents hosts all agents, which are currently on this platform. Each platform offers services to the agents, some of which are indicated in the figure.[5] Agents can be created (transition new) or destroyed (transition destroy). Agents can communicate by message exchange. Two agents of the same platform can communicate by the transition internal communication, which binds two agents, the sender and the receiver, to pass one message over a synchronous channel.[6] External communication (external communication) only binds one agent, since the other agent is

---

[4] This zooming into net tokens should not to be confused with place refining.

[5] Note that only mandatory services are mentioned here. A typical platform will offer more and specialised services, for example implemented by special service agents.

[6] This is just a technical point, since via synchronous channels provided by RENEW asynchronous message exchange is implemented.

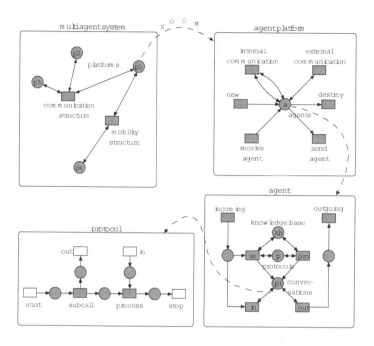

Fig. 8. Agent systems as nets within nets

bound on a second platform somewhere else in the agent system. Also mobility facilities are provided on a platform: agents can leave the platform via the transition **send agent** or enter the platform via the transition **receive agent**.

A platform is therefore quite similar to a location in the general mobility scenario as it was introduced in the beginning of section 2.

Agents are also modelled in terms of nets. They are encapsulated, since the only way of interaction is by message passing. Agents can be intelligent, since they have access to a **knowledge base**. The behaviour of the agent is described in terms of protocols, which are again nets. Protocols are located as templates on the place **protocols**. Protocol templates can be instantiated, which happens for example if a message arrives. An instantiated protocol is part of a conversation and lies in the place **conversations**.

The detailed structure of protocols and their interaction have been addressed before in [21], so we skip the details here and proceed with the modelling of agent migration.

3.2   Mobility as a System View

When modelling a complex system it is often undesirable to see the overall complexity at every stage of modelling and execution (simulation in our case). Therefore the notion of a system view is introduced. Several views on an agent system are possible, for example the history of message transfer (message process), the ongoing conversations and/or active protocols (actual dynamic state) or the distribution of agents among a system of several locations (platforms).

Using nets within nets as a modelling paradigm allows for the direct use of system models at execution time. This can be exploited as follows: The overall system is designed as a system net with places defining locations and transitions representing possible moves from one location to another. This is a direct transformation of the general mobility modelling ideas of section 2. The adaption to host platforms (that encapsulate the mobile agents) instead of agents directly is straightforward (adding one level of indirection) and can be omitted here.

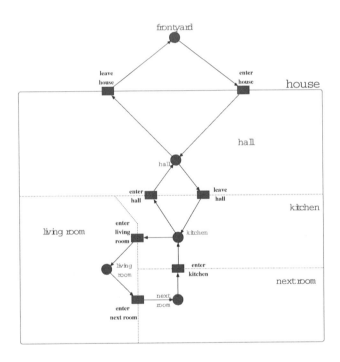

Fig. 9. Example system net

Figure 9 shows an example of such a system net. Places are locations (rooms) in or in front of a house, transitions model possible movements between the rooms. The walls of the house are drawn in for illustrationary purposes only. This example is carried forward in the next section (4).

The interesting point is that it is possible to "decorate" the places in the system net with additional interesting features, for example a characteristic subset of the services of the hosted platform together with some of the pre- and postconditions of these services. The beholder of such an enhanced system net is provided with a complete view of important system activities without having to deal with the underlying complexity.[7]

The difference between this proposal and a visualisation tool that shows some activities of e.g. a program running in background is twofold: First, using our proposal, the modelling process concludes with a running system model. A normal modelling process requires at least three stages to gain a comparable result: (a) model the system, (b) implement the model, and (c) write a visualisation for the program. Second, the visualisation of a system model at execution time is indeed the implementation of the system. This eliminates several potential sources of errors shifting from model to implementation to visualisation in an ordinary software design process.

4   Mobile Robot Case Study

To illustrate the modelling method introduced before, we introduce a small case study, a mobile household robot, unfortunately just a *software* robot.

4.1   Specification

The robot is internally implemented according to the MULAN architecture introduced in subsection 3.1. But that is not what the supervisor of the robot wants to know and see. What he requires is a simple view on his household, the robots location and state, and maybe some supplementary information. That is just what we provide with the extended system net view on the agent system.

The household is represented by the system net in Figure 10.[8] The household consists of several rooms (locations): hall, living room, kitchen, next room, and the front yard (dark places). Each room offers special services to the robot: it can fetch coffee in the kitchen, serve it in the living room, fetch mail in the front yard, open and close the door in the hall, and so on (light transitions). The possible movements from one location to another are displayed as dark transitions. Note that moving from room to room is not symmetric in this scenario. For example it is not possible to move directly from the kitchen to the next room. Service transitions are supplemented with additional information (service state/buffer, light places) showing for instance if new mail has arrived, coffee is available and so on. Extraneous actions not accessible for the robot are displayed as thin-lined transitions: arrival of new mail, new assignments for the robot etc.

The door of the house is used to show another possibility of viewing special parts of the system: the state of the door (open/closed) is directly modelled.

---

[7] This is illustrated in the next section: The mobility system net is just a small part of the overall system consisting of several agents, platforms, technical substructure e.g. for the remote communication and so on. This is hidden from the user as long as the user does not wish to see the implementation details.

[8] The use of colour greatly supports the differentiation of different types of places, transitions, or arcs. Unfortunately this is – even in the adapted form as in the figures – not so obvious in a black and white representation.

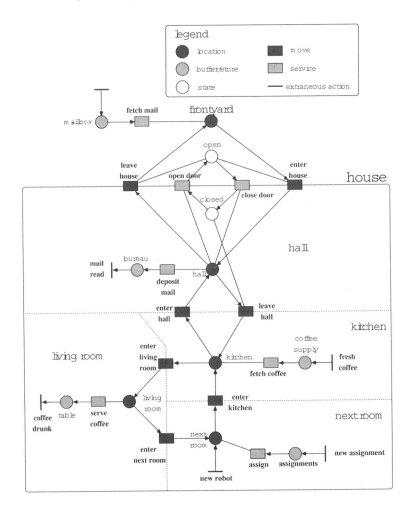

Fig. 10. Household System Net

This system state is not belonging to a single service (as for example the state of the mailbox), but is queried by a couple of service transitions including the movements into and out of the house.

This model of the household is filled with life by implementing an appropriate robot agent and defining the desired services for the platforms. The behaviour modelling for this kind of agents has been introduced elsewhere [21].

Having defined the functionality of platforms and agents the parts are inserted in the household system net, which is a straightforward operation that claims for automation (tool support). After that the system is ready for execution.

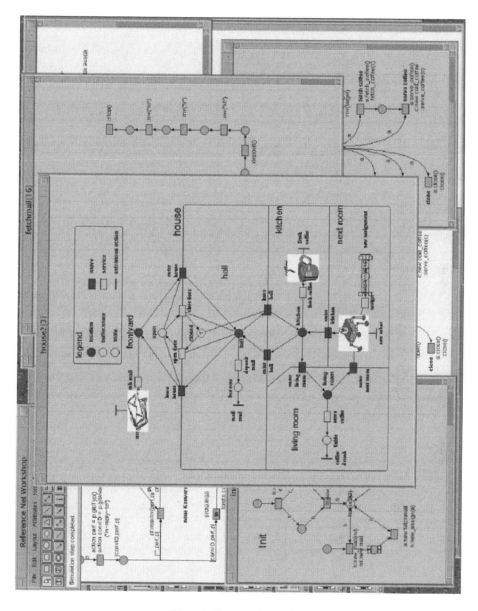

Fig. 11. System in Action

4.2   Execution

Figure 11 shows an early state of a sample simulation of the agent system. For illustration purposes some nets that normally work "behind the scenes" are made visible in this screen shot. An ordinary simulation would just show the net house2 in the middle of the figure. It is outside the scope of this paper to explain the full functionality of the MULAN agent framework, so just the mobility overview net is regarded.[9]

Taking a closer view on the central household system net an additional feature becomes visible: Tokens can be visualised by an intuitive image. In the figure the incoming mail is visible in the upper left, and a fresh coffee on the lower right side. This is a one-to-one matching between token and image – whenever such a token is produced, moved, or deleted, the image follows. The case of the robot in the next room (lower middle) is a little bit more tricky (to implement), because what is really lying on the location place is a platform hosting the robot agent, and not the robot itself. What can be seen in the system net is actually a dynamic token figure of the platforms showing representatives of the agents hosted by the respective platform (empty platform means no image).

It is now easy to see the state of the overall system by looking at the system net. All the simplifications in presentation do not mean a loss of generality in the system's implementation. It is possible to take control of the whole agent system implementing the robot scenario by just double clicking on one of the tokens. The tool RENEW will then display the respective net (similar to those introduced in subsection 3.1), allowing for a complete inspection of the running system without having to interrupt.

In further execution steps the robot will receive one or more assignments, for example to serve fresh coffee in the living room. Its internal representation says that coffee is only available in the kitchen, so it moves there to look for the coffee service of the kitchen platform. In the scenario of Figure 11 this service is available, so the robot fetches the coffee and moves to the living room, where it serves the coffee. The robot ends the assignment by moving back to the next room and waiting for orders.

The behaviour modelling and planning the robot carries out to perform its tasks is outside the scope of this paper. The mobility system net greatly helps in validating a certain robot planning algorithm, for instance by unmasking bad habits of the robot like fetching the coffee, leaving the house for mail, returning to the living room later and serving cold coffee. It is mainly used for debugging and presentation of the system, but it also helps at design time because the developer is forced to think about the structure of the system.

5   Related Work

The work presented in this paper is related to different fields of research from (mobility) calculi to agent implementation frameworks. We will briefly address similarities and differences.

---

[9] Details of the MULAN framework are topic of an ongoing diploma thesis [11] and a technical report [32] as well as other papers, e.g. [10]

5.1   Mobility Calculi

During the last years a variety of process calculi have been introduced or adopted to cover mobility: π-calculus [28], Ambient-calculus [5] and Seal calculus [38] are just three of the more popular ones.

Mobility calculi like the Seal calculus offer about the same modelling power while being (partly) easier to analyse than our extended Petri net formalism. We favour nets within nets for the description of mobility, since Petri net models are much easier to read without losing the exactness. This is the main difference to models based upon calculi or logic, like HiMAT of Cremonini et al. [9]. Nets within nets can be used for modelling purposes in our MULAN architecture as well as for analysis purposes. Additionally, our formalisation allows for an intuitive description of location and concurrency in an integrated way, which – in our opinion – would not have been possible with simple Petri nets or mobility calculi alone.

To justify our argumentation (why we are not using a common mobility calculus), we take the Seal Calculus as an example. In the Seal Calculus a process term P with locality is denoted as $E[P]$. Mobility is described by sending seals over channels. The term $\bar{x}^*\{y\}.P$ describes a process, which can send a seal y over the channel $\bar{x}^*$ and then behaves like P. For receiving the term $x^*\{z\}.Q$ describes a process, which receives the seal z, substitutes this seal for every occurrence of z in Q and then behaves like this new Q.[10] To our opinion, the notation used in the Seal Calculus – which is quite similar e.g. to that of the π-calculus – is not well-suited to model mobile agent systems by "normal" software developers that are interested in a fast development of the software and want the systems properties to be detected automatically. It is not possible to "see" the overall structure of a system described in terms of a mobility calculus in the way that even non-computer scientists can "read" a Petri net.

5.2   Frameworks: UML and Extensions

The use of widespread (graphical) modelling formalisms like UML[30] is not possible, because all included types of diagrams (e.g. class diagrams, activity diagrams, state charts) are static and/or do not offer a notion of (dynamic) embeddedness. Additionally they lack a formal semantics. This is especially the case for the deployment diagram, that was originally used to model the *static* allocation of objects among different computers (clients and servers).

In the majority of cases mobile agents are directly implemented using a programming language like Java [33]. As an example we cite Uhrmacher et al. [34] (representative for other publications): The implementation is done using Java whilst the mobility aspect of the system is presented using simple graphics (without semantics). In contrast to this approach we try to unify illustration and implementation using the same (Petri net) model for both purposes.

Agent oriented extensions to UML like AUML [4] offer no solution to the problem of modelling mobility.

---

[10] The Seal Calculus thus describes a *spontaneous move*. See Figure 4 in subsection 2.3.

5.3   Other Petri Net Approaches

There are some approaches to model agents and/or multi agent systems using varieties of Petri nets: [7,27,29,12,17,13,40]. Narrowing this to mobile agent systems only a minority of proposals remain, some of them will be commented explicitly. One difference between our approach and others is – to our best knowledge – the use of nets-within-nets that allows for a dynamically changing hierarchy of agents and platforms and therefore of the structure of the system.

Some more subtle differences to selected approaches are:

Ferber and Gutknecht [14] do not model the overall system explicitly. They only give informational illustrations of its structure. The same difference holds for the approach of Briot et al. [2], that gives an interesting combination of Petri net models and an Agent platform implemented in Java.

Xu and Deng [39] use Predicate/Transition-nets to model agent systems consisting of components following the MASIF [18] proposals. They allow only two kinds of migrations: *autonomous* and *passive*. This corresponds to subjective moves and objective moves in our nomenclature. Because agents and the structure of the agent system are modelled as nets on the same level, only the names of the agents are moved through the agent system.[11] This indirectedness makes debugging and presentation of the system (a little bit) more difficult.

Buchs and Hulaas [19] use an extended variant of high-level algebraic Petri nets to model mobile (agent) systems. The main difference to our models lies in the type of mobility: Their system uses a π-calculus-like style, e.g. mobility via passing of names. In our opinion it is more intuitive to remove the mobile entity from one location and let it enter a new one. This was shown in subsection 2.2. Nevertheless, it is possible to simulate our approach with π-calculus like mobility and vice versa.

Mobile Petri nets [1] provide a description of mobility by embedding the π-calculus into the formalism of ordinary Petri nets. Since the formalism is far from the general intuition of a token game, it cannot be used here for the proposed goal of supporting the specification process of complex software systems.

6   Conclusion and Outlook

In this article we have presented the paradigm of "nets within nets" for the modelling of mobility. Nets within nets are attractive, since the concept allows for an intuitive representation of mobile entities as well as an operational semantics which is implemented in the RENEW-simulator.

We have shown how this approach can help in modelling and executing mobile agent systems and under-pinned our proposal with a small case study.

Besides the software-technical aspects, the exact semantics of nets within nets – in principal – allows for an analysis of mobile applications. This still needs more work, a starting point is [22]. Formal treatment of mobile (agent)

---

[11] This is the same kind of mobility as it is used for example in the π-calculus.

systems may prevent "classical" problems like deadlocks because of insufficient resources, but also helps in "modern" problems like security.

The use of an intuitive graphical formalism like Petri nets offers the room for a more widespread use as a modelling tool in agent oriented software engineering (AOSE) as opposed to text-based formalisms (mobile calculi), as the use of graphic modelling is an accepted technique in mainstream software engineering (see for example the Unified Modelling Language (UML) [16]).

Additional formal results will be directly integrated in the design of the Petri net based multi agent system architecture MULAN.

References

1. Andrea Asperti and Nadia Busi. Mobile Petri nets. Technical report, Department of computer science, University of Bologna, TR UBLCS-96-10, 1996.
2. Jean-Pierre Briot, Walter Merlat, and Min-Jung Yoo. Modelling and validation of mobile agents on the web. In *1998 International Conference on Web-based Modeling and Simulation*, pages 23–28, 1998.
3. Lorenzo Bettini and Rocco De Nicola. Translating strong mobility into weak mobility. In Gian Pietro Picco (Ed.), *Mobile Agents. 5th International Conference, MA 2001, Atlanta. Proceedings*, volume 2240 of Lecture Notes in Computer Science, page 182 pp., Springer Verlag, Berlin, 2001.
4. Bernhard Bauer, James Odell, and H. van Dyke Parunak. Extending UML for Agents. In *Proceeding of Agent-Oriented Information Systems Workshop*, pages 3–17, 2000.
5. Luca Cardelli, Andrew D. Gordon, and Giorgio Ghelli. Ambient groups and mobility types. Technical report, Microsoft Research and University of Pisa, 2000.
6. Søren Christensen and Niels Damgaard Hansen. Coloured Petri nets extended with channels for synchronous communication. In Robert Valette (Ed.), *Application and Theory of Petri Nets 1994, Proc. of 15th Intern. Conf. Zaragoza, Spain, June 1994*, LNCS, pages 159–178, Springer Verlag, Berlin, June 1994.
7. W. Chainbi, C. Hanachi, and C. Sibertin-Blanc. The multi-agent prey/predator problem: A Petri net solution. In Borne, P., Gentina, J.C., Craye, E., and El Khattabi, S. (Eds.), *Proceedings of the CESA'96 Conference, Computational Engineering in Systems Applications, Lille, France*, pages 291–299. IEEE Society Press, July 1996.
8. G. Cabri, L. Leonardi, and F. Zambonelli. Weak and strong mobility in mobile agent applications. In *Proceedings of the 2nd International Conference and Exhibition on The Practical Application of Java (PA JAVA 2000)*, Manchester (UK), April 2000.
9. Marco Cremonini, Andrea Omicini, and Franco Zambonelli. Modelling network topology and mobile agent interaction: An integrated framework. In *Proceedings of the 1999 ACM Symposium on Applied Computing (SAC'99)*, 1999.
10. Michael Duvigneau, Daniel Moldt, and Heiko Rölke. Concurrent architecture for a multi-agent platform. In *Proceedings of the 2002 Workshop on Agent Oriented Software Engineering (AOSE'02)*, volume 2585 of Lecture Notes in Computer Science, Springer Verlag, Berlin, 2003.
11. Michael Duvigneau. Bereitstellung einer Agentenplattform für petrinetzbasierte Agenten. Master's thesis, University of Hamburg, Computer Science Department, Germany, 2002.

12. Joao M. Fernandes and Orlando Belo. Modeling Multi-Agent Systems Activities Through Colored Petri Nets. In *16th IASTED International Conference on Applied Infomatics (AI'98)*, pages 17–20, Garmisch-Partenkirchen, Germany, Feb. 1998.

13. Jaques Ferber. *Multi-Agent Systems: An Introduction to Distributed Artificial Intelligence*. Addison Wesley Longman, Harlow, UK, 1999.

14. Jaques Ferber and Oliver Gutknecht. A meta-model for the analysis and design of organization in multi-agent systems. In *Proc. of ICMAS*, 1998.

15. Foundation for Intelligent Physical Agents. *FIPA Agent Management Support for Mobility Specification*, 30. June 2000. Available at http://www.fipa.org/specs/fipa00087/.

16. Martin Fowler. *Analysis patterns: reusable object models*. Addison-Wesley series in object-oriented software engineering. Addison-Wesley, 1997.

17. Gustavo M. Gois, Angelo Perkusich, Jorge C. A. de Figueiredo, and Evandro B. Costa. Towards a multi-agent interactive learning environment oriented to the Petri net domain. In *Proc. IEEE Int. Conf. on Systems, Man, and Cybernetics (SMC'98), 11–14 October 1998, San Diego, USA*, pages 250–255, October 1998.

18. OMG (Object Management Group). MASIF – Multi Agent System Interoperability Facility. Technical report, OMG, 1998.

19. Jarle Hulaas and Didier Buchs. An experiment with coordinated algebraic Petri nets as formalism for modeling mobile agents. In *Workshop on Modelling of Objects, Components, and Agents (MOCA'01) / Daniel Moldt (Ed.)*, pages 73–84. DAIMI PB-553, Aarhus University, August 2001.

20. Kurt Jensen. *Coloured Petri nets, Basic Methods, Analysis Methods and Practical Use*, volume 1 of *EATCS monographs on theoretical computer science*. Springer Verlag, Berlin, 1992.

21. Michael Köhler, Daniel Moldt, and Heiko Rölke. Modelling the structure and behaviour of Petri net agents. In J.M. Colom and M. Koutny (Eds.), *Proceedings of the 22nd Conference on Application and Theory of Petri Nets*, volume 2075 of *Lecture Notes in Computer Science*, pages 224–241. Springer Verlag, Berlin, 2001.

22. Michael Köhler and Heiko Rölke. Mobile object net systems: Concurrency and mobility. In H.-D. Burkhard, L. Czaja, G. Lindemann, A. Skowron, and P. Starke (Eds.), *Proceedings of the International Workshop on Concurrency, Specification, and Programming (CS&P 2002)*, Berlin, 2002.

23. Olaf Kummer. Simulating synchronous channels and net instances. In J. Desel, P. Kemper, E. Kindler, and A. Oberweis (Eds.), *Forschungsbericht Nr. 694: 5. Workshop Algorithmen und Werkzeuge für Petrinetze*, pages 73–78. University of Hamburg, Computer Science Department, 1998.

24. Olaf Kummer. *Referenznetze*. Dissertation, University of Hamburg, Computer Science Department, Vogt-Kölln Str. 30, 22527 Hamburg, Germany, 2002.

25. Olaf Kummer and Frank Wienberg. *Reference net workshop (Renew)*. University of Hamburg, http://www.renew.de, 1998.

26. Olaf Kummer, Frank Wienberg, and Michael Duvigneau. *Renew – User Guide*. University of Hamburg, Computer Science Department, Vogt-Kölln Str. 30, 22527 Hamburg, Deutschland, 1.6 edition, 2002.

27. Toshiyuki Miyamoto and Sadatoshi Kumagai. A Multi Agent Net Model of Autonomous Distributed Systems. In *Proceedings of CESA'96, Symposium on Discrete Events and Manufacturing Systems*, pages 619–623, 1996.

28. Robin Milner, Joachim Parrow, and David Walker. A calculus of mobile processes, parts 1–2. *Information and computation*, 100(1):1–77, 1992.

29. Daniel Moldt and Frank Wienberg. Multi-Agent-Systems based on Coloured Petri Nets. volume 1248 in Lecture Notes in Computer Science, pages 82–101, Springer Verlag, Berlin, 1997.
30. Rational Software Corporation. Rational UML Homepage. URL: http://www.rational.com/uml/, 2000.
31. Wolfgang Reisig. *Petri Nets: An Introduction.* Springer Verlag, Berlin, 1985.
32. Heiko Rölke. The Multi Agent Framework Mulan. Technical report, University of Hamburg, 2002.
33. Sunsoft. Java Online Reference Manual. http://www.javasoft.com.
34. Adelinde M. Uhrmacher, Petra Tyschler, and Dirk Tyschler. Modeling and simulation of mobile agents. *Elsevier, Artificial Intelligence,* 2000.
35. Rüdiger Valk. Modelling of task flow in systems of functional units. Technical Report FBI-HH-B-124/87, University of Hamburg, 1987.
36. Rüdiger Valk. Petri nets as token objects: An introduction to elementary object nets. In Jörg Desel and Manuel Silva (Eds.), *Application and Theory of Petri Nets,* volume 1420 of *LNCS,* pages 1–25, June 1998.
37. Rüdiger Valk. Concurrency in communicating object Petri nets. In G. Agha, F. De Cindio, and G. Rozenberg (Eds.), *Concurrent Object-Oriented Programming and Petri Nets,* volume 2001 of Lecture Notes in Computer Science, Springer Verlag, Berlin, 2001.
38. Jan Vitek and Giuseppe Castagna. Seal: A framework for secure mobile computations. In *ICCL Workshop: Internet Programming Languages,* pages 47–77, 1998.
39. Dianxiang Xu and Yi Deng. Modeling mobile agent systems with high level Petri nets. In *Proc. of IEEE International Conference on Systems, Man, and Cybernetics (SMC'00),* pages 3177–3182, Nashville, October 2000.
40. Haiping Xu and Sol M. Shatz. A Framework for Modeling Agent-Oriented Software. In *Proc. of the 21th International Conference on Distributed Computing Systems (ICDCS-21),* Phoenix, Arizona, April 2001.

Modular System Development with Pullbacks[*]

Marek A. Bednarczyk[1], Luca Bernardinello[2], Benoît Caillaud[3],
Wiesław Pawłowski[1], and Lucia Pomello[2]

[1] IPI PAN, Gdańsk, Poland
[2] DISCO, Università degli Studi di Milano-Bicocca, Milano, Italy
[3] IRISA, Rennes, France

Abstract. Two, seemingly different modular techniques for concurrent
system development are investigated from a categorical perspective. A
novel approach is presented in which they turn out to be merely special
instances of *pullback*, a general categorical limit construction. Interest-
ingly, the approach is based on truly concurrent semantics of systems.

1 Mathematical Preliminaries

A *transition system* is a tuple $\mathcal{S} = (S, \hat{s}, A, T)$ where S is a set of *states*, $\hat{s} \in S$ is the *initial state* of $\mathcal{S}$, A is a set of *actions*, and $T \subseteq S \times A \times S$ is a *transition relation*. The transition relation captures the idea of dynamic evolution of systems whereby the execution of an action results in a change of the current state.

In the sequel, various decorations of systems are inherited by their compo-
nents, e.g., the initial state of $\mathcal{S}_3$ is $\hat{s}_3$, etc., and the usual notational conventions apply. Thus, we write $p \xrightarrow{a} q$ whenever $\langle p, a, q \rangle \in T$, and call it an *atomic step* in $\mathcal{S}$. Similarly, $p \not\xrightarrow{a} q$ if $p \xrightarrow{a} q$ does not hold, and $p \xrightarrow{a}$ if $p \xrightarrow{a} q$ holds for some $q \in S$, respectively. The atomic step notation inductively extends to *paths*, i.e., finite sequences of subsequent steps. Thus, for the empty sequence $\varepsilon \in A^*$ we let $p \xrightarrow{\varepsilon} q$ iff $p = q$. For a nonempty sequence $a\varpi$ we let $p \xrightarrow{a\varpi} q$ iff there exists a state r such that $p \xrightarrow{a} r$ and $r \xrightarrow{\varpi} q$.

Morphisms of transition systems were invented to explain how the dynamic behaviour of one system is simulated within another system.

Definition 1. *A morphism $f : \mathcal{S}_1 \to \mathcal{S}_2$ of transition systems is a pair $f = \langle \sigma, \alpha \rangle$ where $\sigma : S_1 \to S_2$ is a total while $\alpha : A_1 \rightharpoonup A_2$ is a partial function which satisfy*

- $\sigma \hat{s}_1 = \hat{s}_2$.
- $p \xrightarrow{a} q$ *in* $\mathcal{S}_1$ *and* αa-*defined implies* $\sigma p \xrightarrow{\alpha a} \sigma q$ *in* $\mathcal{S}_2$.
- $p \xrightarrow{a} q$ *in* $\mathcal{S}_1$ *and* αa-*undefined implies* $\sigma p = \sigma q$ *in* $\mathcal{S}_2$.

---

[*] Partially supported by CATALYSIS, a programme within CNRS/PAN cooperation
framework, and by MIUR and CNR/PAN exchange programme.

W.M.P. van der Aalst and E. Best (Eds.): ICATPN 2003, LNCS 2679, pp. 140–160, 2003.
© Springer-Verlag Berlin Heidelberg 2003

The first condition simply says that morphisms preserve the initial states. According to the second, a step in S_1 caused by an action observable in S_2 via α is mapped into an atomic step of S_2. Finally, according to the third condition, steps caused in S_1 by actions unobservable in S_2 via α do not have effects observable in S_2 via σ. Together, the conditions guarantee that each *computation* of S_1, i.e., a path $\hat{s}_1 \xrightarrow{\varpi} p$ in S_1, gets mapped to a computation in S_2, namely to $\hat{s}_2 \xrightarrow{\alpha\varpi} \sigma p$. One can also consider the *image* of S_1 in S_2 via f, i.e., a subsystem $S = f(S_1)$ of S_2 defined as follows.

- $S = \{\sigma p \in S_2 \,|\, p \in S_1\}$, with $\hat{s} = \hat{s}_2$.
- $A = \{\alpha a \in A_2 \,|\, a \in A_1, \alpha a\text{-defined}\}$.
- $\sigma p \xrightarrow{\alpha a} \sigma q$ in S whenever $p \xrightarrow{a} q$ in S_1.

It is convenient to introduce artificial *empty* steps $p \xrightarrow{\varnothing} p$ for each state p. Then one can think that steps caused in S_1 by α-unobservable actions are mapped to the empty steps in S_2. With this convention we can succinctly rewrite the second and third conditions of Def. 1 as follows.

$$p \xrightarrow{a} q \text{ in } S_1 \quad \text{implies} \quad \sigma p \xrightarrow{\alpha a} \sigma q \text{ in } S_2 \tag{1}$$

The notion of morphism introduced in Def. 1 seems to be the most commonly accepted in the literature. In fact, one often restricts attention to the subclass of morphisms which are total on actions. On the other hand, many models of concurrent systems, Petri nets in particular, offer a broader framework, in which *sets*, or even *bags* of actions contribute to system's evolution in the form of a complex, non-atomic step. Thus, one could consider morphisms more general than those allowed by Def. 1. The simplest would be to allow α to map an action to a set, or a bag of actions so that atomic steps in the source system get mapped to complex steps in the target system.

Formally, this generalization is correct in the sense that computations in the source get mapped to computations in the target. From this perspective an even more general notion of morphism can be considered in which actions are mapped to paths. Yet, only in the area of action refinement one can track these ideas.

The lukewarm support can be attributed to conceptual problems that come along. One of them stems from the apparent difficulty to define the image construction. Consider, for instance, a morphism $f : \mathcal{N}_1 \to \mathcal{N}_2$ of net systems. Nowadays it is uniformly accepted that this should induce a morphism $\mathrm{CG}(f) : \mathrm{CG}(\mathcal{N}_1) \to \mathrm{CG}(\mathcal{N}_2)$, i.e., a simulation of the abstract behaviour of $\mathcal{N}_1$, the *case graph* of $\mathcal{N}_1$, within the abstract behaviour of $\mathcal{N}_2$. Now, what could it mean for f to map a single event a of $\mathcal{N}_1$ to a set $\{a', a''\}$ of events of $\mathcal{N}_2$? Should we consider $\{a', a''\}$ as a new atomic event in the image of $\mathcal{N}_1$ via f? If so, would there be this connection between such complex atom and a', in case $a' = \alpha b$ for another event b of S_1?

To convey the ideas we consider a class of simple concurrent systems. *Elementary net systems* offer a concrete and elegant model of concurrent computations. Ehrenfeucht and Rozenberg, see [7], considered a class of deterministic transition systems called *elementary transition systems*, and showed that they are exactly the case graphs of elementary nets. More formally, they showed that for any elementary transition system S there exists an elementary net system $\mathcal{N} = \text{SN}(S)$ with case graph isomorphic to S, i.e., $\text{CG}(\mathcal{N}) \simeq S$. This result was among the first two constructive solutions to the problem of *synthesis* of a distributed realization $\text{SN}(S)$ of a given abstract behaviour S. The other solution was provided earlier by Zielonka, see [13].

Later, see [10], this result was strengthened, and the correspondence was revealed to take the form of an adjunction between the category $\mathbb{ETS}$ of elementary transition systems, and a category $\mathbb{ENS}$ of elementary net systems. More precisely, Nielsen et al., showed that the process of synthesizing an elementary net is a functor, with $\text{SN} : \mathbb{ETS} \rightarrow \mathbb{ENS}$ being left adjoint and inverse to $\text{CG} : \mathbb{ENS} \rightarrow \mathbb{ETS}$. Therefore, the adjunction is in fact a coreflection.

Such a close correspondence between two categories has important consequences. For instance, universal categorical constructions are preserved by the functors: colimits by the left adjoint functor SN, and limits by the right adjoint functor CG. Due to $\text{CG}(\text{SN}(S)) \simeq S$ one can also identify elementary transition systems as a subcategory of elementary net systems.

Many constructions on nets are specified up to the properties of their case graphs, for instance by providing the specification of a system in temporal logic. It is tempting to perform all the work in the realm of transition systems. Then, if the construction was based on limits, one can call upon the coreflection, and translate the components back to the realm of Petri nets via synthesis. If the category of Petri nets is sufficiently complete, see [3], the same construction can then be performed on nets. Moreover, the case graph of the resulting net will be isomorphic to the construction performed in the category of transition systems.

Consequently, in this note we shall be concerned with Petri nets rather indirectly, especially on the technical side. The main attention will be devoted to the existence of limits in the categories of Petri net systems abstract behaviours.

Elementary transition systems satisfy the following conditions.

No short loops	$p \xrightarrow{a} q$ implies $p \neq q$	(2)
No parallel arrows	$p \xrightarrow{a} q$ and $p \xrightarrow{b} q$ implies $a = b$	(3)
State reachability	$(\exists \varpi)\hat{s} \xrightarrow{\varpi} p$	(4)
Action reachability	$(\exists p, q)p \xrightarrow{a} q$	(5)

The essential notion necessary to define elementary transition systems and to facilitate the adjunction result is that of a region, see [7].

A *region* in S is a set R of states of S which is consistent with respect to action crossing the borders of R. Thus, $R \ni p \xrightarrow{a} q \notin R$ implies that for any other a-step $r \xrightarrow{a} s$ one gets the same picture: $r \in R \not\ni s$. The same holds for actions entering the region. The set of all regions of S is denoted $\mathcal{R}_S$. In what follows we write $R°a$, resp., $a°R$, to indicate that action a leaves, resp., enters, region R. Consequently, $°a = \{R \in \mathcal{R}_S \mid R°a\}$, $R° = \{a \in A \mid R°a\}$, etc.

A transition system is *elementary* if it satisfies conditions (2)-(5) and, additionally, the following regional axioms.

State Separation	$p \neq q$ implies $(\exists R \in \mathcal{R}_S)p \in R \not\ni q$	(6)
State-Action Separation	$p \xrightarrow{a} q$ implies $(\exists R \in \mathcal{R}_S)R^\circ a \wedge p \notin R$	(7)

The first regional axiom states that two different states can be separated by a region. The second axiom states that if all regions exited by a contain p then $p \xrightarrow{a}$ must hold in S.

Transition system S is *deterministic* provided $p \xleftarrow{a} q \xrightarrow{a} r$ implies $p = r$. Note that from the state separation axiom it follows that each elementary transition system is deterministic. In the sequel only deterministic transition systems are considered.

2   Introduction — A Primer on Modular Synthesis of Concurrent Systems

Following Ehrenfeucht and Rozenberg an elementary net system $\text{SN}(S)$ corresponding to a given elementary transition system S can be constructed as follows.

Take the regions of S as places. Let the set of actions of S be the set of events of $\text{SN}(S)$. The flow relation F is given by $F(R, a)$ iff $R^\circ a$ and $F(a, R)$ iff $a^\circ R$. Finally, declare place R to be marked initially iff $\hat{s} \in R$.

The above procedure is global and unstructured — the net is constructed in a single step. Modular approaches concentrate, instead, on gradual and systematic ways of system construction. Here, two such approaches pertaining to elementary net/transition systems are recalled.

2.1   Synthesis via Action Identification

An alternative way to look at elementary net systems was put forward by Bernardinello, see [5]. Namely, one can characterise each elementary net as a *product* of simple components, so called state machines. This observation paves the way to a modular presentation of net synthesis.

A *state machine* is a reachable marked Petri net of a very simple type: each event has exactly one pre-condition and one post-condition, while the initial marking consists of a single place with one token in it. Just like in the case of elementary nets no loops are allowed, and every two elements have different pre- or post-elements. Clearly, in a state machine every reachable marking is a singleton. Consequently, the behaviour of every state machine is purely sequential — no two events can ever be fired concurrently.

The idea of the product of elementary net systems and the decomposition into such sequential components is best explained on a toy example. In the middle of Fig. 1 an elementary net system is presented. It admits decomposition into two state machine components presented on the left and on the right. The net in the middle can be seen as a product of the two state machines. The product is computed by putting the nets side by side, separately, and then identifying events with identical names. To stay within the realm of elementary nets one should, in general, also clean things up. For instance, non-reachable places and/or events should be removed, while indistinguishable places glued together.

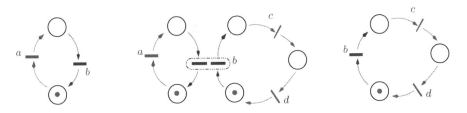

Fig. 1. Decomposition of an elementary net into sequential components.

This product, it turns out, is a categorical product in a category of elementary net systems with *rigid morphisms*, i.e., morphisms in the sense of Def. 1 which can either delete an event or map it to itself, but never rename it, see [4] for examples and details. We have already remarked that the case graph functor, as a right adjoint, preserves all limits that exist in its domain. Moreover, the adjunction cuts down to the subcategories with rigid morphisms. Thus, the case graph of the rigid product of two net systems computed in $\mathbb{ENS}$, is a rigid product of their case graphs in $\mathbb{ETS}$. This is demonstrated on Fig. 2. On the left and on the right the case graphs of the state machines from Fig. 1 are presented. Their rigid product computed in accord to Def. 2, see below, is depicted in the middle.

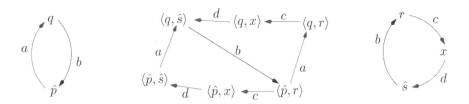

Fig. 2. Product of two elementary transition systems.

A formal definition of this product follows. Note that it applies to all transition systems, not just the elementary, and has been used in many situations since the beginning of Computing Science. One can argue, for instance that it corresponds to one of the CSP operators.

Definition 2. *Consider transition systems* S_i, $i = 1, 2$. *Their* rigid product, *sometimes denoted* $S_1 \prod^r S_2$, *is the transition system* S *defined as follows.*

- $S = S_1 \times S_2$, with projections $\pi_i : S \to S_i$, $i = 1, 2$.
- $\hat{s} = \langle \hat{s}_1, \hat{s}_2 \rangle$.
- $A = A_1 \cup A_2$, with partial projections $\alpha_i : A \rightharpoonup A_i$, for $i = 1, 2$, given by $\alpha_i a = a$ if $a \in A_i$, and $\alpha_i a$ undefined otherwise.
- $p \xrightarrow{a} q$ in S iff $\pi_1 p \xrightarrow{\alpha_1 a} \pi_1 q$ in S_1 and $\pi_2 p \xrightarrow{\alpha_2 a} \pi_2 q$ in S_2.

By taking REACH$(S_1 \prod^r S_2)$, i.e., the reachable subsystem of S, one obtains a rigid product in the full subcategory of reachable systems.

Let us remark here, that for any transition system S there exists its maximal subsystem REACH(S) satisfying the required reachability conditions: either (4) or (5) or both. Let us state without proof the following result.

Proposition 1. *Transition systems with morphisms given in Def. 1 constitute a complete category* TS. *Its subcategory* TS^r *with rigid morphisms is also complete, with binary products as given in Def 2.*

A limit in the subcategories with reachable systems is obtained by taking the reachable subsystem of the limit. This cuts down to the category of elementary systems, and its subcategory with rigid morphisms.

More about the existence of limits in the categories of transition systems follows in Sec. 4. Let us only remark that the operation of computing REACH(S) is functorial for the various reachability criteria imposed by (4), (5) or both. In fact, the functors are right adjoint to the inclusions. Thus, again, it is enough to prove that $S_1 \prod^r S_2$ is a product in the category of all transition systems with rigid morphisms to deduce that REACH$(S_1 \prod^r S_2)$ is a product in the subcategory of reachable systems.

In summary, each elementary transition system can be computed by applying the rigid product constructor to *basic* elementary transition systems — those, in which every action participates in a single transition. The same does not hold for all elementary net systems. There is no problem, however, if one is interested in nets as the concurrent realisations of their abstract behaviours. For instance, the saturated net systems considered by Ehrenfeucht and Rozenberg ([7]), and those constructed from minimal regions following Bernardinello ([5]) can be constructed from state machines by identification of common events, i.e., by repeated application of the rigid product construction.

2.2   Synthesis via Identification of Regions

Recently, a novel modular composition technique applicable to elementary net and transition systems has been proposed, see [6]. The idea is to glue together two systems on pairs of *complementary* places/regions.

We refrain from explaining the details of the construction as defined on nets, consult [6] for details. Instead, the example on Fig. 3 demonstrates the idea at work. Consider the two nets $\mathcal{N}_1$ and $\mathcal{N}_2$ in the upper corners of the figure, each with the distinguished pair of places denoted R and R̄. These places are complementary in the sense that the pre-events of R are the post-events of the

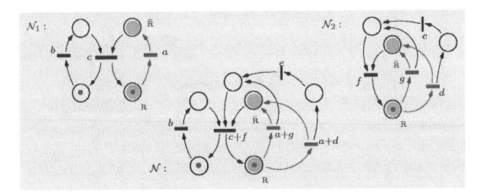

Fig. 3. Composition of two elementary nets by identification of places.

corresponding R̄, and vice versa. It is required that both places R are initially marked, and thus places R̄ are not. Net $\mathcal{N}$ (in the middle) shows the effect of the composition. Again, we start by putting disjoint copies of the two nets side by side. Then, both places R are identified, and so are those labelled R̄.

The events, like b or e, which neither input nor output from R and R̄ are not affected. The flow relation for such events is inherited from the components. Other events, like a, must either empty R and fill R̄, or, like c, fill R and empty R̄. We take all matching pairs of such events, like $a + g$ and $c + f$, with one event from each component. Note that in this way duplicate copies of some events may give rise to several complex events, viz. $a + g$ and $a + d$. The flow relation for the complex events is the union of flows of their elements.

Finally, if one is interested in elementary nets one should restrict attention to reachable places and events.

Looking at Fig. 3 one can easily see how the original nets can be recovered from their composition. This can be formalized by establishing suitable morphisms from the resulting net to the components.

This method of building nets from simpler components is quite intuitive. It is mimicked by a suitable operation on the transition systems as shown on Fig. 4. In the upper corners we see the case graphs of the two component nets presented on Fig. 3. Regions corresponding to the distinguished places are also depicted. The transition system in middle of Fig. 4 is the result of the composition. The details of its construction are described in Def. 3. Note that, following [6], elementary transition systems are composable on regions only under certain assumptions.

Definition 3. *Assume that elementary transition systems* $\mathcal{S}_1$ *and* $\mathcal{S}_2$ *are composable on regions* R_1 *and* R_2, *i.e., that the following conditions are satisfied.*

- $\hat{s}_1 \in R_1$ *and* $\hat{s}_2 \in R_2$.
- $\{a \in A_1 \mid R_1{}^\circ a\} = \emptyset$ *iff* $\{a \in A_2 \mid R_2{}^\circ a\} = \emptyset$.
- $\{a \in A_1 \mid a^\circ R_1\} = \emptyset$ *iff* $\{a \in A_2 \mid a^\circ R_2\} = \emptyset$.

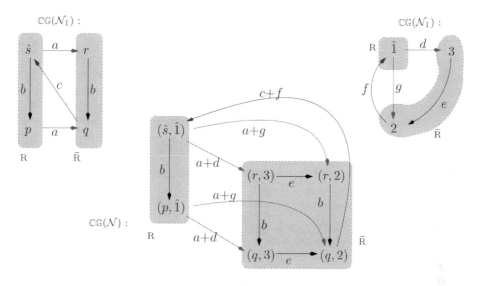

Fig. 4. Composition of two elementary transition systems by identification of places.

Then, consider S defined as follows, where $\bar{R}_i = S_i \setminus R_i$, for $i = 1, 2$.

- $S = (R_1 \times R_2) \cup (\bar{R}_1 \times \bar{R}_2)$. *The corresponding projections $\pi_i : S \to S_i$ are defined by $\pi_i \langle p_1, p_2 \rangle = p_i$, for $i = 1, 2$.*
- $\hat{s} = \langle \hat{s}_1, \hat{s}_2 \rangle$.
- $A = A_1 \setminus (R_1{}^\circ \cup {}^\circ R_1) \cup A_2 \setminus (R_2{}^\circ \cup {}^\circ R_2) \cup (R_1{}^\circ \times R_2{}^\circ) \cup ({}^\circ R_1 \times {}^\circ R_2)$. *The corresponding partial projections $\kappa_i : A \to A_i$ are defined by $\kappa_i a = a$, if $a \in A_i$ and undefined otherwise, and $\kappa_i \langle a_1, a_2 \rangle = a_i$, for $i = 1, 2$.*
- $\langle p_1, p_2 \rangle \xrightarrow{a} \langle q_1, q_2 \rangle$ *iff* $p_1 \xrightarrow{\kappa_1 a} q_1$ *and* $p_2 \xrightarrow{\kappa_2 a} q_2$.

Finally, composition of S_1 and S_2 on regions R_1 and R_2, denoted $S_1[R, \bar{R}]S_2$, is the result of taking the reachable subsystem of S.

One can show that the composition of two composable elementary transition systems is elementary. The main result of [6] is that the two composition operations, of elementary net systems and elementary transition systems agree: $\mathbb{CG}(\mathcal{N}_1[R, \bar{R}]\mathcal{N}_2) \simeq \mathbb{CG}(\mathcal{N}_1)[R, \bar{R}]\mathbb{CG}(\mathcal{N}_2)$, i.e., the case graph of the composed net is isomorphic to the composition of the case graphs. Until now, however, there was no categorical explanation of the result. The reasons for that failure are discussed in the following section.

3   Synchronisation as a Pullback?

Clearly, $\langle \pi_i, \kappa_i \rangle : S_1[R, \bar{R}]S_2 \to S_i$, constitute morphisms of transition systems for $i = 1, 2$. The composability conditions implicitly refer to a transition system **2** presented on the top of Fig. 5, and the two morphisms presented there. System **2** captures the essence of dividing the state space of a transition system into two complementary regions, with two actions corresponding to moving out from the region to its complement and back. The state components σ_i send all states in the corresponding region R to the state R of **2**, and the states in the region $\bar{\text{R}}$ to the state $\bar{\text{R}}$ of **2**. The action components α_1 map all actions crossing from R to R in S_i to action *out*, actions crossing R in the opposite direction are mapped to *in*. All other actions that do not cross R are α_i-undefined.

More generally, the choice of a region R, with $\hat{s} \in$ R, in *any* elementary transition system S uniquely determines a morphism $f : S \rightarrow \mathbf{2}$ defined as above. Conversely, *any* region in a transition system S is the inverse image of the state component of a morphism $f : S \rightarrow \mathbf{2}$. In this sense system **2** is the *type* of elementary nets, see [1].

The square of morphisms on Fig. 5 is commutative. This strengthens the intuition that system $S_1[\text{R}, \bar{\text{R}}]S_2$ is in fact a *synchronisation* of S_1 and S_2 induced by a specific choice of regions. Equivalently, it is a synchronisation induced by a choice of abstractions of S_1 and S_2 within **2** as a common interface.

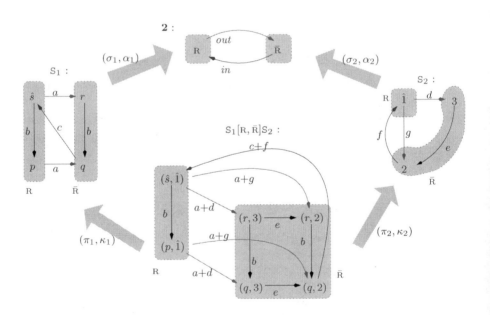

Fig. 5. System $S_1[\text{R}, \bar{\text{R}}]S_2$ as a synchronisation of S_1 and S_2 via interface **2**.

There is an evident analogy between this synchronisation operation, and the construction of a *pullback* — one of the universal categorical limits. To see this consider Fig. 6.

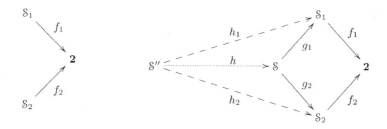

Fig. 6. The interface of synchronisation and the pullback

The starting point of the construction of $S_1[R, \bar{R}]S_2$ is presented on the left, where we are given two morphisms $f_1 : S_1 \to \mathbf{2}$ and $f_2 : S_2 \to \mathbf{2}$ with a common *synchronisation interface* $\mathbf{2}$. Clearly, this presentation generalises the compatibility conditions given in Def 3. The same situation is at the beginning of pullback construction. The difference is that one is not willing to accept *any* solution, i.e., a system S with a pair of *projections* $g_1 : S \to S_1$ and $g_2 : S \to S_2$ which make the square commute. This merely captures that the solution *adheres* to the restrictions imposed by the interface. One strives to find S which is an *optimal* solution. Formally, given any other solution S'' with a pair of morphisms $h_1 : S'' \to S_1$ and $h_2 : S'' \to S_2$ which adhere to the interface, there should exist a unique mediating morphism $h : S'' \to S$ through which both h_1 and h_2 would factorize: $g_1 \circ h = h_1$ and $g_2 \circ h = h_2$. The situation is depicted on the right of Fig. 6

Now, it is natural to ask whether the synchronisation operation proposed in [6] is an instance of a pullback construction. An answer can be given, provided we fix a category in which the pullbacks are sought. A natural choice would be to consider transition systems with the class of morphisms described in Def. 1. This category is complete. This is a good news, since the pullbacks can be constructed for all interfaces. Sadly, it turns out that the pullback construction computed in this category does not coincide with the synchronisation operation proposed by Bernardinello et al., as described in Def. 3. The difference is vividly demonstrated on Fig. 7. The result $S_1[R, \bar{R}]S_2$ of synchronisation of S_1 and S_2, is recalled on the left. Their pullback, with respect to the same interface, computed in this category, is described on the right of Fig. 7. Clearly, $S_1[R, \bar{R}]S_2$ is a proper subsystem of the pullback, and so we have hidden in the pullback the labels of all transitions except the new one. There is one new complex action in the pullback which contributes to one new transition — the one drawn with thick dashed arrow.

Having demonstrated that the synchronisation construction is not universal one is left with two options. First, if one believes that the category is *the* right one, perhaps the best idea is to forget about the construction. Most likely the construction will not have any interesting properties. The second option is to investigate other categories in which the construction might turn out to be universal.

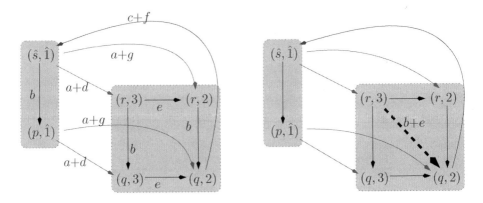

Fig. 7. Synchronisation versus pullback in $\mathbb{TS}$.

In our case one can see that the pullback construction behaves somewhat strangely. The interface does not impose any restrictions between action b from the component $\mathcal{S}_1$, and action e from the component $\mathcal{S}_2$. Thus, it would be most natural to have them implemented as independent events, compare Fig. 3. Yet, additionally, the pullback construction introduces an action $b+e$ which, via projections, corresponds to a *simultaneous* execution of these two, seemingly independent actions.

Thus, one would want to work within a model in which the tight synchronisation $b+e$ could be prohibited. At the same time we would like to keep actions like $a+d$ or $c+f$ which also are tight synchronisations of actions from the two components. These, seemingly contradictory requirements can be met in the category studied in the next Section.

4   Asynchronous Systems with Step Semantics

Let us formally introduce a more general model. We could continue with elementary transition systems, but there would be a price to pay. First, elementary transition systems are conceptually less suitable than *asynchronous systems*, see [12,2], the model we are about to introduce. This is reflected not only in the number of axioms imposed in both cases. More importantly, the crucial notion of *action independence* is brought forward only in the definition of asynchronous systems, whereas it remains hidden in the structure of elementary transition systems.

Definition 4. *An asynchronous system is a tuple* $\mathcal{A} = (S, \hat{s}, A, \|, T)$ *where* $\mathcal{S} = (S, \hat{s}, A, T)$ *is a deterministic transition system underlying* $\mathcal{A}$, *and* $\| \subseteq A \times A$ *is an irreflexive and symmetric binary relation of independence, provided the following swap property holds.*

$$p \xrightarrow{a} q \xrightarrow{b} r \quad and \quad a \parallel b \quad implies \quad p \xrightarrow{b} s \xrightarrow{a} r \quad for \ some \ s \in S. \quad (8)$$

Note that an elementary transition system S gives rise to an asynchronous system when the independence relation is defined as follows.

$$a \parallel b \quad \text{iff} \quad p \xrightarrow{\;a\;} q \xrightarrow{\;b\;} \text{ in } S \quad \text{and} \quad \xleftarrow{\;a\;} r \xrightarrow{\;b\;} \text{ in } S$$

It can be shown that the asynchronous systems obtained by taking $\parallel$ are *concrete*, i.e., they are rigid products of their sequential components, see [4] for details.

For a binary relation like $\parallel\, \subseteq A \times A$ consider the set $\mathsf{Cliques}(\parallel)$ of cliques of mutually $\parallel$-related elements.

$$\mathsf{Cliques}(\parallel) = \{x \subseteq A \mid (\forall a, b \in x)a \neq b \Rightarrow a \parallel b\}$$

Note that $\emptyset, \{a\} \in \mathsf{Cliques}(\parallel)$. Thus, identifying elements of A with the corresponding singleton cliques the relation $\parallel$ can be conservatively extended to $\parallel\, \subseteq \mathsf{Cliques}(\parallel) \times \mathsf{Cliques}(\parallel)$ by letting $x \parallel y$ iff $a \parallel b$ for all $a \in x$ and $b \in y$. Clearly, $x \parallel \emptyset$ holds for any $x \in \mathsf{Cliques}(\parallel)$.

The idea behind the independence of actions, as captured by the swap property (8), is that two independent actions can occur concurrently whenever they can occur one after another. Thus, we can extend the atomic step relation to arbitrary sets of mutually independent actions as follows.

$$p \xrightarrow{\;x\;} q \quad \text{iff} \quad p \xrightarrow{\;\varpi\;} q \quad \text{for some linearization } \varpi \text{ of } x \in \mathsf{Cliques}(\parallel)$$

Note that due to (8), $p \xrightarrow{\;\varpi\;} q$ holds for some linearization ϖ of x iff it holds for *any* linearization of x. Also, $p \xrightarrow{\;\emptyset\;} p$ holds for any p. In the sequel, notation $p \xrightarrow{\;a\;} q$ and $p \xrightarrow{\;a+c\;} q$ is used rather than the formally correct $p \xrightarrow{\;\{a\}\;} q$ and $p \xrightarrow{\;\{a,c\}\;} q$, respectively.

One-safe Petri nets give rise to asynchronous systems. Indeed, the marking graph of any P/T net is a deterministic transition system. Then, if $\mathcal{N}$ is 1-safe, one can associate with each net an independence relation in two universal ways. One way is to choose the *dynamic independence* and say that two events a and b are independent in $\mathcal{N}$ if a step $a + b$ can be fired from a reachable marking. The other is to choose *structural independence* and declare a and b as independent if the pre- and post-conditions of a are disjoint with the pre- and post-conditions of b.

An asynchronous system with empty independence relation is just a deterministic transition system. Let us also look at conservative extensions of morphisms as given in Def. 1.

One such conservative extension has already been proposed, cf. [2]. The idea is to add the requirement that the action part of a morphism weakly preserves concurrency in the sense that if $a \parallel_1 b$ and both αa and αb are defined then $\alpha a \parallel_2 \alpha b$ holds. With our convention extending independence relation to steps the same can be more succinctly described as follows.

$$a \parallel_1 b \quad \text{implies} \quad \alpha a \parallel_2 \alpha b \tag{9}$$

Clearly, if the independence is empty, the above holds trivially.

It turns out that asynchronous systems and morphisms introduced above form a complete category. An example of a product, i.e., the pull-back of two systems with respect to the unique morphisms to the terminal object of the category, is depicted on Fig. 8. The actions from the components of the product

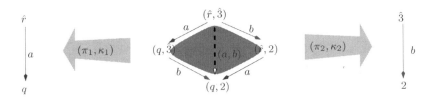

Fig. 8. A product of asynchronous systems in the category with flat morphisms.

are preserved in the product as independent actions. Unfortunately, their synchronisation is also added. So, the example demonstrates the same unwelcome phenomenon that we have faced before, cf. Fig. 7.

The notion recalled above is not, however, the only possible conservative extension of the notion of a morphism of deterministic transition systems. Another, more liberal proposal follows.

Definition 5. *Let $A_i = (S_i, \hat{s}_i, A_i, \|_i, T_i)$ be two asynchronous systems, for $i = 1, 2$. Their synchronising morphism $f : A_1 \to A_2$ is a pair $f = \langle \sigma, \alpha \rangle$ where*

- *$\sigma : S_1 \to S_2$ preserves the initial states:* $\sigma\hat{s}_1 = \hat{s}_2$.
- *$\alpha : A_1 \to$ Cliques($\|_2$) preserves independence:* $a \|_1 b$ *implies* $\alpha a \|_2 \alpha b$

Together they should map atomic steps of A_1 into steps of A_2 as follows.

- *$p \xrightarrow{a} q$ in A_1 implies $\sigma p \xrightarrow{\alpha a} \sigma q$ in A_2.*

Thus, the only difference in reference to the classical definition is that instead of sending an action to null or to another action, a possibility to send it to a set of mutually independent actions is added. Preservation of the independence relation means that not only atomic, but also each multi-step in the source system is mapped to a multi-step in the target system.

The completeness of the category of asynchronous systems equipped with synchronising morphisms is investigated in the next section. Let us finish by stating some basic facts.

Proposition 2. *Asynchronous systems with synchronising morphisms as given in Definition 5 form a category $\mathbb{AS}$.*

The subcategory $\mathbb{AS}^f$ with flat morphisms obtained by imposing that the cardinality of αa is not greater than 1 is isomorphic to the category of asynchronous systems studied in [2].

Let us finally remark that in the richer category in which non-flat morphisms are allowed the product of the systems depicted in Fig. 8 will be the same diamond, but without the extra action (a, b). The details of the constructions of limits in this, and in related categories are studied in the sequel sections.

5   Pullbacks, and Other Limits

Computing a pullback, one of the universal categorical constructions, can be seen as a generalisation of binary products. Indeed, assume that a category admits a *terminal object* $\mathbf{1}$, i.e., that for any object $\mathcal{A}$ there exists a unique morphism $! : \mathcal{A} \rightarrow \mathbf{1}$. Given objects $\mathcal{A}_1$ and $\mathcal{A}_2$ assume that there exists a pullback $p_i : \mathcal{A} \rightarrow \mathcal{A}_i$, $i = 1, 2$, of the morphisms $!_1 : \mathcal{A}_1 \rightarrow \mathbf{1}$ and $!_2 : \mathcal{A}_2 \rightarrow \mathbf{1}$. It follows then easily that $\mathcal{A}$ is a categorical product of $\mathcal{A}_1$ and $\mathcal{A}_2$.

Conversely, products can be used to compute pullbacks, provided the category admits equalizers. Let us remind that given a pair of parallel morphisms $f, g : \mathcal{A}' \rightarrow \mathcal{A}''$ their *equalizer* is a morphism $\mathbf{eq}(f, g) : \mathcal{A} \rightarrow \mathcal{A}'$, which equalizes f and g, i.e., $f \circ \mathbf{eq}(f, g) = g \circ \mathbf{eq}(f, g)$, and does it in the optimal way, i.e., any other morphism equalizing f and g factors uniquely through $\mathbf{eq}(f, g)$. The construction of a pullback via product and equalizer is described on Fig. 9. Thus,

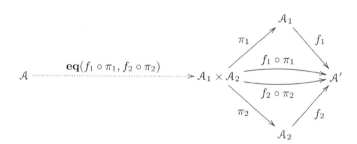

Fig. 9. Pullback constructed from product and equalizer

given $f_i : \mathcal{A}_i \rightarrow' \mathcal{A}'$, $= 1, 2$, first compute the categorical product $\mathcal{A}_1 \times \mathcal{A}_2$ with projections π_1 and π_2. Then, enforce commutation of the square by equalizing the morphisms $f_1 \circ \pi_1$ and $f_2 \circ \pi_2$.

The above observation is valid in any category. In fact, from products and equalizers one can compute arbitrary finite limits, see [8]. Our idea is to apply this characterisation of pullbacks to demonstrate the existence of the pullbacks of asynchronous systems with synchronising morphisms.

An asynchronous system consists of two components, while a morphism comprises two maps for manipulating the components. Consequently, one can view the category defined above as being built over two simpler *component* categories. One, concerning the state part, is the category $\text{SET}_\star$ of pointed sets $(S, \hat{s})$, with $\hat{s} \in S$, and functions between the sets which preserve the distinguished points.

The other category, concerning the action part, is the category $\mathrm{SET}_{\|}$ of sets with independence relation $(A, \|)$, and multi-functions $\alpha : A_1 \to \mathsf{Cliques}(\|_2)$ which preserve the independence relation. Clearly, forgetting about actions, and hence the independence and transition relations, results in a forgetful functor from $\mathbb{AS}$ to $\mathrm{SET}_\star$. Similarly, a forgetful functor from $\mathbb{AS}$ to $\mathrm{SET}_{\|}$ is obtained.

In the context of the present section it is useful to learn how the limits are constructed in the component categories. The reason is simple — as we shall see the limits in the category of asynchronous systems are reflected by these forgetful functors.

5.1   Pullbacks in $\mathrm{SET}_\star$

Binary product of $(S_1, \hat{s}_1)$ and $(S_2, \hat{s}_2)$ in $\mathrm{SET}_\star$ is just the cartesian product $S_1 \times S_2$ with the pair $\langle \hat{s}_1, \hat{s}_2 \rangle$ as the designated element.

The construction of equalizer is also simple. Given $\sigma, \varsigma : (S_1, \hat{s}_1) \to (S_2, \hat{s}_2)$ consider $S = \{s \in S_1 \mid \sigma s = \varsigma s\}$. Clearly, $\hat{s}_1 \in S$. It is immediate that the inclusion $\imath : S \hookrightarrow S_1$ equalizes σ and ς. The inclusion is also optimal in the sense, that any $\rho : S' \to S_1$ that equalizes σ and ς necessarily factors through $\imath$, and does it in a unique way. Namely, the image $\{\rho s' \in S_1 \mid s' \in S'\}$ of S' via ρ is included in S, since $\sigma(\rho s') = \varsigma(\rho s')$ by assumption. Thus, the characterisation of pullbacks in $\mathrm{SET}_\star$ described on Fig. 9 gives the following result.

Proposition 3. *Let* $\sigma_i : (S_i, \hat{s}_i) \to (S', \hat{s}')$, *for* $i = 1, 2$. *Then* $(S, \langle \hat{s}_1, \hat{s}_2 \rangle)$ *where* $S = \{\langle p_1, p_2 \rangle \in S_1 \times S_2 \mid \sigma_1 p_1 = \sigma_2 p_2\}$ *with projections* $\varsigma_i : S \to S_i$ *given by* $\varsigma_i \langle s_1, s_2 \rangle = s_i$, $i = 1, 2$, *is a pullback in* $\mathrm{SET}_\star$.

5.2   Products in $\mathrm{SET}_{\|}$

Characterisation of finite limits in $\mathrm{SET}_{\|}$ is more complicated. Let us start by introducing some notions and notation, and making a number of observations.

Given an object $(A, \|)$ in $\mathrm{SET}_{\|}$ and $x, y \in \mathsf{Cliques}(\|)$ let us say that x and y are *consistent* whenever $x \cup y \in \mathsf{Cliques}(\|)$.

Lemma 1. *Let* $x, y \in \mathsf{Cliques}(\|)$ *and consider* $\alpha : (A, \|) \to (A', \|')$ *in* $\mathrm{SET}_{\|}$. *Then*

1. $x \| y$ *iff* $x \cup y \in \mathsf{Cliques}(\|)$ *and* $x \cap y = \emptyset$.
2. $x \| y$ *implies* $\alpha x, \alpha y \in \mathsf{Cliques}(\|')$ *and* $\alpha x \|' \alpha y$.
3. $x \cup y \in \mathsf{Cliques}(\|)$ *implies* $\alpha(x \cap y) = \alpha x \cap \alpha y$.

Lemma 1(3) says that morphisms in $\mathrm{SET}_{\|}$ preserve *consistent* meets. Without this proviso this is not valid. For instance, let $A = \{a, b, c\}$ with $a \| b \| c$ and $A' = \{d, e\}$ with $d \| e$. Then $\alpha : A \to A'$ given by $\alpha a = \alpha c = e$ and $\alpha b = d$ is a valid morphism. However, $\{d\} = \alpha(\{a, b\} \cap \{b, c\}) \neq \alpha\{a, b\} \cap \alpha\{b, c\} = \{d, e\}$.

Now, with products the situation is rather easy.

Consider $(A_1, \|_1)$ and $(A_2, \|_2)$. Without loss of generality one can assume $A_1 \cap A_2 = \emptyset$. Their product $(A, \|)$ in $\mathrm{SET}_{\|}$ is given by $A = A_1 \cup A_2$ and $\| = \|_1 \cup$

$\|_2 \cup A_1 \times A_2 \cup A_2 \times A_1$. Clearly, $\|$ is symmetric and irreflexive. Define projections $\pi_i : A \to A_i$, for $i = 1, 2$, as follows.

$$\pi_i a = \begin{cases} \emptyset & a \notin A_i \\ \{a\} & a \in A_i \end{cases}$$

Note that $\pi_i x = x \cap A_i$, $i = 1, 2$ follows from the above definition.

Lemma 2. $(A, \|)$ *with projections defined above is a product of* $(A_1, \|_1)$ *and* $(A_2, \|_2)$ *in* $\mathbb{S}\mathrm{ET}_\|$.

Let us also notice that the projections are flat.

5.3 Equalizers in $\mathbb{S}\mathrm{ET}_\|$

With equalizers the situation is more complex. Let $\alpha, \beta : (A_1, \|_1) \to (A_2, \|_2)$.

Again, we would like to take the sub-object of $(A_1, \|_1)$ which is equalized by α and β. The problem is, however, that it is not enough to restrict attention to the elements of A_1 on which α and β agree.

Example 1. Put $A_1 = \{a, b\}$ with $a \|_1 b$, and let $(A_1, \|_1) = (A_2, \|_2)$. Consider α and β given by $\alpha b = \beta a = b$ and $\alpha a = \beta b = a$. One morphism equalizing α and β is obtained by taking the embedding of the empty set, with the empty independence relation. Another is given by taking $A = \{\star\}$ and $\lambda \star = a + b$. The other is, in fact, an equalizer of α and β in $\mathbb{S}\mathrm{ET}_\|$.

Thus, even when α and β agree on no singleton, they may nevertheless agree on some larger cliques, and this leads to non-trivial equalizers.

Let us therefore consider $\mathcal{X}_{\alpha\beta} = \{x \in \mathsf{Cliques}(\|_1) \mid \alpha x = \beta x\}$.

Lemma 3. $\mathcal{X}_{\alpha\beta}$ *is closed with respect to consistent: meets, sums and differences.*

From lemma 3 it follows that *atoms*, i.e., non-empty minimal elements of $\mathcal{X}_{\alpha\beta}$, are irreducible in the following sense, see [3].

Corollary 1. *Let* $x, y, z \in \mathcal{X}_{\alpha\beta}$, x *an atom, satisfy* $x, y \subseteq z$. *Then either* $x \subseteq y$ *or* $x \subseteq z \setminus y$. *Any two different consistent atoms* $x, y \in \mathcal{X}_{\alpha\beta}$ *are disjoint. Thus,* $x \|_1 y$.

As a consequence one obtains that each element y of $\mathcal{X}_{\alpha\beta}$ admits unique decomposition into a sum of disjoint atoms of $\mathcal{X}_{\alpha\beta}$. In what follows $Atoms(\mathcal{X}_{\alpha\beta})$ denotes the set of all atoms of $\mathcal{X}_{\alpha\beta}$. Also, let $\mathcal{X}_{\alpha\beta}^y = \{x \in Atoms(\mathcal{X}_{\alpha\beta}) \mid x \subseteq y\}$ for all $y \in \mathcal{X}_{\alpha\beta}$.

Proposition 4. *Every* $y \in \mathcal{X}_{\alpha\beta}$ *admits* $\mathcal{X}_{\alpha\beta}^y$ *as a unique decomposition into a sum of disjoint atoms of* $\mathcal{X}_{\alpha\beta}$.

The existence of unique decompositions of the elements of $\mathsf{Cliques}(\|_1)$ equalized by α and β paves the way for the construction of equalizers. Namely, consider $(Atoms(\mathcal{X}_{\alpha\beta}), \|)$ with $x \parallel y \Leftrightarrow x \parallel_1 y$.

Thus, the idea is to consider the atoms of $\mathcal{X}_{\alpha\beta}$ as new actions, the hitherto complex structure of which will now be forgotten. Actually, we do keep track of their past in the evaluation mapping $[\alpha, \beta] : Atoms(\mathcal{X}_{\alpha\beta}) \to \mathsf{Cliques}(\|_1)$ defined by $[\alpha, \beta]x = x$.

The independence between new atoms is inherited from independence of the cliques of mutually independent actions in $(A_1, \|_1)$. Therefore, $[\alpha, \beta]$ is in fact a morphism $[\alpha, \beta] : (Atoms(\mathcal{X}_{\alpha\beta}), \|) \to (A_1, \|_1)$ in $\mathbb{S}\mathrm{ET}_\parallel$.

Proposition 5. $[\alpha, \beta]$ *is an equalizer of α and β in* $\mathbb{S}\mathrm{ET}_\parallel$.

Let us note that Example 1 demonstrates that even if both, α and β, are flat morphisms, their equalizer $[\alpha, \beta]$ in the category with synchronising morphisms may fail to be flat.

5.4   Pullbacks in $\mathbb{S}\mathrm{ET}_\parallel$

The construction of products and equalizers can be used, as described on Fig. 9, to compute pullbacks. Putting the constructions together result in the following elementary characterisation of pullbacks in the category $\mathbb{S}\mathrm{ET}_\parallel$. Since the categorical constructions are defined up to isomorphism we assume, without loss of generality, that the objects in the pullback situation are disjoint.

Definition 6. *Let $\alpha_1 : (A_1, \|_1) \to (A', \|') \leftarrow (A_2, \|_2) : \alpha_2$ and assume that $A_1 \cap A_2 = \emptyset$. Define $(A, \|)$ as follows.*

- $A = \{\{a_1\} \mid a_1 \in A_1, \alpha_1 a_1 = \emptyset\} \cup \{\{a_2\} \mid a_2 \in A_2, \alpha_2 a_2 = \emptyset\}$
$$\cup \left\{ x_1 \cup x_2 \;\middle|\; \begin{array}{l} x_i \in \mathsf{Cliques}(\|_i), \text{ for } i = 1, 2, \alpha_1 x_1 = \alpha_2 x_2 \neq \emptyset, \\ \forall y_1, y_2 \text{ such that } x_1 \supseteq y_1 \neq \emptyset \neq y_2 \subseteq x_2 \\ \alpha_1 y_1 = \alpha_2 y_2 \Rightarrow y_1 = x_1 \wedge y_2 = x_2 \end{array} \right\}$$
   with projections $\kappa_i : A \to \mathsf{Cliques}(\|_i)$ given by $\kappa_i(x_1 \cup x_2) = x_i$, for $i = 1, 2$.
- $x \parallel y$ *iff* $\kappa_1 x \parallel_1 \kappa_1 y$ *and* $\kappa_2 x \parallel_2 \kappa_2 y$.

The following result is almost immediate.

Proposition 6. $(A, \|)$ *together with $\kappa_i : (A, \|) \to (A_i, \|_i)$, $i = 1, 2$, is a pullback of $\alpha_i : (A_i, \|_i) \to (A', \|')$, $i = 1, 2$ in $\mathbb{S}\mathrm{ET}_\parallel$. If α_1 and α_2 are flat then so are κ_1 and κ_2.*

5.5   Pullbacks in the Category of Asynchronous Systems

We have shown the existence and characterized the construction of binary products and equalizers in the component categories $\mathbb{S}\mathrm{ET}_\star$ and $\mathbb{S}\mathrm{ET}_\parallel$. In fact, in both categories the construction of binary products easily generalises to arbitrary products. This guarantees the existence of arbitrary limits, see [8], i.e., both categories are complete.

Moreover, the forgetful functors from the category $\mathbb{AS}$ of asynchronous systems to the component categories $\mathbb{SET}_\star$ and $\mathbb{SET}_\parallel$ together *reflect* the limits. That is, to construct a limit of a diagram in $\mathbb{AS}$ one can compute the limits in the component categories, and then equip the results with an appropriate transition relation. Rather than proving such a general result let us verify that it works for pullbacks.

The following provides an elementary characterisation of pullbacks.

Definition 7. *Let $f_i = \langle \sigma_i, \alpha_i \rangle : \mathcal{A}_i \to \mathcal{A}'$, for $i = 1, 2$, be such that $A_1 \cap A_2 = \emptyset$. Consider $\mathcal{A}$ defined as follows.*

- $S = \{ \langle p_1, p_2 \rangle \in S_1 \times S_2 \mid \sigma_1 p_1 = \sigma_2 p_2 \}$,
  together with projections $\varsigma_i : S \to S_i$ given by $\varsigma_i \langle s_1, s_2 \rangle = s_i$ for $i = 1, 2$.
- $\hat{s} = \langle \hat{s}_1, \hat{s}_2 \rangle$.
- $A = \{ \{a_1\} \mid a_1 \in A_1, \alpha_1 a_1 = \emptyset \} \cup \{ \{a_2\} \mid a_2 \in A_2, \alpha_2 a_2 = \emptyset \}$
$$\cup \left\{ x_1 \cup x_2 \;\middle|\; \begin{array}{l} x_i \in \mathsf{Cliques}(\|_i), \text{ for } i = 1, 2, \alpha_1 x_1 = \alpha_2 x_2 \neq \emptyset, \\ \forall y_1, y_2 \text{ such that } x_1 \supseteq y_1 \neq \emptyset \neq y_2 \subseteq x_2 \\ \alpha_1 y_1 = \alpha_2 y_2 \Rightarrow y_1 = x_1 \wedge y_2 = x_2 \end{array} \right\}$$
  with projections $\kappa_i : A \to \mathsf{Cliques}(\|_i)$ given by $\kappa_i(x_1 \cup x_2) = x_i$, $i = 1, 2$.
- $a \parallel b$ *iff* $\kappa_1 a \parallel_1 \kappa_1 b$ *and* $\kappa_2 a \parallel_2 \kappa_2 b$.
- $p \xrightarrow{a} q$ *iff* $\varsigma_1 p \xrightarrow{\kappa_1 a} \varsigma_1 q$ *and* $\varsigma_2 p \xrightarrow{\kappa_2 a} \varsigma_2 q$.

The following result is now straightforward.

Theorem 1. *$\mathcal{A}$ constructed above is an asynchronous system which, together with $g_i = \langle \varsigma_i, \kappa_i \rangle : \mathcal{A} \to \mathcal{A}_i$, for $i = 1, 2$, forms a pullback in the category of asynchronous systems with synchronising morphisms. The pullback in the full subcategory of reachable systems is obtained by taking $\mathrm{REACH}(\mathcal{A})$, i.e., the reachable part of $\mathcal{A}$. Morphisms g_1 and g_2 are flat whenever f_1 and f_2 are both flat.*

6   Applications and Future Work

The remaining part of the paper demonstrates the utility of the construction of pullbacks in the category of asynchronous systems with synchronising morphisms. In particular, we show that the two, seemingly quite different methods of system synthesis discussed in Sec. 2.1 and in Sec. 2.2 are in fact special cases of pullbacks.

6.1   Rigid Product as a Pullback

We have argued in section 2.1 that elementary systems, both transition systems and nets, can be synthesised by conjoining a number of simple sequential systems via rigid product construction, cf. Def. 2 and Prop. 1. Morin was the first to notice that a large class of asynchronous systems can be characterised as rigid products of their sequential components in a category of state-reachable asynchronous systems with rigid flat morphisms, see [9]. This result has been used in the studies of the functorial synthesis of 1-safe Petri nets with labelled transitions from asynchronous systems, see [4]. There, the need to restrict attention to rigid morphisms proved to be a severe limitation. Here, we show that rigid products are instances of pullback construction in the richer category of asynchronous systems with synchronising morphisms. Since Def. 2 deals with deterministic systems, we should start by generalising it to asynchronous systems first. The following extension simply adds the missing description of the independence relation.

Definition 8. *Consider asynchronous systems* $\mathcal{A}_1$ *and* $\mathcal{A}_2$ *Their* rigid product *is the asynchronous system* $\mathcal{A}$ *defined as follows.*

- $S = S_1 \times S_2$, *with projections* $\pi_i : S \to S_i$, $i = 1, 2$.
- $\hat{s} = \langle \hat{s}_1, \hat{s}_2 \rangle$.
- $A = A_1 \cup A_2$, *with partial projections* $\alpha_i : A \rightharpoonup A_i$, *for* $i = 1, 2$, *given by* $\alpha_i a = a$ *if* $a \in A_i$, *and* $\alpha_i a = \emptyset$ *otherwise.*
- $a \parallel b$ *iff* $\alpha_1 a \parallel_1 \alpha_1 b$ *and* $\alpha_2 a \parallel_2 \alpha_2 b$.
- $p \xrightarrow{a} q$ *in* S *iff* $\pi_1 p \xrightarrow{\alpha_1 a} \pi_1 q$ *in* S_1 *and* $\pi_2 p \xrightarrow{\alpha_2 a} \pi_2 q$ *in* S_2.

By taking REACH$_\sigma(\mathcal{A})$, *the state-reachable subsystem of* $\mathcal{A}$, *one obtains a rigid product in the full subcategory of reachable systems.*

Now, consider $\mathcal{A}_1$ and $\mathcal{A}_2$. Their rigid product $\mathcal{A}$ as described in Def. 8 forces synchronisation of the components on common actions. It is also worth noting that as a result of the composition certain actions which were declared as independent in $\mathcal{A}_1$, say, may become dependent in the result, since they were not independent in $\mathcal{A}_2$.

Now, consider $\mathbf{1}_{A_1 A_2} = (\{\star\}, \star, A_1 \cap A_2, \parallel_\star, T_\star)$ defined as follows.

$$a \parallel_\star b \text{ iff } a \neq b \qquad \text{and} \qquad \star \xrightarrow{a} \star, \quad \text{for all } a \in A_1 \cap A_2$$

Clearly, $\mathbf{1}_{A_1 A_2}$ is an asynchronous system. Moreover, the system can be used to synchronise $\mathcal{A}_1$ and $\mathcal{A}_2$ via morphisms $f_i = \langle !_i, \alpha_i \rangle : \mathcal{A}_i \to \mathbf{1}_{A_1 A_2}$ defined as follows for $i = 1, 2$.

$$!_i s = \star \qquad \text{and} \qquad \alpha_i a = \begin{cases} a & a \in A_1 \cap A_2 \\ \emptyset & a \notin A_1 \cap A_2 \end{cases}$$

Proposition 7. *The rigid product of asynchronous systems* $\mathcal{A}_1$ *and* $\mathcal{A}_2$ *defined in Def. 8 is a pullback of* $f_1 : \mathcal{A}_1 \to \mathbf{1}_{A_1 A_2} \leftarrow \mathcal{A}_2 : f_2$.

6.2 Synchronisation as a Pullback

The synchronisation of elementary transition systems via identification of regions was motivated and introduced in section 2.2. In section 3 we have discussed the apparent affinity of this idea to the idea of pullback. We have also demonstrated that the construction put forward by Bernardinello et al. is not a pullback in any category of transition/asynchronous systems with *flat* morphisms. Nevertheless, this construction turns out to be a pullback in the richer category with synchronising morphisms.

Indeed, consider the following pullback situation

$$f_1 : \mathcal{A}_1 \rightarrow \mathbf{2} \leftarrow \mathcal{A}_2 : f_2 \qquad (10)$$

where $\mathbf{2} = (\{R, \bar{R}\}, R, \{in, out\}, \emptyset, T)$ is the two state elementary transition system described on Fig. 5. The interface system $\mathbf{2}$ is sequential, i.e., the independence relation is empty. Thus, morphisms $f_1 = \langle \sigma_1, \alpha_1 \rangle$ and $f_2 = \langle \sigma_2, \alpha_2 \rangle$ are necessarily flat. Moreover, each of them determines a pair of complementary regions $R_i = \sigma_i^{-1}(R)$ and $\bar{R}_i = \sigma_i^{-1}(\bar{R})$, for $i = 1, 2$.

Proposition 8. *Let the transition systems $\mathcal{S}_1$ and $\mathcal{S}_2$ underlying $\mathcal{A}_1$ and $\mathcal{A}_2$ from (10) be elementary and composable on regions R_1 and R_2, cf. Def. 3. Then, the reachable transition system underlying the pullback of f_1 and f_2 is isomorphic to $\mathcal{S}_1[R, \bar{R}]\mathcal{S}_2$.*

6.3   Future Work

The importance of the method of putting systems together by means of synchronising their activities on common actions has been recognized very early. It is one of the basic constructors in Hoare's CSP, as well as in several process algebras. It was also one of the first composition operations studied by Arnold and Nivat. Nowadays, it is commonly found in the area of DES synthesis and control. As noticed by Morin this operation is the corner-stone of synthesis for a large class of concurrent systems ([9]) and their Petri net realisations ([4]).

The above together with another, seemingly quite different method of putting systems together proposed by Bernardinello et al., have been shown here to be instances of *pullback*, a conceptually simple categorical construction.

This observation prompts several natural generalisations. Firstly, one can use the full power of pullbacks and allow arbitrary interfaces for system synchronisation, not just $\mathbf{2}$ or $\mathbf{1}_{A_1 A_2}$. Secondly, one can use limits more general than pullbacks for system development. For instance, one can consider putting together three components $\mathcal{A}_1$, $\mathcal{A}_2$ and $\mathcal{A}_3$, synchronised via two interfaces $\mathcal{A}'$ and $\mathcal{A}''$ given by $f_i : \mathcal{A}_i \rightarrow \mathcal{A}'$ and $g_j : \mathcal{A}_j \rightarrow \mathcal{A}''$, for $i = 1, 2, 3$ and $j = 1, 2$. The task now would be to find a systems $\mathcal{A}$ which would be a simultaneous synchronisation of $\mathcal{A}_1$, $\mathcal{A}_2$ and $\mathcal{A}_3$ which fulfills the restrictions imposed by the choice of f_i and g_j, for $i = 1, 2, 3$ and $j = 1, 2$, *all* at the same time. This, is clearly possible if limits more general than pullbacks are allowed.

As demonstrated here, one can play the same game with pullbacks, and indeed with arbitrary finite limits, in the much larger class of behaviours of asynchronous systems. We also believe that the results of [3] can be easily extended to cope with the more liberal notion of morphisms of Petri nets envisaged here. Therefore, the resulting category of safe Petri nets will also turn out to be finitely complete. Then, continuity of the case graph functor would guarantee that the case graph of a net constructed as a limit from some finite diagram of nets will be isomorphic to the result of the same pullbacks applied to the case graphs of the respective nets.

In the categorical characterisation of rigid products, cf. Prop. 7, and composition by region identification, cf. Prop. 8, the projections and the morphisms involved in the constructions are flat. Yet, within the classical framework with flat morphisms the pullback construction offers results different than expected, see Fig. 8. This observation seems to offer an interesting insight. Namely, one is tempted to believe that an elegant abstract characterisation of synchronisation can only be achieved when *truly concurrent* semantics of systems is called upon, and the synchronising morphisms are allowed into the picture. Better understanding this facet is also worth further investigations.

References

1. BADOUEL, E., and PH. DARONDEAU. Dualities between nets and automata induced by schizophrenic objects. Proc. 6th Intnl. Conf. CTCS, LNCS **953**, pp.: 24–43, Springer-Verlag, 1995.
2. BEDNARCZYK, M. A. *Categories of asynchronous systems.* PhD Thesis, University of Sussex, 1–88, 1988.
3. BEDNARCZYK, M. A, BORZYSZKOWSKI, A. M. and R. SOMLA. Finite completeness of categories of Petri nets. *Fundamenta Informaticae*, **43**, 1–4, pp.: 21–48, 2000.
4. BEDNARCZYK, M. A., and A. M. BORZYSZKOWSKI. On concurrent realization of reactive systems and their morphisms. H. Ehrig et al. (eds.) *Unifying Petri Nets — Advances in Petri nets*, LNCS **2128**, pp.: 346–379, Springer-Verlag, 2001.
5. BERNARDINELLO, L. Synthesis of net systems. Proc. *Application and Theory of Petri Nets*, LNCS **691**, pp.: 89–105, Springer-Verlag, 1993.
6. BERNARDINELLO, L, FERIGATO, C and L. POMELLO. Towards modular synthesis of EN systems. In B. Caillaud et al. (eds.) *Synthesis and Control of Discrete Event Systems*, 103–113, Kluwer Academic Publishers, 2002.
7. EHRENFEUCHT, A and G. ROZENBERG. Partial (set) 2 structures, I & II. *Acta Informatica*, **27**, 4, pp.: 315–368, 1990.
8. MACLANE, S. *Categories for the Working Mathematician.* Graduate Texts in Mathematics, Springer-Verlag, 1971.
9. MORIN, R. Decompositions of asynchronous systems. In *Proc. CONCUR'98*, LNCS **1466**, pp. 549–564. Springer, 1998.
10. NIELSEN, M., ROZENBERG, G. and P.S. THIAGARAJAN. Elementary transition systems. *Theoretical Computer Science*, **96**, 1, pp.: 3–32, 1992.
11. NIELSEN, M., ROZENBERG, G. and P.S. THIAGARAJAN. Elementary transition systems and refinement. *Acta Informatica*, **29**, pp.: 555–578, 1992.
12. SHIELDS, M. *Deterministic asynchronous automata.* In E. J. Neuhold and G. Chroust (Eds.) *Formal Methods in Programming*, pp. 317–345, North-Holland, 1985.
13. ZIELONKA, W. Notes on finite asynchronous automata. *RAIRO, Informatique Théoretique et Applications*, vol. 21, pp.: 99–135, 1987.

Specification and Validation of the SACI-1 On-Board Computer Using Timed-CSP-Z and Petri Nets

Adnan Sherif, Augusto Sampaio, and Sérgio Cavalcante

Federal University of Pernambuco
Center of Informatics
P.O. BOX 7851 Cidade Universitaria
50740-540 Recife - PE Brazil
{ams,acas,svc}@cin.ufpe.br

Abstract. In this paper we focus on the application of integrated formal methods to the specification and validation of a fault tolerant real-time system (the on-board computer of a Brazilian micro-satellite). The work involves the application of a framework which covers from the formal specification to the analysis and use of tools to prove properties of the system. We used Timed-CSP-Z, a combination of Timed CSP and Z, to specify the system behavior, and then a strategy for converting the specification to TER Nets, a high level Petri Nets based formalism with time. The conversion enables us to use the CABERNET tool to analyse the behavior of the system.

Keywords: Real-time Systems, Case Studies, Time CSP, Z

1 Introduction

The use of real-time systems in safety critical applications makes them a serious candidate for the use o formal methods concerning their specification and validation. During the last few years several formal frameworks for the specification of real-time systems have been developed. Formalisms such as Duration Calculus [2] and Real-time Logic [11] have proved useful to express time constraints of a system. However we need to deal with other aspects such as control and abstract data types, which are more adequately captured using languages such as Z [16]. On the other hand Z does not include time operators and lacks the expressive tools to deal with control (such as parallel composition and synchronization). Any attempt to deal with a real case study will show the need to address all these aspects.

In previous work we have proposed an approach to the specification and validation of real time systems [22,21]. The proposed language (Timed-CSP-Z) extends CSP-Z [6,7] with facilities to specify real-time systems . The extension was achieved by substituting the CSP part of the language by Timed CSP (a version of CSP capable of expressing timed behavior). Concerning analysis, the

W.M.P. van der Aalst and E. Best (Eds.): ICATPN 2003, LNCS 2679, pp. 161–180, 2003.

adopted solution proposed the translation of Timed-CSP-Z to a Petri Net formalism called Timed Environment Relational Nets (TER Nets)[9]. Rules were developed to convert a specification written in Timed-CSP-Z to TER Nets, so that one can use a tool (CABERNET) to analyze the Petri Nets resulting from converting the original specification.

In this paper we focus on the application of the strategy mentioned above to formally specify the On-Board Computer (OBC) of the first Brazilian satellite for scientific applications [20,4] using Timed-CSP-Z. We then show how this specification can be translated into TER Nets, and how the resulting nets can be studied in order to check classical properties such as deadlock freedom using the CABERNET tool. The formal development of a realistic industrial application like the SCAI-1 OBC is worthwhile in its own right (due to the criticality of the application). In the context of this research, more specifically, this effort has been crucial to validate the practical applicability of the proposed approach. This seems to be a differential of our approach when contested with closer related approaches. For example, the work reported in [12,23] has not been applied to our knowledge to any realistic and complete case study.

The remainder of this paper is organized as follows. Section 2 briefly describes the SACI-1 On-Board Computer (OBC). In Section 3 we present the formal specification of the Watchdog Timer process and of the Fault Tolerant Router process (which are part of the SACI-1 OBC). In Section 4 we focus on the approach to specification analysis, where we briefly discuss the strategy to convert a Timed-CSP-Z specification into TER Nets, and the use of the CABERNET tool to conduct the analysis automatically, through an example: the Fault Tolerant Router process.

2   The SACI-1 On-Board Computer

The SACI-1 is the first of a series of low cost micro-satellites produced by INPE (the Brazilian Institute for Space Research). The objective of these satellites is to achieve scientific experiments in gravity near zero environment. Each experiment is designated to a dedicated sub-system, which carries out the acquisition and processing of the relevant data, which needs to be stored before transmission to the ground station for analysis [15].

The OBC is the heart of the satellite; it is responsible for the following activities:

- *Collecting and storing data.* The satellite completes its orbit in about every 52 minutes. The communication with the ground station is possible only for a short period (which lasts about 8 minutes). This time is known as the communication window, and occurs once every 72 hours. Therefore the satellite needs to store the results of the experiments until the window is visible and the data can be transmitted to the ground station.
- *Communication with the ground station.* The Communication is initialized by the ground station as soon as the communication window becomes visible. It is a two-way communication. The satellite sends the data gathered

during the period of time elapsed between communications. It also receives commands from the ground station with instructions that affect the satellite behavior.

– *Controlling the satellite orbit.* The satellite speed and altitude is monitored by the OBC to ensure that the satellite continues in orbit.

Based on the above mentioned activities, the OBC can be observed abstractly as a black box with three external interfaces, each one for a given activity. Figure 1 shows the interfaces of the OBC with the rest of the satellite. Observe from Figure 1 that the interfaces are duplicated. There are two copies of each interface, for if one of them fails the other can still take over.

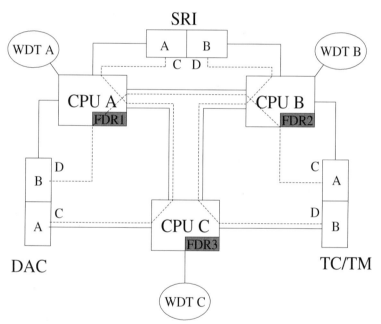

DAC   : Data Acquisition and Control

TC/TM : Telecommand and Telemetry

SRI   : Serial Interface

Fig. 1. Detail Architecture of the OBC.

Since the OBC works unattended and without the possibility of an intervention from the ground station, a fault tolerant mechanism is needed. The mechanism helps to recover from any errors and reconfigures the system if a serious failure occurs. This fault tolerance requirement is achieved by redundancy.

The OBC is composed of a network of three T80 transputers, which act as the processors of the OBC. Each transputer has four serial interfaces used to communicate with the external world. In the design of the OBC we observe that two of the four serial interfaces for each transputer are connected to one of the external interface ports, each on a different interface. For example, CPU_B is connected to the TC/TM interface A, and connected to the SRI interface B, which permits processes running on CPU_B to handle both interface TC/TM and SRI, if needed. This makes each transputer capable of managing two external interfaces. The other two serial interfaces are used to connect each transputer with the other two transputers, and is used for communication among the transputers.

Observe that the dotted lines are alternative connections to be used by the CPUs. If one of them fails the other two are capable of taking over the interfaces (the failed CPU was connected to) without affecting the OBC behavior. Note also that if two CPUs fail the remaining CPU can still take control of the satellite.

An important aspect of the architecture of the SACI-1 is the Watchdog Timers (WDT)s. There are three of them in the OBC, each one designated to a CPU. The objective of the WDT (as its name indicates) is to watch over the CPU to make sure it never engages in a task that will lead it to stop serving the rest of the system.

Each CPU has a Fault Tolerant Router (FTR) process associated with it. The FTR acts as the operating system of the OBC. The main task of this program is to handle communication among the various processes running on the same CPU or on different CPUs. A detailed specification of the FTR can be found in the next section.

The OBC is considered a network of independent active processes. An active process is always ready to act on request, it undergoes its operational cycle, leaving the standby state. When it ends its task it returns to the standby state. Active processes communicate with each other by message exchange, which are routed around the network.

3   Formal Specification of the SACI-1 OBC in Timed-CSP-Z

Timed-CSP-Z is based on CSP-Z [6,7], which is itself a combination of CSP and Z. CSP-Z encapsulates the process description in a specification unit containing two main blocks: a CSP block containing the CSP equations that describe the sequential and concurrent behavior of the events, and a Z block which defines a state and an operation associated with each event of the CSP part. An operation defines the pre-condition for an event to occur and the effect of the event occurrence on the state space. The language also permits combining the behavior of specification units using the CSP operators.

CSP-Z is based on the Failure Divergence model of CSP [18]. The adaptation of the model was proposed by Fischer in [6]. In [14] Mota and Sampaio present a strategy which allows the use of FDR [8] as a model checking tool for CSP-Z.

In this section we show how time aspects are added to CSP-Z in order allow the specification of real-time systems. To achieve this, the CSP part of the specification is substituted by Timed CSP. A new semantic model, Timed Failure Model $\mathcal{M}_{\mathcal{TF}}$ [19], is used as a semantic basis for the proposed language, Timed-CSP-Z.

The syntax of Timed-CSP-Z is not much different from the one of CSP-Z. The original syntax is conserved, as Timed CSP [17,3] is a superset of CSP and the Z part of the language is not effected by this extension.

3.1   Watchdog Timer (WDT)

The task of the WDT is to monitor the FTR and make sure it is functioning properly. The WDT does this by waiting for a reset signal from the FTR within a given amount of time. If the time elapses and the reset signal is not received, the WDT sends an interrupt signal to the observed FTR and waits again for a new reset signal. If it does not receive this signal within a predefined amount of time it will consider the monitored FTR to be out of order. First the WDT will try to recover the FTR by resetting it and then returns to function normally as before. If the FTR does not respond for more than seven consecutive times then it is considered to be failed and so should be cut off from the rest of the system. If this happens then the WDT process itself will stop (behaving like the canonical deadlock process).

A process specification in Timed-CSP-Z starts with the keyword *Spec* followed by the process name.

Spec WDT_i

Observe that the process name WDT is indexed, which indicates that the process is parameterised by this variable. We can parameterise a process for modularity. In our case we will specify WDT_i to represent the behavior of the WDT processes. When defining the complete system we will have three copies of the process, namely WDT_1, WDT_2 and WDT_3; each occurrence of i represents one of the three different CPUs of the satellite.

The next step is to declare the channels used by the process. The channels can be divided into external channels, prefixed by the keyword *channel*, and internal channels, prefixed by the keyword *local channel*. Channels can represent the events in case the channel has no type. If a channel is declared of an appropriate type then the channel is used for communication between processes: the information exchanged will be of the declared type. In the case of our WDT process only external event channels are declared.

channel $resetWDT_i$, $failed_i$, $WDTint_i$, $resetFTR_i$

All Timed-CSP-Z processes as previously mentioned have two main blocks. The first is the specification of the control part (in Timed CSP) and the other is the specification of the data part (in Z). The Timed CSP part starts with the definition of the *main* process. This process is the starting point of the Timed

CSP specification.

$$main_i = (resetWDT_i \rightarrow main_i) \stackrel{WDTPeriod}{\rhd} Interrupt_i$$

$$Interrupt_i = WDTint_i \rightarrow$$

$$(resetWDT_i \rightarrow main_i) \stackrel{WDTIntPeriod}{\rhd} InterruptFailed_i$$

$$InterruptFailed_i = (resetFTR_i \rightarrow main_i) \,\square\, (failed_i \rightarrow STOP)$$

The above specification ensures that the WDT will wait for a reset signal within the time period defined between the start of the process main and the elapse of the amount of *WDTPeriod*. In case the reset is not received then the WDT will send a *WDTint* signal to call the attention of the monitored FTR. It then waits for the reset signal for a new amount of time. If the reset is not received, the WDT sends a *resetFTR* signal re-initialising the FTR or a failed signal indicating the failure of the FTR. The decision of considering the FTR to be failed or not is determined by the Z part of the specification, which begins by defining any constants and related axioms used in the specification of the process. In the case of the WDT process we define *WDTPeriod* and *WDTIntPeriod* as two constants of type natural.

$WDTPeriod : \mathbb{N}$	$WDTIntPeriod : \mathbb{N}$
$WDTPeriod = 100$	$WDTIntPeriod = 100$

Next we define the state affected by each event of the process. This is defined using a Z state declaration. A special schema *Init* is also defined as the initialisation of the process state. The operational view is that it will be executed as soon as the process starts and before any activity is carried out.

State	_Init_
$FailCounter : \mathbb{N}$	$State'$
$FailCounter \leq 7$	$FailCounter' = 0$

Observe that the only variable in the *state* schema is a natural variable which will hold the count of failures at any given instance. It is limited by the restriction that it should hold values less or equal to 7, which is the maximum number of times the CPU can fail. The *Init* schema initialises the *FailCounter* variable with 0. The next step is to define the Z schemas relative to the Timed CSP events. The schema has the same name as the event but prefixed by the keyword com_. Each Z schema will also define the preconditions of the event. It also describes the consequence of the event execution on the data variables of the process state.

```
┌─ com_resetFTR_i ──────────────────┐    ┌─ com_failed_i ─────────────────┐
│ ΔState                            │    │ ΞState                         │
│───────────────────────────────── │    │────────────────────────────────│
│ FailCounter < 7                   │    │ FailCounter = 7                │
│ FailCounter' = FailCounter + 1    │    └────────────────────────────────┘
└───────────────────────────────────┘
```

Observe that $com\_resetFTR_i$ is defined to have the precondition that *Fail-Counter* should be less than 7. The effect of executing this event will be to increment by one the value of the counter.

```
┌─ com_resetWDT_i ─────────────────────────────────────────────────────┐
│ ΔState                                                               │
│──────────────────────────────────────────────────────────────────── │
│ FailCounter' = 0                                                     │
└──────────────────────────────────────────────────────────────────────┘
```

$com\_WDTint_i \mathrel{\widehat{=}} [\,\Xi State\,]$

Observe as well that $com\_resetWDT$ resets the counter to zero. *WDTint* is declared not to affect the state of the process. These events normally represent actions or external interrupts. With this we finish the body of the WDT_i unity.

end Spec WDT$_i$

3.2   Fault Tolerant Router Process (FTR)

This process is responsible for routing the messages between the application processes running on the three satellite CPUs. Before we go further in the specification of our case study we define some data types needed in the remainder of the specification. The specification makes use of a data type representing the block of messages exchanged between the FTR and the other processes. The data type *OBCMessage* is defined globally (that is, not inside a specification unit), so that the data type can be used by all the other process specifications. The following is the definition of some data types using Z.

We start by introducing a data type DATA to hold the data carried by the message. Next we introduce a type LENGTH which represents the length of the data block. Observe that one advantage of using Z is particularly the possibility to define data types and operations on these types without giving details of the implementation of the type. Such types are called *given sets*. We also introduce the free (enumerated) *BOOL* with constants *True* and *False*.

[DATA]
[LENGTH]
BOOL ::= *True* | *False*

Next we define another enumerated type to represent the names of the processes and yet another to represent the Functions or commands exchanged in the messages. A complete list of these sets and a detailed description of its significance can be found in [21].

$$PROCESS ::= TC \mid TM \mid SGC \mid HK \mid D \mid EDA \mid EDAC \mid SDC \mid ADC \mid$$
$$DAC \mid SRI \mid EMM \mid NONE$$
$$FUNCTION ::= UpdateOBCClock \mid TransmitTM \mid UpdateExpTable \mid ...$$

We define the *OBCMessage* as being formed of two state components. The first is of the type *TControl* and holds the information regarding the sender and the receiver of the message. The second holds the message itself, an element of the type *TComm*, which is composed of the function representing the action the sender requests the receiver to take, the data regarding the operation, and the length of the data block. These are defined as Z schemas which are similar to records.

```
┌─ TControl ─────────────        ┌─ TComm ──────────────
│  Sender : PROCESS              │  Action : FUNCTION
│  Reciever : PROCESS            │  Data : DATA
                                 │  Length : LENGTH
```

```
┌─ OBCMessage ──────────────────────────────
│  Control : TControl
│  Message : TComm
```

The specification of the FTR starts with the declaration of the channels used by the process. Note that the process has typed channels indicating the type of data to be transmitted over the channel. The channel direction is determined by the Z part of the specification.

Spec FTR_i
   Channel $getMsg_i$:[x:OBCMessage]
   Channel $sendMsg_i$:[x:OBCMessage]
   Channel $WDTint_i$, $resetWDT_i$, $resetFTR_i$, $OtherInt_i$
   Channel $failed_i$, $failed_{(i\oplus3)+1}$, $failed_{((i+1)\oplus3)+1}$
   local channel $ProcessInt_i$, $ConfigureFTR_i$

Next we define the Timed CSP part of the specification. The main process starts by configuring the process location table according to the current state of the CPUs. After configuration the FTR process starts its normal activity which is to manage the traffic of messages among the system application processes.

Observe, however, that the IO operation can be interrupted by an incoming interrupt, which can be one of the following events:

$\quad\quad$ WDTInt$_i$ $\quad\quad\quad$ failed$_{(i\oplus3)+1}$ $\quad\quad\quad$ failed$_{((i+1)\oplus3)+1}$ $\quad\quad\quad$ Otherint$_i$

The operator $\oplus$ represents a modular summation operator. The occurrence of any of the above mentioned interrupts will suspend the normal activities of the FTR. This is modeled with the help of the untimed CSP interrupt.

$main = ConfigureFTR_i \rightarrow OPERATION$
$OPERATION = (HANDEL\_IO_i \ \hat{} \ INTHANDLER_i)$
$HANDEL\_IO_i = getMsg_i ?x \rightarrow sendMsg_i !x \rightarrow resetWDT_i \rightarrow HANDEL\_IO_i$
$INTHANDLER_i = (WDTInt_i \rightarrow ((resetWDT_i \rightarrow OPERATION)$
$\square (resetFTR_i \rightarrow main)$
$\square (failed_i \rightarrow STOP)))$
$\square (failed_{(i \oplus 3)+1} \rightarrow main)$
$\square (failed_{((i+1) \oplus 3)+1} \rightarrow main)$
$\square (OtherInt_i \rightarrow ProcessInt_i \rightarrow resetWDT_i \rightarrow OPERATION)$

The following step is to define the state space of the process. First we define two enumerated types and a type synonym (abbreviation) that will make our specification more readable.

$PROCSTATUS ::= NOTDEFINED \mid READY \mid BUSY$
$CPUID ::= 1..3$
$ROM\_CONFIG == (CPUID \rightarrow BOOL) \rightarrow (PROCESS \rightarrow CPUID)$

The type $PROCSTATUS$ represents the possible states of a process. The states are $NOTDEFINED$ at initialization, $READY$ when the process is ready to receive a message from another process, and $BUSY$ when it is engaged in a communication. The state of the process affects its behavior. The main reason for using the process status flag (PROCSTATUS) is to ensure that a process will start to communicate with another only if both parts are ready to communicate.

$CPUID$ is used to indicate the current CPU; it ranges from 1 to 3 as there are three CPUs composing the OBC. Finally, the ROM_CONFIG constant (as its name suggests) represents the predefined combinations for the distribution of the application processes among the satellite CPUs. These combinations are represented as a function. The processors status are represented by ($CPUID \rightarrow BOOL$) indicating which CPUs are functioning properly ($true$) and which CPUs are to be considered failed ($false$). The function ROM_CONFIG returns a valid distribution of the application processes over the CPUs. The distribution of the application processes among the CPUs is presented as a table modeled as a function from process identifier to CPU identifier.

___ State _____
| $Buffer : OBCMessage$
| $ProcStatus : PROCESS \rightarrow PROCSTATUS$
| $ProcLocation : PROCESS \rightarrow CPUID$
| $CPUOk : CPUID \rightarrow BOOL$
| $CurrentCPU : CPUID$
|_____
| $CurrentCPU = i$

```
┌─ Init ──────────────────────────────────────────────
│ State'
├──────────────────────────────────────────────────────
│ Buffer' = EmptyMsg
│ ∀ p : PROCESS •
│ ProcStatus(p)' = NOTDEFINED
│ ∀ p : PROCESS •
│ ProcLocation(p)' = CPU'
│ ∀ c : CPUID • CPUFailed(c)' = True
│ CPU' = i
└──────────────────────────────────────────────────────
```

The State schema contains the following elements. *Buffer* is a local storage area used to temporarily store the message before transmitting it to the destination process. *ProcStatus* is a function used to lookup the status of the application process. *ProcLocation* is used to hold the current configuration showing the distribution of the application processes over the CPUs. *CPUOk* is a function which represents the status of the CPUs. Given the CPU identification the function returns *true* to indicate that the CPU did not fail or *false* to indicate that the given CPU was considered failed by its application processes. The component *CurrentCPU* simply refers to the identification number of the current CPU and is equal to the process parameter i. The schema Init states that the buffer is initially empty.

The schema $com\_ConfigureFTR_i$ is used to define the process configuration. This is done by loading the process location table with the appropriate *rom* configuration according to the current active CPUs. The schema also initializes the process status table by setting all the process status to UNDEFINED.

```
┌─ com_ConfigureFTR_i ──────────────────────────────────
│ ΔState
├──────────────────────────────────────────────────────
│ ProcLocation' = ROM_CONFIG(CPUFailed)
│ Buffer' = EmptyMsg
│ ∀ p : PROCESS • ProcStatus(p)' = NOTDEFINED
│ CPUFailed' = CPUFailed
│ CPU' = CPU
└──────────────────────────────────────────────────────
```

Next the Z schema for the *getMsg* and *SendMsg* is presented. Observe that these events are declared as typed channels and so have input or output data according to the variable declaration in the Z schema. The same variable name is used in the corresponding Z schema. In the case of *getMsg* the variable x is redefined in the Z schema as x? indicating that the channel and the corresponding Z schema will receive a data input. In the case of *SendMsg* the output variable x is redefined in the Z schema as x! indicating that the channel and the corresponding Z schema will produce a data output. Observe as well that a channel can only have one direction within the same process. That is, if it is used for input it cannot be used for output within the same specification unit, and vice versa.

$$
\begin{array}{l}
\rule{0.5em}{0pt}\textit{com\_getMsg}_i \rule{8em}{0.4pt} \\
\Delta State \\
x? : OBCMessage \\
\hline
ProcLocation(x?.Control.Sender) \\
\quad = CPU \\
Buffer' = x? \\
ProcLocation' = ProcLocation \\
ProcStatus' = ProcStatus \\
CPUOk' = CPUOk \\
CurrentCPU' = CurrentCPU
\end{array}
\qquad
\begin{array}{l}
\rule{0.5em}{0pt}\textit{com\_sendMsg}_i \rule{8em}{0.4pt} \\
\Delta State \\
x! : OBCMessage \\
\hline
x! = Buffer \\
Buffer' = EmptyMsg \\
ProcLocation' = ProcLocation \\
ProcStatus' = ProcStatus \\
CPUOk' = CPUOk \\
CurrentCPU' = CurrentCPU
\end{array}
$$

EmptyMsg is a constant representing an empty buffer state. The other events have no effect on the state of the variables observed by the process and are omitted for lack of space.

end Spec FTR_i

3.3 The SACI-1 OBC

Here we present the top level specification of the OBC system. To achieve this we use the CSP operators to combine and join specification units. we start with the CPU processes. Each of the three CPUs of the on-board computer is formed of the parallel composition of an FTR, a WDT and the application processes (APP). Observe that each of the application processes is a separate specification unit. In [21] a detailed specification of each of these specification units is presented.

$$
\begin{aligned}
APP_i = {}& TC_i \| TM_i \| SGC_i \| EDAC_i \| EDA_i \| SRI_i \| SDC_i \| ADC_i \| DAC_i \\
& \| HK_i \| D_i \| EMM_i \\
CPU'_i = {}& FTR_i \underset{\{resetWDT_i,\, resetFTR_i,\, WDTint_i,\, failed_i\}}{\|} WDT_i \\
CPU_i = {}& APP_i \underset{\{getMsg\_i,\, putMsg\_i\}}{\|} CPU'_i
\end{aligned}
$$

The set of events used as subscripts of the parallel operator describes the points of synchronization of the processes running in parallel.

Before we introduce the final equation that composes the three CPUs to form the OBC process we define two sets that will be used in the equation.

$$
\begin{aligned}
INTERFACE_{12} &= \{failed_1, failed_2\} \\
INTERFACE_{123} &= INTERFACE_{12} \cup \{failed_3\}
\end{aligned}
$$

Finally we define the equation that defines the complete on-board computer subsystem of the SACI-1.

$$OBC_i = \underset{INTERFACE_{12}}{(CPU_1 \parallel CPU_2)} \underset{INTERFACE_{123}}{\parallel CPU_3}$$

4   The Specification Analysis

As mentioned before no analysis tools exist for Timed-CSP-Z to help in model checking. The language that gave origin to Timed-CSP-Z, CSP-Z, has no tools either. In [14] a strategy was introduced which allows the use of FDR (a tool used in the analysis of CSP processes) to the analysis of CSP-Z processes. Unfortunately, the same method cannot be adapted for Timed-CSP-Z as the observation model of Timed-CSP-Z is different from that of CSP and, currently, FDR does not offer any analysis based on the timed models of CSP.

Based on recent work in the literature [13,1] related to process algebra, we have adopted the solution of translating Timed-CSP-Z to a Petri Net formalism called Timed Environment Relational Nets (TER Nets)[9] for which there exists an analysis tool (CABERNET). TER Nets consist of a high-level Petri Net where tokens are environments, i.e., functions associating values to variables. Furthermore, an action is associated with each transition describing which input tokens can participate in a transition firing and the possible tokens produced by the transition.

The conversion rules for CSP are taken from previou work related to converting CSP to Petri Nets [13,1]. We have extended these conversion rules to meet CSP-Z specifications where a state space has been added. We then proposed new rules to cover the Timed-CSP-Z notation. A complete description of these rules can be found in [21,22]. To better understand the principals behind the conversion mechanism we introduce the basic element rule and a short example.

Basic Elements Rule

The basic element of a Timed-CSP-Z specification is an event, say e, in the CSP part of the specification, and a corresponding Z schema representing the Z part of the specification. Each Z schema predicate is divided into two parts: the precondition for the event to occur (which depend on the value of input variables and current value of the state space of the specification) and the post-condition or the effect of the event occurrence on the state space. The translation of this basic element is shown in Figure 2, where $S1$ represents the input place to the basic element and holds the data space before the event occurrence, and $S2$ is the output place which holds the resulting state of the data space generated by the event occurrence. The event itself is represented by the transition that connects the input places to the output places. The transition firing represents the event occurrence.

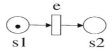

Fig. 2. The PN Representation of an Event e

To better illustrate the rule consider the following portion of a specification:

Spec ex1
    channel e
    main = e → main

```
┌─ State ──────────────────
│ x : Int
│
└──────────────────────────

┌─ Init ───────────────────
│ State'
│
│ x' = 0
└──────────────────────────
```

```
┌─ com_e ──────────────────────────────
│ ΔState
│
│ x < 10
│ x' = x + 1
└──────────────────────────────────────
```

end Spec

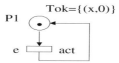

Fig. 3. An Example of Converting Timed-CSP-Z to TER Nets

The example above can be translated to the TER Net shown in Figure 3. The initial token in place P_1 represents the starting point of the specification. The *Init* schema is represented in the initialisation value of the token. The value of the token *tok* at place P_1 represents the state space before the occurrence of the event e. The resulting token represents the result of the occurrence of the event e. The event is represented by the transition e in the Petri Net. The action associated to the transition is defined from the Z schema associated to the corresponding CSP event. The action is defined by specifying the sequence of input places to the transition and the sequence of output places to the transition. Then the action behaviour is defined using a logical predicate to state the conditions on the input places and the result of the transition firing on the tokens produced in the output place.

act_e = $< P_1, P'_1 >| P_1.x < 10$ and $P'_1.x = P_1.x + 1$

All arcs have weight 1 meaning that for each transition firing there will be only one resulting token for each place.

Concerning how the concept of time is represented in TER Nets, each environment contains a variable, called chronos, whose value is of numerical type, representing the time stamp of the token. For example, time stamps would assume natural number values, when dealing with discrete time models; on the other hand they assume real number values in order to deal with a continuous time model. The actions associated with the transitions are responsible for producing timestamps for the tokens that are inserted into the output places based on the values of the environments of the chosen input enabling tuples. The basic idea is that the timestamp represents the time when the token was produced. In order to capture the intuitive concept of time, however, chronos cannot be treated as an unconstrained variable and some axioms must be satisfied. For details of ER Nets and Time in TER Nets see [9,10]. Our main aim is to emphasize how the analysis of our case study has been conducted.

An Analysis Example

We again study the WDT and FTR processes introduced in the last section after been converted to TER based on the conversion rules shown in [21,22]. After reducing the original TER Net using Petri Net reduction rules and by not considering transitions that either are not time critical or do not contribute to our time analysis, we have the resulting TER Net that can be observed in Figure 4.

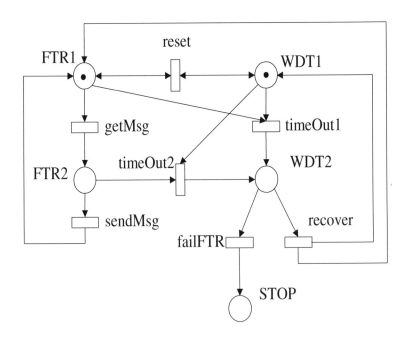

Fig. 4. The FTR and the WDT Processes

We consider that the data space of the net is empty as it can be seen in the declaration of the type *FTR_Token* below. The type *WDT_Token* is reduced to the *failCounter* entry which counts the number of times the FTR process failed to meet its timing requirements.

$$FTR\_Token =<>$$
$$WDT\_Token =<< failCounter, 0 >>$$

The net will act on two state spaces: that of the FTR process, which in this case is empty, and the other is the WDT process state space. The transitions *getMsg* and *sendMsg* simulate the FTR communication process. The actions associated with these transitions are:

$$act\_getMsg =<< FTR1 >, < FTR2 >>|$$
$$FTR2.chronos = FTR1.chronos + StartCommDelay$$
$$act\_sendMsg =<< FTR2 >, < FTR1 >>|$$
$$FTR1.chronos = FTR2.chronos + CommTime$$

From the actions associated with the transitions, observe that the only effect of any transition firing is on the time variable chronos. The time value *StartCommDelay* indicates the time before the FTR process starts a communication operation whereas the time variable *CommTime* contains the amount of time needed for the FTR process to conclude a communication operation.

The *reset* transition is used to reset the WDT process. The result of firing this transition is to set the *failCounter* state variable to zero. The precondition for this transition to fire is that the firing time of the transition should be less than or equal to the time it was armed and the passage of the amount of time indicated by *WDTPeriod*. This can be observed in the following action associated to the transition.

$$act\_reset =<< FTR1, WDT1 >, < FTR1, WDT1 >>|$$
$$FTR1' = FTR1$$
$$\text{and } WDT1'.failcounter = 0$$
$$\text{and } WDT1'.chronos \leq WDT1.chronos + WDTPeriod$$

The transitions *timeout1* and *timeout2* are used to suspend the transition *getMsg* and *sendMsg*, respectively. They impose the condition that if the FTR does not reset the WDT before the time defined by *WDTPeriod* elapses from the last reset then the FTR will be considered failed and therefore should be checked by the WDT to see if it can continue operating. The actions that determine the behavior of the transitions are shown below.

$$act\_timeout1 =<< FTR1, WDT1 >, < WDT2 >>|$$
$$WDT2.failcounter = WDT1.failcounter + 1$$
$$\text{and } WDT2.chronos > WDT1.chronos + WDTPeriod$$

$$act\_timeout2 \; =<< FTR2, WDT1 >, < WDT2 >>|$$
$$WDT2.failcounter = WDT1.failcounter + 1$$
$$and \; WDT2.chronos > WDT1.chronos + WDTPeriod$$

The transition *recover* is fired by the WDT to determine that the FTR should not be considered failed and should continue its operation normally. As previously specified, the WDT considers the FTR to be failed if it does not meet its timing requirements for seven consecutive times. In the action below we can observe clearly the precondition imposed that satisfies this requirement. Observe that the precondition is imposed using a variable which permits us to determine the number of times the FTR process can fail.

$$act\_recover \; =<< WDT2 >, < FTR1, WDT1 >>|$$
$$WDT2.failcounter < 7$$
$$and \; WDT1 = WDT2$$
$$and \; FTR1 = WDT2$$

The action associated to the *failFTR* transition is used to terminate the process when it reaches the maximum number of consecutive failures indicated by the *MaxNumberAttempts*. This is done by sending the active token to a sink place *STOP*.

$$act\_failFTR \; =<< WDT2 >, < STOP >>|$$
$$WDT2.failCounter = MaxNumberAttempts$$
$$and \; STOP = WDT2$$

Consider the following values for the variables used in the specification:

$$StartCommDelay = 1$$
$$CommTime = 1$$
$$WDTDelay = 5$$
$$MaxNumberOfAtempts = 2$$
$$TimePeriod = 15$$

When executing the net using the CABERNET tool to generate the TRT, a large amount of states is generated. For this reason we have limited to show a part of the graph representing the TRT generated by a deadlock analysis of the FTR process. The graph shown in Figure 5(a) represents the worst case of the FTR process functionality where it always engages in a communication operation and never resets the WDT. By studying the generated graph we observe that the states $S3, S8, S18, S23, S33, S38$ all represent the firing of the transition *Task* while the states $S5, S10, S20, S25, S35, S40$ represent the *TaskDone* transition firing. Observe that all substates in the graph which occur after the mentioned states have the possibility of executing the reset and so eliminate the case of a deadlock state. This functionality is expected to occur as the FTR process functions normally but if it does not respond to the WDT, then it is considered failed and then should be cut out of the rest of the system.

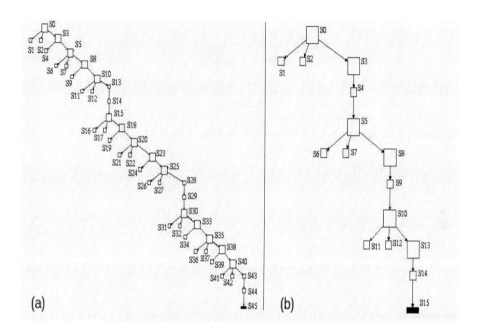

Fig. 5. TRT of The FTR Process

Now consider the case where the timing parameters are as follows:

$$StartCommDelay = 2$$
$$CommTime = 5$$
$$WDTDelay = 3$$
$$MaxNumberOfAtempts = 2$$
$$TimePeriod = 15$$

The TRT (Time Reachable Tree) generated by CABERNET for the above parameters can be seen in Figure 5(b). Note that in this case the process is never capable of executing its principal task represented by the transitions *getMsg* and *sendMsg*. This is because the event *TimeOut2* is enabled to fire before the transition *sendMsg* fires. This type of trace is undesired in the behavior of the SACI-1 OBC as it will lead the CPU to fail immediately and if it happens on all three CPUs then the OBC will fail and the satellite will be lost.

5   Conclusions

In this paper we have concentrated on the application of an approach to the specification and validation of an industial real-time system: the micro satellite SACI-1 On-Board Computer (OBC). The proposed approach suggests the use of Timed-CSP-Z for specification and a Petri Net based formalism for analysis (specification validation). Timed-CSP-Z allows us to capture the several facets of a real-time system in an integrated way: the control flow, possibly concurrent and distributed (in CSP), the data space in terms of abstract data types (in Z) and the time aspects (in Timed-CSP). The case study was essential to consolidate the proposed approach to specification and validation of real-time systems.

The application of formal methods to a real and complex system offers various potential benefits. Apart from validation the proposed approach, it can be used as a precise documentation of the functionality of critical parts of the satellite. It can also be used as a reference for the support group of INPE during the maintenance phase of the project, or the development of new projects based on the SACI-1. The formal description is a solid basis for the verification of the satisfaction of desired properties and the absence of undesirable properties in the system specification, as illustrated in Section 4. The TER Net model, through the CABERNET environment, can be used as a simulation tool to study the effect of the time parameters. As discussed in Section 4, deadlock is very sensitive to the variation of time parameters.

Nevertheless, it is important to justify the use of different formalisms for specification and analysis, since this has required a conversion between them. The use of Petri Nets for analysis was a consequence of the lack of tools to validate the Timed-CSP-Z specification. In this way we could benefit from both formalisms; we developed an initial abstract and structured model of the system (in Timed-CSP-Z) and then systematically translated it (following well-defined rules) into a Petri Net based formalism adequate for mechanical analysis.

There are several similar efforts reported in the literature to combine formalisms. For example, RTL-Z [5] combines Real Time Logic (RTL) and Z,RT-Z [23] combines Timed CSP and Z. Perhaps the closest work to ours is the one carried out by Dong and Mahony in [12] where they introduce a new language based on the combination of Timed CSP and Object-Z. The new language is named Timed Communicating Object-Z. In this approach, objects are modelled as processes with the Object-Z objects, predicates and operations based on the CSP syntax. This has the disadvantage of not having a clear separation between the different aspects of the process, where the behavior of the process is incorporated into the data state of the system, while in our approach the control state is totally independent of the data state and can be studied separately; this makes the specification more readable as well. Furthermore, to our knowledge, no significant specification of a case study has been developed using Timed Communication Object-Z. Concerning analysis, no approach has been suggested.

The main topic for further research is the formalisation of the conversion rules from Timed-CSP-Z to TER Nets.

References

1. G. Balbo. Performance Issues in Parallel Programming. In *13th International Conference on Application and Theory pf Petri Nets*, volume 616, pages 1–23, Sheffield, UK, 1992. Lecture Notes in Computer Science.

2. Zhou Chaochen, C. A. R. Hoare, and Anders P. Ravn. A calculus of durations. *Information Processing Letters*, 40(5):269–276, 1991.

3. J. Davies and S. Schneider. A brief history of timed csp. *Theoretical Computer Science*, 138(2):243–271, 1995.

4. A. R. de Paula Jr. Fault Tolerance Aspects of the SACI-1. *VI Simpósio de Computadores Tolerantes a Falhas*, 1995.

5. C.J. Finge. Specification and verification of real-time behaviors using z and rtl. *Lecture Notes in Computer Science*, 571:393–409, 1992.

6. C. Fischer. Combining CSP and Z. Technical report, University of Oldenburg, 1996.

7. C. Fischer. *Combination and implementation of processes and data: from csp-oz to java*. PhD thesis, University of Oldenburg, 2000.

8. Formal Systems (Europe) Ltd. *FDR: User Manual and Tutorial, version 2.01*, August 1996.

9. C. Ghezzi, D. Mandrioli, S. Morasca, and M. Pezze. A Unified High-level Petri Net Formalism for Time-Critical Systems. *IEEE Transactions on Software Engineering*, 17(2):160–172, 1991.

10. C. Ghezzi and M. Pezze. Cabernet: an environment for the specification and verification of real-time systems. In *In Proceedings of 1992 DECUS Europe Symposium, Cannes (F)*, 1992.

11. F. Jahanian, A. K. Mok, and D. A. Stuart. Formal specification of real-time systems. Technical Report TR-88-25, Department of Computer Science, University of Texas at Austin, June 1988.

12. B. Mahony and J. Song Dong. Blending Object-Z and Timed CSP: An introduction to TCOZ. In *Proceedings of the 1998 Internaltional Conference on Software Engineering*, pages 95–104, 1998.

13. A. Mazzeo, N. Mazzocca, S. Russo, C. Savy, and V. Vittorini. Formal Specification of Concurrent Systems: A Structured Approach. *The Computer Journal*, 41(3):145–162, 1998.

14. A. Mota and A. Sampaio. Model-Checking CSP-Z, Strategy, Tool Support and Industrial Application. *Science of Computer Programing*, 40(1):59–96, 2001.

15. J. A. C. F. Neri. SACI-1: A Cost-Effective Microssatellite Bus for Multiple Mission Payloads. Technical report, Instituto Nacional de Pesquisas Espaciais - INPE, 1995.

16. B. Potter, J. Sinclair, and D. Till. *An Introduction to Formal Specification and Z*. Prentice-Hall, 1991.

17. G. M. Reed and A. W. Roscoe. A timed model for communication sequential processes. In *Proceedings of ICALP '86*, volume 226. Lecture Notes in Computer Science, 1986.

18. A. W. Roscoe. An alternative order for the failures model. *Journal of Logic and Computation*, 2(5):557–578, 1992.

19. A. W. Roscoe. *The Theory and Practice of Concurrency*. Prentice-Hall International, 1998.

20. M. E. Saturno and J. B. Neto. Software Requirement Specification for the OBC/SACI-1 Application Programs. Technical report, Instituto Nacional de Pesquisas Espaciais - INPE, 1995.

21. A. Sherif. Formal Specification and Validation of Real-Time Systems. Master's thesis, Centro de Informática, UFPE, 2000. http://www.di.ufpe.br/~ams/tese.ps.gz.
22. A. Sherif, A. Sampaio, and S. Cavalcante. An Integrated Approach to Specification and Validation of Real-Time Systems. In *Proceedings of Formal Methods Europe 2001: Formal Methods for Increasing Software Productivity, Lecture Notes in Computer Science*, volume 2021, pages 278–299. Springer, 2001.
23. C. Suhl. RT-Z: An Integration of Z and timed CSP. In *Proceedings of the 1st Internaltional Conference on Integrated Formal Methods*, 1999.

On the Use of Petri Nets for the Computation of Completion Time Distribution for Short TCP Transfers

R. Gaeta, M. Gribaudo, D. Manini, and M. Sereno

Dipartimento di Informatica, Università di Torino,
Corso Svizzera, 185, 10149 Torino, Italia

Abstract. In this paper we describe how the completion time distribution for short TCP connections can be computed using Deterministic Stochastic Petri Net (DSPN) models of TCP protocol. A DSPN model of TCP is a representation of the finite state machine description of the TCP transmitter behavior, and provides an accurate description of the TCP dynamics. The DSPN requires as input only the packet loss probability, and the average round trip time for the TCP connections being considered. The proposed model has been validated by comparing it against simulation results in various network scenarios, thus proving that the model is accurate. Numerical results are presented to prove the flexibility and the potentialities of the proposed methodology.

Keywords: Performance Evaluation, Communication Network Planning, Transport Control Protocol, Deterministic Stochastic Petri Nets, Completion Time Distribution

1 Introduction

The performance in the Internet network is mainly dominated by physical network parameters, the IP routing strategies, the incoming traffic parameters, as well as all the aspects related with the used protocols.

In particular, the Quality of Service (QoS) perceived by end users in their access to Internet services is often driven by TCP, the connection oriented transport protocol, whose congestion control algorithm dictates the latency of information transfer.

TCP has been subject to numerous performance studies based on simulations, measurements, and analytical models. Modeling the TCP behavior using analytical paradigms is the key to obtain more general and parametric results and to achieve a better understanding of the TCP behavior under different operating conditions. At the same time, the development of accurate models of TCP is difficult, because of the intrinsic complexity of the protocol algorithms, and because of the complex interactions between TCP and the underlying IP network.

Although numerous papers proposed models and/or methods to obtain the *expected* throughput rates or transfer time, the variability of QoS requirements

W.M.P. van der Aalst and E. Best (Eds.): ICATPN 2003, LNCS 2679, pp. 181–200, 2003.

for different types of services and the need to provide *Service Level Agreements (SLA)* to end users, require the computation of more sophisticated performance metrics, such as completion time distributions time for TCP connections.

In this paper we show how transfer time distributions can be computed (and quantiles can be derived) using Deterministic Stochastic Petri Nets (DSPN). The DSPN model we developed requires two input parameters: the *packet loss probability*, and the *average round trip time*.

The packet loss probability is the probability that a TCP packet is dropped by one of the IP routers along the path from the TCP transmitter to the TCP receiver. Buffers in IP router have large but finite capacity therefore a packet is discarded (lost) either if an IP router has exhausted its buffering capacity for an incoming packet or if the buffer implements an Active Queue Management (AQM) algorithm where early packet drops are used to prevent network congestion (see [7] for an example of AQM scheme).

The round trip time (RTT) is a measure of the time it takes for a packet to travel from the source to the destination plus the time required for the acknowledgement packet to reach the source. The total time it takes for a packet to reach the destination (D) is the sum of three main components, i.e., $D = D_q + D_p + D_t$ where:

- D_q is the queuing delay; at each IP router it has to go through, a packet experiences a queuing delay as it waits its turn to be transmitted onto the proper link.
- D_p is the propagation delay; once a bit is transmitted onto the link, it needs to propagate to its destination. The time required to propagate depends on the link physical medium type and on the link length.
- D_t is the transmission delay; all bytes of a packet must be transmitted onto the link. The time required to push all bytes of a packet on to the link depends on the packet size and on the link capacity.

The two parameters required by the DSPN model, i.e., packet loss probability, and average round trip time, can be obtained in three different ways; interesting relations exist between the method used for deriving packet loss probability and average round trip time and the possible utilization of the proposed DSPN model:

- they can be measured over an actual network or an experimental setup; in this case, completion time distributions can also be measured, although this is significantly more complex, but the advantage of a model like the one described in this paper lies in the possibility of performing a *what-if* analysis, e.g., to assess the effectiveness of network upgrades.
- they can be derived from simulation experiments; in this case, completion time distribution can be also estimated from the same simulation experiments, but the CPU times needed to obtain reliable estimates for distributions from simulation experiments could be exceedingly high. The DSPN model of TCP thus allows simulation to be used only to obtain parameters (packet loss probability, and average round trip time) that are known to

be relatively easy to accurately estimate, and then to exploit the model to obtain completion time distributions.
- they can be estimated with an analytical model of the underlying IP network. In this case, the DSPN model of TCP can be used to compute completion time distributions.

The balance of this paper is outlined as follows: Section 2 provides a short overview of previous works that are closely related to our proposal. Section 3 presents a short description of TCP. In Section 4 we present our modeling assumptions and the description of the DSPN model of TCP. In Section 5 we present numerical results to validate the model and for providing some examples of the possible uses of the proposed methodology. Finally, Section 6 contains some concluding remarks, and outlines possible directions for future developments.

2   Related Works

The literature on analytical models of TCP is vast therefore it is difficult to provide here a comprehensive summary of previous contributions. In this section we only summarize some of the approaches that have been successfully used so far in this field and that are most closely related to our work.

One of the first approaches to the computation of the latency of short file transfers is presented in [3]. In this paper, an analytical model is developed to include connection establishment and slow start to characterize the data transfer latency as a function of transfer size, average round trip time and packet loss rate.

Works in [4] [1] too cope with finite TCP connections; their peculiarity is the analysis of connections that exhibit a on-off behavior, following a Markov model. In these works, the description of the protocol behavior is decoupled from the description of the network behavior and the interaction between the two sub-models is handled by iterating the solution of the submodels until the complete model solution converges according a fixed point algorithm. In particular, [1] uses a GSPN based approach for developing the TCP behavior models.

The approach presented in [14] is based on the use of queueing network models for modeling the behavior of the window protocol. This approach has been extended in [8]. These works are based on the description of the protocol using a queueing network model that allows the estimation of the load offered to the IP network (represented using a different queueing network model). The merit of the queueing network approach lies in the product form solution that these models exhibit thus leading to more efficient computation of the interesting performance indexes.

On the other hand the approach presented in [12], allows the computation of the completion time distribution by using an analysis based on the possible paths followed by a TCP connections. The main problem of this approach is that the number of possible paths grows according to an exponential law with respect to the number of packets to be transferred over the connection. To overcome this problem an approximate method has been proposed. The approach presented in

[12] allows to account for independent losses, but other type of losses cannot be easily considered.

The use of Petri nets for performance evaluation of TCP and/or of issues related with this protocol is not new. In particular, in [16] a model of TCP, based on the Infinite-State Stochastic Petri Nets, is used to show the impact of lower-layer protocols (TCP and its windowing mechanisms) on the perceived performance at the application level (HTTP). In [5] a TCP model, based on Colored Petri Nets, is used to compare the behavior of two different TCP versions (Reno and Tahoe) when they face different packet loss situations, and to evaluate the impact of packet loss on the throughput. In [24] a TCP model, based on Extended Fuzzy-Timing Petri Nets, is used to evaluate the impact of QoS parameters on virtual reality systems.

To appreciate the contribution of the present paper is intersting to observe that the analysis presented in [5] and [24] are performed by simulation. On the other hand, the models presented in [1,5,16] allow to obtain first order measures such as throughput, average buffer occupation, and packet loss probability. It is important to point out that in network planning activities, the variability of QoS requirements for different types of services and the need to provide SLA to end users, require the computation of more sophisticated performance metrics, such as completion time distribution for TCP connections.

The contribution of the present paper is a DSNP model that allows the computation of completion time distribution for TCP connections. In general the computation of completion time distribution by means of simulation requires CPU times that could be exceedingly high. The complexity of the solution of the DSPN model is quite limited (see Table 5). Therefore the presented model can be effectively used in the context of planning and performance evaluation of IP networks.

3   The Transport Control Protocol: A Short Description

In this paper we mainly focus the attention on the TCP variant called *TCP Tahoe*. This version of TCP has been the first modification of the early implementation of TCP [19] and was proposed at time when Internet suffered from congestion. The TCP Tahoe implementation added a number of new algorithms and refinements. The new algorithms include: Slow Start, Congestion Avoidance, and Fast Retransmit [11]. The refinements include a modification to the round trip time estimator used to set retransmission timeout values [11]. The three algorithms provide the core algorithm of congestion control of TCP which is based on additive-increase/multiple-decrease principles. The congestion control of TCP gets the level of congestion from the arriving data-acknowledgement packets (ACK). When ACKs do not arrive at the sender, the TCP source waits for a timeout period: when new data is acknowledged by the TCP receiver, the congestion window (cwd) is increased (additive-increase); when congestion occur (congestion is indicated by packet loss detected by a retransmission timeout or the reception of three duplicate ACKs) one-half (multiple-decrease) of the current window size is saved in the slow start threshold (a congestion estimation value), and additionally the *cwd* is set to one packet. The technique to manage timeout is called *timer backoff* strategy. This technique computes an initial timeout value using the estimated round trip time. However, if the timer expires and causes a retransmission, TCP increases the timeout. In fact, each time it must retransmit a packet, TCP increases the timeout. To keep timeouts from becoming ridiculously long, most implementations limit increases to an upper bound. There are a variety of techniques to compute backoff. Most choose a multiplicative factor γ and set the new value equal to $= \gamma \cdot$ timeout. Typically, γ is equal to 2. When the TCP transmitter does not have yet an estimation of the round trip time (in case of the transmission of the first packet) the initial value of the timeout is set to a given value.

The Slow Start mechanism is initiated both at the start of a TCP connection or after a retransmission timeout. It increases the *cwd* by one packet every ACK until *cwd* reaches the slow start threshold. During the slow start the sender is able to identify the bandwidth available by gradually increasing the number of data packets sent in the network. When *cwd* reaches the slow start threshold TCP uses the Congestion Avoidance mechanism which increases the *cwd* by a fraction of *cwd* ($1/cwd$) for each ACK received until *cwd* reaches the maximum congestion window size. After three duplicate ACKs are received the fast retransmit algorithm infers that a packet has been lost, retransmits the packet (without waiting for a timeout) and sets the slow start threshold to half the current window, and additionally, the current window is set to one packet. The Reno TCP version includes a further algorithm named Fast Recovery [2] which instead of reducing the *cwd* to one packet size after a three-duplicate ACK event, allows Reno to reduce the *cwd* to half its current value.

4   The Modeling Approach and Assumptions

In this section we describe the model assumptions and provide a detailed description of the DSPN model we propose.

4.1   Model Assumptions

The main goal of this work is the study of completion time distribution of finite TCP connections; to this aim we choose to represent in very detailed manner the TCP protocol and in an abstract way the remaining part of the communication network. In particular, our model captures aspects such as congestion window (*cwd*) evolution, slow start and congestion avoidance phases, TCP packet transmissions (and re-transmissions), management of packet losses due to time-out, and due to triple duplicate. As in most TCP modeling proposals, we do not model the connection setup and the connection closing phases. The network aspects are represented by two parameters: *round trip time* (RTT), and *packet loss probability* (*p*).

We model the TCP behavior in terms of *rounds*. A round starts with the back-to-back transmission of *cwd* packets, where *cwd* is the current size of the TCP congestion window. Once all packets falling within the congestion window have been sent in this back-to-back manner, no other packets are sent until the ACKs are received for these *cwd* packets or packet loss/losses are detected. These ACKs receptions mark the end of the current round and the beginning of the next round. In this model, the duration of a round is equal to the round trip time and is assumed to be independent of the window size. Note that we have also assumed that the time needed to send all the packets in a window is smaller than the round trip time. All these assumptions are quite common in the TCP modeling literature (see for instance [17]) and in the most cases are justified by observations and measurements of the TCP behavior.

In our model we assume that the duration of the round trip time is deterministic. This is an assumption that holds in communication networks where the fixed propagation delay is the dominant component of the round trip time. This is a common situation in wide area network.

Loss Packet Assumptions. The selection of a loss model is a key question in designing models of TCP. We assume that packets and ACKs are sent in groups over rounds, and that losses are independent from round to round. This assumption, which is made in most analytical studies (see for instance [3,17]), is partially justifiable with the understanding that TCP tends to send packets in bursts in a manner similar to how our models send packets in rounds. The independence of packet losses occurring in different rounds is especially likely to hold for connections with moderate to high round trip times since the time needed to send all the packets in a window is then much smaller than the round trip time [2]. On the other hand the independence of packet losses occurring within the same round is much stronger and in some cases is not realistic. For instance, this assumption is not realistic when a congested router uses the classical *drop-tail* policy.[1] In our model we focus on the following intra-round loss models:

- *Bernoulli*: Each packet is independently lost with a fixed probability p.
- *Drop-tail*: In each round, we consider the data packets sequentially. The first packet in the round is lost with probability p; for every other packets, if the previous packet was not lost, the packet is lost with probability p; if a packet is lost all subsequent packets in the round are lost.
- *Correlated*: In each round, we consider the data packets sequentially. The first packet in the round is lost with probability p; for every other packets, if the previous packet was not lost, the packet is lost with probability p; otherwise, it is lost with probability q.

The Bernoulli model is arguably the most basic model for packet loss. Owing to its simplicity, it lends to an easier analysis that the other loss models. The Bernoulli model may be appropriate for modeling congestion arising in routers

---

[1] A router that implements a drop-tail policy discards IP packets when its buffer is full.

that implement the *Random Early Detection (RED)* policy [7], since such routers respond to congestion by dropping IP packets uniformly at random. The drop-tail model is an idealization of the packet loss dynamics associated with a FIFO drop-tail router. It is assumed in this model that during congestion, routers drop packets in bursts [18], thus causing packets in the "tail" of a round to be lost. The Correlated model is somewhat less stringent. It characterizes the loss pattern as a Bernoulli distribution of loss episodes, each episode consisting of a group of consecutive packets, the length of which is approximated by a geometric distribution. Recent evidence for such correlated packet loss include [18,23]. We note that the Correlated model actually includes both the Bernoulli case (when $q = p$) and the drop-tail model ($q = 1$) as extreme cases.

4.2   DSPN: Definition and Notation

The modeling assumptions that we use, suggest to use the Stochastic Petri nets variant that allows to use deterministic, immediate, and exponential transitions (DSPNs) [9,13]. We also use other Petri net features such as marking dependent arcs and transition guard functions. For the model we propose, these features are not mandatory but they allow to obtain a more compact model.

In the following model description transition labels are written in italic style lower case *characters* for immediate transitions and upper case letter *CHARACTERS* for timed transitions. Places labels are written in sans serif style and upper case CHARACTERS.

The marking of place P_i is denoted as $\#P_i$, the cardinality of marking dependent arcs is denoted as: $\langle expr \rangle$, where *expr* is the expression that represent the cardinality of the arc.

4.3   Model Description

In this section we illustrate the developed model for the computation of the completion time distribution of finite TCP connections. The specification of the model of Figure 1, requires the definition of guard functions, priorities, and weights, for immediate transitions, definition of delays for the timed transitions, and specification of the initial marking. Tables 1, 2, and 3 reports all these information. The subnet composed by places STEP1, STEP2, STEP3, STEP4, and transitions *s1*, *s2*, *s3*, *RTT* models the *round evolution*, i.e., since our model represents the TCP behavior in terms of round this subnet can be considered "the clock of the model". In this subnet there is one deterministic transition, *RTT*, whose firing delay is equal to the round trip time. Place CWD models the congestion window, i.e., the number of tokens in this place represents the current value of the congestion window. If the protocol is in Slow Start phase place SS_FLAG is marked, while if it is in Congestion Avoidance phase place CA_FLAG is marked. Marking of place SS_THRESH represents the threshold that triggers the change of the TCP state from Slow Start to Congestion Avoidance phase. The tokens in place PCKTOSEND represent the packets that have to be transmitted. The initial marking of places STEP1, SS_FLAG, and CWD is equal to 1,

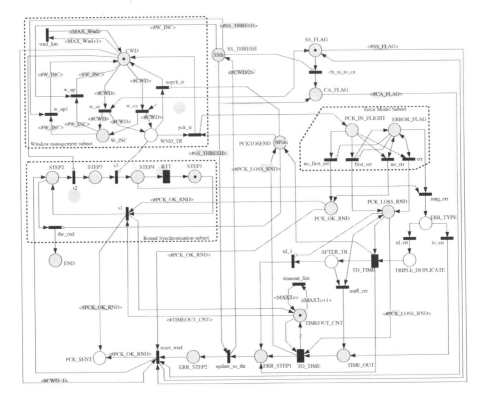

Fig. 1. DSPN model of TCP Tahoe

while sum of markings of places SS_THRESH and CWD represents the threshold
that allows TCP to change from Slow Start to Congestion Avoidance phase.
The subnet composed by immediate transitions w_ss, pck_tr, $nopck\_tr$, w_up,
w_up1, and places W_INC, and WND_TR models the congestion window evolution during the Slow Start phase. In the initial marking only transition $s1$ is
enabled and may fire (note that transition w_ss and transition w_ca are not enabled because their guard functions $f(w\_ss) = [(\#\mathsf{STEP2} = 1) \wedge (\#\mathsf{CWD} > 0)]$
and $f(w\_ca) = [(\#\mathsf{STEP2} = 1) \wedge (\#\mathsf{CWD} > 0)]$). When place STEP2 and place
SS_FLAG are marked transition w_ss becomes enabled and may fire. The cardinality of arcs connecting place CWD to transition w_ss, and w_ss to places W_INC
and WND_TR depends on the marking of place CWD. In this manner, the effect
of the firing of w_ss "moves" the marking of CWD to places W_INC and WND_TR
while place CWD is emptied. In the resulting marking, places W_INC, WND_TR,
and STEP2 are marked, transitions $s2$, pck_tr, $nopck\_tr$, w_up, and w_up1, have
concession; due to guard functions (see Table 1) only transition $s2$ is enabled
and may fire. When place STEP3 is marked there are two transitions that can
be enabled in mutual exclusion, pck_tr, or $nopck\_tr$. These transitions allow to

Table 1. Definition of weights, priority levels, and guard functions for the immediate transitions (p is the packet loss probability)

Label	Weight	Priority	Guard Function
			Immediate transitions
s1	1	2	
s2	1	1	
s3	1	1	
w_ss	1	1	#STEP2 = 1 ∧ #CWD > 0
w_ca	1	1	#STEP2 = 1 ∧ #CWD > 0
pck_tr	1	1	#STEP3 = 1
nopck_tr	1	1	#STEP3 = 1
w_up	1	4	#STEP1 = 1 ∧ #W_INC > 0 ∧ #SS_THRESH ≥ #W_INC
w_up1	1	4	#STEP1 = 1 ∧ #W_INC > 0 ∧ #SS_THRESH < #W_INC
ch_ss_to_ca	1	4	#STEP2 = 1
wnd_lim	1	5	
no_first_err	1 − p	2	#STEP3 = 0
first_err	p	2	#STEP3 = 0
no_err	1 − p	2	#STEP3 = 0
err	p	2	#STEP3 = 0
mng_err	1	1	
to_err	3	1	
td_err	#CWD − 3	1	#CWD > 3
update_ss_thr	1	3	
reset_wnd	1	1	
the_end	1	5	
timeout_lim	1	6	

Table 2. Definition of delays for the deterministic transitions

Label	Delay
	Deterministic transitions
RTT	round trip time
TO_TIME	(see Eq. 1)
TD_TIME	round trip time

move tokens from place WND_TR to place CWD; furthermore, transition *pck_tr* models the transmission of TCP packets. If there is no packet to send, i.e., place PCKTOSEND empty, transition *nopck_tr* is enabled while *pck_tr* is not. In this case, the firings of *nopck_tr* only move the tokens from place WND_TR to place CWD. When place WND_TR is empty, only transition *s3* is enabled and may fire because of the guard functions. When place STEP4 is marked, the deterministic transition *RTT* is enabled and it fires after a round trip time elapses. When the token reaches place STEP1, four transitions have concession (*s1*, *w_up*, *w_up1*, and *w_ss* if place SS_FLAG is marked or *s1*, *w_up*, *w_up1*, and *w_ca* if place CA_FLAG is marked). Due to priorities and guard functions, only *w_up* or *w_up1* can be enabled. These two mutually exclusive transitions, model the doubling of the congestion window size when TCP is in slow start phase and the linearly increasing of the congestion window size when TCP is in congestion avoidance phase.

In particular, when #SS_THRESH ≥ #W_INC transition *w_up* is enabled; its firing empties place W_INC, puts #W_INC tokens in place CWD, and removes #W_INC tokens from place SS_THRESH. When #SS_THRESH < #W_INC tran-

Table 3. Initial Marking Definition

Initial marking	
Label	Marking
STEP1	1
CWD	1
PCKTOSEND	Number of packets to send
SS_FLAG	1
SS_THRESH	Initial Slow Start Threshold value
TIMEOUT_CNT	1

sition *w_up1* is enabled, instead; its firing empties place W_INC, puts #W_INC tokens in place CWD, and empties place SS_THRESH.

When SS_THRESH becomes empty and place STEP2 is marked, transition *ch_ss_to_ca* becomes enabled and may fire. Its firing removes the token from place SS_FLAG and puts it in place CA_FLAG.

The immediate transition *w_ca* models the congestion window evolution during the Congestion Avoidance phase. In this case the behavior is similar to the Slow Start case. When transition *w_ca* fires it puts #CWD tokens in place WND_TR and only one token in place W_INC. In this manner the size of the congestion window size is increased by one each round.

The maximum number of tokens in place CWD is bounded by the transition *wnd_lim* and the by the input and output arcs that connect this transition to place CWD (the constant MAX_WND denotes the maximum congestion window size).

The immediate transition *pck_tr* models the transmission of TCP packets. This transition removes tokens from place PCKTOSEND and puts them in place PCK_IN_FLIGHT. The four immediate transitions *no_first_err*, *first_err*, *no_err*, and *err*, represent different error model behaviors. Due to priorities and guard functions, when the token is in place STEP4 transitions *no_first_err* and *first_err* become enabled. These immediate transitions model the occurrence of the first packet loss in the round. The weights of these transitions are such that the probability that *first_err* fires (resp. *no_first_err* fires) is equal to p (resp. $1-p$), where p is the loss probability of the first loss in a round. If a loss occurs these two transitions are no longer enabled in the same round, while *no_err* and *err* become enabled. These immediate transitions model the occurrence of the packet losses that may occur after the first loss. The firings of transitions *no_first_err* and *no_err* put tokens in place PCK_OK_RND. These tokens represent successful packet transmissions. On the other hand, transitions *first_err* and *err* put tokens in place PCK_LOSS_RND. These tokens represent lost packets that have to be re-transmitted.

When a packet loss occurs in a round, place ERROR_FLAG becomes marked and hence transition *mng_err* becomes enabled. Its firing puts a token in place ERR_TYPE thus enabling transitions *to_err* and *td_err*. These immediate transitions model the two different types of losses that may occur: losses detected by time-out expiration, and losses detected by triple duplicates [11]. There are situations where only the first case may occur; in particular when the size of the congestion size is smaller than three a packet loss can only be detected by time-out expiration (this is accounted by means of a guard function on td_err). The weights of to_err and td_err are computed by using the same criterion that has been used in [17] (see this reference for details). In particular, the probability that a loss is due to time-out is $\max\{1, \frac{3}{\#\mathsf{CWD}}\}$.

We first describe the evolution of the model in case of packet loss detected by time-out expiration. The firing of transition to_err removes the token from place STEP4 (in this manner the round evolution is stopped) and puts a token in place TIME_OUT. In this marking, the deterministic transition TO_TIME is enabled. The firing time of this transition represents the TCP time-out. If the lost packet is the first packet of the connection, the TCP sender does not have an estimate of the round trip time and in this case the value of the time-out is a constant that may be dependent on the TCP implementation, we denote this time by T_{out}. In our case we set $T_{out} = 6$ sec. When the lost packet is not the first one of the connection, the TCP transmitter has an estimate of the round trip time and it can use the round trip time estimate to compute the time-out value, we denote this time by T_{out_RTT}. In both situations it may happen that more than one consecutive time-out expirations occur. In this case, the TCP transmitter exponentially increases the value of the time-out (*time-out backoff*). After a small number of consecutive time-out expirations the value of the time-out is kept constant (in the model, after six consecutive time-out expirations the marking of place TIMEOUT_CNT is not allowed to increase). The number of tokens in place TIMEOUT_CNT represents the number of consecutive time-out that have occurred. The maximum number of tokens in this place is bounded by transition $timeout\_lim$ and the by the input and output arcs that connect this transition to place TIMEOUT_CNT (the constant $MAXTo$ denotes the maximum number of consecutive times that time-out can be increased). The marking of place TIMEOUT_CNT is set equal to one when a successful packet transmission occurs. We manage all possible cases by using a marking dependent firing time for transition TO_TIME defined in the following manner:

$$\text{firing time of } TO\_TIME = \begin{cases} 2^{(\#\mathsf{TIMEOUT\_CNT}-1)}T_{out} & \text{if } \#\mathsf{PCK\_SENT} = 0 \\ 2^{(\#\mathsf{TIMEOUT\_CNT}-1)}T_{out\_RTT} & \text{if } \#\mathsf{PCK\_SENT} > 0. \end{cases} \quad (1)$$

The first case accounts for the loss of the first packet of the connection, while the latter for the loss of a generic packet. When transition TO_TIME fires, tokens from place PCK_LOSS_RND are moved back to place PCKTOSEND. This represents the packets that have been lost and have to be re-transmitted. The firing of TO_TIME puts a token in place ERR_STEP1 and enables transition $update\_ss\_thr$. The firing of this immediate transition puts a token in place ERR_STEP2 and updates the Slow Start threshold to the value $\frac{\#\mathsf{CWD}}{2}$. This is obtained by a pair of marking dependent input and output arcs that connect place SS_THRESH to transition $update\_ss\_thr$. When place ERR_STEP2 is marked transition $reset\_wnd$ is enabled and may fire. The firing of this transition has several effects:

- it moves tokens from place PCK_OK_RND to place PCK_SENT (the packets that have been successfully transmitted);
- it moves the token from place CA_FLAG to place SS_FLAG (after a loss, TCP restarts from the Slow Start phase);
- it sets the current congestion window equal to one;
- it puts the token in place STEP2 (this resumes the round evolution).

The evolution of the model in case of tripe duplicate is quite similar to the time-out occurrence. In this case the firing of the immediate transition td_err enables the timed transition TD_TIME whose firing time is equal to the round trip time. The firing of TD_TIME puts a token in place AFTER_TD. When this place is marked and place PCK_LOSS_RND is empty the immediate transition td_1 is enabled, on the other hand if place PCK_LOSS_RND is marked transition $mult\_err$ is enabled. These two mutually exclusive immediate transitions (td_1 and $mult\_err$) model the cases when in the round only a single triple duplicate loss occurs. On the other hand if in the round two or more losses occur (the first detected by triple duplicate) the second packet loss triggers a time-out (in this case transition $mult\_err$ is enabled and puts a token in place TIME_OUT).

If during the round no losses occur then immediate transition $s1$ removes all tokens from place PCK_OK_RND and puts them in place PCK_SENT.

When place PCKTOSEND is empty and place STEP2 is marked transition the_end is enabled; its firing puts a token in place END. This represents the final absorbing state; the completion time distribution is then defined as the distribution to reach this absorbing state.

5   Results

In this section we first validate the proposed model by comparison against simulation results obtained in different network scenario; following the evaluation of the accuracy of the analytical approach we use the model we developed for analyzing cases of possible TCP scenarios. The completion time distribution is computed resorting to a transient analysis of the stochastic process underlying the DSPN model. In particular, we compute the probability that the TCP connection has completed its transfer of packets at time t by computing the probability that place END is marked at time t.

To perform the numerical experiments we use the tool TimeNet [25]. and an ad-hoc solution algorithm [10] that is faster that the one implemented in the tool TimeNet because it takes advantage from the peculiarities of the proposed model, i.e., there are only immediate and deterministic transitions (in each tangible marking only one deterministic transition is enabled).

5.1   Model Validation

In order to validate the DSPN model, we compare its results against simulation results obtained using the network simulator (ns-2) [15], which provides a detailed description of the dynamics of the Internet protocols.

We choose a simple network scenario where different type of traffic sources co-exist. In particular, we consider a mix of long and short TCP connections, as well as, a mix of TCP and UDP traffic.

Indeed, numerous measurements show that the Internet traffic is now dominated by short flows involving transfer of small Web objects $10 - 20Kb$ in size [21]. While most Internet flows are short-lived, the majority of the packets belongs to long-lived flows, and this property holds across several levels of aggregation [21]. On the other hand, the proliferation of streaming contents, and hence of UDP-based traffic, forms a significant portion of Internet traffic [22].

– *Network Scenario.* Figure 2 illustrates a simple single-bottleneck topology that we use for our simulations. It is a simple network topology with a single bottleneck link whose bandwidth (C) is equal to $2Mbps$, and the two ways propagation delay (D_p) is equal to $120msec$. The bottleneck router has a buffer capacity equal to 100 packets and uses the RED queue management.

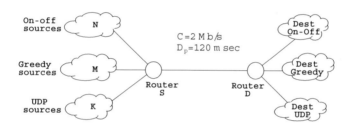

Fig. 2. Simulation topology

– *Traffic Characteristics.* We consider three types of traffic sources in our simulations:

1. A set of N homogeneous *On-off* TCP traffic sources characterized by:
   - and alternating behavior where the silence periods are exponentially distributed with average equal to 1 sec;
   - a packet emission distribution $g = \{g_i\}$, where g_i is the probability that the number of packets that the connection has to transfer is equal to i; in our experiment we have that $g_1 = 0.3$, and and $g_i = 0.1$ for $i = 3, 5, 8, 12, 35, 120$.
2. A set of M long-lived TCP traffic sources (greedy sources) that belong to FTP sessions with an infinite amount of data to transmit. In our simulations $M = N/10$.
3. A set of K UDP traffic sources. In all the experiments the portion of UDP traffic is 5% of the available bandwidth. In our simulations $K = 2$.

The size of TCP data packets is set to 500 bytes. The size of the TCP acknowledgments is set to 40 bytes. The maximum congestion window size is set to 21 packets. The UDP sources are Constant-Bit-Rate with rate equal to 64000 bps.

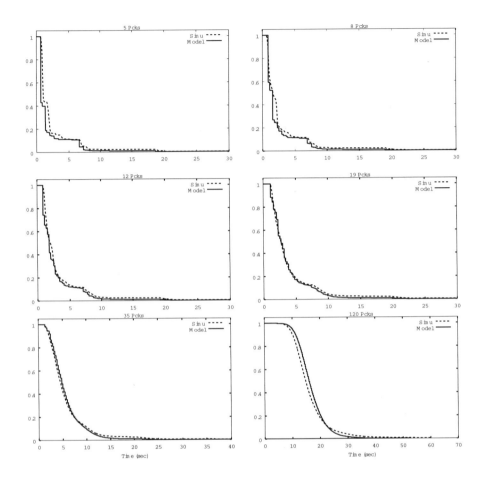

Fig. 3. Analysis vs Simulation, scenario with $N = 90$. 1–CDF of the completion time for connection with: 5 packets to transmit (upper row left column); 8 packets to transmit (upper row right column); 12 packets to transmit (middle row left column); 19 packets to transmit (middle row right column); 35 packets to transmit (lower row left column); 120 packets to transmit (lower row right column)

– *Simulation Methodology.* In our ns simulation the duration of the discared transient period is 300 sec, while the duration of each batch is 120 sec. Every simulations runs a maximum of 3000 batches, corresponding to a maximum simulated time of 360000 sec. Despite this upper bound limit we adopt a criterion to stop the simulations runs which refers to the completion time distribution measure. To estimate the completion time distribution we restrict the analysis to the interval $[0 - 1000 * D_p]$; i.e., 1000 times the propagation delay. We split this time interval into histograms whose width is 10 times the physical delay. We stop the simulation when 70% of the histograms reach the 10% accuracy. We estimate the completion time distribution as well as other measures such as 90%, 95%, and 98% quantile, packet loss probability, and average round trip time. We measure all the metrics with a confidence level of 97.5%.

Figure 3 shows the complement of the cumulative distribution function (CDF), i.e., 1−CDF, of the connection completion time for 5, 8, 12, 19, 35, and 120 packets and $N = 90$. Although the plots of the CDF of the completion time give an intuition of the accuracy of the proposed model, they only allow to conduct a qualitative analysis. To give a quantification of possible differences between model and simulative results we resort to the computation of some parameters of the completion time distributions; in particular, we present results referring to the 90%, 95%, and 98% quantiles of completion time distribution. Table 4 shows the 90%, 95%, and 98% quantiles, the loss probability, and RTT for $N = 80, 90, 100$. Results obtained for different values of N in the range $30, \ldots, 150$ show similar level of accuracy: the relative errors for the 90%, 95%, and 98% quantiles are in the range 2÷20%.

The CPU time required for any simulation run was in the order of magnitude of several hours (at least ten), while the model solution required a few minutes of CPU time except for the case of 120 packets whose solution required one hour of CPU time. Please note that CPU times are drastically reduced when the solution technique proposed in [10] is used.

5.2   Model Exploitation

In general the computation of completion time distribution by means of simulation requires CPU times that could be exceedingly high. On the other hand, the complexity of the solution of the DSPN model is quite limited (as it can be derived by the state space size presented in Table 5). Therefore the DSPN model can be effectively used in the context of planning and performance evaluation of IP networks. Figure 4 shows some results obtained with a loss probability

Table 4. Comparison of simulation vs model results with $N = 80, 90$

n. Pcks	90%		95%		98%	
$N = 80, RTT = 0.223251, p = 0.093475$						
	mod	sim	mod	sim	mod	sim
5	3.35	4.32	6.67	7.08	8.68	8.16
8	4.46	5.52	7.11	7.68	9.79	10.08
12	5.58	6.48	7.78	8.28	11.36	10.81
19	7.14	7.81	8.45	9.12	12.92	12.72
35	8.90	9.36	10.69	11.16	16.07	18.12
120	20.09	19.81	22.30	23.16	29.02	29.88
$N = 90, RTT = 0.225975, p = 0.106652$						
	mod	sim	mod	sim	mod	sim
5	6.68	6.96	6.68	7.81	18.68	18.96
8	6.90	7.21	7.58	8.16	18.90	19.08
12	7.13	7.44	8.03	8.76	19.13	19.44
19	7.68	8.41	9.16	9.84	19.58	20.04
35	9.84	10.44	11.65	12.61	21.39	21.84
120	22.72	23.04	25.21	28.08	32.99	34.81

Table 5. State Space Size of the DSPN model with different number of packet to transfer

Num. of Packets	State Space Size
8	158
10	229
30	1799
50	5052
80	14375
100	23372
120	34479

equal to 0.05 and different values for the round trip time (from 0.45 to 0.60 sec). Figure 5 shows similar results obtained with a round trip time equal to 0.4 sec and different values for the packet loss probability (from 0.01 to 0.15).

Results similar to those presented in Figures 4 and 5 are particular meaningful in the context of network planning. For instance, assume we are interested to plan a network which guarantees with confidence equal to 95% to all TCP connections shorter than 50 packets (in this case, since we assume that the packet size is equal to 500 Bytes, for transfers of files smaller that 25 KBytes) a completion time shorter than 15 sec. From the plots depicted in Figure 5 (right column) we can see that to achieve this target we need to plan a network which guarantees packet loss probability smaller that 0.12 (12%). If instead we are interested to guarantee a confidence level equal to 98%, for the same type of TCP connections, we need to ensure loss probability smaller that 0.08 (8%).

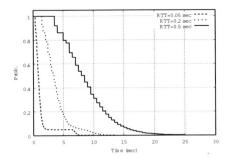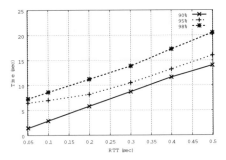

Fig. 4. Analytical results obtained for TCP connections with 50 packets, with fixed loss probability (0.05) and different values for the round trip time (from 0.05 to 0.50 sec): completion time distribution (left column) and quantiles (right column)

The DSPN model can also be used for investigating loss model effects. This could be meaningful to evaluate the impact of different IP router management policies and their effects on the correlation of loss occurrences. Figure 6 shows the completion time distributions (left column) and the 90%, 95%, and 98% quantiles (right column) obtained for a TCP connections of 120 packets, a network scenario

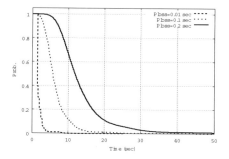
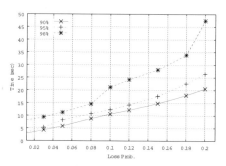

Fig. 5. Analytical results obtained for TCP connections with 50 packets, with fixed loss probability (0.05) and different values for the round trip time (0.2 sec) and different values for the loss packet probability (from 0.01 to 0.2): completion time distribution (left column) and quantiles (right column)

characterized by round trip time equal to 0.2 sec and packet loss probability equal 0.1, and different loss packet models: Bernoulli, drop-Tail, and Correlated with different correlation values ($q = 0.25, 0.5, 0.75$).

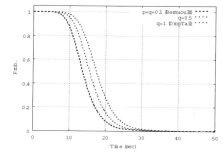
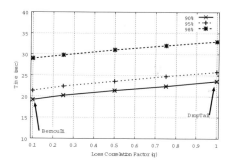

Fig. 6. Analytical results obtained for TCP connections of 120 packets, RTT equal to 0.2 sec, $p = 0.1$, and different loss packet models: Bernoulli, drop-Tail, and Correlated with different correlation values ($q = 0.25, 0.5, 0.75$). Completion time distribution (left column) and quantiles (right column).

All the plots presented so far focus on cumulative distribution functions. These functions are useful for deriving quantitative information (for instance quantiles). In the following we show two graphs that represent probability density functions (pdfs). These pdfs can give other useful information because in a certain sense the can provide a "fingerprint" of the probability distribution and the analysis of their shapes could useful for interesting considerations. Figure 7 shows two plots that present pdfs obtained for TCP connections of 50 packets (left column), and for TCP connections of 100 packets (right column). In both cases, the round trip time is equal to 0.2 sec. Each plot reports two pdfs obtained using two different packet loss probability values: 0.1 and 0.2.

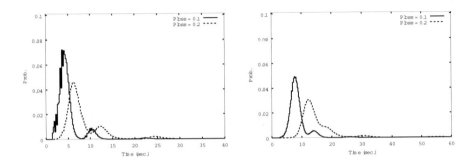

Fig. 7. Probability density functions obtained with a round trip time equal to 0.2 sec, TCP connections of 50 packets (left column), and TCP connections of 100 packets (right column)

One of the most intriguing and, probably, the most ambitious future work of this research topic is the use of the DSPN model for guessing an analytical form of the completion time distribution for TCP connections. In this case the goal would be first the fitting of the completion time distributions with some known distribution and hence the derivation of the parameters for this distribution starting from TCP and/or network parameters (number of packets to transfer, round trip time, loss probability, etc.). This very ambitious goal would yields a drastical change in performance evaluation and planning of IP networks because the TCP modeling could be replaced by the analytical form of the completion time distribution for TCP connections.

6   Conclusions and Further Developments

In this paper we have shown how DSPN models of the TCP protocol can be used to obtain accurate estimates of the distributions of the completion time of short TCP connections. The DSPN models of TCP require as input only the packet loss probability, and the average round trip time for the TCP connections being considered.

The method presented in this paper, which is based on the use of DSPN models, allows the computation of the completion time distribution by using a transient analysis of the model, it easily allows to model different type of losses (independent, correlated, etc.), and, more important, it is amenable to possible extensions, some of which are currently underway. In particular:

– we are developing DSPN models for other, more recent, TCP versions such as Reno [20] and NewReno [6];

– The DSPN models could be used for evaluating the effect of the round trip time distribution. In this case the deterministic transition that represents the round trip time must be replaced by a timed transition with general distributed firing time.

– The DSPN models can be also modified to capture other aspects of the TCP protocol and/or the underlying IP network that have not been modeled so far. In particular, it is not too complicated the modeling of aspects such as losses of ACKs (in general in the TCP modeling literature it is assumed that the ACKs are never lost). Modeling TCP connections with asymmetric data transfer rates, i.e., different forward (data packets per second) and reverse (ACKs per seconds) rates. This feature is quite common in network technology such as ADSL.

– Another possible future work concerns the derivation of a fitting methodology to derive an analytical form for the completion time distribution for TCP connections.

References

1. M. Ajmone Marsan, C. Casetti, R. Gaeta, and M. Meo. Performance analysis of TCP connections sharing a congested internet link. *Performance Evaluation*, 42(2–3), September 2000.

2. M. Allman, V. Paxon, and W. Stevens. TCP Congestion Control. Technical report, RFC 2581, April 1999.

3. N. Cardwell, S. Savage, and T. Anderson. Modeling TCP latency. In *Proc. IEEE Infocom 2000*, pages 1742–1751, Tel Aviv, Israel, March 2000. IEEE Comp. Soc. Press.

4. C. Casetti and M. Meo. A New approach to Model the Stationary Behavior of TCP Connections. In *Proc. IEEE Infocom 2000*, Tel Aviv, Israel, March 2000. IEEE Comp. Soc. Press.

5. J.C.A. de Figueiredo and L. M. Kristensen. Using Coloured Petri nets to investigate behavioural and performance issues of TCP protocols. In *Proc. of the 2nd Workshop on the Practical Use of Coloured Petri Nets and De-sign/CPN*, 1999.

6. S. Floyd and T. Henderson. The NewReno Modification to TCP,s Fast Recovery Algorithm. Technical report, RFC 2582, 1999.

7. S. Floyd and V. Jacobson. Random Early Detection Gateways for Congestion Avoidance. *IEEE/ACM Transaction on Networking*, 1(4), August 1997.

8. M. Garetto, R. Lo Cigno, M. Meo, and M. Ajmone Marsan. A Detailed and Accurate Closed Queueing Network Model of Many Interacting TCP Flows. In *Proc. IEEE Infocom 2001*, Anchorage, Alaska, USA, 2001. IEEE Comp. Soc. Press.

9. R. German. *Performance Analysis of Communication Systems: Modeling with Non-Markovian Stochastic Petri Nets*. John Wiley and Sons, 2000.

10. M. Gribaudo. Transient Solution Methods for Deterministic Stochastic Petri Nets. Technical report, Università di Torino, 2002.
http://www.di.unito.it/ marcog/res.html.

11. V. Jacobson. Congestion avoidance and control. In *Proc. ACM SIGCOMM '88*, 1988. An updated version is avaliable via
ftp://ftp.ee.lbl.gov/papers/congavoid.ps.Z.

12. E. Király, M. Garetto, R. Lo Cigno, M. Meo, and M. Ajmone Marsan. Computation of the Completion Time Time Distribution of Short-Lived TCP Connections. Technical report, Politecnico di Torino, 2002.
13. C. Lindemann. *Performance Modelling with Deterministic and Stochastic Petri Nets*. John Wiley and Sons, 1998.
14. R. Lo Cigno and M. Gerla. Modelling Window Based Congestion Control Protocols with Many Flows. In *Proc. Performance 1999*, Istanbul, Turkey, 1999.
15. S. MCanne and S. Floyd. ns-2 network simulator (ver.2). Technical report, 1997. URL http://www.isi.edu/nsnam/ns/.
16. A. Ost and B. R. Haverkort. Analysis of Windowing Mechanisms with Infinite-State Stochastic Petri Nets. *ACM Performance Evaluation Review*, 26(8):38–46, 1998.
17. J. Padhye, V. Firoiu, D. Towsley, and J. Kurose. Modeling TCP Reno performance: a simple model and its empirical validation. *IEEE/ACM Transaction on Networking*, 8(2):133–145, 2000.
18. V. Paxon. End-to-End Internet Packet Dynamics. *IEEE/ACM Transaction on Networking*, 7(3), June 1999.
19. J. Postel. Transmission Control Protocol. Technical report, RFC 793, September 1981.
20. W. Stevens. TCP Slow Start, Fast retransmit, and Fast Recovery Algorithms. Technical report, RFC 2001, IETF, Jan 1997.
21. K. Thompson, G. Miller, and R. Wilder. Wide-area internet traffic patterns and charateristics. *IEEE Network*, 11(6), Nov-Dec 1997.
22. A. Wolman, G. Voelker, N. Sharma, N. Cardwell, M. Brown, T. Landray, D. Pinnel, A. Karlin, and H. Levy. Organization-Based Analysis of Web-Object Sharing and Caching. In *Proceedings of USENIX Symposium on Internet Technologies and Systems*, October 1999.
23. Y. Zhang, V. Paxson, and S. Shenker. The Stationarity of Internet Path Properties: Routing, Loss, and Throughput. Technical report, AT&T Center for Internet Research at ICSI, http://www.aciri.org/, May 2000.
24. Y. Zhou, T. Murata, and T.A. DeFanti. Modeling and performance analysis using extended fuzzy-timing Petri nets for networked virtual environments. *IEEE Trans. on Systems, Man, and Cybernetics; B: Cybernetics*, 30(5):737–756, 2000.
25. A. Zimmermann, R. German, J. Freiheit, and G. Hommel. TimeNET 3.0 Tool Description. In 8th *Intern. Workshop on Petri Nets and Performance Models*, Zaragoza, Spain, Sep 1999. IEEE-CS Press.

Model Checking Safety Properties in Modular High-Level Nets

Marko Mäkelä[*]

Helsinki University of Technology,
Laboratory for Theoretical Computer Science,
P.O.Box 9205, 02015 HUT, Finland
marko.makela@hut.fi
http://www.tcs.hut.fi/ msmakela/

Abstract. Model checking by exhaustive state space enumeration is one of the most developed analysis methods for distributed event systems. Its main problem—the size of the state spaces—has been addressed by various reduction methods.

Complex systems tend to consist of loosely connected modules, which may perform internal tasks in parallel. The possible interleavings of these parallel tasks easily leads to a large number of reachable global states. In modular state space analysis, the internal actions are explored separately in each module, and the global state space only includes synchronisations. This article introduces nested modular nets, which are hierarchal collections of nets synchronising via shared transitions, and presents a simple algorithm for model checking safety properties in modular systems.

Keywords: modular systems, state space enumeration, model checking, high-level nets

1 Introduction

Complex systems are often divided into modules that can be managed more easily. The internal structure of the modules is hidden behind high-level interfaces, the connection points for composing a complete system out of the components.

Abstracting from implementation details may make it easier to understand how a system works. However, these details may become significant when one wants to assert something about the behaviour of the composed system. To verify whether a desired property holds in the system, one could analyse all its reachable states. The question is whether the easily resulting *state space explosion* [18] can be ameliorated by utilising the division of the system into modules.

In a technique called *compositional reachability analysis* [19] or *modular verification* [7], a model is analysed in multiple phases. Sometimes, it is possible

---

[*] This research was supported by Jenny and Antti Wihuri Fund and by Academy of Finland (Project 47754).

W.M.P. van der Aalst and E. Best (Eds.): ICATPN 2003, LNCS 2679, pp. 201–220, 2003.

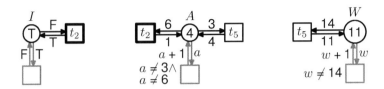

Fig. 1. A partial model of a control system. The gray rectangles denote internal transitions. The three nets synchronise on two labels, t_2 and t_5. The model can be flattened to a single net by fusing the transitions marked with t_2 and t_5.

to transform the property being checked into something that can be checked on each component separately or on a composition of some of the components. Also, the state space can be composed incrementally, collapsing the sequences of internal actions in each intermediate composition.

A modular state space exploration algorithm may save space and time compared to an algorithm for monolithic or flat models. Figure 1 corresponds to part of [16, Figure 1], which models a controller of automated guided vehicles. MARIA [15] constructed the reachability graph of the full model—30,965,760 nodes and 216,489,984 edges—in almost eight hours on a 1 GHz AMD Athlon system equipped with 1 GB of memory. In modular reachability analysis, the edges of the *synchronisation graph* are occurrences of synchronisations. In Figure 2, these are the black edges. Our algorithm constructs the synchronisation graph of the full model—512 nodes and 1,600 edges—in a split second.

Section 3 shows that it is safe to prohibit the occurrences of internal transitions in non-synchronising modules. This reduces the number of synchronisation states of the model in Figure 1 from six in Figure 2 to two.

Compositional or modular state space analysis has been presented for communicating state machines [7] and Petri nets. Input/output nets [10] communicate via dedicated places. Modular place/transition nets [5] use shared transitions. The techniques have also been sketched for high-level nets [1,4], but to our knowledge, no state space exploration algorithm has been presented before.

This paper describes an algorithm for checking safety properties in modular state spaces. Section 2 defines a class of modular high-level nets, and Section 3 defines state spaces for these nets. Section 4 describes our algorithm, and Section 5 reports experimental results. Finally, Section 6 concludes the presentation.

2   Nested Modular Nets

High-level nets are based on net graphs, consisting of the disjoint sets of places P and transitions T and the set of arcs $F \subseteq (P \times T) \cup (T \times P)$. Due to space constraints, we refer the reader to [2] for a definition of high-level nets.

Christensen and Petrucci define modular nets [4, Definition 2.1] as triples (S, PF, TF). The set S contains *modules*, which are high-level nets with no shared places or transitions. The sets $PF \subseteq 2^P$ and $TF \subseteq 2^T$ are the *place*

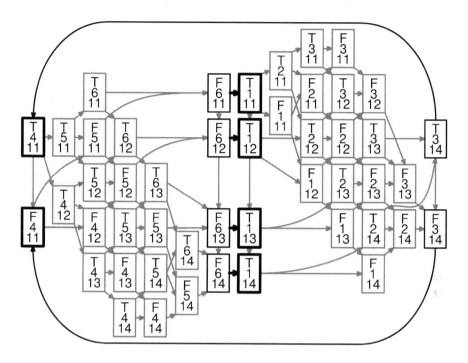

Fig. 2. The complete state space of the model presented in Figure 1, consisting of 48 reachable states and 98 transition occurrences. The initial state is the node at the top left. The gray edges and nodes denote the occurrences of internal transitions and the resulting states. The states immediately before or after a synchronisation are highlighted with a black border. There are four occurrences of t_2, the bold arrows in the centre of the picture, and two occurrences of t_5, the black edges leading from the right to the left of the figure.

fusion sets and *transition fusion* sets, respectively. The sets P and T refer to the combined sets of places and transitions in all modules.

The elements of the transition fusion set $tf \in TF$ are sets of transitions. They can be thought as *synchronisation labels*. If a transition of a module does not belong to any tf, it is an *internal transition*. Otherwise, it is an *external transition* that cannot occur on its own, but only in a synchronous step with other transitions in some tf to which it belongs.

Our definition of modular nets differs from [4, Section 2]. We make the simplifying assumption that there is no place fusion: $PF = \emptyset$. Since communication via shared places can be transformed into synchronisation via shared transitions [4, Section 5], it suffices to support the latter.

Furthermore, instead of defining one top-level structure containing basic nets as modules, we allow modules within modules. Cheung and Kramer motivate the use of subsystem hierarchies by analysing a model of a gas station [3, Section 2.4]. We believe that nested modules could be useful in the verification of multilayered protocols. Figure 3 illustrates how a modular model of a communication protocol can be reused as a module of a distributed application.

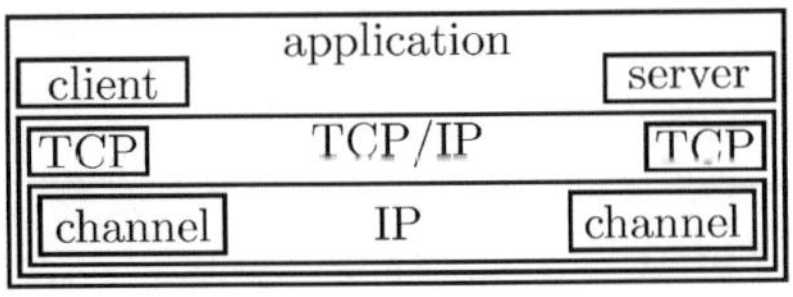

Fig. 3. A possible module hierarchy of a distributed Internet application. The TCP/IP layer consists of two TCP processes that synchronise with primitives provided by the IP layer, which encapsulates some channels. Client–server communication takes place via synchronisations with the TCP/IP module.

We shall define a nested modular net as a *hierarchy tree* of modules S.

Definition 1 (Hierarchy tree). *Let S be a finite set of high-level nets, such that each $s \in S$ contains sets of places P_s and transitions T_s, and $(P_{s_1} \cup T_{s_1}) \cap (P_{s_2} \cup T_{s_2}) = \emptyset$ for all $s_1, s_2 \in S, s_1 \neq s_2$. A hierarchy tree H of S is a connected directed graph $H = (S, C \subset S \times S)$ for which the following hold:*

1. *H has a unique root $s_0 \in S$, such that $C(s) \neq s_0$ for all $s \in S$, and*
2. *other nets in H have a unique ancestor: for each $s \in S \setminus \{s_0\}$, there exists exactly one $s' \in S$ such that $s \in C(s')$.*

A modular net (S, PF, TF) of Christensen and Petrucci can be represented as a nested modular net that has a root module $s_0 \notin S$ with $P_{s_0} = T_{s_0} = \emptyset$ and a hierarchy tree $(\{s_0\} \cup S, \{s_0\} \times S)$.

The fusion sets TF and PF shall be restricted in such a way that only modules with a common parent in the tree can synchronise with each other.

Definition 2 (Hierarchal fusion sets). *Let $H = (S, C)$ be a hierarchy tree, and let P_s and T_s be as in Definition 1. Let $T = \bigcup_{s \in S} T_s$ and $P = \bigcup_{s \in S} P_s$. The sets $TF \subseteq 2^T$ and $PF \subseteq 2^P$ are fusion sets.*

TF (or PF) is H-hierarchal if for all $tf \in TF$ (or $pf \in PF$), the modules of the transitions included in tf (or places in pf) are siblings: for all $(s_1, s_1'), (s_2, s_2') \in C$, either $s_1 = s_2$ or $T_{s_1'} \cap tf = \emptyset \vee T_{s_2'} \cap tf = \emptyset$ (or $P_{s_1'} \cap pf = \emptyset \vee P_{s_2'} \cap pf = \emptyset$).

A transition $t \in T_s$ of a module $s \in S$ is an *internal transition* if there is no $tf \in TF$ such that $t \in tf$. If $t \in tf$ for some $tf \in TF$, t *synchronises on tf*.

The state of a collection of nets is an assignment of markings for each net. In our hierarchy tree, we can define the state of a subtree of nets as an assignment of markings for the root of the subtree and for all its descendants in the tree. These are included in the transitive closure of the child relation:

Definition 3 (Transitive closure). *Let $C \subseteq S \times S$ be a binary relation. Then $C^+ \subseteq S \times S$ is the transitive closure of C, the smallest relation that fulfils the following definition.*

1. $(s, s') \in C \Rightarrow (s, s') \in C^+$,
2. $(s, s') \in C^+ \wedge (s', s'') \in C^+ \Rightarrow (s, s'') \in C^+$.

Next, we shall define a transformation that imports the places of the descendent nets into each ancestor net in the hierarchy tree. For each transition fusion set $tf \in TF$, the transformation also instantiates a *synchronisation transition* by the same name tf. A transition $t \in tf$ is enabled if and only if all transitions $t' \in tf$ are enabled—or tf is enabled in the parent.

Definition 4 (Nested modular net). *Let $H = (S, C)$ be a hierarchy tree, and let TF be a H-hierarchal transition fusion set, and let $PF = \emptyset$ be the place fusion set of S. Let $s \in S$ be a module with the sets of places P_s, transitions T_s and arcs F_s. The modular augmentation $\mathcal{M}(s)$ of s is defined as $\mathcal{M}(P_s) = P_s \cup P'_s$, $\mathcal{M}(T_s) = T_s \cup T'_s$ and $\mathcal{M}(F_s) = F_s \cup F'_s$ as follows.*

1. *a)* $P'_s = \bigcup_{s' \in C^+(s)} P_{s'}$ *(import all places from the descendants)*
   b) $T'_s = \{tf \in TF : \exists s' \in C(s) : T_{s'} \cap tf \neq \emptyset\}$ *(transform the fusions between transitions in child nets into synchronisation transitions in the parent)*
2. *each synchronisation transition $tf \in T'_s$ is a fusion of the transitions $t \in tf$:*
   a) *the set of variables of tf is the union of the sets of variables of $t \in tf$.*
   b) *the guard of tf is the conjunction of the guards of $t \in tf$.*
   c) *for each $t \in tf$, if there is an arc $f = (t, p)$ or $f = (p, t)$ such that $t \in T_{s'}$ and $f \in F_{s'}$ for some $s' \in C(s)$, then there is an arc $f' = (tf, p)$ or $f' = (p, tf)$ in F'_s, respectively. The label of f' is that of f.*

The triple $(H, TF, \mathcal{M})$ is a nested modular net.

Definition 5 (Markings and projected markings). *Let $(H, TF, \mathcal{M})$ be a nested modular net with $H = (S, C)$. Let $s \in S$, and let M be a marking of $\mathcal{M}(s)$. For $s' \in C^+(s)$, the projection of M on $\mathcal{M}(s')$ is $M_{s'} := \bigcup_{p \in \mathcal{M}(P_{s'})} \{(p, M(p))\}$ where s' has the set of places $P_{s'}$.*

Definition 6 (Occurrence rule for nested modular nets). *Let $(H, TF, \mathcal{M})$ be a nested modular net and let $H = (S, C)$. Let $s \in S$ be a net with the transition set T_s. Let $t \in \mathcal{M}(T_s)$ and let m be an assignment for the variables of t. In a given marking M_1 of $\mathcal{M}(s)$, transition t is $\mathcal{M}(s)$-enabled in mode m if*

1. $t \in T_s$: $M^* = M_1$
   $t \notin T_s$: $M^* = M_1$ *or there is a sequence of internal transitions $t_1 \ldots t_n$ such that $M_1 \xrightarrow{t_1} M_2 \cdots \xrightarrow{t_n} M^*$ and each t_i belongs to some $s' \in C(s)$ that synchronises on t—i.e., each s' has a transition $t' \in T_{s'}$ such that $t' \in t$,*
2. t *is enabled in mode m in M^*.*

Then, t may $\mathcal{M}(s)$-occur in mode m, which results in the successor marking M' that is obtained by firing t in mode m in the marking M^*.

Definition 6 modifies the occurrence rule of the underlying high-level nets by defining special treatment of synchronisation transitions. By Definition 4, a transition corresponding to a synchronisation label $t \notin T_s, t \in TF$ is a fusion of the transitions of the child nets $s' \in C(s)$ that synchronise on the label, or belong to the set t. The fused transition t can only be enabled if all its components in the child nets are enabled in some marking M^* reachable from M_1 via a possibly empty sequence of internal transitions.

The synchronisation graph of a modular net can be constructed by exploring its $\mathcal{M}(s_0)$-enabled transitions, where s_0 is the root of the hierarchy tree. This graph only contains the occurrences of synchronisation transitions.

The reachability graph of a modular net can be computed by flattening the hierarchy to an ordinary high-level net and applying the occurrence rule of ordinary high-level nets on the flattened net. The reachability graph may contain more occurrences of synchronisation transitions than the synchronisation graph, since the occurrences of the internal transitions of child nets are not restricted.

Definition 7 (Flattened nested modular net). *Let $(H, TF, \mathcal{M})$ be a nested modular net with the hierarchy tree $H = (S, C)$ whose root is s_0. For each $s \in S$, let there be the set of places P_s, the set of transitions T_s and the set of arcs $F_s \subseteq (P_s \times T_s) \cup (T_s \times P_s)$. Let $P = \bigcup_{s \in S} P_s$ and $T = \bigcup_{s \in S} T_s$. Then $\mathcal{F}(H, TF, \mathcal{M})$ is the* flattened nested modular net *of $(H, TF, \mathcal{M})$, with the following elements:*

1. $P = \mathcal{M}(P_{s_0}) = \bigcup_{s \in S} P_s$ *(all the places of all nets),*
2. $T = \bigcup_{s \in S} \mathcal{M}(T_s) \setminus \bigcup_{tf \in TF} tf$ *(all internal transitions),*
3. $F = \bigcup_{s \in S} \mathcal{M}(F_s) \cap ((P \times T) \cup (T \times P))$ *(the arcs attached to the transitions).*

Definition 7 also imports synchronisation transitions $tf \in TF$ to the flattened net, unless they are external, i.e., $tf \in tf' \in TF$. Only the "outermost" synchronisation transitions tf' (such that $tf' \notin tf''$ for all $tf'' \in TF$) are imported.

3   Modular State Spaces

Next, we shall define the reachability graph of a high-level net and the synchronisation graph of a nested modular high-level net. Both are called *state spaces*. We shall also define the *equivalent state space* of a nested modular net and prove that it is equivalent to the state space of a flattened nested modular net.

Definition 8 (State space). *Let s be a high-level net with the initial marking M_0. The state space of s is a directed rooted graph $G = (V, E, v_0)$, with $E \subseteq V \times V$ and $v_0 \in V$, the smallest graph for which the following hold:*

1. $v_0 = M_0$ *is the initial state,*
2. *for each $M \in V$, if $M \longrightarrow M'$, then $M' \in V$ and $(M, M') \in E$.*

Let $(H, TF, \mathcal{M})$ be a nested modular net with the root net s_0. The state space of $(H, TF, \mathcal{M})$ is defined analogously, with v_0 corresponding to the initial marking of $\mathcal{M}(s_0)$ and with the edges in E corresponding to the occurrences of $\mathcal{M}(s_0)$-enabled transitions.

The three nets s_1, s_2, s_3 in Figure 1 can be interpreted as $(H, TF, \mathcal{M})$, such that $H = (S, C)$, $S = \{s_0, s_1, s_2, s_3\}$, s_0 is empty, and $C = \{s_0\} \times \{s_1, s_2, s_3\}$. The synchronisations are $TF = \{\{t_2^1, t_2^2\}, \{t_5^2, t_5^3\}\}$, where the superscripts identify the

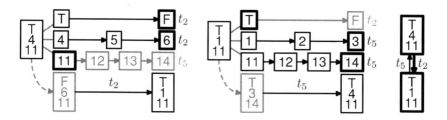

Fig. 4. The construction of the synchronisation graph of the controller model (Figure 1). In the initial marking, only t_2 is enabled, as all modules that synchronise on it can internally reach a state where it is enabled. Similarly, only t_5 is enabled in the successor state. In a synchronisation, the states of non-participating modules do not change.

modules, e.g., $t_2^2, t_5^2 \in T_{s_2}$. Figure 2 represents the state space of $\mathcal{F}(H, TF, \mathcal{M})$. The state spaces of $(H, TF, \mathcal{M})$ are illustrated in Figure 4. We assert that the two synchronisation states correspond to the two groups of states enclosed in a thick border in Figure 2. These groups can be formalised as follows.

Definition 9 (Related markings [4, Definition 3.3.1]). *Let $(H, TF, \mathcal{M})$ be a nested modular net. Let $H = (S, C)$ and $s \in S$. Let $\mathbb{M}_s$ be the set of all markings of $\mathcal{M}(s)$. Let Π_s map each marking $M \in \mathbb{M}_s$ to a set of markings reachable via $\mathcal{M}(s')$-enabled internal transitions of the modules $s' \in C(s)$. Let $R_s \subseteq \mathbb{M}_s \times \mathbb{M}_s$ identify markings with common internal successor states: $(M_1, M_2) \in R_s \Leftrightarrow \Pi_s(M_1) \cap \Pi_s(M_2) \neq \emptyset$. Let R_s^+ be the transitive closure of R_s.*

Similar to Christensen and Petrucci [4], checking whether a state M' is in $\Pi_s(M)$ does not require $\Pi_s(M)$ to be generated—it is sufficient to check that in each module $s' \in C(s)$, the local component $M'_{s'}$ is either reachable from the local component $M_{s'}$ via internal transitions, or $M'_{s'} \in \Pi_{s'}(M_{s'})$.

   The left half of Figure 2 represents the initial state of the model in Figure 1—$M_0 = \{(I, \mathsf{T}), (A, 4), (W, 11)\}$—and the internal states $\Pi_{s_0}(M_0)$. Let $M_1 = \{(I, \mathsf{F}), (A, 4), (W, 11)\}$. Clearly, $\Pi_{s_0}(M_0) \cap \Pi_{s_0}(M_1) \neq \emptyset$, and thus R_{s_0} contains (M_0, M_1). Note that by definition, R_{s_0} is reflexive and symmetric. $R_{s_0}^+$ is also transitive and thus an equivalence relation. In this example, $R_{s_0}^+ = R_{s_0}$.

   Next, we shall define a mapping for the state space of a nested modular net and prove that it equals the state space of the corresponding flattened net.

Definition 10 (Equivalent state space [4, Definition 3.3.3]). *Let $G = (V, E, v_0)$ be the state space of the nested modular net $(H, TF, \mathcal{M})$ with $H = (S, C)$. Let s_0 be the root of H, and let $\mathbb{M}$ and R^+ be as in Definition 9. The equivalent state space of G is $G' = (V', E', v_0)$, defined inductively as follows:*

1. *$V' = \bigcup_{M \in V} R_{s_0}^+(M)$*
2. *for all states $v, v' \in V'$, $(v, v') \in E$ if $v \xrightarrow{t, m} v'$ where*
    a) *t is $\mathcal{M}(s_0)$-enabled in mode m in the marking v, or*
    b) *t is an internal transition of some $s \in C(s_0)$, and t is $\mathcal{M}(s)$-enabled in mode m in the marking v.*

Proposition 1 (Equivalence of state spaces [4, Theorem 3.3.4]). *Let $(H, TF, \mathcal{M})$ be a nested modular net and $\mathcal{F}(H, TF, \mathcal{M})$ be a flattened nested modular net. Let $G_{\mathcal{F}} = (V_{\mathcal{F}}, E_{\mathcal{F}}, v_0)$ be the state space of $\mathcal{F}(H, TF, \mathcal{M})$. Let $G = (V, E, v_0)$ be the state space of $(H, TF, \mathcal{M})$, and let $G' = (V', E', v_0)$ be the equivalent state space of G. Then $G_{\mathcal{F}} = G'$.*

By Definitions 7 and 10, each state space has the same initial state v_0. The sets of potential markings coincide for the modular and the flattened net, as they both contain the same set of places, by Definition 7.

Compared to the flattened net (Definition 7), the occurrence rule for nested modular nets (Definition 6) hides the occurrences of internal transitions. Thus, for each $(M, M') \in E$, there is a path $(M, M_1), (M_1, M_2), \ldots, (M_n, M') \in E_{\mathcal{F}}$ where the intermediate states $M_1 \ldots M_n$ result from the occurrences of internal transitions. Thus, $V \subseteq V_{\mathcal{F}}$ and $E \subseteq E_{\mathcal{F}}^{+}$. By Definitions 9 and 10, V' and E' are extended from V and E by adding states and edges corresponding to the occurrences of internal transitions. Since $G_{\mathcal{F}}$ and G' only differ by these occurrences, we have $V_{\mathcal{F}} = V'$ and $E_{\mathcal{F}} = E'$.

4   Checking Safety Properties

4.1   Algorithm for Exhaustive Modular State Space Exploration

Figure 5 presents a basic algorithm for checking safety properties by exhaustive modular state space enumeration. If we ignore the right column of this figure and the invocation of MODULES in the procedure EXPLORE, we have the basic exploration algorithm for flat state spaces.

The procedure EXPLORE constructs the state space of a net $\mathcal{M}(s)$ by exploring all enabled transitions in each state reachable from M_1. It invokes TRANSITIONS in each state $M \in V$. For each enabled transition of s, TRANSITIONS invokes REPORT in order to insert previously unknown states $M' \notin V$ into the search queue Q and into the set of reachable states V.

While the procedure TRANSITIONS computes the successor states of M by exploring the enabled transitions of s, the procedure MODULES explores the synchronising transitions of each module $s' \in C(s)$ and invokes SYNC to compute the synchronisations. The transition enabling test invoked by TRANSITIONS and SYNC has been described in [13].

By invoking EXPLORE on each s', the procedure MODULES constructs a synchronisation relation $\mathcal{S}$ that maps child nets to synchronisation labels and local states where these synchronisations are enabled. If $(s', tf, M^{s'}) \in \mathcal{S}$, the marking $M^{s'}$ of s' is reachable from $M_{s'}$—the projection of M on $\mathcal{M}(s')$—via internal transitions, and a transition synchronising on tf is enabled in $M^{s'}$. Each invocation of MODULES is associated with such a relation $\mathcal{S}$. The relation is extended by TRANSITIONS and explored by SYNC.

The procedure SYNC iterates over certain subsets of the synchronisation relation $\mathcal{S}$. MARIA implements the relation with some mappings and arrays. The iteration has been implemented as a recursive loop over those modules $s' \in C(s)$

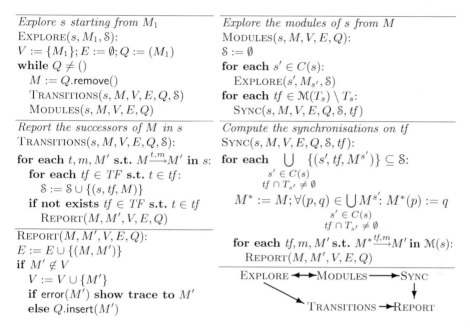

Fig. 5. The basic algorithm for modular state space analysis and its call graph. All parameters are passed by reference. The synchronisation information S is associated with each invocation of MODULES. V is a set of encountered states, E is a set of edges and Q is a queue of unexplored states. TRANSITIONS assumes that there is at most one $t \in T_s \cap \mathit{tf}$. If $t_1, \ldots, t_n \in T_s \cap \mathit{tf}$, the transitions $t_1, \ldots, t_n$ will have to be fused together in a pre-processing step. The algorithm is invoked for a nested modular net $(H = (S, C), \mathit{TF}, \mathcal{M})$ as EXPLORE(s_0, M_0, S) where s_0 is the root net, M_0 is the initial marking of $\mathcal{M}(s_0)$ and $S = \emptyset$.

that synchronise on tf. On each round, a marking $M^{s'}$ of s' is assigned to M^*. That is, M^* is first initialised to M, and the markings of the synchronising modules $\mathcal{M}(s')$ will be substituted in M^*.

An Example Run. We shall demonstrate the algorithm with the model presented in Figure 1 and discussed near Definition 9. The root net s_0 of the model contains no places or transitions. The modular augmentation $\mathcal{M}(s_0)$ is depicted in Figure 6. Its initial marking is $M_0 = \{(I, \mathsf{T}), (A, 4), (W, 11)\}$.

The algorithm is invoked as EXPLORE$(s_0, M_0, \emptyset)$. It initialises the set V and the queue Q with the initial marking M_0. It removes the marking from the queue and passes it as a parameter to TRANSITIONS. Since s_0 has no transitions (but $\mathcal{M}(s_0)$ has), invoking TRANSITIONS does not change anything.

Next, EXPLORE invokes the procedure MODULES. The left part of Figure 4 shows how MODULES splits the marking into the markings of child nets and invokes EXPLORE on each of them. Let us look at the EXPLORE call for s_1, the leftmost net in Figure 1. On the first round of the **while** loop, TRANSITIONS calls

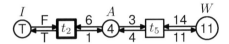

Fig. 6. The modular augmentation $\mathcal{M}(s_0)$ of the empty root net s_0 of the control system presented in Figure 1.

REPORT($\{(I, \mathsf{T})\}, \{(I, \mathsf{F})\}, V, E, Q$), which inserts $\{(I, \mathsf{F})\}$ into the local search queue of EXPLORE($s_1, \{(I, \mathsf{T})\}, \mathcal{S}$). Thus, the **while** loop of EXPLORE enters a second round, with $M = \{(I, \mathsf{F})\}$. For this marking, TRANSITIONS finds out that the transition synchronising on t_2 is enabled. Thus, it adds $(s_1, t_2^1, \{(I, \mathsf{F})\})$ to the synchronisation set $\mathcal{S}$. After this, EXPLORE returns to MODULES.

By invoking EXPLORE on all nets $C(s_0) = \{s_1, s_2, s_3\}$, MODULES constructs the set $\mathcal{S} = \{(s_1, t_2^1, \{(I, \mathsf{F})\}), (s_2, t_2^2, \{(A, 6)\}), (s_3, t_5^3, \{(W, 14)\})\}$. Next, it invokes SYNC on t_2 and t_5. For t_2, SYNC finds exactly one subset of $\mathcal{S}$, namely $\{(s_1, t_2^1, \{(I, \mathsf{F})\}), (s_2, t_2^2, \{(A, 6)\})\}$. The markings $M = M_0$, $M^{s_1} = \{(I, \mathsf{F})\}$ and $M^{s_2} = \{(A, 6)\}$ are combined to $M^* = \{(I, \mathsf{F}), (A, 6), (W, 11)\}$. In M^*, the transition t_2 is enabled, and a successor marking $\{(I, \mathsf{T}), (A, 1), (W, 11)\}$ of M is reported. For t_5, there is no subset $\{(s_2, t_5^2, M^{s_2}), (s_3, t_5^3, M^{s_3})\}$ of $\mathcal{S}$, since there is no $(s_2, t_5^2, M^{s_2}) \in \mathcal{S}$. This can be compared to the left part of Figure 4.

So, MODULES returns to EXPLORE after having recorded one successor marking. The second and last round of the **while** loop in EXPLORE($s_0, M_0, \emptyset$) proceeds in a similar fashion; see the middle part of Figure 4.

It should be noted that the parameter M of REPORT and the sets E do not affect the control flow of the algorithm, and thus they need not be implemented.

Correctness. We assert that the algorithm presented in Figure 5 computes state spaces (Definition 8) of nested modular nets.

Definition 6, the occurrence rule for nested modular nets, has one essential addition to the occurrence rule of the underlying nets. The case $t \notin T_s$ deals with synchronisations, preceded by a sequence of internal transitions. Similarly, the algorithm in Figure 5 extends the basic state space exploration algorithm [14, Algorithm 1] with the procedures MODULES and SYNC that explore the internal transitions of modules and make all possible synchronisations occur.

The subroutine EXPLORE terminates when all states have been explored, or the search queue Q runs out. States are added to Q by REPORT, which is the only routine altering Q, the set of reachable states V and the set of edges E.

When EXPLORE is invoked on a net s that contains no module, $C(s) = \emptyset$, the invocation of MODULES in the **while** loop of EXPLORE does not affect anything. Clearly, each invocation of TRANSITIONS augments V and E with the successor states of M—those states that result from the occurrences of internal transitions. Since each reachable state is inserted into Q exactly once and since EXPLORE invokes TRANSITIONS on each state in Q, it is easy to see that after the **while** loop terminates, the sets V and E correspond to the state space of s, with

$v_0 = M$. Provided that at most one transition of s synchronises on any given tf (see the caption of Figure 5), the set S will contain a tuple for each enabled external transition of the net and for each reachable state where it is enabled.

The procedure MODULES becomes significant when a net s contains modules. It explores the reachable states in each module $s' \in C(s)$, starting from $M_{s'}$, the current marking M projected on s'. Once all invocations of EXPLORE have returned to MODULES, the set S contains an item for each module and for each state where a synchronisation is possible. The outermost **for each** loop in SYNC iterates over all possible synchronisation points on tf and initialises a marking M^* on each iteration. The markings M^* are reachable from M by the occurrences of internal transitions of those modules that synchronise on tf. This is equivalent to the case $t \notin T_s$ of Definition 6. Finally, SYNC generates the successor states of M by making each enabled instance of tf occur in each M^*.

We conclude that EXPLORE($s_0, M_0, \emptyset$) constructs the state space of a modular algebraic system net $(H, TF, \mathcal{M})$ with the root s_0 and the initial state M_0.

4.2 Specifying Safety Properties

The procedure REPORT in Figure 5 checks a safety property on a newly generated state. Erroneous states are not explored further—they are reported to the user.

The safety model checker in MARIA recognises three kinds of erroneous states:

– states that satisfy a "reject" or "deadlock" formula,
– states that cannot be compacted due to a constraint violation, and
– states whose successors cannot be computed due to an evaluation error.

A "reject" formula is a Boolean condition on reachable markings. A "deadlock" formula is a condition on reachable markings where no transition is enabled.

More generic properties can be specified in linear temporal logic. It covers infinite executions, but its "safety" subset [12] is equivalent to finite state automata, which deal with finite execution sequences. The property "whenever A becomes 2, it will remain less than 5" does not hold in the net s_2 of Figure 1, since its place A acts a counter from 1 to 6. In the automata-theoretic approach to verification, a desired property of a system is negated and translated into an automaton. The system is in error if an accepting state (the dashed one in Figure 7) is reachable in the product automaton of the system's state space and the automaton corresponding to the negated property.

We would like to synchronise a modular state space with a property automaton. In the left part of Figure 4, $A \neq 2$ and the property automaton of Figure 7 remains in its initial state 0. In the middle part, the markings of A are 1, 2 and 3. In the state $A = 2$, the automaton moves to the state 1, but it cannot move to its accepting state 2, even though the system clearly violates the property!

Obviously, the property automaton must be fused with the relevant module, as in Figure 8. Figure 9 shows the state space of the fused model and the local states that are reachable from the last synchronisation. The property is violated in the dashed state. We see that the product of a property and a model may have more states than the plain model (Figure 4). Thus, it may be wise to check this kind of properties one at a time to avoid a combinatorial explosion.

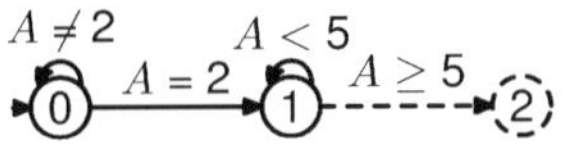

$$A = 2$$

Fig. 7. A finite automaton for the formula $\neg\Box\,((A = 2) \Rightarrow \Box(A < 5))$.

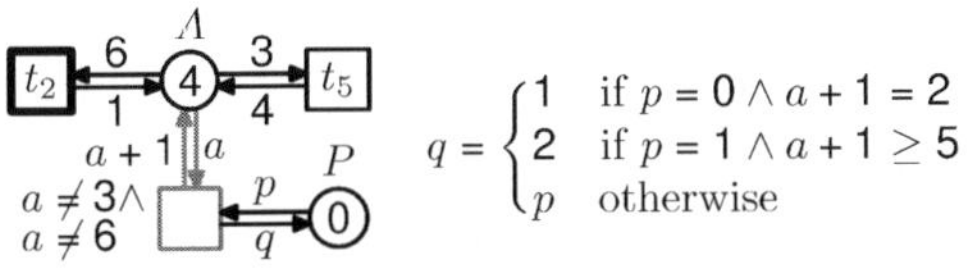

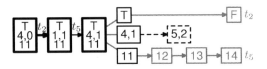

$$q = \begin{cases} 1 & \text{if } p = 0 \wedge a + 1 = 2 \\ 2 & \text{if } p = 1 \wedge a + 1 \geq 5 \\ p & \text{otherwise} \end{cases}$$

Fig. 8. The automaton of Figure 7 composed with the net s_2 of Figure 1. Only the occurrences of the internal transition can change the state of the automaton.

Fig. 9. An error trace of the property (Figure 7) and the model (Figures 1 and 8).

4.3   Constructing Error Traces

An *error trace* is a sequence of model actions that leads from the initial state to an erroneous state or transition. The algorithm presented in Figure 5 does not construct error traces—it is only aware of the last state on the trace.

The information needed for producing error traces should be stored in as little space as possible, so that more memory is available for accommodating the sets of encountered states and the collections of unprocessed states. Some information can be omitted and recomputed when an error is detected. This allows the verification of larger systems and more complex properties. For instance, the complete set of edges E may need much more storage space than the set of reachable states V if the state space contains many cycles and branches, or the edges are labelled with the names and firing modes of the occurring transitions.

Efficient production of an error trace requires a function ancestor : $\mathbb{M} \rightarrow \mathbb{M}$ that maps each new state to the state from which it was obtained (the parameters M and M' of REPORT in Figure 5, $M' \notin V$). All states on an error trace can be enumerated by repetitively applying this function on the error state until the first state M_1 is reached. Once all the states $M_1, \ldots, M_n$ of the trace are known, the transitions can be obtained by computing the successor states of each state M_i in the trace and displaying a transition leading from M_i to M_{i+1}. There might not be a unique shortest error trace—this method produces one of them.

The function ancestor could be defined as something that follows the edge relation E backwards. Alas, we cannot store E, as we want to preserve memory. Stern and Dill [17] propose an addition to the algorithm: whenever a state

M' is inserted into Q, it is also appended to an auxiliary file together with the position of its ancestor M in the said file. The collection Q must associate each unprocessed state with these file positions. This file provides the mapping ancestor.

Writing the counterexample recovery information to a file does not significantly affect the performance, since sequential file access is fast. Only when a counterexample trace is produced, random (slow) access is needed. Even at that point, the input/output overhead may be insignificant.

Our implementation of MODULES (Figure 5) shows an error trace to M if any of the EXPLORE invocations reports an error. The tool can be told to stop after reporting a specified number of errors. When the model in Figure 1 is checked for the property in Figure 7, MARIA reports the trace (part of Figure 9) in two parts: from $\{(A, 4), (P, 1)\}$ to $\{(A, 5), (P, 2)\}$, and from the initial state of the modular system to the synchronisation state $\{(I, \mathsf{T}), (A, 4), (P, 1), (W, 11)\}$.

This arrangement produces short error traces—in fact, the produced traces are as short as possible if the state spaces are constructed in breadth-first order.

5   Experiments

5.1   Automated Guided Vehicles

The first system [16, Figure 1] that was analysed with our algorithm models the coordination of automated guided vehicles on a factory floor. The state space of the flattened model consists of 30,965,760 nodes and 216,489,984 edges. The model is distributed with MARIA in the file modular.pn.

For the modular model, the algorithm in Figure 5 produces a state space of 836 nodes and 2,644 edges. This state space consists of 325 strongly connected components, one of which is terminal. This is somewhat surprising, since the state space of the flattened model consists of a single strongly connected component. Each of the remaining 324 components consists of a single state. Thus, there are no cycles between the 324 states—all edges lead towards the terminal component.

In the initial marking in [16, Figure 1], several modules are in an *intermediate state* in the sense that some internal transitions have occurred after synchronisations. Starting from a marking where the occurrences of these internal transitions have been undone, we obtained 512 nodes and 1,600 edges. This corresponds to the terminal strongly connected component of the original state space.

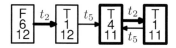

Fig. 10. The state space of the modular model (Figure 1) with an initial marking corresponding to [16, Figure 1]. The state space in Figure 4 is smaller.

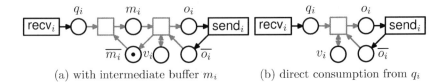

(a) with intermediate buffer m_i (b) direct consumption from q_i

Fig. 11. A schematic view of two alternative implementations of the modules s_i of the leader election protocol, $1 \leq i \leq n$. The initial markings of v_i and o_i and the arc inscriptions have been omitted, and q_i should be accessed as a queue. The composition is $(H, TF, \mathcal{M})$ with $H = (S, C)$, $S = \{s_0, \ldots, s_n\}$ (s_0 is empty), $C = \{s_0\} \times \{s_1, \ldots, s_n\}$ and $TF = \bigcup_{i=1}^{n}\{\{\mathsf{send}_i, \mathsf{recv}_{(i \bmod n)+1}\}\}$.

Table 1. State space sizes and exploration times on a 1.67 GHz AMD Athlon XP for the leader election protocol (Figure 11) for different numbers of processes.

n	flat (a)			flat (b)			modular		
	nodes	edges	time	nodes	edges	time	nodes	edges	time
3	155	299	0.0 s	69	126	0.0 s	33	63	0.0 s
4	712	1,847	0.0 s	240	588	0.0 s	90	227	0.0 s
5	3,428	11,194	0.1 s	870	2,693	0.0 s	251	800	0.0 s
6	16,788	66,039	0.8 s	3,213	12,013	0.1 s	713	2,746	0.1 s
7	82,663	380,263	5.3 s	11,949	52,310	0.6 s	2,041	9,210	0.2 s
8	407,695	2,146,961	31.6 s	44,544	223,338	2.9 s	5,863	24,267	0.9 s

In Figure 1, if the internal transition in the middle occurs twice and the ones at the sides occur once, the result corresponds to a subset of [16, Figure 1]. Figure 10 shows the state space starting from this marking.

5.2   Leader Election in Unidirectional Ring

One of the examples distributed with SPIN [8] is the leader election protocol in a unidirectional ring [6]. Figure 11 depicts the operation of the modules. Each module s_i has an input queue q_i, from which it takes messages m_i that are processed, affecting the local variables v_i. The modules also contain an output buffer o_i that can hold at most one message. In the initial state, the output buffers o_i are filled with the initiating messages of the protocol.

According to Table 1, the modules sketched in Figure 11(a) generate much bigger state spaces for the flattened net than those shown in Figure 11(b). The state space of the root net is the same for both variations.

Karaçalı and Tai [11] have modelled the system with extended finite state machines. The flat state spaces reported in [11, Table 1] are an order of magnitude bigger than those in Table 1. However, their reduction algorithm appears to outperform modular analysis by generating only $8n + 13$ states and $8n + 12$ events for systems consisting of $3 \leq n \leq 6$ processes.

The algorithm of Karaçalı and Tai [11, Section 5] makes use of a transition dependency relation [11, Section 4]. Such relations are difficult to derive for high-

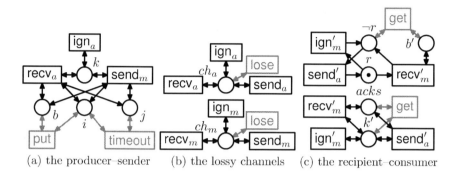

(a) the producer–sender   (b) the lossy channels   (c) the recipient–consumer

Fig. 12. The modules of the sliding window protocol. The inscriptions have been omitted. The message and acknowledgement channels support the actions send (send a message wr), recv (receive a message rd) and ign (ignore rd). The parameters rd and wr are shared variables among the synchronising transitions.

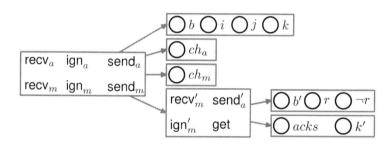

Fig. 13. The hierarchy tree of the sliding window protocol model (Figure 12). The modules s are denoted by gray rectangles, and gray arrows represent the child relation C. The synchronisation labels and the places are indicated for each module. The recipient–consumer module defines four synchronisation labels, three of which synchronise on the similarly named labels of the root module.

level Petri nets, which do not have separate control and data flows. Karaçalı and Tai have improved partial order reduction with something that resembles modular state space exploration. It would be interesting to see if partial order reduction could be efficiently implemented on top of modular analysis.

5.3   Sliding Window Protocol

Figures 12 and 13 show our modular model of the sliding window protocol, which is distributed with MARIA in the file swn-m.pn. There are two communication channels, which are connected to a sending and a receiving process. The producer entity that feeds the sending process is modelled by the internal transition put, and the consumer is modelled by the transitions get. If the channels lose messages, a sender timeout will eventually occur.

Both channel modules in Figure 12 contain one place, which holds one token that represents the contents of the queue. The producer–sender has a transmission buffer b for tw items and three indices to it: i and k to the bounds of the transmission window and j to the next item to be sent ($i \leq j \leq k$). The receiver has a buffer b' for rw items, the index number k' of the next awaited message and an array $acks$ that indicates which messages have been acknowledged.

The recipient–consumer module is divided further into two modules. Since both modules are free of internal transitions, they could be fused into a single module, a behaviour-equivalent replacement of their parent. The synchronisation labels of the recipient–consumer module are invisible to the root module. We must explicitly define the synchronisation transitions for $recv'_m$, ign'_m and $send'_a$ to synchronise on the labels $recv_m$, ign_m and $send_a$ of the root module.

Compared to previous examples, our sliding window protocol model is not very well suited for modular analysis. The internal and external transitions are rarely in conflict, except when the channels have very limited capacity. In fact, synchronisations are possible in almost all local states of the modules.

Table 2. State space sizes (numbers of nodes and edges) for the protocol in Figure 12. The parameters are the transmission and reception window sizes tw and rw and the type of the channel. Each state space is strongly connected. The third columns for modular state spaces indicate the numbers of edges obtained by caching the pre-synchronisation states.

$tw,$	reliable channel					lossy channel			
rw	flat		modular			flat		modular		
1,1	12	12	8	8	8	108	310	64	172	96
1,2	18	18	12	12	12	462	1,686	348	1,734	636
1,3	24	24	16	16	16	1,336	5,372	1,104	7,928	2,192
2,1	54	90	39	72	54	2,118	8,349	1,422	9,408	2,853
2,2	72	120	52	96	72	9,388	40,256	7,080	65,692	15,156
2,3	90	150	65	120	90	27,265	122,555	22,115	268,185	49,140
3,1	160	336	116	292	192	25,292	113,036	18,412	216,304	40,792
3,2	200	420	145	365	240	109,550	508,790	84,775	1,235,830	193,835
3,3	240	504	174	438	288	323,724	1,537,638	262,026	4,629,366	611,754

Caching Pre-Synchronisation States. Table 2 indicates that applying modular analysis to our model of the sliding window protocol slightly reduces the number of reachable states but increases the number of edges.

The number of edges can be reduced by slightly modifying the procedure $SYNC(s_0, \ldots)$ in Figure 5. Before generating the successors of M^*, it would add the item (M^*, tf) to a set and return if the set already contained that item.

Table 2 shows three columns of numbers for modular analysis. The leftmost two columns indicate the numbers of states and edges in the state space of the root net. The third column shows the number of edges reduced by applying this

Table 3. Processor time consumption in seconds and peak heap memory consumption in kilobytes of a 1.67 GHz AMD Athlon XP system running GNU/Linux when exploring the protocol in Figure 12 on lossy channels.

$tw,$ rw	nodes	edges	plain mem	plain time	local cache mem	local cache time	pre-cache edges	pre-cache mem	pre-cache time	pre&local mem	pre&local time
1,1	64	172	309	0.0	311	0.0	96	343	0.0	342	0.0
1,2	348	1,734	309	0.1	307	0.0	636	339	0.1	382	0.0
1,3	1,104	7,928	309	0.3	523	0.1	2,192	339	0.3	494	0.1
2,1	1,422	9,408	309	0.3	451	0.1	2,853	339	0.3	422	0.1
2,2	7,080	65,692	309	1.8	2,007	0.5	15,156	343	1.8	2,314	0.5
2,3	22,115	268,185	309	6.0	7,735	2.0	49,140	343	6.0	9,270	1.9
3,1	18,412	216,304	309	4.8	4,475	1.4	40,792	339	4.7	5,754	1.3
3,2	84,775	1,235,830	473	24.2	23,763	8.4	193,835	2,335	24.2	29,914	8.4
3,3	262,026	4,629,366	1,021	82.7	100,679	32.5	611,754	535	84.4	125,290	33.1

method on the root net. The method can only be applied on the root net, since it could prevent synchronisations from occurring in the parent net.

Clearly, this modification does not affect the set of reachable states. However, it is questionable whether the modification is useful in practice. The number of edges obtained by caching the pre-synchronisation states could serve as a benchmark for a more intelligent reduction method.

The times in Table 3 exclude the time needed for invoking a C compiler and linker. The peak memory usage was determined by making MARIA allocate everything via `malloc` and by tuning the memory allocator of GNU libc 2.3.1 with `mallopt`. The figures include some non-constant overhead due to pooling. A comparison of the columns "plain" and "pre-cache" reveals only slight differences in resource usage, even though our implementation does not make efficient use of memory, since it stores each pair (M^*, tf) separately.

The effect of this modification depends on implementation aspects and the model. For the interpreter option of MARIA, this modification reduces the time needed for exploring the sliding window protocol by 25 %. The model of the automated guided vehicles (Section 5.1) is unaffected by this modification. In the leader election protocol (Table 1), the number of edges is reduced by 20 %.

Caching Local States. Our implementation of the algorithm presented in Figure 5 of Section 4.1 includes an optional synchronisation state cache. Instead of associating a set S with each invocation of MODULES, this option associates a set S' with each invocation of EXPLORE.

If $(M_{s'}, s', tf, M') \in S'$, the marking M' of s' is reachable via internal transitions from $M_{s'}$ and a transition synchronising on tf is enabled in M'. The cache maps a local state $(M_{s'}, s')$ to pairs of synchronisation labels and states (tf, M').

The procedure MODULES invokes EXPLORE only once for each pair $(M_{s'}, s')$. This caching may save a considerable amount of time if only few of the local successors of $M_{s'}$ are possible synchronisation points. But it will also consume more memory than the original algorithm, because the cache S' associated with the root net s_0 must be preserved until the whole model has been explored.

The columns "local cache" and "pre&local" of Table 3 indicate up to 70 % shorter execution times and 200-fold increase in memory usage compared to the columns "plain" and "pre-cache." A more efficient data structure for managing the cache would be needed to apply this modification in practice.

6   Conclusion and Future Work

We presented a slightly more general version of modular high-level nets than Christensen and Petrucci [4] and an algorithm for checking safety properties in these nets by exhaustive enumeration of modular state spaces. Unlike the algorithm sketch of Christensen and Petrucci [5, Section 8.1], our algorithm does not compute any strongly connected components of state space graphs, and thus does not need to store the transition relation. Our implementation consumes only slightly more memory per state than the algorithm for exploring flat state spaces, and it is compatible with the parallel state space exploration option [14].

From the theoretical point of view, it may be difficult to see the benefits of modular analysis. An experienced modeller would eliminate the internal transitions in Figure 1 by rewriting the arc inscriptions of the synchronising transitions and by adapting the initial marking. The flat state space of the resulting model is identical to the modular state space of the root net.

However, just like the precise modelling of complex systems is easier with high-level nets or other high-level languages than with place/transition systems or state machines, we believe that utilising modular analysis can improve productivity. Some of the tedious work of preparing models for verification [18, Section 7.1] can be avoided and shifted to the state space exploration algorithm. It is unnecessary to reduce the number of internal states or to avoid interleavings between local actions in different processes, as the algorithm takes care of them. Unoptimised models are likely to be easier to maintain than optimised ones. Finally, modules or entire models can be reused in other models.

Although we have described the algorithm in terms of Petri nets, we believe that it can be applied to exploring any system that has a notion of processes or modules that communicate via shared actions. It would be interesting to try the algorithm on some real-world high-level models—such as communication protocol specifications written in SDL [9]—and to see whether the results could be improved by applying partial order reduction methods, as in [11].

Section 4.2 lacks an example of a property covering multiple modules, but we believe that such properties can be checked with "assertion" or "fact" synchronisation transitions that should never be enabled. Extending our algorithm to model check liveness properties is the subject of future research.

The specification of modular systems deserves further research. A system can be modularised in several ways. Ideally, the decomposition of the model should produce greatly reduced state spaces and cater for reuse. Further experiments are needed in order to come up with recommended modularisation strategies.

Acknowledgements. The author would like to thank Charles Lakos for suggesting this research topic, and Laure Petrucci and Charles for valuable feedback during this work. The helpful comments of the anonymous referees are acknowledged.

References

1. Eugenio Battiston, Fiorella De Cindio, and Giancarlo Mauri. Modular algebraic nets to specify concurrent systems. *IEEE Transactions on Software Engineering*, 22(10):689–705, October 1996.
2. Jonathan Billington et al. High-level Petri nets—concepts, definitions and graphical notation, version 4.7.3. Final Draft International Standard ISO/IEC 15909, ISO/IEC JTC1/SC7, Genève, Switzerland, May 2002.
3. Shing Chi Cheung and Jeff Kramer. Checking safety properties using compositional reachability analysis. *ACM TOSEM*, 8(1):49–78, January 1999.
4. Søren Christensen and Laure Petrucci. Modular state space analysis of coloured Petri nets. In Giorgio De Michelis and Michel Diaz, editors, *Application and Theory of Petri Nets 1995, 16th International Conference*, volume 935 of *Lecture Notes in Computer Science*, pages 201–217, Turin, Italy, June 1995. Springer-Verlag.
5. Søren Christensen and Laure Petrucci. Modular analysis of Petri nets. *The Computer Journal*, 43(3):224–242, 2000.
6. Danny Dolev, Maria Klawe, and Michael Rodeh. An $O(n \log n)$ unidirectional distributed algorithm for extrema finding in a circle. *Journal of Algorithms*, 3(3):245–260, September 1982.
7. Orna Grumberg and David E. Long. Model checking and modular verification. *ACM TOPLAS*, 16(3):843–871, May 1994.
8. Gerard J. Holzmann. Spin—formal verification. http://spinroots.com/.
9. Specification and description language (SDL). Recommendation Z.100 (08/02), International Telecommunication Union, Geneva, Switzerland, September 2002.
10. Eric Y. T. Juan, Jeffrey J. P. Tsai, and Tadao Murata. Compositional verification of concurrent systems using Petri-net-based condensation rules. *ACM TOPLAS*, 20(5):917–979, September 1998.
11. Bengi Karaçalı and Kuo-Chung Tai. Model checking based on simultaneous reachability analysis. In Klaus Havelund, John Penix, and Willem Visser, editors, *SPIN Model Checking and Software Verification, 7th International SPIN Workshop*, volume 1885 of *Lecture Notes in Computer Science*, pages 34–53, Stanford, CA, USA, August 2000. Springer-Verlag.
12. Orna Kupferman and Moshe Y. Vardi. Model checking of safety properties. In Nicolas Halbwachs and Doron Peled, editors, *Computer Aided Verification 1999, 11th International Conference (CAV99)*, volume 1633 of *Lecture Notes in Computer Science*, pages 172–183, Trento, Italy, July 1999. Springer-Verlag.
13. Marko Mäkelä. Optimising enabling tests and unfoldings of algebraic system nets. In José-Manuel Colom and Maciej Koutny, editors, *Application and Theory of Petri Nets 2001, 22nd International Conference*, volume 2075 of *Lecture Notes in Computer Science*, pages 283–302, Newcastle upon Tyne, England, June 2001. Springer-Verlag.

14. Marko Mäkelä. Efficiently verifying safety properties with idle office computers. In Charles Lakos, Robert Esser, Lars M. Kristensen, and Jonathan Billington, editors, *Formal Methods in Software Engineering and Defence Systems 2002*, volume 12 of *Conferences in Research and Practice in Information Technology*, pages 11–16, Adelaide, Australia, June 2002. Australian Computer Society Inc.

15. Marko Mäkelä. Maria: Modular reachability analyser for algebraic system nets. In Javier Esparza and Charles Lakos, editors, *Application and Theory of Petri Nets 2002, 23rd International Conference*, volume 2360 of *Lecture Notes in Computer Science*, pages 434–444, Adelaide, Australia, June 2002. Springer-Verlag.

16. Laure Petrucci. Design and validation of a controller. In *Proceedings of the 4th World Multiconference on Systemics, Cybernetics and Informatics*, volume VIII, pages 684–688, Orlando, FL, USA, July 2000. International Institute of Informatics and Systemics.

17. Ulrich Stern and David L. Dill. Parallelizing the Murφ verifier. In Orna Grumberg, editor, *Computer Aided Verification 1997, 9th International Conference (CAV97)*, volume 1254 of *Lecture Notes in Computer Science*, pages 256–267, Haifa, Israel, June 1997. Springer-Verlag.

18. Antti Valmari. The state explosion problem. In Wolfgang Reisig and Grzegorz Rozenberg, editors, *Lectures on Petri Nets I: Basic Models*, volume 1491 of *Lecture Notes in Computer Science*, pages 429–528. Springer-Verlag, 1998.

19. Wei Jen Yeh and Michal Young. Compositional reachability analysis using process algebra. In *Proceedings of the Symposium on Software Testing, Analysis, and Verification*, pages 49–59, Victoria, British Columbia, October 1991. ACM Press.

On Reachability in Autonomous Continuous Petri Net Systems

Jorge Júlvez*, Laura Recalde**, and Manuel Silva

Dep. Informática e Ingeniería de Sistemas, Centro Politécnico Superior de Ingenieros,
Universidad de Zaragoza, María de Luna 3, E-50015 Zaragoza, Spain
{julvez,lrecalde,silva}@unizar.es

Abstract. Fluidification is a common relaxation technique used to deal in a more friendly way with large discrete event dynamic systems. In Petri nets, fluidification leads to continuous Petri nets systems in which the firing amounts are not restricted to be integers. For these systems reachability can be interpreted in several ways. The concepts of *reachability* and lim-*reachability* were considered in [7]. They stand for those markings that can be reached with a finite and an infinite firing sequence respectively. This paper introduces a third concept, the δ-*reachability*. A marking is δ-reachable if the system can get arbitrarily close to it with a finite firing sequence. A full characterization, mainly based on the state equation, is provided for all three concepts for general nets. Under the condition that every transition is fireable at least once, it holds that the state equation does not have spurious solutions if δ-reachability is considered. Furthermore, the differences among the three concepts are in the border points of the spaces they define. For mutual lim-reachability and δ-reachability among markings, i.e., reversibility, a necessary and sufficient condition is provided in terms of liveness.

1 Introduction

Discrete systems with large populations or heavy traffic appear frequently in many fields: manufacturing processes, logistics, telecommunication systems, traffic systems,... It becomes, therefore, interesting to develop adequate formalisms and tools for the analysis and verification of such systems. The "natural" approach to study the above mentioned kind of systems consists in using discrete models. The main drawback is that often an exploration of the state space is needed for the verification of properties. Unfortunately, the size of the state space can grow exponentially with respect to the size of the population of the system, and so many properties are computationally too heavy to be verified.

An interesting approach to study discrete systems with large populations is based on the fluidification of the model. Thus, it is not discrete any more but continuous. This is a classical relaxation technique that can also be applied in

---

* Supported by a grant from D.G.A. ref B106/2001
** Partially supported by project CICYT and FEDER TIC2001-1819

W.M.P. van der Aalst and E. Best (Eds.): ICATPN 2003, LNCS 2679, pp. 221–240, 2003.

the context of Petri nets. Usually, but not always [9], the greater the population of the discrete system the better the continuous approximation.

In PNs, fluidification has been introduced independently from three different perspectives:

- At the net level fluidification was introduced and developed by R. David and coauthors since 1987 [3,1]. In this case, the fluidification of timed discrete systems generates deterministic continuous models, and also hybrid models if there is a partial fluidification.
- Analogously, fluidifying the firing count vector (thus also the marking) in the state equation allows the use of convex geometry and linear programming instead of integer programming, making possible the verification of some properties in polynomial time. The systematic use of linear programming on autonomous and timed system was proposed also in 1987 [8,10].
- K. Trivedi and his group introduced [11,2] a partial fluidification on some stochastic models. The fluidification only affects one or a limited number of places originating stochastic hybrid systems.

Like in [7], in this paper autonomous Petri net models will be considered. In particular, this means that no time interpretation will be applied on the firing of the transitions. A total nondeterminism on the evolution of the system exists. Notice, however, that if the transitions are timed, the evolution/behaviour of the system will always be constrained to some of the possible evolutions/behaviours of the autonomous system.

The paper is organized as follows: in Section 2 reachability in continuous systems is introduced formally and by means of examples. Three different ways of understanding (interpreting) reachability will be considered: reachability in a finite number of steps or simply reachability, reachability in an infinite number of steps or lim-reachability, and δ-reachability that has to do with the capacity of the system to get arbitrarily close to a given "continuous" marking. In order to make the paper more readable, a preview of the main results will be given in that section. Sections 3, 4 and 5 are devoted to the characterization of the sets of reachable markings according to the different concepts: reachability, lim-reachability and δ-reachability respectively. Moreover, it will be seen that it is decidable whether a given "continuous" marking belongs to any of those three concepts. Finally, Section 6 studies reversibility in continuous systems.

2   Definitions and Preview

In the following it is assumed that the reader is familiar with Petri nets (PNs) (see [6,4] for example). The usual PN system will be denoted as $\langle \mathcal{N}, \mathbf{m_0} \rangle$, where $\mathcal{N} = \langle P, T, \mathbf{Pre}, \mathbf{Post} \rangle$. If not explicitly said, all the Petri nets systems considered here are *continuous*. A continuous system is understood as a relaxation of a *discrete* system. The main difference between continuous and discrete PNs is in the firing count vector and consequently in the marking, which in discrete PNs are restricted to be in the naturals, while in continuous PNs are relaxed into the non-negative real numbers. The marking of a place can be seen as an amount of fluid being stored, and the firing of a transition can be considered as a flow of this fluid going from a set of places (input places) to another set of places (output places). Thus, instead of tokens and discrete firings, it is more convenient to talk of levels in the places (deposits/reservoirs) and flows through transitions (valves).

The firing of a transition is also modified and brought to the non-negative real domain. A transition t is *enabled* at $\mathbf{m}$ iff for every $p \in {}^{\bullet}t$, $\mathbf{m}[p] > 0$. In other words, the enabling condition of continuous systems and that of discrete ordinary systems can be expressed in an "analogous" way: every input place should be marked. Notice that to decide whether a transition in a continuous system is enabled or not it is not necessary to consider weights of the arcs going from the input places to the transition. However, the arc weights are important to compute the enabling degree of a transition and to obtain the new marking after a firing. As in discrete systems, the *enabling degree* at $\mathbf{m}$ of a transition measures the maximal amount in which the transition can be fired in a single occurrence, i.e., $\mathrm{enab}(t, \mathbf{m}) = \min_{p \in {}^{\bullet}t}\{\mathbf{m}[p]/\mathbf{Pre}[p, t]\}$. The firing of t in a certain amount $\alpha \leq \mathrm{enab}(t, \mathbf{m})$ leads to a new marking $\mathbf{m}'$, and it is denoted as $\mathbf{m} \xrightarrow{\alpha t} \mathbf{m}'$. It holds $\mathbf{m}' = \mathbf{m} + \alpha \cdot \mathbf{C}[P, t]$, where $\mathbf{C} = \mathbf{Post} - \mathbf{Pre}$ is the token flow matrix (incidence matrix if $\mathcal{N}$ is self-loop free). Hence, as in discrete systems, $\mathbf{m} = \mathbf{m_0} + \mathbf{C} \cdot \sigma$, the state (or fundamental) equation summarizes the way the marking evolves. Right and left natural annullers of the token flow matrix are called T- and P-semiflows, respectively. As in discrete systems, when $\mathbf{y} \cdot \mathbf{C} = \mathbf{0}$, $\mathbf{y} > \mathbf{0}$ the net is said to be *conservative*, and when $\mathbf{C} \cdot \mathbf{x} = \mathbf{0}$, $\mathbf{x} > \mathbf{0}$ the net is said to be *consistent*. A set of places Θ is a *trap* iff $\Theta^{\bullet} \subseteq {}^{\bullet}\Theta$. Similarly, a set of places Σ is a *siphon* iff ${}^{\bullet}\Sigma \subseteq \Sigma^{\bullet}$. The support of a vector $\mathbf{x} \geq \mathbf{0}$ will be denoted as $\|\mathbf{x}\|$ and represents the set of positive elements of $\mathbf{x}$.

In order to illustrate the firing rule in a continuous system, let us consider the system in Figure 1(a). The only enabled transition at the initial marking is t_1 whose enabling degree is 1. Hence, it can be fired in any real quantity going from 0 to 1. For example, firing by 0.5 would yield marking $\mathbf{m}_1 = (0.5, 0.5, 1, 0)$. At $\mathbf{m}_1$ transition t_2 has enabling degree equal to 0.5; if it is fired in this amount the resulting marking is $\mathbf{m}_2 = (0.5, 0.5, 0, 0.5)$. Both $\mathbf{m}_1$ and $\mathbf{m}_2$ are reachable markings with finite firing sequences, or simply reachable markings.

The set of all reachable markings for a given system $\langle \mathcal{N}, \mathbf{m_0} \rangle$ is denoted as $\mathrm{RS}(\mathcal{N}, \mathbf{m_0})$:

Definition 1 $\mathrm{RS}(\mathcal{N}, \mathbf{m_0}) = \{ \mathbf{m} |$ *a finite fireable sequence* $\sigma = \alpha_1 t_{a_1} \ldots \alpha_k t_{a_k}$ *exists such that* $\mathbf{m_0} \xrightarrow{\alpha_1 t_{a_1}} \mathbf{m}_1 \xrightarrow{\alpha_2 t_{a_2}} \mathbf{m}_2 \ldots \xrightarrow{\alpha_k t_{a_k}} \mathbf{m}_k = \mathbf{m}$ *where* $t_{a_i} \in T$ *and* $\alpha_i \in \mathbb{R}^+\}$.

An interesting property of $\mathrm{RS}(\mathcal{N}, \mathbf{m_0})$ is that it is a *convex* set (see [7]). That is, if two markings $\mathbf{m}_1$ and $\mathbf{m}_2$ are reachable, then for any $\alpha \in [0, 1]$ $\alpha \mathbf{m}_1 + (1 - \alpha)\mathbf{m}_2$ is also a reachable marking.

Let us consider again the system in Figure 1(a) with initial marking $\mathbf{m_0} = (0.5, 0.5, 0, 0.5)$. At this marking either transition t_1 or transition t_3 can be fired. The firing of t_3 in an amount of 0.5 makes the system evolve to marking $(0.5, 0.5, 0.5, 0)$ from which t_2 can be fired in an amount of 0.25 leading to

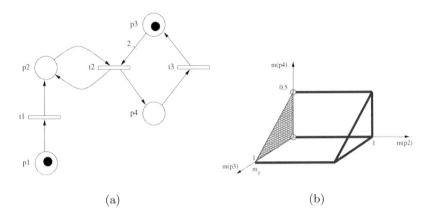

(a)   (b)

Fig. 1. (a) Autonomous continuous system (b) Lim-Reachability space

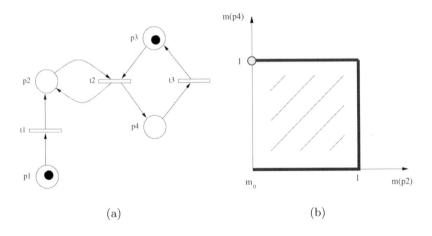

(a)   (b)

Fig. 2.   (a) Autonomous continuous system (b) Reachability space and Lim-Reachability space coincide

marking (0.5, 0.5, 0, 0.25). Now, the markings of places p_1, p_2 and p_3 are the same that those of the system at $\mathbf{m}_0$, but the marking of p_4 is half of its marking at $\mathbf{m}_0$. The continuous firing of transitions t_2 and t_3 by its maximum enabling degree causes the elimination of half of the marking of p_4. Assume that we go on firing transitions t_2 and t_3. Then, as the number of firings increases the marking of p_4 approaches 0, value that will only be reached in the limit. Notice that the marking reached in the limit (0.5, 0.5, 0, 0) corresponds to the emptying of an initially marked trap ($\Theta = \{p_3, p_4\}, \Theta^\bullet = {}^\bullet\Theta = \{t_2, t_3\}$), fact that does not occur in discrete systems. From the point of view of the analysis of the behaviour of the system, it is interesting to consider this marking as limit-reachable, since it is the one to which the state of the system may converge. We will define the set of such markings that are reachable with a finite/infinite firing sequence:

Definition 2 *[7] Let $\langle \mathcal{N}, \mathbf{m_0} \rangle$ be a continuous system. A marking $\mathbf{m} \in (\mathbb{R}^+ \cup \{0\})^{|P|}$ is lim-reachable, iff a sequence of reachable markings $\{\mathbf{m}_i\}_{i \geq 1}$ exists such that*

$$\mathbf{m_0} \xrightarrow{\sigma_1} \mathbf{m}_1 \xrightarrow{\sigma_2} \mathbf{m}_2 \cdots \mathbf{m}_{i-1} \xrightarrow{\sigma_i} \mathbf{m}_i \cdots$$

and $\lim_{i \to \infty} \mathbf{m}_i = \mathbf{m}$. *The* lim-*reachable space is the set of* lim-*reachable markings, and will be denoted* lim-RS$(\mathcal{N}, \mathbf{m_0})$.

Figure 1(b) depicts the lim-reachability space of system in Figure 1(a). It is not necessary to represent the marking of place p_1 since $\mathbf{m}_1 = 1 - \mathbf{m}_2$. The set of lim-reachable markings is composed of the points inside the prism, the points in the non shadowed sides, the points in the thick edges and the points in the non circled vertices.

For some systems, the sets RS$(\mathcal{N}, \mathbf{m_0})$ and lim-RS$(\mathcal{N}, \mathbf{m_0})$ are identical. This means that in this case, with regard to the set of reachable markings, there is no difference between considering sequences of finite or infinite length. See Figure 2 for an example. Only $\mathbf{m}_2$ and $\mathbf{m}_4$ are represented since $\mathbf{m}_1 = 1 - \mathbf{m}_2$ and $\mathbf{m}_3 = 1 - \mathbf{m}_4$. The innner points of the square defined by the vertices $(0, 0)$, $(0, 1)$, $(1, 1)$ and $(1, 0)$, and the thick lines in Figure 2(b) are part of the reachability and the lim-reachability space, while the points going from $\mathbf{m_0}$ to $(0, 1)$ (including $(0, 1)$) do not belong to these sets.

However, in general, the set of reachable markings, RS$(\mathcal{N}, \mathbf{m_0})$ is a subset of the set of lim-reachable markings, lim-RS$(\mathcal{N}, \mathbf{m_0})$. For the system in Figure 3(b), neither p_1 nor p_2 can be emptied with a finite firing sequence because every time a transition is fired some marks are put in both places. For that system the set of reachable markings is $(\alpha, 2 - \alpha)$, $0 < \alpha < 2$. Nevertheless, considering the sequence $\frac{1}{2}t_1, \frac{1}{4}t_1, \frac{1}{8}t_1, \ldots$, in the k-th step, the system reaches the marking $(2^{-k}, 2-2^{-k})$. When k tends to infinity the marking of the system tends to $(0, 2)$. Therefore the infinite firing of t_1 (t_2) will converge to a marking in which p_1 (p_2) is empty. Thus the set of markings reachable in the limit is $(\alpha, 2-\alpha)$, $0 \leq \alpha \leq 2$. Notice that the only difference between both sets lim-RS$(\mathcal{N}, \mathbf{m_0})$ and RS$(\mathcal{N}, \mathbf{m_0})$ is in the markings $(0, 2)$ and $(2, 0)$. Observe that even under consistency and conservativeness RS$(\mathcal{N}, \mathbf{m_0}) \neq$ lim-RS$(\mathcal{N}, \mathbf{m_0})$.

For the system in Figure 3(a), p_1 (p_2) can be emptied with the firing of t_1 (t_2) in an amount of 1. Hence, although the systems in Figure 3 have the same incidence matrix, their sets of finitely reachable markings are not the same.

Both RS$(\mathcal{N}, \mathbf{m_0})$ and lim-RS$(\mathcal{N}, \mathbf{m_0})$ are not in general closed sets. For example in Figure 2(b) the points on the segment going from $(0, 0)$ (initial marking) to $(0, 1)$ do neither belong to RS$(\mathcal{N}, \mathbf{m_0})$ nor to lim-RS$(\mathcal{N}, \mathbf{m_0})$. Nevertheless, any point on the right of this segment does belong to both sets RS$(\mathcal{N}, \mathbf{m_0})$ and lim-RS$(\mathcal{N}, \mathbf{m_0})$. For a given set A, the closure of A is equal to the points in A plus those points which are infinitely close to points in A, but are not contained in A. In the case of the set depicted in Figure 2(b) its closure is equal to the

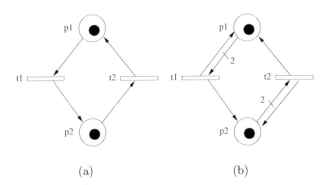

(a)   (b)

Fig. 3. Continuous systems that have the same incidence matrix and whose reachability spaces do not coincide.

inner and edge points of the square defined by the vertices (0, 0), (0, 1), (1, 1) and (1, 0), that is, it is obtained by adding the segment $[(0, 0), (0, 1)]$.

Focusing on the spaces defined by $\mathrm{RS}(\mathcal{N}, \mathbf{m_0})$ and lim-$\mathrm{RS}(\mathcal{N}, \mathbf{m_0})$ and closing them, it will be noticed that the points limiting both spaces are exactly the same. This is because if the system can get as close as desired to a given point with an infinite sequence, it can also get as close as desired with a finite sequence and vice versa. Hence, the following property can be stated:

Property 3 *The closure of* $\mathrm{RS}(\mathcal{N}, \mathbf{m_0})$ *is equal to the closure of* lim-$\mathrm{RS}(\mathcal{N}, \mathbf{m_0})$.

Assume that, given a system, $\mathrm{RS}(\mathcal{N}, \mathbf{m_0})$ and lim-$\mathrm{RS}(\mathcal{N}, \mathbf{m_0})$ are not identical sets, i.e., $\mathrm{RS}(\mathcal{N}, \mathbf{m_0}) \not\subseteq$ lim-$\mathrm{RS}(\mathcal{N}, \mathbf{m_0})$. This means that for every $\mathbf{m}$ in lim-$\mathrm{RS}(\mathcal{N}, \mathbf{m_0}) \setminus \mathrm{RS}(\mathcal{N}, \mathbf{m_0})$, $\mathbf{m}$ is a *border* point of lim-$\mathrm{RS}(\mathcal{N}, \mathbf{m_0})$; that is, there are markings in $\mathrm{RS}(\mathcal{N}, \mathbf{m_0})$ infinitely close to $\mathbf{m}$ that do not belong to lim-$\mathrm{RS}(\mathcal{N}, \mathbf{m_0})$. Let us make a final consideration on the system of Figure 1(a). It has been seen that the initial firing of t_1 enables t_2 and that an infinite sequence consisting on firing t_2 and t_3 will empty p_3 and p_4, reaching marking (0.5, 0.5, 0, 0). In that example t_1 was fired in an amount of 0.5. Nevertheless, p_3 and p_4 can be emptied also if t_1 is fired in an amount α such that $0 < \alpha \leq 1$. For example, if we take $\alpha = 0.1$, we fire t_1 in an amount of 0.1 and then fire t_2 five times in an amount of 0.1. Now we can fire completely, in an amount of 0.5, transition t_3. Repeating this procedure, in the limit p_3 and p_4 become empty. Thus, it can be said that the marking $(1 - \alpha, \alpha, 0, 0)$ is lim-reachable for any α such that $0 < \alpha \leq 1$. Hence, marking (1, 0, 0, 0) is not lim-reachable but the system can get as close as desired to it by taking a small enough α. This marking can then be interpreted as the fact that a little leak of fluid from p_1 to p_2 can cause the emptying of p_3 and p_4. In some situations, it may be useful to consider those markings like (1, 0, 0, 0), that are not reachable, but for which the system can get as close as desired.

Let us consider a norm in order to determine the proximity of two markings. Let $|\mathbf{x}|$ denote the norm of vector $\mathbf{x} = (x_1, \dots, x_n)$ defined as: $|\mathbf{x}| = |x_1| + \dots +$

$|x_n|$. A new reachability concept for continuous systems will be introduced: the δ-reachability. The set of δ-reachable markings will be written as δ-RS$(\mathcal{N}, \mathbf{m_0})$ and accounts for those markings to which the system can get as close as desired firing a finite sequence. Formally:

Definition 4 δ-RS$(\mathcal{N}, \mathbf{m_0})$ *is the closure of* RS$(\mathcal{N}, \mathbf{m_0})$: δ-RS$(\mathcal{N}, \mathbf{m_0}) = \{ \mathbf{m} \mid$ *for every* $\epsilon > 0$ *a marking* $\mathbf{m'} \in$ RS$(\mathcal{N}, \mathbf{m_0})$ *exists such* $|\mathbf{m'} - \mathbf{m}| < \epsilon\}$.

Since the closure of RS$(\mathcal{N}, \mathbf{m_0})$ is equal to the closure of lim-RS$(\mathcal{N}, \mathbf{m_0})$, δ-RS$(\mathcal{N}, \mathbf{m_0})$ is also equal to the set of markings to which the system can get as close as desired firing an infinite sequence. RS$(\mathcal{N}, \mathbf{m_0})$ and lim-RS$(\mathcal{N}, \mathbf{m_0})$ are, therefore, subsets of δ-RS$(\mathcal{N}, \mathbf{m_0})$.

Therefore, till now three different kinds of reachability concepts have been defined:

- Markings that are reachable with a finite firing sequence, RS$(\mathcal{N}, \mathbf{m_0})$.
- Markings to which the system converges, eventually, with an infinitely long sequence, lim-RS$(\mathcal{N}, \mathbf{m_0})$.
- Markings to which the system can get as close as desired with a finite sequence, δ-RS$(\mathcal{N}, \mathbf{m_0})$.

Let us finish this section by defining the linearized reachability set with respect to the state equation:

Definition 5 LRS$(\mathcal{N}, \mathbf{m_0}) = \{\mathbf{m} \mid \mathbf{m} = \mathbf{m_0} + \mathbf{C} \cdot \boldsymbol{\sigma} \geq \mathbf{0}$ *with* $\boldsymbol{\sigma} \in (\mathbb{R}^+ \cup \{0\})^{|T|}\}$.

Notice that given a consistent system (i.e., $\exists\ \mathbf{x} > \mathbf{0} | \mathbf{C} \cdot \mathbf{x} = \mathbf{0}$) it holds: LRS$(\mathcal{N}, \mathbf{m_0}) = \{\mathbf{m} \mid \mathbf{m} = \mathbf{m_0} + \mathbf{C} \cdot \boldsymbol{\sigma} \geq \mathbf{0}$ with $\boldsymbol{\sigma} \in \mathbb{R}^{|T|}\}$. In [7] it was shown that for consistent systems in which every transition is fireable at least once, the sets LRS$(\mathcal{N}, \mathbf{m_0})$ and lim-RS$(\mathcal{N}, \mathbf{m_0})$ are the same. This result will be generalized by describing the set of lim-reachable markings of a general system.

By definition LRS$(\mathcal{N}, \mathbf{m_0})$ is a closed set. $\mathbf{m}$ is a *border* point of LRS$(\mathcal{N}, \mathbf{m_0})$ iff for every $\epsilon > 0$ there exists $\mathbf{m'}$, $|\mathbf{m'} - \mathbf{m}| < \epsilon$ such that $\mathbf{m'} \notin$ LRS$(\mathcal{N}, \mathbf{m_0})$.

The *open* set of LRS$(\mathcal{N}, \mathbf{m_0})$ is the result of removing every border point from LRS$(\mathcal{N}, \mathbf{m_0})$ and will be denoted as]LRS$(\mathcal{N}, \mathbf{m_0})[$.

Notice that given a system $\langle \mathcal{N}, \mathbf{m_0} \rangle$ if there exists $\mathbf{y} \neq \mathbf{0}$ such that $\mathbf{y} \cdot \mathbf{C} = \mathbf{0}$ then every $\mathbf{m} \in$ LRS$(\mathcal{N}, \mathbf{m_0})$ is a border point of LRS$(\mathcal{N}, \mathbf{m_0})$, and so in this case]LRS$(\mathcal{N}, \mathbf{m_0})[= \emptyset$. If such $\mathbf{y}$ exists all the points in LRS$(\mathcal{N}, \mathbf{m_0})$ are contained in a hyperplane of smaller dimension than the number of places. In particular, if a system has a p-semiflow, every marking in LRS$(\mathcal{N}, \mathbf{m_0})$ is a border point. Those markings having null components are also border points of LRS$(\mathcal{N}, \mathbf{m_0})$.

Since all reachable, lim-reachable and δ-reachable markings are solution of the state equation, the following relation is satisfied:
RS$(\mathcal{N}, \mathbf{m_0}) \subseteq$ lim-RS$(\mathcal{N}, \mathbf{m_0}) \subseteq \delta$-RS$(\mathcal{N}, \mathbf{m_0}) \subseteq$ LRS$(\mathcal{N}, \mathbf{m_0})$.

Along the paper this relationship among the different sets will be completed showing that the open linearized set,]LRS$(\mathcal{N}, \mathbf{m_0})[$, is contained in RS$(\mathcal{N}, \mathbf{m_0})$ and that δ-RS$(\mathcal{N}, \mathbf{m_0}) =$ LRS$(\mathcal{N}, \mathbf{m_0})$ if every transition is fireable at least once.

3   RS($\mathcal{N}$, $\mathbf{m_0}$)

The goal of this section is first to provide a full characterization of the set of reachable markings (Subsection 3.1) and then to show a computation algorithm that decides the reachability of a given target marking (Subsection 3.2).

3.1   Reachability Characterization

Before showing the main result (Theorem 12), some intermediate lemmas will be presented in order to ease the final characterization. First, let us introduce an algorithm to compute the sets of transitions fireable from the initial marking, and some interesting results dealing with continuous systems.

Let $FS(\mathcal{N}, \mathbf{m_0})$ be the set of sets of transitions for which there exists a sequence fireable from $\mathbf{m_0}$ that contains those and only those transitions in the set. Formally,

Definition 6 $FS(\mathcal{N}, \mathbf{m_0}) = \{\ \theta |\ there\ exists\ a\ sequence\ fireable\ from\ \mathbf{m_0},\ \sigma,\ such\ that\ \theta = \|\boldsymbol{\sigma}\|\}$.

Algorithm 7 (Computation of the set $FS(\mathcal{N}, \mathbf{m_0})$)

1. *Let V be the set of transitions enabled at $\mathbf{m_0}$*
2. *$FS := \{v | v \subseteq V\}$ % all the subsets of V including the empty set*
3. *Repeat*
   - **3.1.** *take $f \in FS$ such that it has not been taken before*
   - **3.2.** *fire sequentially from $\mathbf{m_0}$ every transition in f without disabling any enabled transition. Let $\mathbf{m}$ be the reached marking.*
   - **3.3.** *$V := \{t | t$ is enabled at $\mathbf{m}$ and $t \notin f\}$*
   - **3.4.** *$FS := FS \cup \{f \cup v | v \subseteq V\}$*
4. *until FS does not increase*

Notice that step **3.2.** can always be achieved since for any element $f \in FS(\mathcal{N}, \mathbf{m_0})$ there exists a fireable sequence that contains every transition in f. Algorithm 7 accounts for all possible subsets of transitions that can become enabled, and so its complexity is exponential on the number of transitions and so is the size of the set $FS(\mathcal{N}, \mathbf{m_0})$. As an example, considering the net in Figure 4 with initial marking $\mathbf{m_0} = (1, 0, 1, 1, 0)$ the result of Algorithm 7 is $FS(\mathcal{N}, \mathbf{m_0}) = \{\ \{\}, \{t_2\}, \{t_3\}, \{t_4\}, \{t_2,\ t_3\}, \{t_2,\ t_4\}, \{t_3,\ t_4\}, \{t_2,\ t_3,\ t_4\}, \{t_1, t_2\}, \{t_4, t_5\}, \{t_1, t_2, t_3\}, \{t_1, t_2, t_4\}, \{t_2, t_4, t_5\}, \{t_1, t_2, t_4, t_5\}, \{t_3, t_4, t_5\}, \{t_1, t_2, t_3, t_4\}, \{t_2, t_3, t_4, t_5\}, \{t_1, t_2, t_3, t_4, t_5\}\}$.

Now let us introduce four lemmas that will help to characterize the set of reachable markings. The first one simply states that continuous systems are homothetic w.r.t. the scaling of $\mathbf{m_0}$.

Lemma 8 *[7] Let $\langle \mathcal{N}, \mathbf{m_0} \rangle$ be a continuous system. If σ is a fireable sequence yielding marking $\mathbf{m}$, then for any $\alpha \geq 0$, $\alpha\sigma$ is fireable at $\alpha\mathbf{m_0}$ yielding marking $\alpha\mathbf{m}$, where $\alpha\sigma$ represents a sequence that is equal to σ except in the amount of each firing, that is multiplied by α.*

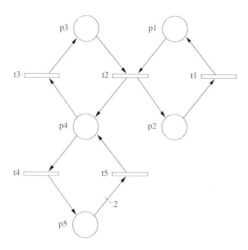

Fig. 4. Non-consistent continuous system

Although this section deals with those markings that are reachable with a finite firing sequence, a lemma that has to do with the markings that can be reached in the limit will be presented. Lemma 9 establishes that if all the transitions in the support of a given firing vector σ are enabled, then $\mathbf{m} = \mathbf{m_0} + \mathbf{C} \cdot \sigma \geq 0$ is reachable in the limit, whatever the value of σ is. Furthermore, there exists a sequence of reachable markings that are "in the direction" of $\mathbf{m}$.

Lemma 9 *Let $\langle \mathcal{N}, \mathbf{m_0} \rangle$ be a continuous system. Let $\mathbf{m} = \mathbf{m_0} + \mathbf{C} \cdot \sigma \geq 0$, $\sigma \geq \mathbf{0}$ and $\mathbf{m_0}$ such that for every $t \in \|\sigma\|$ $\mathrm{enab}(t, \mathbf{m_0}) > 0$. Then, there exists a succession of reachable markings $\mathbf{m_1}, \mathbf{m_2}, \ldots$ fulfilling $\mathbf{m_1} = \mathbf{m_0} + \beta_1 \mathbf{C} \cdot \sigma$, $\mathbf{m_2} = \mathbf{m_0} + \beta_2 \mathbf{C} \cdot \sigma, \ldots$ with $0 < \beta_1 < \beta_2 < \ldots$ that converges to $\mathbf{m}$.*

Proof. Since at $\mathbf{m_0}$ every transition of $\|\sigma\|$ is enabled, α and σ' exist such that σ' is fireable from $\mathbf{m_0}$ and $\sigma' = \alpha\sigma$, i.e., a sequence proportional to the vector leading from $\mathbf{m_0}$ to $\mathbf{m}$ can be fired. If $\alpha \geq 1$, it is clear that $\mathbf{m}$ can be reached from $\mathbf{m_0}$. Otherwise, the firing of σ' leads to $\mathbf{m_0} + \mathbf{C} \cdot \alpha\sigma = (1-\alpha)\mathbf{m_0} + \alpha\mathbf{m_0} + \mathbf{C} \cdot \alpha\sigma = (1-\alpha)\mathbf{m_0} + \alpha\mathbf{m}$. By Lemma 8, if σ' was fireable from $\mathbf{m_0}$, then $(1-\alpha)\sigma'$ is fireable from $(1-\alpha)\mathbf{m_0}$. In this way, we have

$$\alpha\mathbf{m} + (1-\alpha)\mathbf{m_0} \xrightarrow{(1-\alpha)\sigma'} \alpha\mathbf{m} + \alpha(1-\alpha)\mathbf{m} + (1-\alpha)^2\mathbf{m_0}$$

Repeating this procedure, in the iteration n we reach the marking

$$\alpha\mathbf{m}(1 + (1-\alpha) + (1-\alpha)^2 + \ldots + (1-\alpha)^n) + (1-\alpha)^n\mathbf{m_0}$$

Thus, the marking of the system as n goes to infinity converges to $\mathbf{m}$. □

Based on this result a part of the set of reachable markings can be described.

Lemma 10 *Let $\langle \mathcal{N}, \mathbf{m_0} \rangle$ be a continuous system. Let $\mathbf{m} = \mathbf{m_0} + \mathbf{C} \cdot \sigma \geq 0$, $\sigma \geq \mathbf{0}$ and for every $t \in \|\sigma\|$ $\mathrm{enab}(t, \mathbf{m_0}) > 0$ and $\mathrm{enab}(t, \mathbf{m}) > 0$. Then $\mathbf{m} \in RS(\mathcal{N}, \mathbf{m_0})$.*

Proof. Every $t \in \|\sigma\|$ is enabled at $\mathbf{m}$. This means that for every $t \in \|\sigma\|$ all its input places are positively marked at $\mathbf{m}$. Then, we can define an $\mathbf{m}'$ such that $\mathbf{m}' = \mathbf{m_0} + \mathbf{C} \cdot (1 + \alpha)\sigma \geq 0$ with $\alpha > 0$. According to Lemma 9 there is a succession of markings that converges to $\mathbf{m}'$. Since $\mathbf{m}$ is in the line that goes from $\mathbf{m_0}$ to $\mathbf{m}'$ we can stop that sequence at a given step and reach exactly $\mathbf{m}$ in a finite number of firings. □

The following last lemma imposes a necessary and sufficient condition for the fireability of a transition in terms of siphons.

Lemma 11 *Let* $\mathbf{m} \in \mathrm{RS}(\mathcal{N}, \mathbf{m_0})$. *Transition* t *is not fireable for any successor of* $\mathbf{m}$ *iff there exists an empty siphon at* $\mathbf{m}$ *containing a place* p *such that* $p \in {}^{\bullet}t$.

Proof.
($\Leftarrow$)
If there exists such an empty siphon Θ, no transition in $\Theta^{\bullet}$ is fireable.
($\Rightarrow$)
Assume t is not fireable for any successor of $\mathbf{m}$. Then there exists a place p such that $p \in {}^{\bullet}t$ and $\mathbf{m}(p) = 0$. Furthermore, no input transition of p, t', can ever be fired. Hence, for every t' there exists an empty input place p'. Repeating this reasoning we obtain a set of empty places Q. This set Q has the property that all its input transitions $({}^{\bullet}Q)$ are output transitions $(Q^{\bullet})$. Hence Q is an empty siphon. □

Before going on with the characterization of the set of reachable markings let us make some considerations on the conditions a given marking $\mathbf{m}$ should fulfill in order to be reachable. First of all, it is clear that a necessary condition for $\mathbf{m}$ to be reachable is that it has to be solution of the state equation, that is, there must exist σ such that $\mathbf{m} = \mathbf{m_0} + \mathbf{C} \cdot \sigma$. Furthermore, $\|\sigma\|$ must be in $FS(\mathcal{N}, \mathbf{m_0})$ in order to have a fireable sequence. In Section 2 it has been seen that some marked traps can be emptied in a continuous system with the firing of an infinite sequence. If only finite firing sequences are considered, no marked trap can be emptied. Since now the interest lies in finite firing sequences those σ's that correspond to a firing count vector that empties (or fills and then empties) a trap have to be explicitly forbidden. As it will be seen, these necessary conditions are also sufficient for a marking to be reachable.

Given a net $\mathcal{N}$ and a firing sequence σ, let us denote as $\mathcal{N}_\sigma$ the net obtained removing from $\mathcal{N}$ the transitions not in the support of σ and the resulting isolated places. In other words, $\mathcal{N}_\sigma$ is the net composed of the transitions of $\mathcal{N}$ in the support of σ and their input and output places. Using the previous lemmas a full characterization of the set of reachable markings is obtained.

Theorem 12 *A marking* $\mathbf{m} \in \mathrm{RS}(\mathcal{N}, \mathbf{m_0})$ *iff*

1. $\mathbf{m} = \mathbf{m_0} + \mathbf{C} \cdot \sigma \geq 0$, $\sigma \geq 0$
2. $\|\sigma\| \in FS(\mathcal{N}, \mathbf{m_0})$
3. there is no empty trap in $\mathcal{N}_\sigma$ *at* $\mathbf{m}$

Proof.

⊆

Let $\mathbf{m} \in RS(\mathcal{N}, \mathbf{m_0})$. Then, there exists $\boldsymbol{\sigma} \geq \mathbf{0}$ such that $\mathbf{m} = \mathbf{m_0} + \mathbf{C} \cdot \boldsymbol{\sigma}$ and $\|\boldsymbol{\sigma}\| \in FS(\mathcal{N}, \mathbf{m_0})$. Furthermore, there cannot be an empty trap in $\mathcal{N}_{\boldsymbol{\sigma}}$ at $\mathbf{m}$ since it would mean that the trap was emptied with a finite firing sequence.

⊇

Let $\mathbf{m}$ be such that $\mathbf{m} = \mathbf{m_0} + \mathbf{C} \cdot \boldsymbol{\sigma} \geq \mathbf{0}$, $\boldsymbol{\sigma} \geq \mathbf{0}$, $\|\boldsymbol{\sigma}\| \in FS(\mathcal{N}, \mathbf{m_0})$ and there is no empty trap in $\mathcal{N}_{\boldsymbol{\sigma}}$ at $\mathbf{m}$. It will be shown that $\mathbf{m}$ can be reached from $\mathbf{m_0}$ by a finite firing sequence. This will be done in three steps: from $\mathbf{m_0}$ we will reach a marking $\mathbf{m'}$ at which every transition $t \in \|\boldsymbol{\sigma}\|$ is enabled. From $\mathbf{m'}$ we will make the system evolve to a marking $\mathbf{m''}$ at which also every transition $t \in \|\boldsymbol{\sigma}\|$ is enabled and is as closed as desired to $\mathbf{m}$. Finally, due to the way $\mathbf{m''}$ is defined it is shown that $\mathbf{m}$ is reachable from $\mathbf{m''}$. Although the order of the sequence of reachable markings is $\mathbf{m_0}$, $\mathbf{m'}$, $\mathbf{m''}$ and $\mathbf{m}$, we will start by defining $\mathbf{m''}$ and showing how it can be reached.

If every $t \in \|\boldsymbol{\sigma}\|$ is enabled at $\mathbf{m}$ then $\mathbf{m''} = \mathbf{m}$. Otherwise, we will consider the system with marking $\mathbf{m}$ and we will fire backwards and sequentially transitions in $\|\boldsymbol{\sigma}\|$ until we reach a marking $(\mathbf{m''})$ at which every transition in $\|\boldsymbol{\sigma}\|$ is enabled. Notice that this backward firing is equivalent to a forward firing in the reverse net (changing directions of arcs). We will reason that such a firing from $\mathbf{m}$ to $\mathbf{m''}$ is always feasible. Notice that in the reverse net traps have become siphons (structural deadlocks) and the forward firing in the reverse net of transitions in $\|\boldsymbol{\sigma}\|$ never involves the filling of empty siphons of $\mathcal{N}_{\boldsymbol{\sigma}}$ at $\mathbf{m}$. This is because according to the initial condition 3, "there is no empty trap in $\mathcal{N}_{\boldsymbol{\sigma}}$ at $\mathbf{m}$". Therefore, by Lemma 11 we can assure that every transition $t \in \|\boldsymbol{\sigma}\|$ can be fired in the reverse net. Let us denote $\widehat{\boldsymbol{\sigma}}$ the firing count vector such that $\mathbf{m} = \mathbf{m''} + \mathbf{C} \cdot \widehat{\boldsymbol{\sigma}}$ with $\|\widehat{\boldsymbol{\sigma}}\| = \|\boldsymbol{\sigma}\|$.

Now we will define $\mathbf{m'}$ as a marking reached from $\mathbf{m_0}$ at which every transition $t \in \|\boldsymbol{\sigma}\|$ is enabled, $\mathbf{m'} = \mathbf{m_0} + \mathbf{C} \cdot \boldsymbol{\sigma'}$ and $\boldsymbol{\sigma'} \leq \boldsymbol{\sigma}$. This is always possible since $\|\boldsymbol{\sigma}\|$ belongs to $FS(\mathcal{N}, \mathbf{m_0})$ and so we can fire small amounts of the transitions in $\|\boldsymbol{\sigma}\|$ until every transition in $\|\boldsymbol{\sigma}\|$ is enabled. We will define $\boldsymbol{\sigma''}$ as $\boldsymbol{\sigma''} = \boldsymbol{\sigma} - \boldsymbol{\sigma'} - \widehat{\boldsymbol{\sigma}}$, then $\mathbf{m''} = \mathbf{m'} + \mathbf{C} \cdot \boldsymbol{\sigma''}$. Notice that since $\boldsymbol{\sigma'}$ and $\widehat{\boldsymbol{\sigma}}$ can be taken as small as wanted and their supports are contained in the support of $\boldsymbol{\sigma}$, it can always be verified that $\boldsymbol{\sigma''} \geq \mathbf{0}$ and $\|\boldsymbol{\sigma}\| = \|\boldsymbol{\sigma''}\|$. Moreover, Lemma 10 can be directly applied on $\mathbf{m'}$ and $\boldsymbol{\sigma''}$ obtaining that $\mathbf{m''}$ is reachable from $\mathbf{m'}$. And finally, we can conclude that $\mathbf{m}$ is reachable from $\mathbf{m_0}$.   □

Figure 5 sketches the trajectory built by the proof of Theorem 12 to reach $\mathbf{m}$.

As an example, let us take the system in Figure 6. The marking $\mathbf{m} = (0, 0, 0, 0, 1)$ is solution of the state equation and can be obtained with vectors: $\boldsymbol{\sigma}_1 = (1, 0, 1, 1, 0, 0)$ and $\boldsymbol{\sigma}_2 = (0, 1, 0, 0, 1, 0)$. Obviously, $\boldsymbol{\sigma}_2$ fulfills the conditions of Theorem 12, and so it can be concluded that $\mathbf{m}$ can be reached. However, if we consider the system that results of removing transitions t_2, t_5 and the place p_4, then the only possibility to reach $\mathbf{m}$ is with $\boldsymbol{\sigma}_1$ or with $\boldsymbol{\sigma}_1 + \mathbf{x}$ where $\mathbf{x}$ is a T-semiflow. Notice that the nets $\mathcal{N}_{\boldsymbol{\sigma}_1}$ and $\mathcal{N}_{\boldsymbol{\sigma}_1+\mathbf{x}}$ have an empty trap at $\mathbf{m}$ composed of $\{p_2, p_3\}$. Hence, the third condition of Theorem 12 is violated and $\mathbf{m}$ cannot be reached with a finite sequence.

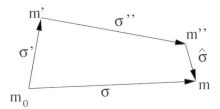

Fig. 5. Trajectory to reach **m** with a finite firing sequence

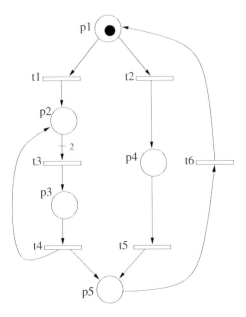

Fig. 6. Marking $(0, 0, 0, 0, 1)$ can be reached with a finite and with an infinite firing sequence

3.2   Deciding Reachability

Based on Theorem 12, an algorithm that decides whether a given marking **m** is reachable or not is introduced. A necessary condition for **m** to be reached is that there must exist a $\sigma \geq 0$ such that $\mathbf{m} = \mathbf{m_0} + \mathbf{C} \cdot \sigma \geq 0$. Given a marking **m** the number of $\sigma \geq 0$ fulfilling the state equation can be infinite. However, as stated in Theorem 12, it is only interesting to consider those σ's such that $\|\sigma\| \in FS(\mathcal{N}, \mathbf{m_0})$. Furthermore, it is not necessary to consider two different σ's that are solution of the state equation and have the same support, since clearly one of those σ's fulfills the condition on the traps of Theorem 12 iff the other one also fulfills it, and the support of one belongs to $FS(\mathcal{N}, \mathbf{m_0})$ iff the other one also belongs to it. This reasoning reduces the number of σ's to be considered to a finite number.

Now let us take into account a set Σ of $\boldsymbol{\sigma}$'s that are solution of the state equation, have different supports and the support of all of them is in $FS(\mathcal{N}, \mathbf{m_0})$. To decide reachability it is only necessary to consider those $\boldsymbol{\sigma}$'s with minimal support. This is because if there is a non minimal $\boldsymbol{\sigma} \in \Sigma$ fulfilling the condition on the traps of Theorem 12, then its support contains the support of a $\boldsymbol{\sigma}' \in \Sigma$ that also fulfills this condition.

Summing up, to decide reachability it is only necessary to consider the $\boldsymbol{\sigma}$'s in a set $\Sigma = \{\boldsymbol{\sigma}_1, \dots, \boldsymbol{\sigma}_k\}$ that fulfills the following conditions:

1. $\mathbf{m} = \mathbf{m_0} + \mathbf{C} \cdot \boldsymbol{\sigma}_i \geq 0$ and $\|\boldsymbol{\sigma}_i\| \in FS(\mathcal{N}, \mathbf{m_0})$
2. $\|\boldsymbol{\sigma}_i\|$ is minimal, i.e., for every $j \neq i$ $\|\boldsymbol{\sigma}_j\| \nsubseteq \|\boldsymbol{\sigma}_i\|$
3. for every $\boldsymbol{\gamma}$ such that $\mathbf{m} = \mathbf{m_0} + \mathbf{C} \cdot \boldsymbol{\gamma} \geq 0$ and $\|\boldsymbol{\gamma}\| \in FS(\mathcal{N}, \mathbf{m_0})$ there exists $i \in \{1, \dots, k\}$ such that $\|\boldsymbol{\sigma}_i\| \subseteq \|\boldsymbol{\gamma}\|$

The third condition guarantees that every $\boldsymbol{\sigma}$ that verifies the first two conditions is included in Σ.

The following algorithm takes as inputs a continuous system, a target marking $\mathbf{m}$, and a set Σ verifying the above conditions for the target marking. The output of the algorithm is the boolean variable *answer* that takes the value *YES* iff $\mathbf{m}$ is reachable. The general idea of the algorithm is first checking whether all traps are marked at $\mathbf{m}$, step 2, and whether there is an empty trap at $\mathbf{m}$ that was marked at $\mathbf{m_0}$, step 3. In these both cases a quick answer to reachability can be given. Otherwise, it is required to iterate on the elements of Σ.

Algorithm 13
    INPUT: $\langle \mathcal{N}, \mathbf{m_0} \rangle$, $\mathbf{m}$, $\Sigma = \{\boldsymbol{\sigma}_1, \dots, \boldsymbol{\sigma}_k\}$
    OUTPUT: answer
    1. *If $\Sigma = \emptyset$ then answer:=NO; exit; end if*
    2. *If there is no empty trap in $\mathcal{N}$ at $\mathbf{m}$ then answer:=YES; exit; end if*
    3. *If there is an empty trap in $\mathcal{N}$ at $\mathbf{m}$ that was not empty at $\mathbf{m_0}$ then*
        answer:=NO; exit; end if
    4. *i:=0*
    5. *loop*
        5.1. *i:=i+1*
        5.2. *If there is no empty trap in $\mathcal{N}_{\boldsymbol{\sigma}_i}$ at $\mathbf{m}$ then answer:=YES;*
            exit; end if
    6. *until i=k*
    7. *answer:=NO*

In [5] a method to compute traps based on the solution of a system of linear equations was proposed. According to this method the support of a solution of that system represents the places of a trap. In steps 2 and 5.2 of Algorithm 13, we are interested only in empty traps at $\mathbf{m}$, therefore only the subnet composed of empty places at $\mathbf{m}$ has to be considered. In step 3, the focus is on the empty traps at $\mathbf{m}$ that were marked at $\mathbf{m_0}$. The only thing that has to be included in the system of inequalities proposed in [5] is forcing that a solution of the system must have at least one non null component corresponding to a non empty place at $\mathbf{m_0}$.

4   lim-RS($\mathcal{N}, \mathbf{m_0}$)

As it has been shown, some traps (for example the one composed of p_3 and p_4 in the system of Figure 1(a)) can be emptied with an infinite firing sequence. Hence when facing the problem of describing the set of lim-reachable markings, it is not necessary to exclude those markings that are result of the state equation and that have empty traps that were previously filled. In this way, the characterization of the lim-RS($\mathcal{N}, \mathbf{m_0}$) is easier and it is only necessary to care about the fireability of the firing count vector $\boldsymbol{\sigma}$ (conditions 1. and 2. of Theorem 12).

Theorem 14 *A marking* $\mathbf{m} \in$ lim-RS($\mathcal{N}, \mathbf{m_0}$) *iff*

$$1.\ \mathbf{m} = \mathbf{m_0} + \mathbf{C} \cdot \boldsymbol{\sigma} \geq 0,\ \boldsymbol{\sigma} \geq 0$$
$$2.\ \|\boldsymbol{\sigma}\| \in FS(\mathcal{N}, \mathbf{m_0})$$

Proof.

$\subseteq$

Let $\mathbf{m} \in$ lim-RS($\mathcal{N}, \mathbf{m_0}$). Since $\mathbf{m}$ is reached by a finite or infinite firing sequence there must exist a firing count vector, $\boldsymbol{\sigma} \geq \mathbf{0}$, corresponding to this sequence such that $\mathbf{m} = \mathbf{m_0} + \mathbf{C} \cdot \boldsymbol{\sigma}$. If the sequence was fireable then $\|\boldsymbol{\sigma}\| \in FS(\mathcal{N}, \mathbf{m_0})$.

$\supseteq$

Let $\mathbf{m}$ be such that $\mathbf{m} = \mathbf{m_0} + \mathbf{C} \cdot \boldsymbol{\sigma} \geq 0$, $\boldsymbol{\sigma} \geq 0$ and $\|\boldsymbol{\sigma}\| \in FS(\mathcal{N}, \mathbf{m_0})$. From $\mathbf{m_0}$ it is possible to fire sequentially a subset of transitions in $\|\boldsymbol{\sigma}\|$, since it belongs to $FS(\mathcal{N}, \mathbf{m_0})$, reaching marking $\mathbf{m'} = \mathbf{m_0} + \mathbf{C} \cdot \boldsymbol{\sigma}'$ at which every transition in $\|\boldsymbol{\sigma}\|$ is enabled. Since $\boldsymbol{\sigma}'$ can be taken arbitrarily small, it can always fulfill $\boldsymbol{\sigma} - \boldsymbol{\sigma}' \geq 0$. Lemma 9 can be applied on the system $(\mathcal{N}, \mathbf{m'})$ and therefore marking $\mathbf{m}$ can be reached in the limit. □

According to Theorem 14 checking whether a given marking is reachable in the limit is a *decidable* problem. For the system in Figure 6 without transitions t_2, t_5 and place p_4 it can can be assured that the marking $\mathbf{m} = (0,\ 0,\ 0,\ 0,\ 1)$ is lim-reachable (but not reachable) since it is solution of the state equation with $\boldsymbol{\sigma} = (1,\ 0,\ 1,\ 1,\ 0,\ 0)$ and $\|\boldsymbol{\sigma}\| \in FS(\mathcal{N}, \mathbf{m_0})$.

If the system fulfills some initial conditions, then the set lim-RS($\mathcal{N}, \mathbf{m_0}$) can be described without the use of $FS(\mathcal{N}, \mathbf{m_0})$. Furthermore, those conditions can be checked in *polynomial* time. For example, for a system, $\langle \mathcal{N}, \mathbf{m_0} \rangle$, in which every transition is enabled at $\mathbf{m_0}$, it holds $FS(\mathcal{N}, \mathbf{m_0}) = \{q | q \subseteq T\}$ and therefore every $\boldsymbol{\sigma} \geq \mathbf{0}$ belongs to $FS(\mathcal{N}, \mathbf{m_0})$.

Corollary 15 *If for every transition* t *enab*$(t, \mathbf{m_0}) > 0$ *then* lim-RS($\mathcal{N}, \mathbf{m_0}$) = LRS($\mathcal{N}, \mathbf{m_0}$).

Let $\langle \mathcal{N}, \mathbf{m_0} \rangle$ be a consistent system in which every transition is fireable at least once, i.e., for every transition t there exists $\mathbf{m'} \in$ RS($\mathcal{N}, \mathbf{m_0}$) such that enab$(t, \mathbf{m'}) > 0$. Clearly $T \in FS(\mathcal{N}, \mathbf{m_0})$. Since the system is consistent it has a T-semiflow $\mathbf{x} > \mathbf{0}$ that can be added to a given $\boldsymbol{\sigma}$, $\mathbf{m} = \mathbf{m_0} + \mathbf{C} \cdot \boldsymbol{\sigma} \geq 0$, fulfilling $\boldsymbol{\sigma} + \mathbf{x} > \mathbf{0}$. It is obvious that $\mathbf{C} \cdot \boldsymbol{\sigma} = \mathbf{C} \cdot (\boldsymbol{\sigma} + \mathbf{x})$ and that $\|\boldsymbol{\sigma} + \mathbf{x}\| = T$. Therefore, $\mathbf{m}$ is lim-reachable.

Corollary 16 ([7]) *If* $(\mathcal{N}, \mathbf{m_0})$ *is consistent and every transition is fireable at least once, then* lim-RS$(\mathcal{N}, \mathbf{m_0}) = $ LRS$(\mathcal{N}, \mathbf{m_0})$.

5 δ-RS$(\mathcal{N}, \mathbf{m_0})$

Let us now assume that given a system, $\langle \mathcal{N}, \mathbf{m_0} \rangle$, every transition is fireable at least once. That is for every transition t there exists $\mathbf{m} \in$ RS$(\mathcal{N}, \mathbf{m_0})$ such that enab$(t, \mathbf{m}) > 0$. The existence of transitions that do not fulfill this condition can be easily detected (see [7]): it is sufficient to iterate on the enabled transitions firing them in half its enabling degree until no more transitions become enabled. Those transitions that are not enabled after the iteration can never be fired. Notice that this assumption does not imply a loss of generality in the following results, since if a transition can never be enabled it can be removed without affecting any possible evolution of the system or changing the set of reachable markings.

In this section the set of markings to which the system can get as close as desired is described. For example, in Figure 4 with $\mathbf{m_0} = (1, 0, 0, 0, 1)$, $\mathbf{m} = (0, 1, 0, 0, 1)$ does not belong neither to RS$(\mathcal{N}, \mathbf{m_0})$ nor to lim-RS$(\mathcal{N}, \mathbf{m_0})$, however $\mathbf{m} = (0, 1, 0, \alpha, 1 - 2 \cdot \alpha)$ belongs to RS$(\mathcal{N}, \mathbf{m_0})$ (hence also to lim-RS$(\mathcal{N}, \mathbf{m_0})$) for every α fulfilling $0 < \alpha \leq 0.5$.

For this set of markings, that will be called δ-reachable, there are no spurious solutions of the state equation.

Theorem 17 *If every transition is fireable at least once from the initial marking, then a marking* $\mathbf{m} \in \delta$-RS$(\mathcal{N}, \mathbf{m_0})$ *iff*

$$1.\ \mathbf{m} = \mathbf{m_0} + \mathbf{C} \cdot \boldsymbol{\sigma} \geq 0,\ \boldsymbol{\sigma} \geq 0$$

i.e., δ-RS$(\mathcal{N}, \mathbf{m_0}) = $ LRS$(\mathcal{N}, \mathbf{m_0})$.

Proof.
$\subseteq$
δ-RS$(\mathcal{N}, \mathbf{m_0}) \subseteq$ LRS$(\mathcal{N}, \mathbf{m_0})$ since LRS$(\mathcal{N}, \mathbf{m_0})$ is a closed set that includes the RS$(\mathcal{N}, \mathbf{m_0})$.
$\supseteq$
Let $\mathbf{m}$ be a solution of the state equation, i.e., $\mathbf{m} = \mathbf{m_0} + \mathbf{C} \cdot \boldsymbol{\sigma} \geq 0$. Since every transition is fireable at least once, let us consider a sequence, $\boldsymbol{\sigma}'$, that reaches a marking, $\mathbf{m}'$, at which every transition in the support of $\boldsymbol{\sigma}$ is enabled. Let us consider the real quantity α determined by $\alpha = \min\{1, \max\{\beta | \mathbf{m}' + \mathbf{C} \cdot \beta \cdot \boldsymbol{\sigma} \geq 0\}\}$. Then, according to Theorem 14, the marking $\mathbf{m}'' = \mathbf{m}' + \mathbf{C} \cdot \alpha \cdot \boldsymbol{\sigma}$ is reachable in the limit from $\mathbf{m}'$. And clearly, it is also reachable in the limit from $\mathbf{m_0}$ ($\mathbf{m}'' \in$ lim-RS$(\mathcal{N}, \mathbf{m_0})$). Notice that if $|\boldsymbol{\sigma}'|$ tends to zero, then the value of α goes to one and $\mathbf{m}''$ approaches $\mathbf{m}$. Thus, firing a finite sequence we can get as close to $\mathbf{m}$ as desired. $\square$

Establishing a bridge to discrete systems, it can be said that if the system is highly populated and it is not necessary to exactly determine the marking at places, then the system can evolve to any marking that is solution of the state equation.

Summarizing on reachability, the following relationship among the different sets of reachable markings can be stated. It asserts that the only differences among the described sets of reachable markings are in the border points of the space defined by the state equation.

Corollary 18 *If every transition is fireable then:*

1. $]\mathrm{LRS}(\mathcal{N}, \mathbf{m_0})[\subseteq \mathrm{RS}(\mathcal{N}, \mathbf{m_0}) \subseteq \mathrm{lim}\text{-}\mathrm{RS}(\mathcal{N}, \mathbf{m_0}) \subseteq \delta\text{-}\mathrm{RS}(\mathcal{N}, \mathbf{m_0}) = \mathrm{LRS}(\mathcal{N}, \mathbf{m_0})$.
2. *Under consistency of* $\mathcal{N}$: $\mathrm{lim}\text{-}\mathrm{RS}(\mathcal{N}, \mathbf{m_0}) = \delta\text{-}\mathrm{RS}(\mathcal{N}, \mathbf{m_0}) = \mathrm{LRS}(\mathcal{N}, \mathbf{m_0})$.

Proof. 1. is a direct consequence of the fact that $\delta\text{-}\mathrm{RS}(\mathcal{N}, \mathbf{m_0})$ is the closure of $\mathrm{RS}(\mathcal{N}, \mathbf{m_0})$ and $\mathrm{lim}\text{-}\mathrm{RS}(\mathcal{N}, \mathbf{m_0})$, and $\delta\text{-}\mathrm{RS}(\mathcal{N}, \mathbf{m_0}) = \mathrm{LRS}(\mathcal{N}, \mathbf{m_0})$.

2. is immediate from Corollary 16 and Theorem 17.   □

6   Reversibility and Liveness

Reversibility is a basic property that has to do with mutual reachability among all markings of the system, or equivalently with the ability to reach the initial marking from any reachable one. Liveness is the capacity of the system of potentially firing any transition from any reachable marking. In discrete systems *if every transition is fireable at least once, then reversibility implies liveness and consistency*: if a system is reversible, it can always get back to the initial marking, therefore it is live because from the initial marking every transition is fireable at least once. Moreover, if a system can always return to the initial marking after every transition has fired, it means that a T-semiflow covering every transition has been fired, that is, the system is consistent.

However, liveness and consistency are not sufficient conditions for reversibility in discrete systems. For example, the system in Figure 7 is consistent and live as discrete, however once t_1 has fired it is impossible to get back to the initial marking. Thus the system is not reversible as discrete.

In continuous systems, assuming that every transition is fireable at least once, it can be observed that reversibility also implies consistency and liveness. As in discrete systems, if reachability with finite sequences is considered, liveness and consistency are not sufficient conditions for reversibility. The system in Figure 6 is consistent and live as continuous considering finite firing sequences. If transition t_1 is fired in any amount, the trap $\{p_2, p_3, p_4\}$ becomes marked, and cannot be emptied with a finite firing sequence. Hence, once t_1 has fired it is not possible to go back to the initial marking, and therefore it can be said that the system is not reversible. Nevertheless, as it will be seen, the system is reversible if lim-reachability and δ-reachability are considered.

In [7] lim-liveness was defined in order to extend the liveness concept to continuous systems regarding lim-reachability. Let us now define also δ-liveness and lim-$(\delta$-$)$reversibility as the natural extensions of the classical definitions for the concepts of lim-reachability and δ-reachability respectively:

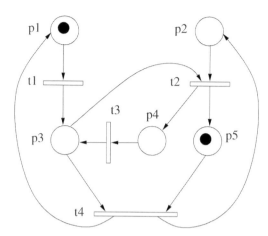

Fig. 7. Non reversible system as discrete or continuous with finite number of firings, but lim-reversible and δ-reversible

Definition 19 $\langle \mathcal{N}, \mathbf{m_0} \rangle$ *is* lim-$(\delta$-$)$*live iff for every* $\mathbf{m} \in$ lim-$(\delta$-$)$RS$(\mathcal{N}, \mathbf{m_0})$ *and for every* $t \in T$ *there exist* $\mathbf{m'} \in$ RS$(\mathcal{N}, \mathbf{m})$ *such that* enab$(t, \mathbf{m'}) > 0$.

Definition 20 $\langle \mathcal{N}, \mathbf{m_0} \rangle$ *is* lim-$(\delta$-$)$*reversible iff for every* $\mathbf{m} \in$ lim-$(\delta$-$)$RS$(\mathcal{N}, \mathbf{m_0})$, $\mathbf{m_0} \in$ lim-$(\delta$-$)$RS$(\mathcal{N}, \mathbf{m})$.

The following theorem states that under lim-reachability and δ-reachability, if every transition can be fired at least once, consistency and lim-$(\delta$-$)$liveness are not only necessary conditions for lim-$(\delta$-$)$reversibility but also sufficient.

Theorem 21 *Let* $\langle \mathcal{N}, \mathbf{m_0} \rangle$ *be such that every transition is fireable at least once.* $\langle \mathcal{N}, \mathbf{m_0} \rangle$ *is consistent and* lim-$(\delta$-$)$*live iff* $\langle \mathcal{N}, \mathbf{m_0} \rangle$ *is* lim-$(\delta$-$)$*reversible.*

Proof.
$(\Rightarrow)$
Since the system is consistent and every transition is fireable at least once, it holds by Corollary 18 that lim-RS$(\mathcal{N}, \mathbf{m_0}) = \delta$-RS$(\mathcal{N}, \mathbf{m_0}) =$ LRS$(\mathcal{N}, \mathbf{m_0})$. Let us consider the lim-$(\delta$-$)$reachable marking $\mathbf{m}$, $\mathbf{m} = \mathbf{m_0} + \mathbf{C} \cdot \boldsymbol{\sigma}$. It will be seen that $\mathbf{m_0}$ is lim-$(\delta$-$)$reachable from $\mathbf{m}$. Since the system is lim-$(\delta$-$)$live, every transition is fireable from $\mathbf{m}$, and therefore a strictly positive marking, $\mathbf{m'} > \mathbf{0}$, can be reached, $\mathbf{m'} = \mathbf{m_0} + \mathbf{C} \cdot (\boldsymbol{\sigma} + \boldsymbol{\sigma'})$. The net is consistent, hence a T-semiflow, $\mathbf{x} > \mathbf{0}$, exists such that $\mathbf{x} - \boldsymbol{\sigma} - \boldsymbol{\sigma'} \geq \mathbf{0}$. By Corollary 15, $\mathbf{m_0} = \mathbf{m'} + \mathbf{C} \cdot (\mathbf{x} - \boldsymbol{\sigma} - \boldsymbol{\sigma'})$ is lim-(δ)reachable from $\mathbf{m'}$.
$(\Leftarrow)$
If the system is reversible and every transition can be fired at least once, then it clearly cannot lim-$(\delta$-$)$reach a marking in which one transition is not fireable any more. It would mean that it cannot get back to the initial marking. Moreover, if after the firing of every transition the system always can return to the initial marking, it means that it is consistent. $\qquad \square$

For example, the system in Figure 7 is consistent and lim-(δ-)live, therefore according to Theorem 21 it is lim-(δ-)reversible. If from the initial marking t_1 is fired in an amount of 1, the marking $(0, 0, 1, 0, 1)$ is reached. Applying the infinite firing sequence $\frac{1}{2}, t_4\frac{1}{2}t_2, \frac{1}{2}t_3, \frac{1}{4}, t_4\frac{1}{4}t_2, \frac{1}{4}t_3, \ldots$ from $(0, 0, 1, 0, 1)$ the system converge to the initial marking.

From Theorem 21, the following Corollary is immediate:

Corollary 22 *Let* $\langle \mathcal{N}, \mathbf{m_0} \rangle$ *be* lim-(δ-)live. $\langle \mathcal{N}, \mathbf{m_0} \rangle$ *is* lim-(δ-)reversible iff $\mathcal{N}$ *is consistent.*

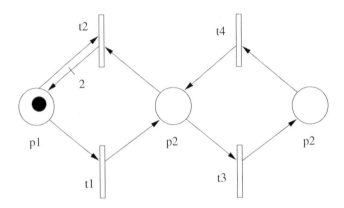

Fig. 8. A lim-(δ-)deadlock-free and not lim-(δ-)live system

Notice that the lim-(δ-)liveness condition in Theorem 21 and Corollary 22 cannot be relaxed to lim-(δ-)deadlock-freeness, where lim-(δ-)deadlock-freeness means that the system cannot lim-(δ-)reach a marking in which no transition is fireable. In other words, as in discrete systems, lim-(δ-)deadlock-freeness does not imply lim-(δ-)liveness, even under consistency and conservativeness. For example, the consistent and conservative system in Figure 8 is lim-(δ-)deadlock-free but not lim-(δ-)live: transitions t_3 and t_4 are potentially fireable from any lim-(δ-)reachable marking, but once t_1 is fired in an amount of 1, neither t_1 nor t_2 will ever be fireable.

7   Conclusions

In continuous nets the concept of "reachable marking" can be interpreted in three different ways:

    a) *reachability*, a marking can be reached with a finite firing sequence.

    b) lim-*reachability*, a marking can be reached with a finite or with an infinite firing sequence.

    c) δ-*reachability*, the system can get as close as desired to a marking with a finite firing sequence.

Each of the three concepts has its own reachability space. These reachability spaces can be fully characterized using, among other elements, the state equation. Moreover, it is decidable whether a marking is reachable according to each concept. Furthermore, there is an inclusion relationship among the sets of markings associated to each concept: $a \subseteq b \subseteq c$. The only differences among these sets are in the border points of the spaces (i.e., the convex hull).

Moreover, as the level of "exigency" regarding reachability decreases (a is the "strongest" and c the "weakest") the characterization of the reachability space becomes progressively easier. In particular, if every transition is fireable at least once, a very weak condition because otherwise unfireable transitions can be simply removed, the set of the markings in c is equal to the solutions of the state equation. In other words, for this last case there exists no spurious solution of the state equation.

Finally, a necessary and sufficient condition for reversibility with respect to lim-reachability and δ-reachability has been provided.

Acknowledgements. We want to thank the referees of this paper for their useful suggestions. Especially referee number 4, who helped us to improve the characterization of the set of reachable markings given in Theorem 12.

References

1. H. Alla and R. David. Continuous and hybrid Petri nets. *Journal of Circuits, Systems, and Computers*, 8(1):159–188, 1998.
2. G. Ciardo, D. Nicol, and K. S. Trivedi. Discrete-event simulation of fluid stochastic Petri nets. *IEEE Trans. on Software Engineering*, 25(2):207–217, 1999.
3. R. David and H. Alla. Continuous Petri nets. In *Proc. of the 8th European Workshop on Application and Theory of Petri Nets*, pages 275–294, Zaragoza, Spain, 1987.
4. F. DiCesare, G. Harhalakis, J. M. Proth, M. Silva, and F. B. Vernadat. *Practice of Petri Nets in Manufacturing*. Chapman & Hall, 1993.
5. J. Ezpeleta, J. M. Couvreur, and M. Silva. A new technique for finding a generating family of siphons, traps and ST-components. application to coloured Petri nets. In G. Rozenberg, editor, *Advances in Petri Nets 1993*, volume 674 of *Lecture Notes in Computer Science*, pages 126–147. Springer, 1993.
6. T. Murata. Petri nets: Properties, analysis and applications. *Proceedings of the IEEE*, 77(4):541–580, 1989.
7. L. Recalde, E. Teruel, and M. Silva. Autonomous continuous P/T systems. In J. Kleijn S. Donatelli, editor, *Application and Theory of Petri Nets 1999*, volume 1639 of *Lecture Notes in Computer Science*, pages 107–126. Springer, 1999.
8. M. Silva and J. M. Colom. On the computation of structural synchronic invariants in P/T nets. In G. Rozenberg, editor, *Advances in Petri Nets 1988*, volume 340 of *Lecture Notes in Computer Science*, pages 387–417. Springer, 1988.
9. M. Silva and L. Recalde. Petri nets and integrality relaxations: A view of continuous petri net models. *IEEE Trans. on Systems, Man, and Cybernetics*, 32(4):314–327, 2002.

10. M. Silva, E. Teruel, and J. M. Colom. Linear algebraic and linear programming techniques for the analysis of net systems. In G. Rozenberg and W. Reisig, editors, *Lectures in Petri Nets. I: Basic Models*, volume 1491 of *Lecture Notes in Computer Science*, pages 309–373. Springer, 1998.
11. K. Trivedi and V. G. Kulkarni. FSPNs: Fluid stochastic Petri nets. In M. Ajmone Marsan, editor, *Application and Theory of Petri Nets 1993*, volume 691 of *Lecture Notes in Computer Science*, pages 24–31. Springer, 1993.

On the Siphon-Based Characterization of Liveness in Sequential Resource Allocation Systems*

Spyros A. Reveliotis

School of Industrial & Systems Engineering
Georgia Institute of Technology
Atlanta, GA 30332, USA
spyros@isye.gatech.edu

Abstract. One of the most interesting developments from, both, a theoretical and a practical perspective, in the emerging theory of resource allocation systems (RAS), is the characterization of the non-liveness of many RAS classes through the Petri net (PN)-based structural object of empty, or more generally, deadly marked siphon. The work presented in this paper seeks to develop a general theory that provides a unifying framework for all the relevant existing results, and reveals the key structures and mechanisms that connect the RAS non-liveness to the concept of deadly marked – and in certain cases, empty – siphon. By taking this generalizing approach, the developed results allow also the extension of the siphon-based characterization of non-liveness to broader RAS classes, and provide a clear and intuitive explanation for the cases where the RAS non-liveness cannot be attributed to such a siphon-based construct.

Keywords: sequential resource allocation systems, deadlock resolution, Petri net structural analysis, siphons

1 Introduction

One of the major breakthroughs underlying our capability to systematically evaluate the liveness of various resource allocation system (RAS) configurations, and to synthesize effective and computationally efficient liveness-enforcing supervisors for non-live RAS, is the formal characterization of the non-liveness of the Petri net (PN) sub-classes modelling the behavior of these environments, through the formation of a particular PN structural object, known as *empty* or, more generally, *deadly marked siphon*[1] [16]. The first results of this type can be found in the seminal work of Ezpeleta and his colleagues [5], which established that in the case of simple sequential processes requesting a single unit from a single resource at each processing stage, non-liveness can be attributed to the

---

* This work has been partially supported by the Keck Foundation and the Logistics Institute - Asia Pacific.
[1] All technical concepts are systematically introduced in the later parts of this paper.

W.M.P. van der Aalst and E. Best (Eds.): ICATPN 2003, LNCS 2679, pp. 241–255, 2003.
© Springer-Verlag Berlin Heidelberg 2003

development of an empty siphon in the dynamics of the PN modeling the considered RAS behavior. Subsequently, a series of publications by various research groups addressed the relationship of empty and/or deadly marked siphons to the liveness of other, more complicated RAS-modelling PN structures, and established the practical significance of this relationship in the development of PN-based liveness enforcing supervisors for these environments. Most of this work is reported in [1,2,3,7,8,12,13,15,17,20,21,24], while some discussion on the historical evolution, interconnections, and significance of these results can be found in [16].

The perusal of the aforementioned literature, however, will also reveal that all the existing results on siphon-based characterization of liveness in sequential RAS have been developed in a rather fragmented fashion, each of them pertaining to a RAS sub-class characterizing a particular type of RAS behavior. Moreover, while they are based on formal and rigorous technical arguments, they fail to provide an explicit and intuitive characterization of the underlying key mechanism that links the non-liveness of the considered RAS classes to the presence of some empty or deadly marked siphons. Hence, the work presented in this paper seeks to develop a general theory that

- will provide a systematic explanation of the relationship between the RAS non-liveness and the presence of deadly marked – and in the case of some simpler RAS structures, empty – siphons;
- will offer, thus, a unifying framework for interpreting all the relevant results existing in the literature;
- will eventually allow the extension of the existing results to broader and/or other RAS behaviors, and it will reveal RAS structures and behaviors for which such a siphon-based characterization of non-liveness will not be possible.

Our approach is based on the identification of a minimal set of requirements for the structure of the RAS processes and their behavior, which when met, will allow the attribution of any experienced RAS non-liveness to the development of deadly marked siphons. More specifically, these siphons are detected in a *"modified"* reachability space of the RAS-modelling PN that constitutes a projection of its original reachability space to an appropriately selected sub-space, and it is formally defined in the following section. Two concepts that will play a central role in the derivation of the presented results are those of the process *quasi-liveness* and *reversibility*. These properties essentially imply that the execution logic underlying the various process flows is inherently consistent, and therefore, any non-liveness of the PN modelling the overall RAS behavior can be attributed to the competition of the concurrently executing processes for the finite system resources. A third concept that appears in the subsequent results, and it is necessary in order to connect the non-liveness of the process-resource net to the presence of deadly marked siphons, is that of *acyclic process flows*, i.e., the developed results pertain to RAS in which the various processes do not present re-circulating loops among their different stages.[2]

The rest of the paper is organized as follows: Section 2 first presents the PN fundamentals that are necessary for the modelling and analysis of the considered RAS structure and behavior, and subsequently it proceeds to the systematic characterization of this RAS class through a series of definitions and assumptions. Section 3 develops the main results of this work, and Section 4 concludes the paper, briefly discussing also the practical significance of the presented results for assessing the RAS liveness and synthesizing, if necessary, appropriate PN-based liveness enforcing supervisors.

2   The Considered RAS Class and Its Petri Net Model

This section first overviews the Petri net (PN) related concepts that are necessary for the formal modelling of the considered RAS class and the analysis of its properties, and subsequently, it provides a detailed characterization of the PN structure modelling the considered resource allocation environments.

2.1   Petri Net Preliminaries

Following the standard Petri net literature, e.g., [11,4], in this work a *marked Petri Net (PN)* is defined by a quadruple $\mathcal{N} = (P, T, W, M_0)$, where P is the set of *places*, T is the set of *transitions*, $W : (P \times T) \cup (T \times P) \rightarrow Z^+$ is the *flow relation*, and $M_0 : P \rightarrow Z^+$ is the net *initial marking*, assigning to each place $p \in P$, $M_0(p)$ *tokens*. In the special case that the flow relation W maps onto $\{0, 1\}$, the Petri net is said to be *ordinary*. If only the restriction of W to $(P \times T)$ maps on $\{0, 1\}$, the PN is said to be *PT-ordinary*. The set of input (resp., output) transitions of a place p is denoted by $\bullet p$ (resp., $p \bullet$). Similarly, the set of input (resp., output) places of a transition t is denoted by $\bullet t$ (resp., $t \bullet$). This notation is also generalized to any set of places or transitions, X, e.g. $\bullet X = \bigcup_{x \in X} \bullet x$. The ordered set $X = < x_1 \ldots x_n > \in (P \cup T)^*$ is a *path*, if and only if (iff) $x_{i+1} \in x_i \bullet$, $i = 1, \ldots, n - 1$. Furthermore, a path X is characterized as a *circuit* iff $x_1 \equiv x_n$. Finally, an ordinary PN such that (s.t.) $\forall t \in T$, $|t \bullet| = |\bullet t| = 1$ (resp., $\forall p \in P$, $|p \bullet| = |\bullet p| = 1$), is characterized as a *state machine* (resp., *marked graph*).

Given a marking M, a transition t is *enabled* iff $\forall p \in \bullet t$, $M(p) \geq W(p, t)$, and this is denoted by $M[t\rangle$. $t \in T$ is said to be *disabled* by $p \in \bullet t$ at M iff $M(p) < W(p, t)$. Furthermore, a place $p \in P$ for which $\exists t \in p \bullet$ s.t. $M(p) < W(p, t)$ is said to be a *disabling* place at M. Firing an enabled transition t results in a new marking M', which is obtained by removing $W(p, t)$ tokens from each place $p \in \bullet t$, and placing $W(t, p')$ tokens in each place $p' \in t \bullet$. The set of markings reachable from M_0 through any fireable sequence of transitions is denoted by $R(\mathcal{N}, M_0)$. A marked PN $\mathcal{N}$ with initial marking M_0 is said to be *bounded* iff

---

[2] This requirement can be relaxed under certain conditions; c.f. [8,12] for details.

all markings $M \in R(\mathcal{N}, M_0)$ are bounded, while $\mathcal{N}$ is said to be *structurally bounded* iff it is bounded for any initial marking M_0. $\mathcal{N}$ is said to be *reversible* iff $\forall M \in R(\mathcal{N}, M_0)$, $M_0 \in R(\mathcal{N}, M)$.

In case that a marked PN is *pure* (i.e., $\forall (x, y) \in (P \times T) \cup (T \times P)$, $W(x, y) > 0 \Rightarrow W(y, x) = 0$), the flow relation can be represented by the *flow matrix* $\Theta = \Theta^+ - \Theta^-$ where $\Theta^+[p, t] = W(t, p)$ and $\Theta^-[p, t] = W(p, t)$. A *p-semiflow* y is a $|P|$-dimensional vector satisfying $y^T \Theta = 0$ and $y \geq 0$, and a *t-semiflow* x is a $|T|$-dimensional vector satisfying $\Theta x = 0$ and $x \geq 0$. A p-semiflow y (t-semiflow x, resp.) is said to be *minimal* iff $\nexists$ a p-semiflow y' (t-semiflow x', resp.) such that $\|y'\| \subset \|y\|$ ($\|x'\| \subset \|x\|$, resp.), where $\|y\| = \{p \in P \mid y(p) > 0\}$ ($\|x\| = \{t \in T \mid x(t) > 0\}$, resp.).

Given a marked PN $\mathcal{N} = (P, T, W, M_0)$, a transition $t \in T$ is *live* iff $\forall M \in R(\mathcal{N}, M_0), \exists M' \in R(\mathcal{N}, M)$ s.t. $M'[t\rangle$, and $t \in T$ is *dead* at $M \in R(\mathcal{N}, M_0)$ iff $\nexists$ marking $M' \in R(\mathcal{N}, M)$ s.t. $M'[t\rangle$. A marking $M \in R(\mathcal{N}, M_0)$ is a (total) *deadlock* iff $\forall t \in T$, t is dead. A marked PN $\mathcal{N}$ is *quasi-live* iff $\forall t \in T, \exists M \in R(\mathcal{N}, M_0)$ s.t. $M[t\rangle$, it is *weakly live* iff $\forall M \in R(\mathcal{N}, M_0), \exists t \in T$ s.t. $M[t\rangle$, and it is *live* iff $\forall t \in T$, t is live. Of particular interest for the liveness analysis of marked PN is a structural element known as *siphon*, which is a set of places $S \subseteq P$ such that ${}^{\bullet}S \subseteq S^{\bullet}$. A siphon S is *minimal* iff $\nexists$ a siphon S' s.t. $S' \subset S$. A siphon S is said to be *empty* at marking M iff $M(S) \equiv \sum_{p \in S} M(p) = 0$, and it is said to be *deadly marked* at marking M, iff $\forall t \in {}^{\bullet}S$, t is disabled by some $p \in S$ [15]. Obviously, empty siphons are deadly marked siphons. It is easy to see that, if S is a deadly marked siphon at some marking M, then (i) $\forall t \in {}^{\bullet}S$, t is a dead transition in M, and (ii) $\forall M' \in R(\mathcal{N}, M)$, S is deadly marked. Furthermore, it can be shown that if marking $M \in R(\mathcal{N}, M_0)$ is a total deadlock, then the set S of disabling places in M constitutes a deadly marked siphon [15]. This last result constitutes the generalization of a well-established relationship between total deadlocks and empty siphons in ordinary PN's [4].

Finally, given two PN's $\mathcal{N}_1 = (P_1, T_1, W_1, M_{01})$ and $\mathcal{N}_2 = (P_2, T_2, W_2, M_{02})$ with $T_1 \cap T_2 = \emptyset$ and $P_1 \cap P_2 = Q \neq \emptyset$ s.t. $\forall p \in Q$, $M_{01}(p) = M_{02}(p)$, the PN $\mathcal{N}$ resulting from the *merging* of the nets $\mathcal{N}_1$ and $\mathcal{N}_2$ *through* the place set Q, is defined by $\mathcal{N} = (P_1 \cup P_2, T_1 \cup T_2, W_1 \cup W_2, M_0)$ with $M_0(p) = M_{01}(p), \forall p \in P_1 \backslash P_2$; $M_0(p) = M_{02}(p), \forall p \in P_2 \backslash P_1$; $M_0(p) = M_{01}(p) = M_{02}(p), \forall p \in P_1 \cap P_2$.

2.2   The Considered RAS Class and the Associated PN Model

For the purposes of the liveness analysis considered in this work, a *(sequential) resource allocation system (RAS)* is formally defined by a set of *resource types* $\mathcal{R} = \{R_l, l = 1, \ldots, m\}$, each of them available at some finite *capacity* $C_l \in Z^+$, and a set of *process types* $\mathcal{J} = \{J_j, j = 1, \ldots, n\}$, that execute sequentially, through a number of *tasks* or *stages*, $J_{jk}, k = 1, \ldots, \lambda_j$, and with each stage J_{jk} engaging a specific subset of the system resources for its execution. More specifically, it is assumed that a process instance advances to the execution of a certain stage, J_{jk}, only after it has secured the required resources, and upon its advancement, it releases the resources held for the execution of the previous stage $J_{j,k-1}$. Furthermore, the set of *tasks* or *stages*, $\{J_{jk}, k = 1, \ldots, \lambda_j\}$, corresponding to process type J_j, presents some additional structure that expresses the associated *process-defining logic* and characterizes the potential process *routings*. Typical structures involved in the definition of the process logic include linear, parallel, conditional and iterative schemes, as well as more complex constructs resulting from the nested combination of the basic ones. Most of the past research on RAS liveness and liveness-enforcing supervision has focused on simpler process structures that allow the modelling of simple linear process flows, potentially enhanced with some routing flexibility (e.g., [5,6,9,10,15,18]).

This work does not make any explicit assumptions about the specific structure of the considered RAS processes, but it only requires that the involved process logic is "inherently consistent", and therefore, any non-liveness arising in the behavior of the resulting RAS and its associated PN model can be attributed to the (mis-)management of the allocation of the finite set of the system resources to the concurrently executing processes. A formal characterization of this notion of *"inherent process consistency"* is provided by the following definition of the considered *process subnet* and its assumed properties.

Definition 1. *For the purposes of this work, a* process (sub-)net *is a Petri net* $\mathcal{N}_P = (P, T, W, M_0)$ *such that:*

i. $P = P_S \cup \{i,\ o\}$ *with* $P_S \neq \emptyset$;
ii. $T = T_S \cup \{t_I,\ t_F,\ t^*\}$;
iii. $i^\bullet = \{t_I\}$; $^\bullet i = \{t^*\}$;
iv. $o^\bullet = \{t^*\}$; $^\bullet o = \{t_F\}$;
v. $t_I^\bullet \subseteq P_S$; $^\bullet t_I = \{i\}$;
vi. $t_F^\bullet = \{o\}$; $^\bullet t_F \subseteq P_S$;
vii. $(t^*)^\bullet = \{i\}$; $^\bullet (t^*) = \{o\}$;
viii. *the underlying digraph is* strongly connected;
ix. $M_0(i) > 0 \ \wedge \ M_0(p) = 0, \ \forall p \in P \backslash \{i\}$;
x. $\forall M \in R(\mathcal{N}_P, M_0), \ M(i) + M(o) = M_0(i) \Longrightarrow M(p) = 0, \ \forall p \in P_S$.

In the PN-based process representation introduced by Definition 1, process instances waiting to initiate processing are represented by tokens in place i, while the initiation of a process instance is modelled by the firing of transition t_I. Similarly, tokens in place o represent completed process instances, while the event of a process completion is modelled by the firing of transition t_F. Transition t^* allows the token re-circulation – i.e., the token transfer from place o to place i – in order to model *repetitive* process execution. Finally, the part of the net between transitions t_I and t_F that involves the process places P_S, models the sequential logic defining the considered process type, and, as it can be seen in Definition 1, it can be quite arbitrary. However, in order to capture the notion of the "inherent process consistency" introduced at the beginning of this subsection, we further qualify the considered process sub-nets through the following two assumptions:

Assumption 1 *The process (sub-)nets considered in this work are assumed to be quasi-live for* $M_0(i) = 1$.

Assumption 2 *The process (sub-)nets considered in this work are assumed to be* reversible *for every* initial marking M_0 *that satisfies Condition (ix) of Definition 1.*

Assumption 1 stipulates that every transition in the considered process subnet models a meaningful event that can actually occur during the execution of some process instance, and therefore, it is not redundant. On the other hand, Assumption 2 essentially stipulates that, at any point in time, all *active* process instances can proceed to completion, and this completion can occur without the initiation of any additional process instances.[3] When taken together, Assumptions 1 and 2 imply also the *liveness* of the considered process nets; we state this result as a lemma, but we skip its proof, since it is a rather well-known result in the PN-research community.

Lemma 1. *Under Assumptions 1 and 2, the considered process nets are also* live.

Since the emphasis of this work is on the characterization and establishment of live resource allocation, the complete characterization of the class of process nets satisfying Assumptions 1 and 2 lies beyond its scope. We notice, however, that all the RAS classes for which there exist results connecting their non-liveness to the development of deadly marked / empty siphons, involve PN-based process models that satisfy the aforementioned assumptions.

Another assumption that is necessary for the development of the analytical results of the next section, is that the various process (sub-)nets are *acyclic*. This concept is defined as follows:

Assumption 3 *The process sub-nets considered in this work are assumed to be* acyclic, *i.e., the removal of transition* t^* *from them renders them acyclic digraphs.*

The modelling of the resource allocation associated with each process stage, $p \in P_S$, necessitates the augmentation of the process sub-net $\mathcal{N}_P$, defined above, with a set of *resource* places $P_R = \{r_l,\ l = 1, \ldots, m\}$, of initial marking $M_0(r_l) = C_l,\ i = 1, \ldots, m$, and with the corresponding flow sub-matrix, Θ_{P_R}, expressing the allocation and de-allocation of the various resources to the process instances as they advance through their processing stages. Notice that the interpretation of the role of transitions t^*, t_I and t_F implies that $(t^*)^\bullet \cap P_R = {}^\bullet(t^*) \cap P_R =$

---

[3] It is noticed, for completeness, that the requirement for process reversibility introduced by Assumption 2 when combined with Definition 1 and Assumption 1, subsumes the notion of process *soundness*, introduced in Workflow theory (c.f., [22, 23]) in order to characterize well-defined process (sub-)nets, for the case where only a single process instance re-circulates in the considered process net. However, Assumption 2 further stipulates that when more than one process instances have been activated, still they will always be able to complete, in spite of any additional effects arising from their interaction through the defining process logic.

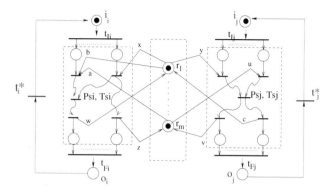

Fig. 1. The process-resource net structure considered in this work

$(t_I)^\bullet \cap P_R = {}^\bullet(t_F) \cap P_R = \emptyset$. The resulting net will be called the *resource-augmented process (sub-)net* and it will be denoted by $\overline{\mathcal{N}_P}$. The *reusable* nature of the system resources is captured by the following assumption regarding the resource-augmented process net $\overline{\mathcal{N}_P}$:

Assumption 4 *Let* $\overline{\mathcal{N}_P} = (P_S \cup \{i, o\} \cup P_R, T, W, M_0)$ *denote a resource-augmented process (sub-)net. Then,* $\forall l \in \{1, \ldots, |P_R|\}$, *there exists a p-semiflow* y_{r_l}, *s.t.:* (i) $y_{r_l}(r_l) = 1$; (ii) $y_{r_l}(r_j) = 0$, $\forall j \neq l$; (iii) $y_{r_l}(i) = y_{r_l}(o) = 0$; (iv) $\forall p \in P_S$, $y_{r_l}(p) = $ *number of units from resource* R_i *required for the execution of stage* p.

While the *p*-semiflows introduced by Assumption 4 characterize the resource allocation taking place at each process stage and the conservative nature of the system resources, they do not reveal anything regarding the adequacy of the available resource set for supporting the execution of the various processing stages, under the sequencing constraints implied by the process-defining logic. This additional concern underlying the correct definition of the various RAS process-types is captured by extending the requirement for *quasi-liveness* of the process net $\mathcal{N}_P$, introduced by Assumption 1, to the resource-augmented process net $\overline{\mathcal{N}_P}$:

Assumption 5 *The resource-augmented process (sub-)nets considered in this work are assumed to be* quasi-live *for* $M_0(i) = 1$ *and* $M_0(r_l) = C_l$, $\forall l \in \{1, \ldots, |P_R|\}$.

The complete PN-based model, $\mathcal{N} = (P, T, W, M_0)$, of any given instance from the considered RAS class is obtained by *merging* the resource-augmented process nets $\overline{\mathcal{N}_{P_j}} = (P_j, T_j, W_j, M_{0_j})$, $j = 1, \ldots, n$, modelling its constituent process types, through their common resource places. The resulting PN class is characterized as the class of *process-resource nets with quasi-live, reversible and acyclic process sub-nets*, and its basic structure is depicted in Figure 1.

Let $P = \bigcup_j P_j$, $P_S = \bigcup_j P_{S_j}$; $I = \bigcup_j \{i_j\}$; $O = \bigcup_j \{o_j\}$. Then, $P = P_S \cup I \cup O \cup P_R$. Furthermore, the re-usable nature of the resource allocation taking place in the entire process-resource net is characterized by a p-semiflow y_{r_l} for each resource type R_l, $l = 1, \ldots, m$, defined by: (i) $y_{r_l}(r_l) = 1$; (ii) $y_{r_l}(r_j) = 0$, $\forall j \neq l$; (iii) $y_{r_l}(i_j) = y_{r_l}(o_j) = 0$, $\forall j$; (iv) $\forall p \in P_S$, $y_{r_l}(p) = y_{r_l}^{(j*)}(p)$, where $\mathcal{N}_{P_{j*}}$ denotes the resource-augmented process sub-net containing place p, and $y_{r_l}^{(j*)}()$ denotes the corresponding p-semiflow for resource R_l. Finally, it is easy to see that Assumption 5 regarding the quasi-liveness of the constituent resource-augmented process sub-nets $\overline{\mathcal{N}_{P_j}}$ implies also the quasi-liveness of the entire process-resource net $\mathcal{N}$.

The next definition extends to the class of process-resource nets, considered in this work, the notion of the *modified* marking, originally introduced in [14, 15] for analyzing the liveness of a more restricted PN subclass, modelling the behavior of sequential RAS with multi-unit resource allocation per stage and routing flexibility.

Definition 2. *Given a process-resource net* $\mathcal{N} = (P_S \cup I \cup O \cup P_R, T, W, M_0)$ *and* $M \in R(\mathcal{N}, M_0)$, *the* modified marking $\overline{M}$ *is defined by*

$$\overline{M}(p) = \begin{cases} M(p) & \text{if } p \notin I \cup O \\ 0 & \text{otherwise} \end{cases} \tag{1}$$

Furthermore, the set of all modified markings induced by the net reachable markings is defined by $\overline{R(\mathcal{N}, M_0)} = \{\overline{M} \mid M \in R(\mathcal{N}, M_0)\}$

We conclude this section by noticing that, from a completely practical standpoint, the main requirement underlying the classical RAS deadlock avoidance theory, is that every process activated in the system will be able to run to completion, without getting entangled in a deadlock situation [19]. In the PN-modelling framework, this requirement is explicitly modelled by stipulating that the corresponding process-resource net is *reversible*. However, in the considered class of process-resource nets, reversibility is equivalent to liveness. This result is also established in Section 3, and it provides the formal link between the theory developed herein and the aforementioned more "practical" considerations of the RAS deadlock avoidance theory (i.e., establishing and maintaining a process flow in which all jobs can run to completion).

3   Liveness Analysis of Process-Resource Nets with Quasi-Live, Reversible, and Acyclic Process Sub-nets

The main result of this section links the non-liveness arising in the class of process-resource nets with quasi-live, reversible and acyclic process sub-nets,[4] to

---

[4] We clarify that in the subsequent development of this paper, a process (sub-)net is characterized *quasi-live* if it satisfies Assumption 1 and the corresponding resource-augmented process sub-net satisfies Assumption 5; it is characterized *reversible* if it satisfies Assumption 2; and it is characterized *acyclic* if it satisfies Assumption 3.

the development of a special type of deadly marked siphon in the net modified reachability space. It is also discussed how this new result encompasses and explains all existing similar results, pertaining to more restricted RAS classes. The last part of the section establishes that, for the considered class of process-resource nets, liveness and reversibility are equivalent concepts.

The result connecting the non-liveness arising in the class of process-resource nets with quasi-live, reversible and acyclic process sub-nets to the presence of deadly marked siphons is developed in a three-step argument, that further reveals the fundamental structures and mechanisms behind it. Hence, its derivation provides also the intuitive explanation requested in the opening discussion of this paper. The first step in this development is established by the following lemma:

Lemma 2. *Consider a process-resource net* $\mathcal{N} = (P_S \cup I \cup O \cup P_R, T, W, M_0)$ *with quasi-live and reversible process sub-nets. If* $\exists M \in R(\mathcal{N}, M_0)$ *s.t.* $\exists$ *a process sub-net* $\mathcal{N}_{P_{j*}}$ *with* $M(i_{j*}) + M(o_{j*}) \neq M_0(i_{j*})$ *and* $\overline{M}$ *is a total deadlock, then* $\exists$ *siphon* S *s.t.*

i.   *S is deadly marked at* $\overline{M}$;
ii. $S \cap P_R \neq \emptyset$;
iii. $\forall p \in S \cap P_R$, *p is a disabling place at* $\overline{M}$.

Proof. Let S denote the set of disabling places in modified marking $\overline{M}$. Since $\overline{M}$ is a total deadlock, $S^{\bullet} = T \supseteq {}^{\bullet}S$. Therefore, S is a siphon, while the definition of S implies also that it is deadly marked. This establishes part (i) in the above lemma.

To establish that $S \cap P_R \neq \emptyset$, consider the process sub-net $\mathcal{N}_{P_{j*}}$. The fact that $M(i_{j*}) + M(o_{j*}) \neq M_0(i_{j*})$ implies that there are active process instances in the sub-net $\mathcal{N}_{P_{j*}}$. But then, Assumptions 2 and 1 imply that sub-net $\mathcal{N}_{P_{j*}}$ remains live in spite of any token removal from places i_{j*} and o_{j*} requested by Definition 2. Hence, the occurrence of the system deadlock at $\overline{M}$ must involve insufficiently marked resource places.

Finally, part (iii) of Lemma 2 is an immediate consequence of the above definition of set S.

In the following, a deadly marked siphon S satisfying also the conditions (ii) and (iii) in Lemma 2, will be called a *resource-induced* deadly marked siphon. Lemma 2 essentially specializes the more general connection between total deadlocks and deadly marked siphons (c.f., Section 2.1), to the subclass of process-resource nets with quasi-live and reversible active processes. From a methodological standpoint, it provides a vehicle for connecting the liveness – and, in certain cases, even the quasi-liveness – of resource allocation taking place in process-resource nets, to resource-induced deadly marked siphons, as long as it can be established that the lack of (any of) these properties implies the existence a reachable marking M s.t. (i) there exists a process sub-net $\mathcal{N}_{P_{j*}}$ with $M(i_{j*}) + M(o_{j*}) \neq M_0(i_{j*})$ and (ii) the corresponding modified marking $\overline{M}$ is a total deadlock. The next lemma establishes that this is the case for the class of process-resource nets with quasi-live, reversible and acyclic process sub-nets.

Lemma 3. *Consider a process-resource net* $\mathcal{N} = (P_S \cup I \cup O \cup P_R, T, W, M_0)$ *with quasi-live, reversible and acyclic process sub-nets. If $\mathcal{N}$ is not live, then, $\exists M \in R(\mathcal{N}, M_0)$ s.t. (i) $\exists$ process sub-net $\mathcal{N}_{P_{j^*}}$ with $M(i_{j^*}) + M(o_{j^*}) \neq M_0(i_{j^*})$ and (ii) $\overline{M}$ is a total deadlock.*

Proof. Since $\mathcal{N}$ is not live, $\exists M' \in R(\mathcal{N}, M_0)$ and $t' \in T$ s.t. t' is dead in M'. We claim that $\exists M \in R(\mathcal{N}, M')$ s.t. (i) $\exists$ process sub-net $\mathcal{N}_{P_{j^*}}$ with $M(i_{j^*}) + M(o_{j^*}) \neq M_0(i_{j^*})$ and (ii) every transition $t \notin (I \cup O)^{\bullet}$ is disabled in M. Indeed, the acyclic structure of the process sub-nets $\mathcal{N}_{P_j}$, $j = 1, \ldots, n$, implies that every transition sequence σ s.t. $M'[\sigma\rangle$ and $\forall t \in \sigma$, $t \notin (I \cup O)^{\bullet}$, will be of finite length. Consider such a maximal transition sequence $\hat{\sigma}$ and let $M'[\hat{\sigma}\rangle M$. Then, at marking M there must exist a process sub-net $\mathcal{N}_{P_{j^*}}$ with $M(i_{j^*}) + M(o_{j^*}) \neq M_0(i_{j^*})$, since otherwise the initial marking M_0 is reachable from M, and then, the quasi-liveness of $\mathcal{N}$ implies that t' is not dead at M'. To see that $\overline{M}$ is a total deadlock for $\mathcal{N}$, simply notice that the specification of $\overline{M}$, by setting $\overline{M}(i_j) = \overline{M}(o_j) = 0$, $\forall j$, essentially disables all transitions $t \in (I \cup O)^{\bullet}$, that, by construction, are the only transitions potentially enabled in M.

The next theorem completes the aforementioned three-step development of the key result of this section, by stating and proving, by means of Lemmas 2 and 3, that in the class of process-resource nets with quasi-live, reversible and acyclic processes, there is a direct relationship between the RAS non-liveness and the presence of resource-induced deadly marked siphons in the modified reachability space of the RAS-modelling PN.

Theorem 1. *Let $\mathcal{N} = (P_S \cup I \cup O \cup P_R, T, W, M_0)$ be a process-resource net with quasi-live, reversible and acyclic processes. $\mathcal{N}$ is live if and only if the space of modified reachable markings, $\overline{R(\mathcal{N}, M_0)}$, contains no resource-induced deadly marked siphons.*

Proof. To show the necessity part, suppose that $\exists M \in R(\mathcal{N}, M_0)$ s.t. $\overline{M}$ contains a resource-induced deadly marked siphon S. Let $r \in S \cap P_R$ be one of the disabling resource places, and consider $t \in r^{\bullet}$ s.t. $\overline{M}(r) < W(r, t)$. The definition of deadly marked siphon implies that $\forall t' \in {}^{\bullet}r$, t' is dead in $R(\mathcal{N}, \overline{M})$. This remark, when combined with Definition 2 and Assumption 4, further imply that $\forall M' \in R(\mathcal{N}, M)$, $M'(r) \leq M(r) = \overline{M}(r)$, since the re-introduction of the tokens removed from places $p \in I \cup O$ and their potential loading in the system, can only decrease the resource availabilities. Therefore, t is a dead transition at M, which contradicts the assumption of net liveness.

To show the sufficiency part, suppose that $\mathcal{N}$ is not live. Then, Lemma 3 implies that $\exists M \in R(\mathcal{N}, M_0)$ s.t. (i) $\exists$ process sub-net $\mathcal{N}_{P_{j^*}}$ with $M(i_{j^*}) + M(o_{j^*}) \neq M_0(i_{j^*})$, and (ii) $\overline{M}$ is a total deadlock. But then, Lemma 2 implies that $\overline{R(\mathcal{N}, M_0)}$ contains a resource-induced deadly marked siphon, which contradicts the working hypothesis.

The following corollary results immediately from Theorem 1; its original statement (and a formal proof) can be found in [14,15].

Corollary 1. *Let $\mathcal{N} = (P_S \cup I \cup O \cup P_R, T, W, M_0)$ be a process-resource net where (i) the process sub-nets $\mathcal{N}_{P_j}$, $j = 1, \ldots, n$, are strongly connected state machines with each circuit containing the places i_j and o_j, and (ii) the resource-augmented process nets $\overline{\mathcal{N}_{p_j}}$ are quasi-live. Then, $\mathcal{N}$ is live if and only if the space of modified reachable markings, $\overline{R(\mathcal{N}, M_0)}$, contains no resource-induced deadly marked siphons.*

The next corollary specializes Theorem 1 to the sub-class of process-resource nets where the process sub-nets $\mathcal{N}_{P_j}$ are acyclic marked graphs. A stronger version of this result, that connects also the lack of quasi-liveness to the presence of resource-induced deadly marked siphons, is presented in [17].

Corollary 2. *Let $\mathcal{N} = (P_S \cup I \cup O \cup P_R, T, W, M_0)$ be a process-resource net where (i) the process sub-nets $\mathcal{N}_{P_j}$, $j = 1, \ldots, n$, are strongly connected marked graphs with each circuit containing the places i_j and o_j, and (ii) the resource-augmented process nets $\overline{\mathcal{N}_{p_j}}$ are quasi-live. Then, $\mathcal{N}$ is live if and only if the space of modified reachable markings, $\overline{R(\mathcal{N}, M_0)}$, contains no resource-induced deadly marked siphons.*

The next result states that for the case of *PT-ordinary* PN's, the problematic siphons interpreting the RAS non-liveness are, in fact, *empty* siphons, and they can also be identified in the *original* net reachability space $R(\mathcal{N}, M_0)$ (besides the modified reachability space $\overline{R(\mathcal{N}, M_0)}$). It encompasses all the relevant results appearing in [3,5,21] and some results appearing in [24].

Corollary 3. *Let $\mathcal{N} = (P_S \cup I \cup O \cup P_R, T, W, M_0)$ be a PT-ordinary process-resource net with quasi-live, reversible and acyclic process sub-nets. $\mathcal{N}$ is live if and only if the space of reachable markings, $R(\mathcal{N}, M_0)$, contains no empty siphons.*

Proof. According to Theorem 1, under the assumptions of Corollary 3, net $\mathcal{N}$ is non-live, iff there exists a marking $M \in R(\mathcal{N}, M_0)$, s.t. $M \neq M_0$ and its modified marking $\overline{M}$ contains a resource-induced deadly marked siphon, S. Furthermore, the development of the result of Theorem 1 (c.f., Lemmas 2 and 3) indicates that S is defined by the set of disabling places of a total deadlock contained in $\overline{M}$. Since every place $p \in S$ is a disabling place in $\overline{M}$, and net $\mathcal{N}$ is PT-ordinary, $\overline{M}(p) = 0$, $\forall p \in S$. Hence, S is an empty siphon in $\overline{M}$. It remains to be shown that the presence of the resource-induced empty siphon S in the modified marking $\overline{M}$ implies the presence of an empty siphon S' in the original marking M. For that, let $S' = \{r_i : r_i \in S\} \cup \{p \in P_S : M(p) = \overline{M}(p) = 0 \land \exists r_i \text{ s.t. } (r_i \in S \land y_{r_i}(p) > 0)\}$. Notice that $S' \neq \emptyset$, since S is a resource-induced empty siphon. We show that S' is a siphon (which is empty, by construction), by considering the next two main cases:

   Case I – $t \in {}^{\bullet}r_k$ **for some** $r_k \in S$: Then, $\exists q \in S$ s.t. $t \in q^{\bullet}$. If $q \in P_R$, then $q \in \{r_i : r_i \in S\} \subset S'$. On the other hand, if $q \notin P_R$, then $q \in P_S$, since

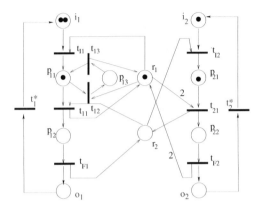

Fig. 2. Example 1 – A case of RAS non-liveness which cannot be attributed to the development of resource-induced deadly marked siphons, due to enabled internal process cycles

$(q^\bullet)^\bullet \cap P_R \neq \emptyset$. Furthermore, $y_{r_k}(q) > 0$ and $M(q) = 0$ (since $q \in S$). Therefore, $q \in \{p \in P_S : M(p) = \overline{M}(p) = 0 \ \wedge \ \exists r_i \ \text{s.t.} \ (r_i \in S \wedge y_{r_i}(p) > 0)\} \subset S'$. In both cases, $t \in (S')^\bullet$.

Case II – $t \in {}^\bullet q$ for some $q \in P_S$ with $M(q) = \overline{M}(q) = 0 \wedge \exists r_k$ s.t. $(r_k \in S \wedge y_{r_k}(q) > 0)$: Then, if $\exists r_l$ s.t. $r_l \in S \wedge t \in r_l^\bullet$, $t \in \{r_i : r_i \in S\}^\bullet \subseteq (S')^\bullet$. Otherwise, $\exists q' \in (I \cup O \cup P_S) \cap {}^\bullet t$ with $\overline{M}(q') = 0$. Furthermore, since $y_{r_k}(q) > 0$ and, by the sub-case assumption, $\forall r_l \in {}^\bullet t$, $M(r_l) > 0$, it must be that $y_{r_k}(q') > 0$. But then, $t \in \{p \in P_S : M(p) = \overline{M}(p) = 0 \ \wedge \ \exists r_i \ \text{s.t.} \ (r_i \in S \wedge y_{r_i}(p) > 0)\}^\bullet \subseteq (S')^\bullet$.

The next example demonstrates that for the case of process-resource nets where the process flows can present internal cycles, the structural concept of resource-induced deadly marked siphon might not be sufficient for interpreting the non-liveness of resource allocation in process-resource nets, even under the assumptions of quasi-live and reversible process sub-nets. The works of [8, 12] identify some special structure on the resource allocation requests, that allows the attribution of the net non-liveness to resource-induced deadly marked siphons, even for the case of process-resource nets with cyclic process routes. However, the complete characterization of the dynamics and the liveness-related properties of process-resource nets with cyclic process routes is an issue open to future investigation.

Example 1: Consider an RAS with two resource types, R_1 and R_2, available at 2 and 1 units, respectively, and two process types, J_1 and J_2. Process type J_1 involves three stages, J_{11}, J_{12} and J_{13}, with corresponding resource allocation requests $[1, 0]$, $[0, 1]$ and $[2, 0]$. Process type J_2 involves two stages, J_{21} and J_{22}, with corresponding resource allocation requests $[0, 1]$ and $[2, 0]$. The RAS-modelling process-resource net, $\mathcal{N}$, that expresses also the flow logic defining the possible process transitions among their stages, is depicted in Figure 2.

In particular, Figure 2 depicts a marking $M \in R(\mathcal{N}, M_0)$, in which the two active processes are deadlocked (notice that transitions t_{11} and t_{21} are dead in the depicted marking M). Yet, the reader can verify that the corresponding modified marking $\overline{M}$, as well as all the modified markings $\overline{M'} \in R(\mathcal{N}, \overline{M})$, contain no resource-induced deadly marked siphon. This results from the fact that the deadlocked process in place p_{11} can circulate freely in the circuit $< p_{11}, t_{12}, p_{13}, t_{13} >$, and therefore, the RAS deadlock of the two processes does not translate to a total deadlock in $\overline{R(\mathcal{N}, \overline{M})}$.

We conclude this section by formally stating and proving that in the considered class of process-resource nets, liveness and reversibility are equivalent concepts.

Theorem 2. *A process-resource net* $\mathcal{N} = (P_S \cup I \cup O \cup P_R, T, W, M_0)$ *with quasi-live, reversible and acyclic process sub-nets is reversible if and only if it is live.*

Proof. The necessity ("only-if") part of this theorem results immediately from Lemma 3, since otherwise $\exists M \in R(\mathcal{N}, M_0)$ s.t. $\exists$ process sub-net $\mathcal{N}_{P_{j^*}}$ with $M(i_{j^*}) + M(o_{j^*}) \neq M_0(i_{j^*})$, and $\overline{M}$ is a total deadlock. In order to establish the sufficiency ("if") part of the theorem, consider a marking $M \in R(\mathcal{N}, M_0)$ s.t. $M \neq M_0$. Then, if for every process sub-net $\mathcal{N}_{P_j}$ it holds that $M(i_j) + M(o_j) = M_0(i_j)$, it should be obvious from the structure of net $\mathcal{N}$ that $M_0 \in R(\mathcal{N}, M)$. Otherwise, using an argument similar to that in the proof of Lemma 3, one can construct a maximal-length firing sequence σ leading to a marking M' s.t. every transition $t \notin (I \cup O)^\bullet$ is disabled in M'. We claim that at M', $\forall$ process sub-net $\mathcal{N}_{P_j}$ it holds that $M'(i_j) + M'(o_j) = M_0(i_j)$, and therefore, $M_0 \in R(\mathcal{N}, M')$, which further implies that $M_0 \in R(\mathcal{N}, M)$. Indeed, by construction, $\overline{M'}$ is a total deadlock of $\mathcal{N}$, and if $\exists$ process sub-net $\mathcal{N}_{P_{j^*}}$ s.t. $M'(i_{j^*}) + M'(o_{j^*}) \neq M_0(i_{j^*})$, Lemma 2 implies that M' contains a resource-induced deadly marked siphon. But then, Theorem 1 implies that $\mathcal{N}$ is not live, which contradicts the working hypothesis.

4 Conclusions

The work presented in this paper extended the currently existing results regarding the siphon-based characterization of (non-)liveness in sequential RAS. Specifically, it provided a unifying framework for interpreting all the relevant results currently existing in the literature, and even more importantly, it extended the theory applicability to broader RAS classes, and it revealed its limitations. It was shown that some key RAS properties that facilitate the interpretation of its non-liveness through the concept of resource-induced deadly marked siphon, are the quasi-liveness, reversibility and acyclicity of its constituent processes.

This finding further suggests that, for RAS with acyclic processes, live behavior can be established through a *two-stage decomposition* procedure, where the first stage seeks to establish inherently consistent – i.e., quasi-live and reversible–

process behaviors, and the second stage seeks to develop, if necessary, a control policy – to be referred to as a liveness enforcing supervisor (LES) – that will ensure the RAS liveness, by preventing the development of resource-induced deadly marked siphons during the process enactment phase. In order to support the first stage of this decomposition, more work is necessary towards developing a more profound understanding of the emerging concepts of process quasi-liveness and reversibility, as defined by Assumptions 1, 2 and 5. On the other hand, to take further advantage of the results presented in this work, the second stage of the aforementioned decomposition must focus on a class of LES's that (i) will admit a PN-based representation, and (ii) their super-imposition on the PN modelling the uncontrolled RAS will result in a PN system that preserves the structural and behavioral assumptions established in Section 2.2. Currently, this is an open research issue, but we expect that our past experience with the synthesis of such supervisors for more restricted RAS classes, will offer useful guidance in this line of research.

Understanding and controlling the dynamics of RAS with internal process cycles is another issue that was shown to lie beyond the boundary of the theory developed in this work, and therefore, it stands open to further investigation. Finally, from an application standpoint, the successful implementation of such a research program will extend our capability towards the effective deployment and (re-)configuration of flexible automation in a broad scope of applications, ranging from automated (e.g., 300mm semiconductor) manufacturing, to driverless urban mono-rail and railway systems, to web-based workflow management systems.

References

1. K. Barkaoui and I. Ben Abdallah. Analysis of a resource allocation problem in fms using structure theory of petri nets. In *Proc. of the 1st Intl Workshop on Manufacturing and Petri Nets*, pages 1–15, 1996.
2. K. Barkaoui, A. Chaoui, and B. Zouari. Supervisory control of discrete event systems based on structure theory of petri nets. In *Proc. of the IEEE Intl Conf. on Systems, Man and Cybernetics*, pages 3750–3755. IEEE, 1997.
3. F. Chu and X-L. Xie. Deadlock analysis of petri nets using siphons and mathematical programming. *IEEE Trans. on R&A*, 13:793–804, 1997.
4. J. Desel and J. Esparza. *Free Choice Petri Nets*. Cambridge Univerrsity Press, 1995.
5. J. Ezpeleta, J. M. Colom, and J. Martinez. A petri net based deadlock prevention policy for flexible manufacturing systems. *IEEE Trans. on R&A*, 11:173–184, 1995.
6. M. P. Fanti, B. Maione, S. Mascolo, and B. Turchiano. Event-based feedback control for deadlock avoidance in flexible production systems. *IEEE Trans. on Robotics and Automation*, 13:347–363, 1997.
7. M. P. Fanti, B. Maione, and T. Turchiano. Comparing digraph and petri net approaches to deadlock avoidance in fms modeling and performance analysis. *IEEE Trans. on Systems, Man and Cybernetics, Part B*, 30:783–798, 2000.
8. M. Jeng and X. Xie. Modeling and analysis of semiconductor manufacturing systems with degraded behaviors using petri nets and siphons. *IEEE Trans. on Robotics and Automation*, 17:576–588, 2001.

9. M. Lawley, S. Reveliotis, and P. Ferreira. The application and evaluation of banker's algorithm for deadlock-free buffer space allocation in flexible manufacturing systems. *Intl. Jrnl. of Flexible Manufacturing Systems*, 10:73–100, 1998.

10. M. A. Lawley. Deadlock avoidance for production systems with flexible routing. *IEEE Trans. Robotics & Automation*, 15:497–509, 1999.

11. T. Murata. Petri nets: Properties, analysis and applications. *Proceedings of the IEEE*, 77:541–580, 1989.

12. J. Park. *Structural Analysis and Control of Resource Allocation Systems using Petri nets*. PhD thesis, Georgia Institute of Technology, Atlanta, GA, 2000.

13. J. Park and S. Reveliotis. Algebraic synthesis of efficient deadlock avoidance policies for sequential resource allocation systems. *IEEE Trans. on R&A*, 16:190–195, 2000.

14. J. Park and S. A. Reveliotis. A polynomial-complexity deadlock avoidance policy for sequential resource allocation systems with multiple resource acquisitions and flexible routings. In *Proc. of CDC 2000*. IEEE, 2000.

15. J. Park and S. A. Reveliotis. Deadlock avoidance in sequential resource allocation systems with multiple resource acquisitions and flexible routings. *IEEE Trans. on Automatic Control*, 46:1572–1583, 2001.

16. S. Reveliotis. Liveness enforcing supervision for sequential resource allocation systems: State of the art and open issues. In B. Caillaud, X. Xie, P. Darondeau, and L. Lavagno, editors, *Synthesis and Control of Discrete Event Systems*, pages 203–212. Kluwer Academic Publishers, 2002.

17. S. A. Reveliotis. Structural analysis of assembly/disassembly resource allocation systems. Technical report, School of Industrial & Systems Eng., Georgia Tech, accessible at: *http://www.isye.gatech.edu/ spyros*, 2001.

18. S. A. Reveliotis and P. M. Ferreira. Deadlock avoidance policies for automated manufacturing cells. *IEEE Trans. on Robotics & Automation*, 12:845–857, 1996.

19. S. A. Reveliotis, M. A. Lawley, and P. M. Ferreira. Structural control of large-scale flexibly automated manufacturing systems. In C. T. Leondes, editor, *The Design of Manufacturing Systems*, pages 4-1 – 4-34. CRC Press, 2001.

20. F. Tricas, J. M. Colom, and J. Ezpeleta. A solution to the problem of deadlock in concurrent systems using petri nets and integer linear programming. In *Proceedings of the 11th Eurpoean Simulation Symposium*, pages 542–546, 1999.

21. F. Tricas, F. Garcia-Valles, J. M. Colom, and J. Ezpeleta. A structural approach to the problem of deadlock prevention in processes with resources. In *Proceedings of the 4th Workshop on Discrete Event Systems*, pages 273–278. IEE, 1998.

22. W. Van der Aalst. Structural characterizations of sound workflow nets. Technical Report Computing Science Reports 96/23, Eindhoven University of Technology, 1996.

23. W. Van der Aalst. Verification of workflow nets. In P. Azema and G. Balbo, editors, *Lecture Notes in Computer Science, Vol. 1248*, pages 407–426. Springer Verlag, 1997.

24. X. Xie and M. Jeng. Ercn-merged nets and their analysis using siphons. *IEEE Trans. on R&A*, 13:692–703, 1999.

Coloured Petri Nets in Development of a Pervasive Health Care System

Jens Bæk Jørgensen

Centre for Pervasive Computing
Department of Computer Science, University of Aarhus
Aabogade 34, DK-8200 Aarhus N, Denmark
jbj@daimi.au.dk

Abstract. *Pervasive computing* implies new challenges for software developers. In addition to tackling common problems pertaining to IT systems in general, new issues like mobility and context-awareness must be dealt with. The contribution of this paper is to demonstrate that *Coloured Petri Nets (CPN)* have potential as an aid in the development of pervasive systems. On a case study of a pervasive health care system for the hospitals in Aarhus, Denmark, we describe how CPN are applied in the everyday software development disciplines of *requirements engineering* and *design*. A number of the observations made in the case study are of a nature making them applicable to use of CPN in development of pervasive systems in general.

Topics. Experience with using nets, case studies; higher-level net models (CPN); application of nets to health and medical systems; requirements engineering; system design; pervasive computing.

1 Introduction

Petri nets are adequate and well-proven to model the behaviour of systems, see, e.g., modelling of workflows in [38] and specification of communication protocols in [29]. However, Petri nets are not in large-scale use in the software industry [41]. Part of the reason for this may be that despite the many virtues of Petri nets, they are not seen as highly needed. For development of traditional administrative systems, the main behavioural issues are handled by a database management system. Moreover, for large distributed enterprise systems, application-level software developers take advantage of well-established middleware architectures like J2EE [46] and CORBA [44] to hide many complex behavioural issues relating to concurrent execution, communication, synchronisation, and resource sharing. Thus, the problems that Petri nets address well are today to some extent not an everyday concern in industrial software development.

This situation could change soon because, arguably, we are on the verge to a paradigm change in software development [40], from familiar distributed systems [10] to *pervasive systems (ubiquitous systems)* [35], which have characteristics that we do not know how to tackle very well today, e.g., mobility and context-awareness [34]. In this paper, we will argue that *Coloured Petri Nets*

W.M.P. van der Aalst and E. Best (Eds.): ICATPN 2003, LNCS 2679, pp. 256–275, 2003.

(CPN) [17,24] have potential to be used successfully for development of pervasive systems. The main reason is that modelling of behaviour may play a more prominent role than previously for application-level developers. We will sustain this conjecture in a case study on a pervasive health care system for the hospitals in Aarhus, Denmark, by describing use of CPN and the Design/CPN tool [45] in two common software development disciplines: *requirements engineering* and *design of middleware*.

The pervasive health care system is introduced in Sect. 2. Requirements engineering is discussed in Sect. 3 and design of middleware in Sect. 4. Section 5 compares with related work and the conclusions are drawn in Sect. 6. The last two sections include a discussion on how to establish CPN more strongly in the software industry, in the context of the Unified Modeling Language (UML) [28, 32,12] and the the new Model Driven Architecture (MDA) vision [47].

2   The Pervasive Health Care System (PHCS)

The *pervasive health care system (PHCS)* [8,48] is being envisioned in a joint project between Aarhus County Hospital, the software company Systematic Software Engineering A/S [50], and the Centre for Pervasive Computing [43] at the University of Aarhus.

PHCS is meant to improve the comprehensive new IT system, the electronic patient record (EPR) [42], which is deployed at the hospitals in Aarhus in the autumn of 2002. EPR solves obvious problems occurring with paper-based patient records such as being not always up-to-date, mislaid, or even lost. However, EPR also has its drawbacks and potentially induces at least two central problems for its users. The first problem is *immobility*: in contrast to a paper-based record, an electronic patient record accessed only from stationary desktop PCs cannot be easily transported. The second problem is *time-consuming login and navigation*: EPR requires user identification and login to ensure information confidentiality and integrity, and to start using the system for clinical work, a logged-in user must navigate, e.g., to find a specific document for a given patient. The motivation for PHCS is to address these problems. The basis is to recognise firstly that users like nurses and doctors are away from their offices and on the move a lot of the time, and secondly that users are frequently interrupted. In the ideal situation, the users should have access to the IT system wherever they go, and it should be easy to resume a suspended work process. The aim of PHCS is to ensure smooth access to and use of EPR by supporting pervasive computing.

The PHCS vision states three general design principles. The first principle is *context-awareness*. This means that the system is able to register and react upon certain changes of context. More specifically, nurses, patients, beds, medicine trays, and other items are equipped with radio frequency identity (RFID) tags [49], enabling presence of such items to be detected automatically by involved context-aware computers, e.g., located by the medicine cabinet and by the patient beds. The second design principle is that the system is *propositional*, in the sense that it makes qualified propositions, or guesses. Context changes may result in automatic generation of buttons, which appear at the task-bar of computers. Users must explicitly accept a proposition by clicking a button – and implicitly ignore or reject it by not clicking. The presence of a nurse holding a medicine tray for patient P in front of the medicine cabinet is a context that triggers automatic generation of a button `Medicine plan:P`, because in many cases, the intention of the nurse is now to navigate to the medicine plan for P. If the nurse clicks the button, she is logged in and taken to P's medicine plan. It is, of course, impossible always to guess the intention of a user from a given context, and without the propositional principle, automatic shortcutting could become a nuisance, because of guesses that would sometimes be wrong. The third design principle is that the system is *non-intrusive*, i.e., not interfering with or interrupting hospital work processes in an undesired way. Thus, when a nurse approaches a computer, it should react on her presence in such a way that a second nurse, who may currently be working on the computer, is not disturbed or interrupted. The last two design principles cooperate to ensure satisfaction of a basic mandatory user requirement: important hospital work processes have to be executed as conscious and active acts by responsible human personnel, not automatically by a computer.

Figure 1 outlines PHCS (with an interface that is simplified and translated into English for the purpose of this paper). The current context of the system is that nurse Jane Brown is engaged in pouring medicine for patient Bob Jones for the giving to take place at 12 a.m. The medicine plan on the display shows which medicine has been prescribed (indicated by 'Pr'), poured ('Po'), and given ('G') at the current time. In this way, it can be seen that Advil and Tylenol have been poured for the 12 a.m. giving, but Comtrex not yet. Moreover, the medicine tray for another patient, Tom Smith, stands close to the computer, as can be seen from the task-bar buttons.

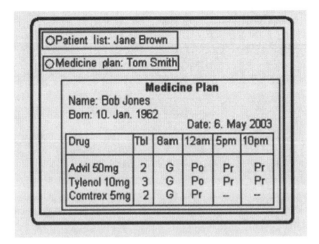

Fig. 1. PHCS – outline

3 Requirements Engineering for PHCS

To aid requirements engineering for PHCS, CPN models of envisioned new work processes and of their proposed computer support were created (described in an earlier version in [20]). The scope of the present paper is the work process *medicine administration*, which is described below.

3.1 Medicine Administration

Assume that nurse N wants to pour medicine into a medicine tray and give it to patient P. First, N goes to the room containing the medicine cabinet (the medicine room). Here is a context-aware computer on which the buttons Login:N and Patient list:N appear on the task-bar, when N approaches. Assume that the second button is clicked. Then, N is logged in and a list of those patients of which she is in charge is displayed on the computer. A medicine tray must be associated with each patient. If a medicine tray is already associated with P, the button Medicine plan:P will appear on the task-bar of the computer, when N takes the tray nearby, and a click will make P's medicine plan appear on the display. However, if P is a newly admitted patient, it is necessary to associate a medicine tray to him. N does so by taking an empty tray from a shelf and making the association. In either case, N pours medicine into the tray and acknowledges this in PHCS. When N leaves the medicine room, she is automatically logged out. N now takes P's medicine tray and goes to the ward where P lies in a bed, which is supplied with a context-aware computer. When N approaches, the buttons Login:N, Patient list:N, and Medicine plan:P will appear on the task-bar. If the last button is clicked, the medicine plan for P is displayed. Finally, N gives the medicine tray to P and acknowledges this in PHCS. When N leaves the bed area, she is automatically logged out again.

The given description captures just one specific combination of sub work processes. There are numerous other scenarios to take into account, e.g., medicine may be poured for one or more patients, for only one round of medicine giving, all four regular rounds of a 24 hours period, or for ad hoc giving; a nurse may have to fetch trays left at the wards prior to pouring; a nurse may approach the medicine cabinet without intending to pour medicine, but only to log into EPR (via PHCS) or to check an already filled medicine tray; two or more nurses may do medicine administration at the same time. To support a smooth medicine administration work process, the requirements for PHCS must deal with all these scenarios and many more. A CPN model, with its fine-grained and coherent nature, is able to support that.

3.2 Medicine Administration CPN Model

The medicine administration CPN model consists of 11 modules (pages) with a total of 54 places and 29 transitions. An overview of the model in terms of a graph with a node for each module and arcs showing the relationship between the modules (the so-called hierarchy page from Design/CPN) is given in Fig. 2.

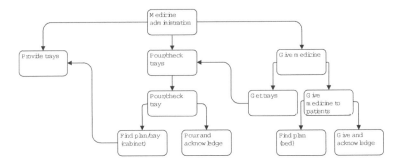

Fig. 2. Medicine administration – CPN model overview

The graph shows how the work process medicine administration is decomposed in sub work processes. An arc between two nodes indicates that the module of the source node contains a so-called substitution transition, whose detailed behaviour is described on the module of the destination node. In this relationship, the source node is called the super-module and the destination the sub-module.

Space does not allows us to present the entire CPN model. Instead, we describe the representative module shown in Fig. 3 (technical remark: substitution transitions are designated by a HS tag, port places by a P tag, and fusion places by an FG tag; port and fusion places are the interfaces of a module to other modules of the model – they are conceptually merged with other places in the model).

The module models the pouring and checking of trays and is represented by the node Pour/check trays in Fig. 2. The medicine cabinet computer is in focus. It is modelled by a token on the Medicine cabinet computer place having colour set COMPUTER, which is a 4-tuple (compid,display,taskbar,users) consisting of a computer identification, its display (main screen), its task-bar, and its current users. In the initial state, the computer has a blank display, no task-bar buttons, and no users. The colour set NURSE is used to model nurses. A nurse is represented as a pair (nurse,trays), where nurse identifies the nurse and trays is a container data structure holding the medicine trays that this nurse currently has in possession. Initially, the nurses Jane Brown and Mary Green are ready (represented as tokens in the Ready place) and have no trays. Occurrence of the Approach medicine cabinet transition models that a nurse changes from being ready to being busy nearby the medicine cabinet. At the same time, two buttons are added to the task-bar of the medicine cabinet computer, namely one login button for the nurse and one patient list button for the nurse. In the CPN model, these task-bar buttons are added by the function addMedicineCabinetButtons appearing on the arc from the transition Approach medicine cabinet to the place Medicine cabinet computer.

The possible actions for a nurse who is by the medicine cabinet are modelled by the three transitions Pour/check tray, Enter EPR via login button, and Leave medicine cabinet. Often, a nurse by the medicine cabinet wants to pour

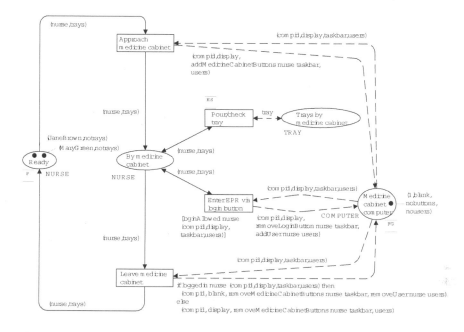

Fig. 3. Medicine administration CPN model – `Pour/check trays` module

and/or check some trays. How this pouring and checking is carried out is modelled on the sub-module `Pour/check tray`, which is bound to the substitution transition having the same name (as can be seen from Fig. 2). The `Enter EPR via login button` transition models that a nurse clicks on the login button and makes a general-purpose login to EPR. It is outside the scope of the model to describe what the nurse subsequently does – the domain of the model is specifically medicine administration, not general EPR use. The transition has a guard which ensures that only a user who is not currently logged into EPR can do so. When a nurse logs in, the login button for that nurse is removed from the task-bar of the computer, modelled by the `removeLoginButton` function on the arc from `Enter EPR via login button` to the `Medicine cabinet computer` place. Moreover, the nurse is added to the set of current users by the function `addUser` appearing on the same arc. The `Leave medicine cabinet` transition models the effect of a nurse leaving: it is checked whether the nurse is currently logged in, modelled by the function `loggedIn` appearing in the if-then-else expression on the arc going from `Leave medicine cabinet` to the `Medicine cabinet computer` place. If the nurse is logged in, the medicine cabinet computer automatically blanks off the screen, removes her task-bar buttons (`removeMedicineCabinetButtons`), and logs her off (`removeUser`). If she is not logged in, the buttons generated because of her presence are removed, but the state of the computer is otherwise left unaltered. When a nurse leaves the medicine cabinet, the corresponding token is put back on the `Ready` place.

3.3   CPN in Requirements Engineering for PHCS

Having described the medicine administration CPN model, the actual use of CPN in requirements engineering for PHCS is now discussed. In the terminology of [39], we will argue that CPN models may be effective means in conjunction with *specification, specification analysis, elicitation, and negotiation and agreement.*

Specification and specification analysis. Specification has a sound foundation because of the formality and unambiguity of the CPN model. From the CPN model of medicine administration, requirements are induced and precisely described by the transitions modelling manipulation of the involved computers. Each transition connected to the places modelling computers, e.g., the place `Medicine cabinet computer` shown in Fig. 3, must be taken into account. The following are examples of requirements induced by the transitions on the module of Fig. 3: (R1) When a nurse approaches the medicine cabinet, the medicine cabinet computer must add a login button and a patient list button for that nurse to the task-bar (transition `Approach medicine cabinet`). (R2) When a nurse leaves the medicine cabinet, if she is logged in, the medicine cabinet computer must blank off its display, remove the nurse's login button and patient list button from the task-bar, and log her out (transition `Leave medicine cabinet`). (R3) When a nurse selects her login button, she must be added as a user of EPR, and the login button must be removed from the task-bar of the computer (transition `Enter EPR via login button`).

Specification analysis is well supported through simulation that allows experiments and trial-and-error investigations of various scenarios for the new envisioned work process. Specification analysis may also be supported through formal verification. However, the CPN model of medicine administration is too large and complex to make, e.g., verification by exploration of the full state space possible in practice. In general, we believe that the full state space of a CPN model made to support requirements engineering typically will be very large. The reason is that often, in the view of the users who should be actively involved in the requirements engineering process, a representation of a work process and its proposed computer support must include many details. This conflicts with modelling the work process in a more coarse-grained, abstract way, with a corresponding smaller state space. Therefore, verification of CPN models supporting requirements engineering is an application area where strong methods for state space reduction, condensation, and exploration are highly needed.

Elicitation. Elicitation is, like specification analysis, well-supported through simulation. We will argue that simulation spurs elicitation by trigging many questions to be asked. Simulation of a CPN model typically catalyses the participants' cognition and generates new ideas. Interaction with an executable model that is a coherent description of multiple scenarios most likely brings up questions, and issues appear that the participants had not thought about earlier.

Examples of questions (Qs) that have appeared during simulation of the CPN model for medicine administration and corresponding answers (As) are: (Q1) What happens if two nurses both are close to the medicine cabinet computer? (A1) The computer generates login buttons and patient list buttons for both of them. (Q2) What happens when a nurse carrying a number of medicine trays approaches a bed? (A2) In addition to a login button and a patient list button for that nurse, only one medicine plan button is generated – a button for the patient associated with that bed. (Q3) Is it possible for one nurse to acknowledge pouring of medicine for a patient while another nurse at the same time acknowledges giving of medicine for that same patient? (A3) No, that would require a more fine-grained concurrency control exercised over the patient records.

Questions like Q1, Q2, and Q3 may imply changes to be made to the CPN model, because emergence of a question indicates that the current version of the CPN model does not reflect the work process properly. As a concrete example, in an early version of the medicine administration CPN model, the leaving of any nurse from the medicine cabinet resulted in the computer display being blanked off. To be compliant with the non-intrusive design principle for PHCS, the leaving of a nurse who is not logged in, should of course not disturb another nurse who might be working at the computer, and the CPN model had to be changed accordingly.

In order to actively involve users like nurses and doctors in elicitation (and specification analysis), the CPN model should be extended with an animation interface. In this way, the simulation of the model could be presented and interacted with by visualising work processes at a hospital department with nurses, medicine trays, medicine cabinets, wards, beds, computers, etc. This would enable nurses and doctors to investigate a future work situation without having to know or care about unfamiliar, technical Petri nets concepts like places, transitions, tokens, enabling, occurrence, etc.

Negotiation and agreement. Negotiation and agreement may be eased via CPN models. Since not all requirements are known and specified before a development project is started, negotiation about requirements must often take place during the project. In many cases, this has strong economical consequences, because a requirements specification for a software system may be an essential part of a legal contract between a customer, e.g., a hospital, and a software company. Therefore, it is important to be able to determine which requirements were included in the initial agreement. Questions like Q1, Q2, and Q3 above may easily be subject to dispute. However, if the involved parties have the agreement that medicine administration should be supported, and have the overall stipulation that the formal and unambiguous CPN model is the authoritative description, many disagreements can quickly be settled.

In summary, CPN models are able to support various common requirements engineering activities, but are of course not a panacea. Use of CPN models does not address, e.g., how to carry out the necessary initial domain analysis, interviews with users, etc. Moreover, the purpose of the presented CPN model is solely to describe the requirements of an IT system, relative to the work processes to be supported. A number of other requirements issues are not addressed properly by the CPN model, e.g., performance and availability issues.

3.4   CPN in Requirements Engineering for Pervasive Systems

It may be argued that the problems solved by use of CPN in requirements engineering for PHCS are not particular to pervasive systems, but applicable to IT systems in general. After all, a CPN model of a work process might also well be beneficial in requirements engineering for other kinds of systems. However, a main objective of a pervasive system is to tightly *integrate* computer support into human work processes [40]. Consequently, with pervasive computing, creation of models of the combined behaviour of users and the computer systems, i.e., work process models, seems to be an inevitable, key activity that should receive more attention than previously.

Moreover, with pervasive computing, requirements engineering must deal with new issues such as mobility and context-awareness. Both issues are accommodated in a natural way in a CPN model. Objects like users (e.g., nurses) and things (e.g., medicine trays) are naturally modelled as CPN tokens, and the various locations of interest can be captured as CPN places. A CPN state as a distribution of tokens on places is a straightforward modelling of a context effecting the appearance of a pervasive system. Mobility in terms of movements of users and things are described by transition occurrences (to model mobility of computational objects like agents, CPN are probably not a proper choice of modelling language – reference nets [22,23] exploiting the nets within nets approach might be a better alternative; another approach to modelling mobility is the ambient calculus [7]).

4   Design of Middleware for PHCS

PHCS as presented above is an ambitious vision addressing both the immobility problem and the time-consuming login and navigation problem discussed in Sect. 2. The latter problem is solved through novel principles of pervasive computing, in particular context-awareness. Application of such leading-edge, potentially immature technological principles involves a risk that the hospitals in Aarhus have not yet decided to take. Therefore, the time-consuming login and navigation problem may persist for some years.

On the other hand, there is an urgent need to solve the immobility problem, i.e., make the current version of EPR, which is accessible from desktop PCs only, mobile. In the first place, this will be done by having nurses and doctors to carry small personal digital assistants (PDAs), with which they can, e.g., access EPR and control other devices such as TV sets.

To make EPR mobile, appropriate middleware must be provided. It is catered for by the *pervasive health care middleware (PHCM)* [3], which is a distributed system consisting of a number of components running in parallel on various mobile and stationary computing devices, and communicating over a wireless network. Some components run on a central background server, while others are deployed on the mobile devices. Figure 4 shows selected components of PHCM – the names hint their functionality.

Fig. 4. PHCM – selected components

The scope of this section is design of the session manager component, shown with thick border in Fig. 4.

4.1 Session Management

A *session* comprises a number of devices that are joined together, sharing data, and communicating in support of some specific work process. A session is appropriate, e.g., if a nurse wants to use her PDA to control a TV set in a ward in order to show an X-ray picture to a patient. In this case, the TV and the PDA must be joined in a session. Another example is a nurse who wishes to discuss some data, e.g., EPR data, or audio and video in a conference setting, with doctors who are in remote locations. Here, the relevant data must be shown simultaneously on a number of devices joined in a session, one device for the nurse and one device for each doctor. In general, session data is viewed and possibly edited by the users through their devices. The PHCM architecture is based on the Model-View-Controller pattern [6]. The model part administers the actual data being shared and manipulated in a session. Each participating device has both a viewer and a controller component which are used as, respectively, interface to and manipulator of the session data. Model, viewer, and controller components communicate over the wireless network.

Sessions are managed by the *session manager*. As can be seen from the class diagram of Fig. 5, the session manager manages zero to any number of sessions, a session comprises one or more devices, and from the point of view of the session manager, a device is either inactive, i.e., not currently participating in any session, or active, i.e., participating in some session. A device participates in at most one session at a time.

The operations that the session manager must provide can be grouped into three main functional areas:

1. *Configuration management*: initiation, reconfiguration (i.e., supporting devices dynamically joining and leaving), and termination of sessions.
2. *Lock management*: locking of session data. Session data is shared and must be locked by a device that wants to edit it.

Fig. 5. Session management – class diagram

3. *Viewer/controller management*: change of viewers and controllers for active devices, e.g., if a nurse enters a room containing a TV, she may wish to view something on the large TV screen instead of on her small PDA display. In this case, viewer and controller replacement on the PDA and the TV is needed.

Devices interact with the session manager by invoking its operations. One interaction scenario is shown in Fig. 6, which illustrates the communication between the session manager and two devices, de(1) and de(2).

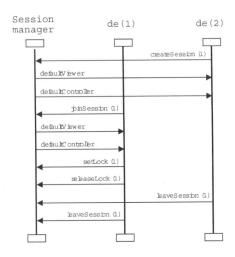

Fig. 6. Session management – session manager / device communication

First, de(2) creates a session, which gets the session identifier 1. The session manager responds to the creation request by providing de(2) with a default viewer and a default controller for the new session. Then, de(1) joins the session, and also gets a default viewer and a default controller from the session manager. At some point, de(1) locks the session, probably does some editing, commits, and later releases the lock. Finally, de(2) and then de(1) leave the session.

Figure 6 is an example illustrating one single possible sequence of interactions between the session manager and some devices. In the session manager design, of course much more is needed. It is crucial to be able to specify and investigate

the general behavioural properties of session management. We now present a
CPN model, whose purpose is to constitute an initial design proposal for the
session manager by identifying the operations to be provided, and describing
their behaviour.

4.2   Session Management CPN Model

The session management CPN model (described in an earlier version in [19],
and much abbreviated in [21]) consists of four modules. The top-level of the
model is the SessionManager module, shown in Fig. 7, where the three main
functional areas are represented by means of substitution transitions. In this way,
each main functional area is modelled by an appropriately named sub-module,
ConfigurationManager, LockManager, and ViewCtrManager, respectively.

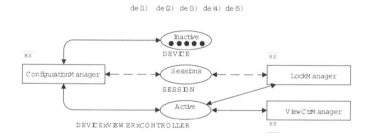

Fig. 7. Session management CPN model – SessionManager module

The model contains declarations of colour sets for devices (DEVICE), viewers
(VIEWER), and controllers (CONTROLLER). In addition, there are Cartesian product colour sets used to model when devices are associated with viewers and
controllers. Sessions are modelled using the SESSION colour set, whose elements
are triples (s,dl,l), where s is a session identifier, dl is a list of devices, and l
is a lock indicator.

In the following, we describe the model modules corresponding to the three
main functional areas of the session manager. The operations of the session
manager correspond to the transitions of the CPN model. The detailed behaviour
of the operations may immediately be derived from the arc expressions and
guards. The latter corresponds to checks that the session manager must carry out
before allowing the corresponding operation to be executed. We have chosen to
use function calls consistently as arc expressions and guards, e.g., a function call
like "createSession s d" instead of the expression "(s,[d],nolock)" (where
s is a session id and d a device). In this way, the sub-routines of the session
manager operations are explicitly identified.

Configuration management. The `ConfigurationManager` module is shown in Fig. 8.

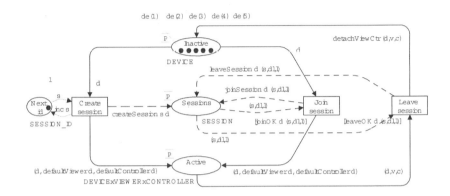

Fig. 8. Session management CPN model – `ConfigurationManager` module

The `ConfigurationManager` module has four places, `Inactive`, `Active`, `Sessions`, and `Next id`. Initially, all devices are inactive, corresponding to all the `DEVICE` tokens initially being at the `Inactive` place. The only transition which is enabled in the initial marking is `Create session`. When it occurs, triggered by an initiating device d, a new session is created. The session gets a fresh session id from the `Next id` place (the session id is incremented each time a new session is created, and starts over from 1, when the largest session id available has been assigned), starts in unlocked mode, and a token corresponding to the new session is put on the `Sessions` place. The token corresponding to the initiating device is augmented with a default viewer and controller, and that triple is put on the `Active` place.

The transition `Join session` adds devices to sessions. Any join must be preceeded by a permission check, modelled by the guard function `joinOK`. When a device d joins a session (s,dl,l), the appropriate token residing on the `Sessions` place is updated accordingly, by use of the `joinSession` function. Moreover, the d token is removed from the `Inactive` place, augmented with a default viewer and controller, and put on the `Active` place. The transition `Leave session` works reversely to `joinSession`. Upon leaving, the applicable viewer and controller for the device are detached.

Lock management. The `LockManager` module is shown in Fig. 9.

`LockManager` contains the two places `Active` and `Sessions`, which are already described. In addition, the module contains the two transitions `Set lock` and `Release lock`. When `Set lock` occurs, the selected session (s,dl,l) is locked by the requesting device, which is identified by the d part of the (d,v,c)

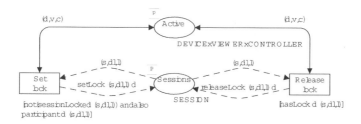

Fig. 9. Session management CPN model – **LockManager** module

token. The session can only be locked if it is not locked already and if the requesting device is currently a participant in that session. These two conditions are checked by the guard of Set lock. The effect of the transition Release lock is to release the lock of the current session. This is only possible if the requesting device is the lock holder, modelled by the **hasLock** function of the guard.

Viewer/controller management. The **ViewCtrManager** module is shown in Fig. 10.

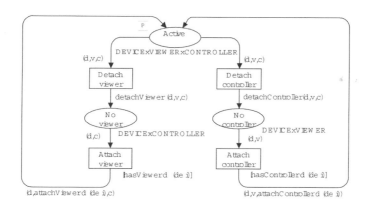

Fig. 10. Session management CPN model – **ViewCtrManager** module

In addition to the place **Active** already described, the **ViewCtrManager** module contains the two places **No viewer** and **No controller**. A token on either **No viewer** or **No controller** models that the corresponding device is suspended – even though the device is participating in a session. This means that the device is temporarily not able to read and write data (only those devices whose tokens are currently on the **Active** place are able to read and, if an appropriate lock is set, write data).

Viewers and controllers are not replaced in one atomic action. The replacement mechanism is modelled by four transitions. The two Detach transitions do, as the names indicate, detach the viewer or controller, respectively. There is no precondition for allowing this. The Attach viewer transition checks that the viewer that some device requests actually can be provided. The request is checked by the guard "hasViewer d (de i)", which evaluates to true if it is possible to equip device d with a viewer for device number i, de i (the devices are indexed and i is a free variable over the index range). The guard function hasViewer enforces that some viewers fit with some devices and not with others. The Attach controller transition works in a similar fashion.

4.3   CPN in Design of Middleware for PHCS

The perspective of the CPN model is the communication between, and thus combined behaviour of, the objects of concern, i.e., a session manager, devices, sessions, etc. Occurrence of any transition corresponds to invocation of an operation of a session manager by some device. The result of the occurrence reflects the corresponding state change for the invoking device and involved session, e.g., a device changes from inactive to active, and a session is extended with an additional device; a lock is set on session data; or a device changes from active to suspended. In general, when using a CPN model as a design of a software component, the relationship with the structural architecture, e.g., represented as a class diagram like in Fig. 5 should be made clear. After all, when the design is to be realised as a running system, today that is typically done by implementing classes in an object-oriented programming language.

Having the session management design represented as a CPN model supports evaluation and comparison of the behavioural consequences of alternative design proposals. The CPN model reflects a number of design decisions for session management, many of which may be argued, e.g., would it be better to allow a device to have more than one viewer and one controller at a time instead of just one of each? Should a session always be explicitly terminated by the initiating device instead of being implicitly terminated when the last participating device leaves? How fine-grained should the locking scheme be? What happens if the session id counter wraps around? Alternatives can be modelled and investigated by making modifications and simulations of the CPN model.

The presented CPN model is a relatively abstract view of session management. Some modelling decisions have been made in order to keep the model relatively simple, but still serving its purpose as constituting an initial design of the session manager. In particular, we have assumed reliable communication and that devices do not crash, i.e., errors in the communication between devices and the session manager are not modelled, and a question like what should happen if the device that is currently having a lock on session data crashes cannot be answered from the current version of the CPN model. In a hectic and busy work day with the PHCM system at the hospitals, communication errors will happen on the wireless network, and devices will crash, e.g., be turned off accidentally or run out of battery. To investigate such problems and strategies for coping with them, the CPN model could be extended.

With respect to design of the session manager, the scope of this paper is construction and simulation of the CPN model. However, formal verification might be an attractive, supplementary option. The session manager is a protocol, which is an application domain where formal verification of CPN models has previously been carried out successfully, see, e.g., [15].

4.4   CPN in Design of Pervasive Systems

The problems addressed by using CPN for design of middleware as described above are not particular to pervasive computing, but applicable to communication protocols in general. However, the problems *arise*, indeed, because of a need to support mobility, a central aspect of pervasive computing.

With pervasive computing, it is likely that application-level developers for some years to come will have to be concerned with designing higher-level protocols (in the OSI seven-layer protocol model sense), because well-established standard off-the-shelf, commercial components will not be available in the near future.

5   Related Work

Given the everyday nature of the software development disciplines of requirements engineering and design addressed in this paper, there is an abundance of related work. We will restrict ourselves to comparing our approaches with use of the *Unified Modeling Language (UML)* [28,32,12], which is almost universally accepted as *the* modelling language in the software industry today. Despite its high industrial acceptance and wide-spread use, UML is certainly not perfect, see, e.g., [31,36]. If we had used UML exclusively in the two development tasks described in this paper, we would have encountered severe problems. Both with respect to requirements engineering and design, our need has been to model behaviour, and in this area, UML strongly needs improvements [2,21,30].

The CPN model of the medicine administration work process can be seen as an alternative to or supplement to a UML use case [9,16]. Use cases model work processes to be supported by a new IT system, and a set of use cases is interpreted as requirements for that system. However, UML use cases have several general and known shortcomings, see, e.g., [36] which points out a number of problems under headlines like *use case modelling misses long-range logical dependency* and *use case dependency is non-logical and inconsistent*. Representing a work process and its proposed computer support as a CPN model, with its precise and coherent nature, alleviates many of the problems inherent to UML use cases. Moreover, having an executable representation of a work process in terms of a CPN model, instead of a static representation in terms of a UML use case, gives better support for specification analysis and elicitation as discussed in Sect. 3.3. In general, using CPN in requirements engineering is not a new idea, see, e.g., [11], where CPN are used to prototype user interfaces. Petri nets have also previously been used for modelling of work processes or workflows, see, e.g., [38]. However, use of Petri nets in requirement engineering for pervasive systems has, to the best of our knowledge, not been described previously.

CPN applied in the design of the middleware component can be seen as a use of CPN to supplement UML [19] – UML is the language used in the class diagram of Fig. 5 and the sequence diagram (message sequence chart) of Fig. 6. If we would have wanted to describe the general behaviour of session management in UML, we should have used state machines or activity diagrams, which are the UML diagram types targeted at modelling behaviour in a general way (in contrast to, e.g., sequence diagrams which merely visualise examples). However, there are a number of substantial reasons to prefer CPN over the candidate UML diagrams, e.g., UML state machines lack a well-defined execution semantics, do not support modelling of multiple instances of classes, and do not scale well to large systems [11,21].

UML supports object-oriented system development – [1] describes a number of ways to combine Petri nets and object-orientation. With respect to using UML and Petri nets together, much of the work done is concerned with automatic translation from certain UML diagram types into Petri nets, often aimed at formal verification, e.g., [33]. Also, Petri nets have been used to give precise execution semantics to different UML diagram types, e.g., [4].

6   Conclusions

With requirements engineering and design of middleware for PHCS, and the generalisations discussed in Sects. 3.4 and 4.4, as evidence, we have argued that CPN with its strong support for modelling of behaviour, have potential to be used successfully for development of pervasive systems.

The emergence of pervasive computing can therefore be seen as a chance to establish CPN more strongly in the software industry. A number of technical arguments for considering CPN a candidate to be used for development of pervasive systems are given in this paper. However, choices of techniques and tools in the software industry are not always done solely based on technical arguments – marketing also has a strong influence. A viable starting point if we want to promote CPN further is to recognise that with the success of UML, the software industry has in large scale adopted modelling as a valuable discipline, and many software developers use in particular UML class diagrams in their everyday work. Those who have tried also to model behavioural aspects in UML might have encountered frustrating shortcomings, and therefore, the motivation to use a supplementary modelling language together with UML may be quite high.

The success of UML is also one of the driving forces behind the new *Model Driven Architecture (MDA)* vision [47] from the Object Management Group (OMG), which is being aggressively marketed as *the* platform for future software development. The overall goal of MDA is to let models be the primary artifacts of software development. If the MDA vision comes true, source code as we know it today will disappear as an everyday concern of software developers, in the same way as assembly code disappeared decades ago. Many issues must be addressed properly for the vision to be realised. Two key aspects are *executable models* and *automatic code generation from models*.

With respect to executable models, MDA is from the beginning hampered because MDA, as marketed by the OMG, is as default tied to use of UML, and in the current version of the UML standard [28], no UML diagram type has a well-defined execution semantics. Many proposals to alleviate this problem have been published, e.g., [5,13], but even though major improvements should be expected in forthcoming versions of UML, a general and widely accepted solution seems far away. With respect to automatic code generation from models, MDA has a feasible starting point in UML. Code generation from UML class diagrams is already widely used, and code generation from (proprietary semantic interpretations of) UML behavioural diagrams is possible, see, e.g., [37]. A severe drawback of using CPN for design of a system is that the step from design to implementation of, e.g., a distributed system typically must be done manually. Automatic code generation from CPN models is in general difficult, although not impossible in some special cases, see, e.g., [26] which describes automatic code generation for an embedded (non-distributed) system.

MDA is boosted with slogans like *the model is the implementation*. Leaving practical issues such as being widely accepted and understandable for involved stakeholders aside, we believe that using CPN in requirements engineering is a significant step towards realising that *the model is the requirements specification*. On the other hand, realising a *the CPN model is the implementation* vision requires considerable advances to be made with respect to automatic code generation from CPN models to different platforms, including construction of various levels of models as envisioned in MDA, ranging from abstract platform independent models (where many CPN models already fit in) to detailed platform specific models, to make it up for running code. In summary, the success of UML and the current commercial boosting of MDA (how distant the vision may seem) is a good chance to establish CPN, and Petri nets in general, more strongly in the software industry – in particular, because of the potential to be used successfully for the development of pervasive systems.

Acknowledgements. This paper benefits from the efforts of many participants in the pervasive health care research project [48]. In particular, Claus Bossen contributed significantly to the work presented in Sects. 2 and 3.

References

1. G. Agha, F.D. Cindio, and G. Rozenberg, editors. *Concurrent Object-Oriented Programming and Petri Nets*, volume 2001 of *LNCS*. Springer-Verlag, 2001.
2. C. Andre, M.A. Peraldi-Frati, and J.P. Rigault. Integrating the Synchronous Paradigm into UML: Application to Control-Dominated Systems. In Jézéquel et al. [18].

3. J. Bardram and H.B. Christensen. Middleware for Pervasive Healthcare, A White Paper. In *Workshop on Middleware for Mobile Computing*, Heidelberg, Germany, 2001.
4. L. Baresi and M. Pezzé. On Formalizing UML with High-Level Petri Nets. In Agha et al. [1].
5. M.v.d. Beeck. Formalization of UML-Statecharts. In Gogolla and Kobryn [14].
6. F. Buschmann, R. Meunier, H. Rohnert, and P. Sommerlad. *Pattern-Oriented Software Architecture*. John Wiley and Sons, 1996.
7. L. Cardelli and A.D. Gordon. Mobile Ambients. In D. Le Métayer, editor, *Theoretical Computer Science, Special Issue on Coordination*, volume 240/1. Elsevier, 2000.
8. H.B. Christensen and J.E. Bardram. Supporting Human Activities – Exploring Activity-Centered Computing. In G. Borriello and L.E. Holmquist, editors, *4th Ubicomp Conference*, volume 2498 of *LNCS*, Gothenborg, Sweden, 2002. Springer-Verlag.
9. A. Cockburn. *Writing Effective Use Cases*. Addison-Wesley, 2000.
10. G. Coulouris, J. Dollimore, and T. Kindberg. *Distributed Systems – Concepts and Design*. Addison-Wesley, 2001.
11. M. Elkoutbi and R.K. Keller. User Interface Prototyping Based on UML Scenarios and High-Level Petri Nets. In Nielsen and Simpson [27].
12. G. Engels, R. Heckel, and S. Sauer. UML – A Universal Modeling Language? In Nielsen and Simpson [27].
13. R. Eshuis and R. Wieringa. An Execution Algorithm for UML Activity Graphs. In Gogolla and Kobryn [14].
14. M. Gogolla and C. Kobryn, editors. *Proceedings of the 4th UML Conference*, volume 2185 of *LNCS*, Toronto, Canada, 2001. Springer-Verlag.
15. S. Gordon and J. Billington. Analysing the WAP Class 2 Wireless Transaction Protocol Using Coloured Petri Nets. In Nielsen and Simpson [27].
16. I. Jacobson, M. Christerson, P. Jonsson, and G. Övergaard. *Object-Oriented Software Engineering: A Use Case Driven Approach*. Addison-Wesley, 1992.
17. K. Jensen. *Coloured Petri Nets – Basic Concepts, Analysis Methods and Practical Use. Volume 1, Basic Concepts*. Monographs in Theoretical Computer Science. An EATCS Series. Springer-Verlag, 1992.
18. J.M. Jézéquel, H. Hussmann, and S. Cook, editors. *Proceedings of the 5th UML Conference*, volume 2460 of *LNCS*, Dresden, Germany, 2002. Springer-Verlag.
19. J.B. Jørgensen. Coloured Petri Nets in UML-Based Software Engineering – Designing Middleware for Pervasive Healthcare. In K. Jensen, editor, *Proceedings of the 3rd CPN Workshop*, Aarhus, Denmark, 2002. Technical report, Department of Computer Science, University of Aarhus.
20. J.B. Jørgensen and C. Bossen. Executable Use Cases for Pervasive Healthcare. In Moldt [25].
21. J.B. Jørgensen and S. Christensen. Executable Design Models for a Pervasive Healthcare Middleware System. In Jézéquel et al. [18].
22. M. Köhler, D. Moldt, and H. Rölke. Modelling the Structure and Behaviour of Petri Net Agents. In J.-M. Colom and M. Koutny, editors, *Proceedings of the 22nd PN Conference*, volume 2075 of *LNCS*, Newcastle, UK, 2001. Springer-Verlag.
23. M. Köhler and H. Rölke. Modelling Mobility and Mobile Agents using Nets within Nets. In Moldt [25].
24. L.M. Kristensen, S. Christensen, and K. Jensen. The Practitioner's Guide to Coloured Petri Nets. *International Journal on Software Tools for Technology Transfer*, 2(2), 1998.

25. D. Moldt, editor. *Proceedings of the 2nd MOCA Workshop*, Aarhus, Denmark, 2002. Technical report, Department of Computer Science, University of Aarhus.
26. K.H. Mortensen. Automatic Code Generation Method Based on Coloured Petri Net Models Applied on an Access Control System. In Nielsen and Simpson [27].
27. M. Nielsen and D. Simpson, editors. *Proceedings of 21st PN Conference*, volume 1825 of *LNCS*, Aarhus, Denmark, 2000. Springer-Verlag.
28. OMG Unified Modeling Language Specification, Version 1.4. Object Management Group (OMG); UML Revision Taskforce, 2001.
29. C. Ouyang, L.M. Kristensen, and J. Billington. A Formal Service Specification for the Internet Open Trading Protocol. In J. Esparza and C. Lakos, editors, *Proceedings of the 23rd PN Conference*, volume 2360 of *LNCS*, Adelaide, Australia, 2002. Springer-Verlag.
30. S. Pllana and T. Fahringer. On Customizing the UML for Modeling Performance-Oriented Applications. In Jézéquel et al. [18].
31. J. Rumbaugh. The Preacher at Arrakeen. In Gogolla and Kobryn [14].
32. J. Rumbaugh, I. Jacobson, and G. Booch. *The Unified Modeling Language Reference Manual*. Addison-Wesley, 1999.
33. J. Saldhana and S.M. Shatz. UML Diagrams to Object Petri Net Models: An Approach for Modeling and Analysis. In *Proceedings of the International Conference on Software Engineering and Knowledge Engineering*, Chicago, Illinois, 2000.
34. M. Satyanarayanan. Challenges in Implementing a Context-Aware System. In *Pervasive Computing – Mobile and Ubiquitous Systems*, volume 1(3). IEEE, 2002.
35. M. Satyanarayanan, editor. *Pervasive Computing – Mobile and Ubiquitous Systems*, volume 1(1). IEEE, 2002.
36. A.J.H. Simons and I. Graham. 30 Things That Go Wrong in Object Modelling with UML 1.3. In H. Kilov, B. Rumpe, and I. Simmonds, editors, *Behavioral Specifications of Businesses and Systems*. Kluwer Academic Publishers, 1999.
37. T. Sturm, J. von Voss, and M. Boger. Generating Code from UML with Velocity Templates. In Jézéquel et al. [18].
38. W.M.P. van der Aalst and K. van Hee. *Workflow Management: Models, Methods, and Systems*. MIT Press, 2002.
39. A. van Lamsweerde. Requirements Engineering in the Year 00: A Research Perspective. In *Proceedings of the 22nd International Conference on Software Engineering*, Limerick, Ireland, 2000. ACM Press.
40. M. Weiser. The Computer for the 21st Century. In *Scientific American*, volume 265 (3). Scientific American, Inc., 1991.
41. G. Wirtz. Application of Petri Nets in Modelling Distributed Software Systems. In D. Moldt, editor, *Proceedings of the 1st MOCA Workshop*, Aarhus, Denmark, 2001. Technical report, Department of Computer Science, University of Aarhus.
42. Aarhus Amt Electronic Patient Record. www.epj.aaa.dk.
43. Centre for Pervasive Computing. www.pervasive.dk.
44. CORBA. www.corba.org.
45. Design/CPN. www.daimi.au.dk/designCPN.
46. J2EE. www.java.sun.com/j2ee.
47. MDA. www.omg.org/mda.
48. Pervasive Healthcare. www.healthcare.pervasive.dk.
49. Radio Frequency Identification. www.rfid.org.
50. Systematic Software Engineering A/S. www.systematic.dk.

Logical Reasoning and Petri Nets

Kurt Lautenbach

University of Koblenz-Landau, Universitätsstr. 1, 56075 Koblenz, Germany,
laut@uni-koblenz.de,
http://www.uni-koblenz.de/ laut

Abstract. The main result of the paper states that a set F of propositional-logic formulas is contradictory iff in all net representations of F the empty marking is reproducible.

1 Introduction

This paper brings some rather different approaches together:

- C.A. Petri's fact theory [GenThi76],
- a theorem by Peterka and Murata [PetMur89],
- a theorem about reproducing the empty marking [Laut02].

In Petri's approach sets of propositional-logical formulas are represented by condition/event nets, where the models correspond to dead $(0, 1)$-markings. Petri called the dead transitions "facts", thus, emphasizing the role of models as factual truths [GenThi76].

Now, an interesting question is how the net representations behave if the sets of formulas are contradictory. Peterka and Murata showed that in the net representations of Horn formulas the empty marking is reproducible iff there exists a non-negative t-invariant with a positive entry for at least one goal transition [PetMur89]. Since the reproductions of the empty marking can be interpreted as proof processes, there exists an obvious connection to contradictory sets of formulas.

As to reproducing the empty marking, there a necessary and sufficient condition that uses only structural net components such as t-invariants, traps, and co-traps (structural deadlocks) [Laut02].

The present paper now is, on the one hand, a completion of the result in [PetMur89] since the formulas need no longer to be Horn formulas. On the other hand, the paper shows an application of the reproduction theorem in [Laut02], because the main result states that a set F of formulas is contradictory

W.M.P. van der Aalst and E. Best (Eds.): ICATPN 2003, LNCS 2679, pp. 276–295, 2003.

- iff in all net representations of F the empty marking is reproducible (theorem 3)
- iff in the "union" of all net representations of F no co-trap of a specific kind exists (theorem 4).

(Among the several different net representations of F, Petri's representation will be called "canonical").

Even though the result looks rather theoretical, it is quite valuable for checking Petri net representations of rule based systems. It was used so far for checking the rules of deductive databases and for proving correctness in net representations of time-logical formulas.

The paper is organized as follows. Chapter 3 is an introduction into propositional logic. Chapter 4 deals with net representations of logical formulas. In particular, one finds connections between logical interpretations and structural concepts in nets. Chapter 5 shows that refutation processes can be represented as reproductions of the empty marking. Chapter 6 finally deals with Horn formulas and integrates the results of [PetMur89] into our approach.

I am greatly indebted to Jörg Müller and Carlo Simon for listening to me and for valuable comments.

2   Petri Nets

As to preliminaries of place/transition nets we refer to [Laut02].

In addition we need the following definition and theorem:

Definition 1. *Let $\mathcal{N} = (S, T, F)$ be a p/t-net and M_0 an initial marking;*

t *is a* fact transition *(goal transition) iff* $^\bullet t = \emptyset$ ($t^\bullet = \emptyset$). *(This concept of a fact transition should not be mixed up with Petri's facts).*

t *is* M_0-firable *iff t is enabled for some follower marking $M' \in [M_0\rangle$.*

Structural deadlocks are called *co-traps*.

Theorem 1. *Let $\mathcal{N} = (S, T, F)$ be a p/t-net.*

$\mathcal{N}$ *is* **0**-reproducing *iff there exists a non-negative t-invariant R of $\mathcal{N}$ whose net representation $\mathcal{N}_R$ neither contains a co-trap nor a trap.*

Proof. See [Laut02].

3   Propositional Logic

Definition 2. *The* alphabet *of propositional logic consists of* atoms $a, b, c, \cdots$, operators $\neg, \vee, \wedge$, *and* brackets *(and)*.

The formulas α *are exactly the words which can be constructed by means of the following rules:*

- *all* atoms *are formulas;*
- *if α is a formula, the* negation $\neg\alpha$ *is a formula, too;*
- *if α and β are formulas, the* conjunction $(\alpha \wedge \beta)$ *and the* disjunction $(\alpha \vee \beta)$ *are formulas, too.*

The implication $(\alpha \rightarrow \beta)$ *is an abbreviation for* $((\neg\alpha) \vee \beta)$; *the* biimplication $(\alpha \leftrightarrow \beta)$ *is an abbreviation for* $((\alpha \rightarrow \beta) \wedge (\beta \rightarrow \alpha))$.

For omitting brackets, we stick to the usual operator hierarchy.

$\mathbb{A}(\alpha)$ *denotes the* set of atoms *contained in a formula α.*

Definition 3. *A* literal *is an atom or the negation of an atom.*

A clause *is a disjunction of literals. Usually, $\square$ denotes the* empty clause.

Let $\tau = \neg a_1 \vee \ldots \vee \neg a_m \vee b_1 \vee \ldots \vee b_n$ be a clause. Often a set notation is used: $\tau = \{\neg a_1, \ldots, \neg a_m, b_1, \ldots, b_n\}$ *or* $\tau = \neg A \cup B$ *for* $A = \{a_1, \ldots, a_m\}$ *and* $B = \{b_1, \ldots, b_n\}$

If in $\tau = (\neg A \cup B)$ *A is empty, τ is a* fact clause; *if B is empty, τ is a* goal clause.

A formula α is in conjunctive normal form *(is a CNF-formula) if it is a conjunction of clauses.* $\mathbb{C}(\alpha)$ *denotes the* set of clauses *contained in a CNF-formula α.*

Definition 4. *Let α be a formula. A mapping $I : \mathbb{A}(\alpha) \rightarrow \{true, false\}$ is an* interpretation *of α.*

An interpretation of α such that α is true is a model *of α.*

If α has no models, α is contradictory. *If α has at least one model, α is* satisfiable. *If all interpretations of α are models, α is* tautological.

If for two formulas α and β $(\alpha \leftrightarrow \beta)$ is tautological, α and β are equivalent.

If for some formulas $\alpha_1, \ldots, \alpha_n, \beta$ $((\alpha_1 \wedge \ldots \wedge \alpha_n) \rightarrow \beta)$ is tautological β is a logical consequence *of $\alpha_1, \ldots, \alpha_n$.*

Definition 5. *Let a be an atom and K_1, K_2 clauses with*

$$K_1 = \{a\} \cup \neg A_1 \cup B_1, \quad \mathbb{A}(K_1) \backslash \{a\} = A_1 \cup B_1$$
$$K_2 = \{\neg a\} \cup \neg A_2 \cup B_2, \quad \mathbb{A}(K_2) \backslash \{a\} = A_2 \cup B_2.$$

Then the clause $\varrho = \neg A_1 \cup \neg A_2 \cup B_1 \cup B_2$ is the resolvent *of K_1 and K_2 w.r.t. a. Constructing ϱ is a* resolution step.

Theorem 2. *A CNF-formula α is contradictory iff the empty clause $\square$ can be reached by finitely many resolution steps starting with the clauses of α.*

Proof. See [Gall87].

Example 1.

$$\alpha = \underset{(1)}{(p \vee q)} \wedge \underset{(2)}{(\neg p \vee q)} \wedge \underset{(3)}{(\neg q \vee r)} \wedge \underset{(4)}{(\neg q \vee \neg r \vee s)} \wedge \underset{(5)}{(\neg q \vee \neg s)}$$

α is contradictory because $\square$ can be derived by finitely many resolution steps starting with the clauses (1) - (5).

In order to demonstrate that we use a *refutation tree*.

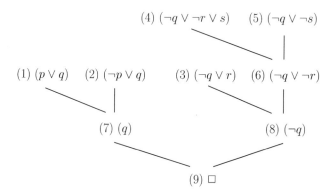

Remark 1. When deriving the empty clause $\square$, tautological clauses as intermediate results can be avoided.

4   The Net Representation of CNF-Formulas

Definition 6 (canonical net representation). *Let α be a CNF-formula and let $\mathcal{N}_\alpha = (S_\alpha, T_\alpha, F_\alpha)$ be a p/t-net;*

$\mathcal{N}_\alpha$ is the canonical p/t-net representation of α iff

- *$S_\alpha = \mathbb{A}(\alpha)$ (set of atoms of α) and $T_\alpha = \mathbb{C}(\alpha)$ (set of clauses of α)*
- *for all $\tau = \neg a_1 \vee \ldots \vee \neg a_m \vee b_1 \vee \ldots \vee b_n \in \mathbb{C}(\alpha)$ where $\{a_1, \ldots, a_m, b_1, \ldots, b_n\} \subseteq \mathbb{A}(\alpha)$, F_α is determined by ${}^\bullet\tau = \{a_1, \ldots, a_m, \tau^\bullet = \{b_1, \ldots, b_n\}$.*

In case $\mathcal{N}_\alpha$ is a c/e-net, $\mathcal{N}_\alpha$ is the canonical c/e-net representation.

Whenever it is not necessary to distinguish between the canonical p/t-net representation and the canonical c/e-net representation, we speak of the *canonical net representation.*

Definition 7 (markings of net representations). *Let α be a CNF-formula, I an interpretation, and $\mathcal{N}_\alpha = (S_\alpha, T_\alpha, F_\alpha)$ the canonical net representation of α; then for all $p \in S_\alpha = \mathbb{A}(\alpha)$*

$$M_I(p) := \begin{cases} 1, & \text{if } I(p) = \text{true} \\ 0, & \text{if } I(p) = \text{false} \end{cases}$$

is the marking of $\mathcal{N}_\alpha$ w.r.t. I.

Example 2 (cf. example 1).

$$\alpha = \underset{(1)}{(p \vee q)} \wedge \underset{(2)}{(\neg p \vee q)} \wedge \underset{(3)}{(\neg q \vee r)} \wedge \underset{(4)}{(\neg q \vee \neg r \vee s)} \wedge \underset{(5)}{(\neg q \vee \neg s)}$$

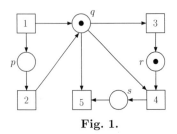

Fig. 1.

Figure 1 shows $\mathcal{N}_\alpha = (S_\alpha, T_\alpha, F_\alpha)$ and M_I with

$S_\alpha = \mathbb{A}(\alpha) = \{p, q, r, s\}$
$T_\alpha = \mathbb{C}(\alpha) = \{1, 2, 3, 4, 5\}$
$F_\alpha = \{(1, p), (p, 2), (1, q), (2, q), (q, 3), (q, 4),$
$\qquad (q, 5), (3, r), (r, 4), (4, s), (s, 5)\}$

I is an interpretation with $I(q) = I(r) = \text{true}$ and $I(p) = I(s) = \text{false}$, so $M_I(q) = M_I(r) = 1$ and $M_I(p) = M_I(s) = 0$.

The equivalence of α and $\neg\neg\alpha$ in propositional logic allows further net representations of α.

Definition 8 (net representation). *Let α be a CNF-formula, $\mathcal{N}_\alpha = (S_\alpha, T_\alpha, F_\alpha)$ the canonical net representation of α, and let $D \subseteq S_\alpha = \mathbb{A}(\alpha)$ be a set of atoms; then the net $\mathcal{N}_{\alpha,D} = (S_{\alpha,D}, T_{\alpha,D}, F_{\alpha,D})$ where*

- $S_{\alpha,D} := \overline{D} \cup \neg D$ *where* $\overline{D} := \mathbb{A}(\alpha) \backslash D$ *and* $\neg D := \{\neg a \mid a \in D\}$
- $T_{\alpha,D} := T_\alpha = \mathbb{C}(\alpha)$
- $F_{\alpha,D} := F_\alpha \backslash (\{(p, t) \mid (p, t) \in F_\alpha \wedge p \in D\} \cup \{(t, p) \mid (t, p) \in F_\alpha \wedge p \in D\})$
  $\qquad \cup (\{(t, \neg p) \mid (p, t) \in F_\alpha \wedge p \in D\} \cup \{(\neg p, t) \mid (t, p) \in F_\alpha \wedge p \in D\})$

is the net representation of α w.r.t. D.

$\mathcal{N}_{\alpha,D}$ equals $\mathcal{N}_\beta$ iff β results from α by replacing $\neg(\neg a)$ for all $a \in D$. Clearly, $\mathcal{N}_\alpha = \mathcal{N}_{\alpha,\emptyset}$.

Definition 9 (markings of net representations). *Let α be a CNF-formula, I an interpretation of α, $\mathcal{N}_\alpha = (S_\alpha, T_\alpha, F_\alpha)$ the canonical net representation, and M_I the marking of $\mathcal{N}_\alpha$ w.r.t. I; let further $D \subseteq \mathbb{A}(\alpha)$ and $\mathcal{N}_{\alpha,D} = (S_{\alpha,D}, T_{\alpha,D}, F_{\alpha,D})$ the net representation of α w.r.t. D; then $M_{I,D}$ is the marking of $\mathcal{N}_{\alpha,D}$ w.r.t. I if the following holds:*

$$M_{I,D}(a) = M_I(a) \text{ if } a \in \overline{D} \text{ and } M_{I,D}(\neg a) = 1 - M_I(a) \text{ if } a \in D$$

Example 3 (cf. example 2).

$$\alpha = \underset{(1)}{(p \vee q)} \wedge \underset{(2)}{(\neg p \vee q)} \wedge \underset{(3)}{(\neg q \vee r)} \wedge \underset{(4)}{(\neg q \vee \neg r \vee s)} \wedge \underset{(5)}{(\neg q \vee \neg s)}$$

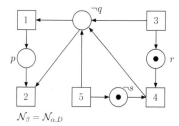

$$D = \{q, s\}$$
$$\overline{D} = \{p, r\}$$
$$\neg D = \{\neg q, \neg s\}$$
$$I(r) = I(\neg s) = true$$
$$I(p) = I(\neg q) = false$$

$\mathcal{N}_\beta = \mathcal{N}_{\alpha,D}$

Fig. 2.

$$\beta = \underset{(1)}{(p \vee \neg(\neg q))} \wedge \underset{(2)}{(\neg p \vee \neg(\neg q))} \wedge \underset{(3)}{((\neg q) \vee r)} \wedge \underset{(4)}{((\neg q) \vee \neg r \vee \neg(\neg s))} \wedge \underset{(5)}{((\neg q) \vee (\neg s))}$$

Lemma 1. *Let α be a CNF-formula and I an interpretation of α; let $\mathcal{N}_\alpha$ be the canonical c/e–net representation of α and $\mathcal{N}_{\alpha,D}$ the c/e–net representation of α w.r.t. D for some $D \subseteq \mathbb{A}(\alpha)$; let M_I be the marking of $\mathcal{N}_\alpha$ w.r.t. I and $M_{I,D}$ the marking of $\mathcal{N}_{\alpha,D}$ w.r.t. I; let τ be a transition of $\mathcal{N}_\alpha$ and τ_D the corresponding transition of $\mathcal{N}_{\alpha,D}$. Then:*

$$\tau \text{ is false for } I \text{ iff } \tau \text{ is enabled in } \mathcal{N}_\alpha \text{ for } M_I$$
$$\text{iff } \tau_D \text{ is enabled in } (N_{\alpha,D} \text{ for } M_{I,D}).$$

Proof. Let $\tau = \neg a_1 \vee \ldots \vee \neg a_m \vee b_1 \vee \ldots \vee b_n$ be the clause form of τ.

Then: ${}^\bullet\tau = \{a_1, \ldots, a_m\}$, $\tau^\bullet = \{b_1, \ldots, b_n\}$,
  ${}^\bullet\tau_D = ({}^\bullet\tau \cap \overline{D}) \cup (\neg({}^\bullet\tau) \cap \neg D)$,
  $\tau_D{}^\bullet = (\tau^\bullet \cap \overline{D}) \cup (\neg({}^\bullet\tau) \cap \neg D).$

τ is false for I

iff for all $a_i \in {}^\bullet\tau$	$: I(a_i)\ \ = true$ and
for all $b_j \in \tau^\bullet$	$: I(b_j)\ \ = false$

iff for all $a_i \in {}^\bullet\tau$	$: M_I(a_i) = 1$ and
for all $b_j \in \tau^\bullet$	$: M_I(b_i) = 0$ (def. 7)

iff τ is enabled in N_α for M_I (def. 6)

iff for all $a_i \in {}^\bullet\tau \cap \overline{D}$	$: M_{I,D}(a_i)\ \ = 1$ and
for all $a_i \in {}^\bullet\tau \cap D$	$: M_{I,D}(\neg a_i) = 0$ and
for all $b_j \in \tau^\bullet \cap \overline{D}$	$: M_{I,D}(b_j)\ \ = 0$ and
for all $b_j \in \tau^\bullet \cap D$	$: M_{I,D}(\neg b_j) = 1$ (def. 9)

iff for all $a_i \in {}^\bullet\tau \cap \overline{D}$	$: M_{I,D}(a_i)\ \ = 1$ and
for all $(\neg b_j) \in \neg(\tau^\bullet) \cap \neg D$	$: M_{I,D}(\neg b_j) = 1$ and
for all $b_j \in \tau^\bullet \cap \overline{D}$	$: M_{I,D}(b_j)\ \ = 0$ and
for all $(\neg a_i) \in \neg({}^\bullet\tau) \cap \neg D$	$: M_{I,D}(\neg a_i) = 0$

iff τ_D is enabled in $N_{\alpha,D}$ for $M_{I,D}$

Corollary 1. *Let $\mathcal{N}_\alpha$ be the canonical c/e-net representation of a CNF-formula α and $\mathcal{N}_{\alpha,D}$ the c/e-net representation of α w.r.t. D for some $D \subseteq \mathbb{A}(\alpha)$. I is a model of α*

$$\text{iff } (\mathcal{N}_\alpha, M_I) \text{ is dead iff } (\mathcal{N}_{\alpha,D}, M_{I,D}) \text{ is dead}$$

While lemma 1 deals with c/e-net representations, the net representation in lemma 2 can be both a c/e-net representation or a p/t-net representation of a CNF-formula α.

Definition 10. *Let α be a CNF-formula, $\mathcal{N}_\alpha$ the canonical net representation of α, and M_I the marking of $\mathcal{N}_\alpha$ w.r.t I;*

$$I^{-1}(true) := \{a \in \mathbb{A}(\alpha) \mid I(a) = true\} = \{a \in \mathbb{A}(\alpha) \mid M_I(a) = 1\}$$
$$I^{-1}(false) := \{a \in \mathbb{A}(\alpha) \mid I(a) = false\} = \{a \in \mathbb{A}(\alpha) \mid M_I(a) = 0\}$$

Lemma 2. *Let α be a CNF-formula and I an interpretation of α; then I is a model of α iff $(\mathcal{N}_{\alpha,I^{-1}(true)}, \mathbf{0})$ is dead.*

$$S_{\alpha,I^{-1}(true)} = \overline{I^{-1}(true)} \cup \neg I^{-1}(true) \qquad \text{(def. 8)}$$
$$M_{I,I^{-1}(true)}(a) = M_I(a)$$

Proof.
$$\text{iff } a \in \overline{I^{-1}(true)} = I^{-1}(false) \text{ (def. 9)}$$
$$\text{iff } M_I(a) = 0 \qquad \text{(def. 10)}$$
$$M_{I,I^{-1}(true)}(\neg a) = 1 - M_I(a)$$
$$\text{iff } a \in I^{-1}(true) \qquad \text{(def. 9)}$$
$$\text{iff } M_I(a) = 1$$

So, $M_{I,I^{-1}(true)}$ is the empty marking $\mathbf{0}$.

If $\mathcal{N}_{\alpha,I^{-1}(true)}$ is a c/e-net representation, I is a model of α

iff $(\mathcal{N}_{\alpha,I^{-1}(true)}, M_{I,I^{-1}(true)})$ is dead (cor. 1).
iff $(\mathcal{N}_{\alpha,I^{-1}(true)}, \mathbf{0})$ is dead.

For the special case of the empty marking all transitions of the c/e-net $\mathcal{N}_{\alpha,I^{-1}(true)}$ are not enabled because their input places are not marked (and not because their output places are marked!). So, $\mathcal{N}_{\alpha,I^{-1}(true)}$ is dead for the empty marking also as a p/t-net. If $\mathcal{N}_{\alpha,I^{-1}(true)}$ is not dead for the empty marking as a c/e-net it is likewise not dead as a p/t net.

Corollary 2. *Let α be a CNF-formula and I an interpretation of α; then I is a model of α iff*

$$I^{-1}(false) \cup \neg I^{-1}(true) \text{ is a co-trap of } \mathcal{N}_{\alpha,I^{-1}(true)}.$$

Proof. If I is a model of α, $(\mathcal{N}_{\alpha,I^{-1}(true)}, \mathbf{0})$ is dead. Then, according to a well known theorem, the set of empty places is a co-trap. So, $S_{\alpha,I^{-1}(true)} = I^{-1}(false) \cup \neg I^{-1}(true)$ is a co-trap.

If $S_{\alpha,I^{-1}(true)} = I^{-1}(false) \cup \neg I^{-1}(true)$ is a co-trap, no place in $(\mathcal{N}_{\alpha,I^{-1}(true)}, \mathbf{0})$ can be marked. So, $(\mathcal{N}_{\alpha,I^{-1}(true)}, \mathbf{0})$ is dead and I is a model of α.

Remark 2. An important consequence of lemma 2 is that among all p/t-net representations $\mathcal{N}_{\alpha,D}$ of a CNF-formula α there exists at least one in which the empty marking cannot be reproduced if I is a model of α, namely $\mathcal{N}_{\alpha,I^{-1}(true)}$.

Example 4.

$$\alpha = \underset{(1)}{(a \vee b)} \wedge \underset{(2)}{(b \vee c)} \wedge \underset{(3)}{(c \vee a)} \wedge \underset{(4)}{(\neg a \vee \neg b \vee \neg c)}$$

The interpretation I with $I(a) = I(b) = true$, $I(c) = false$ is a model of α.

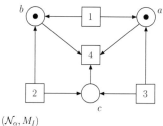

$(\mathcal{N}_\alpha, M_I)$

M_I with $M_I(a) = M_I(b) = 1$, $M_I(c) = 0$ is the marking of $\mathcal{N}_\alpha$ w.r.t. I. (def. 7)

Fig. 3.

$(\mathcal{N}_\alpha, M_I)$ is live if $\mathcal{N}_\alpha$ is a p/t-net, but $(\mathcal{N}_\alpha, M_I)$ is dead if $\mathcal{N}_\alpha$ is a c/e-net.

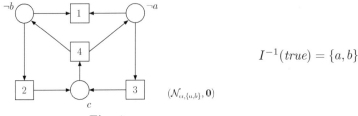

$$I^{-1}(true) = \{a, b\}$$

Fig. 4.

$(\mathcal{N}_{\alpha,\{a,b\}}, \mathbf{0})$ is dead in either case.

Definition 11 (augmented net representation). *Let α be a CNF-formula and $\mathcal{N}_\alpha = (S_\alpha, T_\alpha, F_\alpha)$ the canonical net representation of α; let $\mathcal{N}^*_\alpha = (S^*_\alpha, T^*_\alpha, F^*_\alpha)$ be a p/t-net;*

$\mathcal{N}^*_\alpha$ *is the* augmented net representation *of α iff*

$$S^*_\alpha = \mathbb{A}(\alpha) \cup \neg\mathbb{A}(\alpha) \text{ and } T^*_\alpha = T_\alpha,$$
$$F^*_\alpha = F_\alpha \cup \{(\neg p, t) \mid (t, p) \in F\} \cup \{(t, \neg p) \mid (p, t) \in F\}$$

Let $L \subseteq S^*_\alpha$ be a set of places of $\mathcal{N}^*_\alpha$;

L is called singular *iff $\{a, \neg a\} \subseteq L$ for some $a \in \mathbb{A}(\alpha)$. L is called* non-singular *iff for all $a \in \mathbb{A}(\alpha)$ $\{a, \neg a\} \not\subseteq L$. L is called* semi-covering *iff for all $a \in \mathbb{A}(\alpha)$ either $a \in L$ or $\neg a \in L$.*

Lemma 3. *Let α be a CNF-formula, let L be a non-singular set of places in $\mathcal{N}^*_\alpha$, and $D := \{a \in \mathbb{A}(\alpha) \mid \neg a \in L\}$; then:*

$$L \text{ is a co-trap (trap) in } \mathcal{N}^*_\alpha \text{ iff } L \text{ is a co-trap (trap) in } \mathcal{N}_{\alpha,D}$$

Proof.

$$L = \{a \in \mathbb{A}(\alpha) \mid a \in L\} \cup \neg\{a \in \mathbb{A}(\alpha) \mid \neg a \in L\}$$
$$\subseteq \{a \in \mathbb{A}(\alpha) \mid a \notin D\} \cup \neg\{a \in \mathbb{A}(\alpha) \mid \neg a \in L\}$$
$$= \overline{D} \cup \neg D$$
$$= S_{\alpha, D}$$

So, L is a set of places in $\mathcal{N}_{\alpha, D}$.

$$\bullet L = \{t \in T^*_\alpha \mid ((t, a) \in F^*_\alpha \wedge a \in \overline{D} \cap L) \vee$$
$$((t, \neg a) \in F^*_\alpha \wedge \neg a \in \neg D \cap L)\} \text{ in } N^*_\alpha$$
$$\bullet L = \{t \in T_{\alpha, D} \mid ((t, a) \in F_{\alpha, D} \wedge a \in \overline{D} \cap L) \vee$$
$$((t, \neg a) \in F_{\alpha, D} \wedge \neg a \in \neg D \cap L)\} \text{ in } N_{\alpha, D}$$

$$(t, a) \in F_\alpha^* \wedge a \in \overline{D} \cap L \qquad \text{iff } (t, a) \in F_\alpha \wedge a \in \overline{D} \cap L \qquad \text{(def. 11)}$$
$$\text{iff } (t, a) \in F_{\alpha, D} \wedge a \in \overline{D} \cap L \qquad \text{(def. 8)}$$
$$(t, \neg a) \in F_\alpha^* \wedge \neg a \in \neg D \cap L \text{ iff } (a, t) \in F_\alpha \wedge a \in D \wedge \neg a \in L \qquad \text{(def. 11)}$$
$$\text{iff } (t, \neg a) \in F_{\alpha, D} \wedge a \in D \wedge \neg a \in L \text{ (def. 8)}$$

So, $\cdot L$ (and also $L^\cdot$) is the same set in $\mathcal{N}_\alpha^*$ and $\mathcal{N}_{\alpha, D}$.

Corollary 3. *Let α be a CNF-formula and I an interpretation of α; then*

I *is a model of α iff $\overline{I^{-1}(true)} \cup \neg I^{-1}(true)$ is a semi-covering co-trap of N_α^*.*

Proof. Let I be a model of α;

then $\overline{I^{-1}(true)} \cup \neg I^{-1}(true)$ is a co-trap of $\mathcal{N}_{\alpha, I^{-1}(true)}$ (cor. 2)

then $\overline{I^{-1}(true)} \cup \neg I^{-1}(true)$ is a co-trap of N_α^* (lem. 3)

Let L be a semi-covering co-trap of N_α^*, and let J be the following interpretation of α:

$$J(a) := true \text{ for all } \neg a \in L \cap \neg \mathbb{A}(\alpha)$$
$$J(a) := false \text{ for all } a \in L \cap \mathbb{A}(\alpha).$$

$$\text{Then } L = J^{-1}(false) \cup \neg J^{-1}(true)$$
$$= \overline{J^{-1}(true)} \cup \neg J^{-1}(true).$$

According to lemma 3 L is a co-trap in $N_{\alpha, J^{-1}(true)}$, and according to corollar 2 J is a model of α.

Example 5 (cf. example 4).

$$\alpha = \underset{(1)}{(a \vee b)} \wedge \underset{(2)}{(b \vee c)} \wedge \underset{(3)}{(c \vee a)} \wedge \underset{(4)}{(\neg a \vee \neg b \vee \neg c)}$$

α has three models: $I_1(a) = I_1(b) = true, I_1(c) = false$ $I_2(b) = I_2(c) = true, I_2(a) = false, I_3(c) = I_3(a) = true, I_3(b) = false$

The corresponding semi-covering co-traps are $: \overline{I_1^{-1}(true)} \cup \neg I_1^{-1}(true) = \{c, \neg a, \neg b\}$, $\overline{I_2^{-1}(true)} \cup \neg I_2^{-1}(true) = \{a, \neg b, \neg c\}$ and $\overline{I_3^{-1}(true)} \cup \neg I_3^{-1}(true) = \{b, \neg c, \neg a\}$.

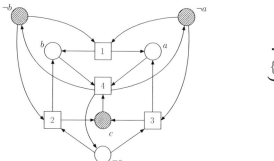

$$\cdot\{c, \neg a, \neg b\} = \{2, 3, 4\}$$
$$\{c, \neg a, \neg b\}^\cdot = \{1, 2, 3, 4\}.$$

$$\underbrace{\phantom{\{c, \neg a, \neg b\}^\cdot = \{1, 2, 3, 4\}}}_{\text{co-trap}}$$

Fig. 5.

5   The Reproduction Semantics of Refutations

In order to introduce the key point of this section we will show an example.

Example 6.

$$\alpha = \underset{(1)}{(a)} \land \underset{(2)}{(\neg a \lor b)} \land \underset{(3)}{(\neg a \lor c)} \land \underset{(4)}{(\neg b \lor \neg c)}$$

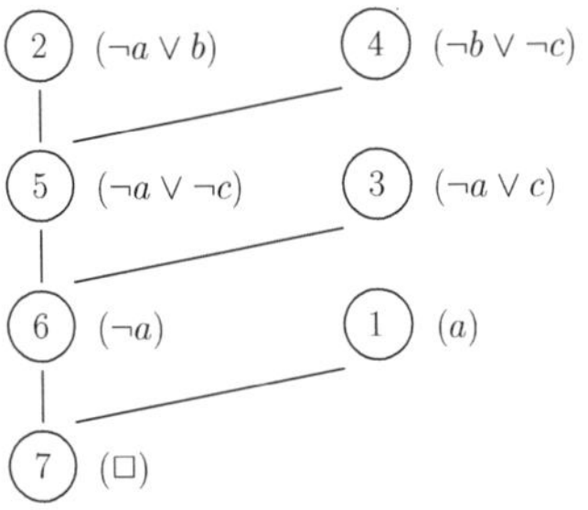

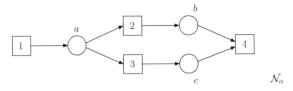

Fig. 6. 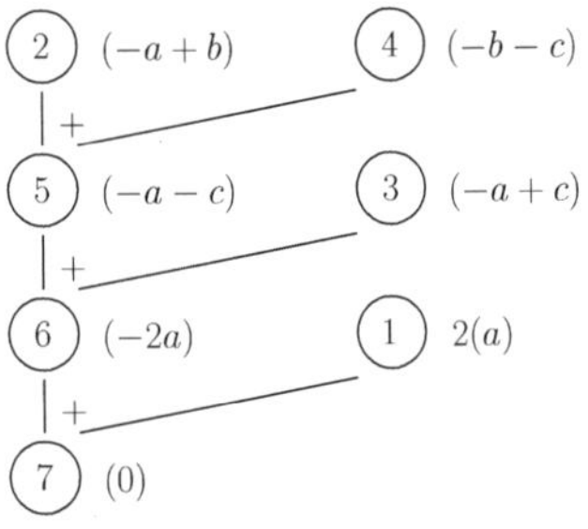   Fig. 7.

Figure 6 shows the refutation tree. Replacing the clauses by formal sums and the resolution steps by summing up two formal sums yields the version in figure 7 of this refutation tree.

These operations can be represented by means of a non-negative linear combination of the formal sums belonging to the clauses of α.

$$\alpha = \underset{①}{2(a)} + \underset{②}{(-a+b)} + \underset{③}{(-a+c)} + \underset{④}{(-b-c)} = 0$$

In order to approach the concepts of p/t-nets, we now choose a vector representation for this linear combination.

$$\begin{matrix} a \\ b \\ c \end{matrix} \quad 2\begin{bmatrix}1\\0\\0\end{bmatrix} + \begin{bmatrix}-1\\1\\0\end{bmatrix} + \begin{bmatrix}-1\\0\\1\end{bmatrix} + \begin{bmatrix}0\\-1\\-1\end{bmatrix} = \begin{bmatrix}1&-1&-1&0\\0&1&0&-1\\0&0&1&-1\end{bmatrix} \cdot \begin{bmatrix}2\\1\\1\\1\end{bmatrix} = \begin{bmatrix}0\\0\\0\end{bmatrix}$$

Obviously, this matrix is the incidence matrix of the canonical net representation N_α of α.

Fig. 8.

The vector $(2\ 1\ 1\ 1)^\top$ of coefficients is a t-invariant of N_α.

Two occurrences of 1 and one of 2, 3, and 4 in any feasible sequence reproduce the empty marking. For example, 1 2 1 3 4 is such a sequence. It represents a proof procedure which first provides the knowledge "a holds" for being used by rule 2 "if a then b". Then, "a holds" has to be provided again since rule 3 "if a then c" uses it, too. Finally, the knowledge "b holds" and the knowledge "c holds" are used by 4, thus showing "b and c hold".

So, using goals means consuming the corresponding tokens.

Definition 12 (non-tautological clause). *Let* $\tau = \neg a_1 \vee \ldots \vee \neg a_m \vee b_1 \vee \ldots \vee b_n$ *be a clause;*

τ *is* non-tautological *iff* $\{a_1, \ldots, a_m\} \cap \{b_1, \ldots, b_n\} = \emptyset$

Definition 13 (columns over a clause). *Let* α *be a CNF-formula, and let* $\tau = \neg a_1 \vee \ldots \vee \neg a_m \vee b_1 \vee \ldots \vee b_n$ *be a non-tautological clause of* α *where*

$$\mathbb{A}(\tau) = \{a_1, \ldots, a_m\} \cup \{b_1, \ldots, b_n\} \subseteq \mathbb{A}(\alpha) \text{ and } \{a_1, \ldots, a_m\} \cap \{b_1, \ldots, b_n\} = \emptyset;$$

then every column vector $C : \mathbb{A}(\alpha) \to \mathbb{Z}$ *where*

$$C(a_i) < 0 \text{ if } a_i \in \mathbb{A}(\tau), 1 \le i \le m$$
$$C(b_j) > 0 \text{ if } b_j \in \mathbb{A}(\tau), 1 \le j \le n$$
$$C(d) = 0 \text{ if } d \in \mathbb{A}(\alpha) \backslash \mathbb{A}(\tau)$$

is called an α-column *of* τ. $|C| := |\tau| := m + n$.

Tautological clauses are not interesting for refutations because every interpretation is a model. So, we will concentrate on non-tautological clauses.

Lemma 4. *Let* α *be a CNF-formula, let* $\tau_1 = \lambda_1 \vee a$ *and* $\tau_2 = \lambda_2 \vee \neg a$ *be two non-tautological clauses with* $a \in \mathbb{A}(\alpha)$, *let* C_1 *and* C_2 *be* α-clauses of τ_1 *and* τ_2, *respectively, and let* $\varrho = \lambda_1 \vee \lambda_2$ *be the non-tautological resolvent of* τ_1 *and* τ_2 *w.r.t.* a; *then: there exists an* α-column C *of* ϱ *that is a positive linear combination of* C_1 *and* C_2.

Proof.

$$\lambda_1 = \neg a'_1 \vee \ldots \vee \neg a'_{m_1} \vee b'_1 \vee \ldots \vee b'_{n_1} \text{ and } \lambda_2 = \neg a''_1 \vee \ldots \vee \neg a''_{m_2} \vee b''_1 \vee \ldots \vee b''_{n_2}$$
$$\varrho = \lambda_1 \vee \lambda_2 \text{ with } \{a'_1, \ldots, a'_{m_1}, a''_1, \ldots, a''_{m_2}\} \cap \{b'_1, \ldots, b'_{n_1}, b''_1, \ldots, b''_{n_2}\} = \emptyset$$
$$\text{and } a \notin \{a'_1, \ldots, a'_{m_2}\} \cup \{b'_1, \ldots, b''_{n_2}\}.$$

For $C_1(a) = k_1, C_2(a) = -k_2, (k_1, k_2 \in \mathbb{N})$, $C := k_2 C_1 + k_1 C_2$ with $C(a) = 0$ is an α-column of ϱ.

Definition 14 (α_D). *Let α be a CNF-formula, $\lambda = \neg a_1 \vee \ldots \vee \neg a_m \vee b_1 \vee \ldots \vee b_n$ a clause of α, and let be $D \subseteq \mathbb{A}(\alpha)$;*

$$\varphi_D(\lambda) := \varphi_D(\neg a_1) \vee \ldots \varphi_D(\neg a_m) \vee \varphi_D(b_1) \vee \ldots \vee \varphi_D(b_n)$$

is defined by

$$\varphi_D(\neg a_i) := \begin{cases} (\neg a_i) & \text{if } a_i \in D \\ \neg a_i & \text{if } a_i \notin D \end{cases} \quad (1 \leq i \leq m)$$

$$\varphi_D(b_j) := \begin{cases} \neg(\neg b_j) & \text{if } b_j \in D \\ b_j & \text{if } b_j \notin D \end{cases} \quad (1 \leq j \leq n)$$

α_D arises from α by replacing all clauses λ by $\varphi_D(\lambda)$.

Lemma 5. *Let α be a contradictory CNF-formula; then: all net representations $\mathcal{N}_{\alpha,D}$ of α contain a non-negative t-invariant.*

Proof. Let be $D \subseteq \mathbb{A}(\alpha)$; then α_D is contradictory. So, starting from α_D, the empty clause is derivable in a sequence R of resolution steps. We may assume that all clauses used in R are non-tautological.

Next, we construct for all those clauses α-columns. Appropriate α-columns for the clauses of α_D are the corresponding columns of the incidence matrix $[\mathcal{N}_{\alpha,D}]$. According to lemma 4, all resolvents have α-columns that are positive linear combinations of columns of $[\mathcal{N}_{\alpha,D}]$. This is in particular true for the zero α-column $\mathbf{0}$ belonging to the empty clause at the end of R. So, the non-negative vector of coefficients of the linear combination yielding $\mathbf{0}$ is a t-invariant of $\mathcal{N}_{\alpha,D}$. ∎

Example 7 (cf. example 2).

$$\alpha = (p \vee q) \wedge (\neg p \vee q) \wedge (\neg q \vee r) \wedge (\neg q \vee \neg r \vee s) \wedge (\neg q \vee \neg s)$$

For $D = \{q, s\}$ we get

$$\alpha_D = \underset{①}{(p \vee \neg(\neg q))} \wedge \underset{②}{(\neg p \vee \neg(\neg q))} \wedge \underset{③}{((\neg q) \vee r)} \wedge \underset{④}{((\neg q) \vee \neg r \vee \neg(\neg s))}$$

$$\wedge \quad \underset{⑤}{((\neg q) \vee (\neg s))}$$

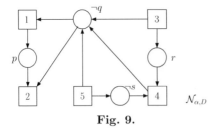

Fig. 9.

	1	2	3	4	5
p	1	-1	0	0	0
$\neg q$	-1	-1	1	1	1
r	0	0	1	-1	0
$\neg s$	0	0	0	-1	1
	3	3	2	2	2

$\mathcal{N}_{\alpha,D}$

$[\mathcal{N}_{\alpha,D}]$

Refutation tree:

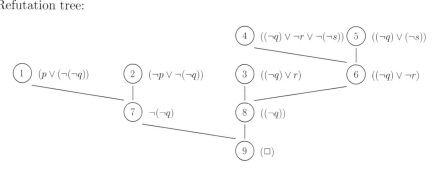

The linear combinations belonging to the resolution steps according to lemma 4:

$$
\text{④}\begin{bmatrix}0\\1\\-1\\-1\end{bmatrix} + \text{⑤}\begin{bmatrix}0\\1\\0\\1\end{bmatrix} = \text{⑥}\begin{bmatrix}0\\2\\-1\\0\end{bmatrix} ; \qquad \text{①}\begin{bmatrix}1\\-1\\0\\0\end{bmatrix} + \text{②}\begin{bmatrix}-1\\-1\\0\\0\end{bmatrix} = \text{⑦}\begin{bmatrix}0\\-2\\0\\0\end{bmatrix} ;
$$

$$
\text{③}\begin{bmatrix}0\\1\\1\\0\end{bmatrix} + \text{⑥}\begin{bmatrix}0\\2\\-1\\0\end{bmatrix} = \text{⑧}\begin{bmatrix}0\\3\\0\\0\end{bmatrix} ; \qquad 3\cdot \text{⑦}\begin{bmatrix}0\\-2\\0\\0\end{bmatrix} + 2\cdot \text{⑧}\begin{bmatrix}0\\3\\0\\0\end{bmatrix} = \text{⑨}\begin{bmatrix}0\\0\\0\\0\end{bmatrix}
$$

Altogether, we have

$$
3\cdot \text{①}\begin{bmatrix}1\\-1\\0\\0\end{bmatrix} + 3\cdot \text{②}\begin{bmatrix}-1\\-1\\0\\0\end{bmatrix} + 2\cdot \text{③}\begin{bmatrix}0\\1\\1\\0\end{bmatrix} + 2\cdot \text{④}\begin{bmatrix}0\\1\\-1\\-1\end{bmatrix} + 2\cdot \text{⑤}\begin{bmatrix}0\\1\\0\\1\end{bmatrix} = \text{⑨}\begin{bmatrix}0\\0\\0\\0\end{bmatrix} ,
$$

thus the coefficient vector $(3\ 3\ 2\ 2\ 2)^\top$ is a t-invariant of $N_{\alpha,D}$.

Lemma 6. *Let α be a contradictory CNF-formula; then:*

in all p/t-net representations $N_{\alpha,D}$ of α the empty marking $\mathbf{0}$ is reproducible.

Proof. Let $N_{\alpha,D}$ be an arbitrary p/t-net representation of α, and let R be a sequence of resolution steps ending with the empty clause $\square$. R may be determined by the sequence $\alpha^0 = \alpha_D, \alpha^1, \alpha^2, \ldots, \alpha^{n-1}, \alpha^n = \square$ of contradictory CNF-formulas where $\alpha^i, 0 \leq i \leq n$, is the conjunction of all clauses that are still used in the remaining resolution steps of R. Let $\mathcal{N}^0 = \mathcal{N}_{\alpha,D}, \mathcal{N}^1, \mathcal{N}^2, \ldots, \mathcal{N}^{n-1}, \mathcal{N}^n$ be the corresponding sequence of p/t-net representations. In the last two of these p/t-nets the empty marking is obviously reproducible:

$\mathcal{N}^n$ consists of a single transition $\square$, $\mathcal{N}^{n-1}$ has the form $\square\!\!\longrightarrow\!\!\bigcirc\!\!\longrightarrow\!\!\square$.

We now want to show indirectly that in $\mathcal{N}^{i-1}$ the empty marking $\mathbf{0}$ is reproducible if that is the case in $\mathcal{N}^i$, $1 \le i \le n$.

So, we assume that $\mathbf{0}$ is *not* reproducible in $\mathcal{N}^{i-1}$. The reason for that is the non-existence of a non-negative T-Invariant whose net representation contains no co-traps or traps according to theorem 1. According to lemma 5, however, there is a non-negative t-invariant in $\mathcal{N}^{i-1}$ because α^{i-1} is contradictory. So, we have to assume that in all net representations of the non-negative t-invariants in $\mathcal{N}^{i-1}$ there exist co-traps or traps.

Let $\mathcal{N}_r^{i-1}$ be the net representation of the t-invariant $r \ge \mathbf{0}$ in $\mathcal{N}^{i-1}$ that is constructed as the t-invariant in the proof of lemma 5. So, r belongs to a resolution procedure. Without restricting generality, we assume $r(\tau) = 1$ iff $r(\tau) > 0$. In case $r(\tau) = k > 1$, $k - 1$ copies of τ are added - in the net and in the resolution procedure as well. Finally, let D be a co-trap of $\mathcal{N}_r^{i-1}$.

Let $\mathcal{N}_{r'}^i$ be the net that results from $\mathcal{N}_r^{i-1}$ as follows: We assume that one resolution step in the resolution belonging to r is

$$\tau_1 = \varrho_1 \vee a, \tau_2 = \varrho_2 \vee \neg a, \text{ resolvent } \varrho = \varrho_1 \vee \varrho_2.$$

On the net level 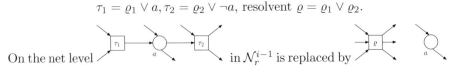 in $\mathcal{N}_r^{i-1}$ is replaced by

Here we benefit from the fact that the non-zero entries of r are equal to 1, since that expresses that every clause/transition is used exactly once in the resolution procedure. Consequently, τ_1 and τ_2 are really replaced by ϱ. It is not necessary to save them for further resolution steps.

Since D is a co-trap ${}^\bullet D \subseteq D^\bullet$ holds. In detail, let A and B be (maybe empty) sets of transitions of $\mathcal{N}_r^{i-1}$ with $A \cap B = \emptyset$ and $\{\tau_1, \tau_2\} \cap (A \cup B) = \emptyset$. Then the following cases are conceivable:

$$
\begin{array}{lll}
(1) & {}^\bullet D = A \cup \{\tau_1, \tau_2\} \subseteq D^\bullet = A \cup \{\tau_1, \tau_2\} \cup B & \\
(2) & {}^\bullet D = A \cup \{\tau_j\} \subseteq D^\bullet = A \cup \{\tau_1, \tau_2\} \cup B & \\
(3) & {}^\bullet D = A \cup \{\tau_j\} \subseteq D^\bullet = A \cup \{\tau_j\} \cup B & \\
(4) & {}^\bullet D = A \subseteq D^\bullet = A \cup \{\tau_1, \tau_2\} \cup B & \\
(5) & {}^\bullet D = A \subseteq D^\bullet = A \cup \{\tau_j\} \cup B & \\
(6) & {}^\bullet D = A \subseteq D^\bullet = A \cup B, & j \in \{1, 2\}.
\end{array}
$$

Now let be

$$D' := \begin{cases} D - \{a\} & \text{,if } {}^\bullet a = \{\tau_1\} \wedge a^\bullet = \{\tau_2\} \\ D & \text{,if } {}^\bullet a \supset \{\tau_1\} \vee a^\bullet \supset \{\tau_2\} \end{cases}$$

Since ϱ inherits the elementship w.r.t. ${}^\bullet D$ and $D^\bullet$ from τ_1 and τ_2, the above cases are transformed as follows thus guaranteeing ${}^\bullet(D') \subseteq (D')^\bullet$

$$(1) - (3) \quad {}^{\bullet}(D') = A \cup \{\varrho\} \, , \, (D')^{\bullet} = A \cup \{\varrho\} \cup B$$
$$(4) - (5) \quad {}^{\bullet}(D') = A \qquad , \, (D')^{\bullet} = A \cup \{\varrho\} \cup B$$
$$(6) \qquad \quad {}^{\bullet}(D') = A \qquad , \, (D')^{\bullet} = A \cup B$$

So $\mathcal{N}_{r'}^{i}$, which is the net representation of a non-negative t-invariant r' in $\mathcal{N}^i$ contains the co-trap D'.

Finally, we have to check whether $\mathcal{N}^i$ has essentially new t-invariants. Since $[\mathcal{N}^i]$ is developed from $[\mathcal{N}^{i-1}]$ by means of a column operation, every non-negative t-invariant in $\mathcal{N}^i$ is either the unchanged copy of a t-invariant in $\mathcal{N}^{i-1}$ or it is developed as shown above. In either case, its net representation contains a co-trap.

Since a similar argument holds for traps, we have shown that the empty marking is not reproducible in $\mathcal{N}^i$ if it is not reproducible in $\mathcal{N}^{i-1}(1 \leq i \leq n)$.

Altogether, we have shown that the empty marking is reproducible in $\mathcal{N}^0 = \mathcal{N}_{\alpha,D}$ since that is trivially the case in $\mathcal{N}^{n-1}$ and $\mathcal{N}^n$.

Example 8 (cf. example 6).

In example 6 the t-invariant is $(2\ 1\ 1\ 1)^{\top}$.

"Standardizing" leads to the following p/t-net

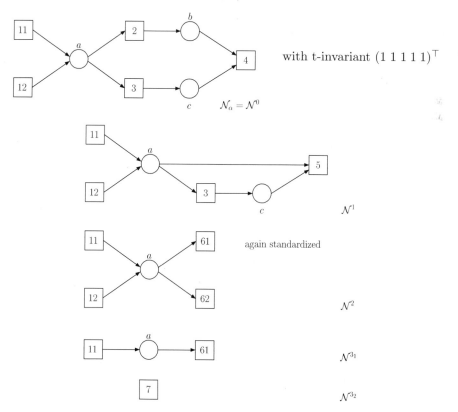

with t-invariant $(1\ 1\ 1\ 1\ 1)^{\top}$

Theorem 3. *Let α be a CNF-formula; α is contradictory iff in all p/t-net representations $\mathcal{N}_{\alpha,D}$ of α the empty marking $\mathbf{0}$ is reproducible.*

Proof. If α is contradictory, in all p/t-net representations $\mathcal{N}_{\alpha,D}$ of α the empty marking $\mathbf{0}$ is reproducible according to lemma 6.

If α is not contradictory, a model I of α exists. Then $(\mathcal{N}_{\alpha,I^{-1}(true)}, \mathbf{0})$ is dead according to lemma 2.

Theorem 4. *Let α be a CNF-formula; α is contradictory iff in $\mathcal{N}_\alpha^*$ no semi-covering co-trap exists.*

Proof. If α is contradictory, in all p/t-net representations $\mathcal{N}_{\alpha,D}$ of α the empty marking $\mathbf{0}$ is reproducible according to theorem 3. Consequently, no $\mathcal{N}_{\alpha,D}$ can have a covering co-trap $C = S_{\alpha,D}$. So, $\mathcal{N}_\alpha^*$ cannot have a semi-covering co-trap C.

If α has a model I then $\overline{I^{-1}(true)} \cup \neg I^{-1}(true)$ is a semi-covering co-trap in $\mathcal{N}_\alpha^*$ according to corollary 3.

Theorem 5. *Let α be a CNF-formula;*

1. *if in $\mathcal{N}_\alpha^*$ no non-singular co-trap exists α is contradictory;*
2. *if in $\mathcal{N}_\alpha^*$ a non-singular co-trap E exists, and if at least one of the net representations $\mathcal{N}_{\alpha,D}$ of α with $E \subseteq S_{\alpha,D}$ is a non-negative one-t-invariant net, then α is not contradictory.*

Proof.   1. If there is no non-singular co-trap in $\mathcal{N}_\alpha^*$ then there is no semi-covering co-trap in $\mathcal{N}_\alpha^*$. In that case, α is contradictory according to theorem 4.
2. If E is a co-trap $\mathcal{N}_\alpha^*$ E is also a co-trap in $S_{\alpha,D}$ (lemma 3). If $\mathcal{N}_{\alpha,D}$ is a one-t-invariant net the empty marking $\mathbf{0}$ is not reproducible in $\mathcal{N}_{\alpha,D}$. Then α is not contradictory according to theorem 3.

Remark 3. If E is a co-trap in a net $\mathcal{N}_{\alpha,D}$ consisting of more than one non-negative one-t-invariant, it should be checked whether all non-negative t-invariants of $\mathcal{N}_{\alpha,D}$ contain co-traps.

6   Horn Formulas

In this section we want to deal with the special case of Horn formulas. In particular we want to show how the corresponding logical simplifications appear in terms of Petri nets. Moreover, we want to show a result by [PetMur89] and integrate it in our approach.

Definition 15. *A clause κ is a* Horn clause *iff it contains at most one positive literal.*

A CNF-formula is a Horn formula *iff its clauses are Horn clauses.*

The canonical representation of a Horn clause is a Horn transition.

Lemma 7. *Let α be a Horn formula and $\mathcal{N}_\alpha = (S_\alpha, T_\alpha, F_\alpha)$ be its canonical p/t-net representation, then: α is contradictory iff in N_α a goal transition is* **0**-*firable.*

Proof. If α is contradictory then according to lemma 6 $\mathcal{N}_\alpha$ is **0**-reproducing. Consequently, at least one goal transition is **0**-*firable.*

Let now $g \in T_\alpha$ be a **0**-firable goal transition.

In order to execute an induction on $|T_\alpha|$ we first assume $|T_\alpha| = 2$.

For $T_\alpha = \{f, g\}$, where f is a fact transition and a Horn transition and g is a goal transition, $\mathcal{N}_\alpha$ has the following form:

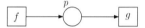

In particular, p is the only input place of g since otherwise g would not be **0**-firable. Consequently, α equals $p \wedge \neg p$ i.e. α is contradictory.

We now assume that all Horn formulas α are contradictory if $|T_\alpha| = n - 1$ holds and g is a **0**-firable goal transition.

And we claim that the same holds if $|T_\alpha| = n$.

In order to prove that, we assume that α' is a Horn formula with $|T_{\alpha'}| = n$. For the **0**-firable goal transition $g' \in T_\alpha$ we assume $\cdot g' = \{p_1, \dots, p_m\}$.

Since g is **0**-firable there exist m **0**-firable transitions $t_1, \dots, t_m$ with $t_i \in \cdot p_i, 1 \leq i \leq m$.

Let furthermore be $\cdot t_1 = \{q_1, \dots q_e\}$. Then we have the following relevant subnet of $\mathcal{N}_{\alpha'}$:

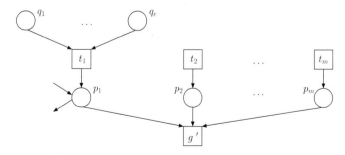

We now join t_1 and g' to one new goal transition g^+

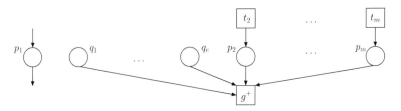

Of course, in the new p/t-net $\mathcal{N}^+$ the goal transition g^+ is **0**-firable, too. More-over, $\mathcal{N}^+$ is the canonical p/t-net representation $\mathcal{N}_\beta = (S_\beta, T_\beta, F_\beta)$ of a Horn formula β that results from α' by replacing the Horn clauses t_1 and g' by the resolvent g^+ of t_1 and g'.

So we have $\mathcal{N}^+ = \mathcal{N}_\beta = (S_\beta, T_\beta, F_\beta)$ with $|T_\beta| = n - 1$.

According to the induction assumption, β is contradictory. So there exists a sequence R_β of resolution steps ending with the empty clause $\square$. If we augment R_β by the additional first resolution step where g^+ is the resolvent of t_1 and g' we get a refutation sequence $R_{\alpha'}$ for α'.

So, α' is contradictory, too.

Lemma 8 (Peterka, Murata). *Let $\mathcal{N} = (S, T, F)$ be a p/t-net with $|t^\bullet| \leq 1$ for all $t \in T$. Let $g \in T$ be a goal transition; then: $\mathcal{N}$ is **0**-reproducing thereby executing g iff $\mathcal{N}$ has a T-invariant $R \geq 0$ where $R(g) > 0$.*

Proof. See [PetMur89].

Lemma 9. *Let $\mathcal{N} = (S, T, F)$ be a p/t-net with $|t^\bullet| \leq 1$ for all $t \in T$. Let $g \in T$ be a goal transition; then: $\mathcal{N}$ is **0**-reproducing thereby executing g iff there exists a set Y of paths between fact transitions and g such that all $p \in {}^\bullet t \cup t^\bullet$ are nodes of a path of Y if $t \in T$ is a node of a path of Y.*

Proof. When $\mathcal{N}$ reproduces **0** thereby executing g the fact transitions throw to-kens onto the net which then flow to the goal transitions where they are removed. So, there exists a path system Y as described above ending with g.

If, on the other hand, such a path system Y exists one can execute a **0**-reproduction in backward direction starting at g and executing only transitions belonging to Y.

Of course, this reproduction implies the corresponding reproduction in forward direction.

Theorem 6. *Let α be a Horn formula and $\mathcal{N}_\alpha = (S_\alpha, T_\alpha, F_\alpha)$ be its canonical p/t-representation; then the following statements are equivalent:*

(1) α is contradictory.
*(2) $\mathcal{N}_\alpha$ is **0**-reproducing.*

(3) $\mathcal{N}_\alpha$ has a t-invariant $R \geq 0$ with $R(g) > 0$ for some goal transition g.

(4) In $\mathcal{N}_\alpha$ a goal transition g is **0**-firable.

(5) In $\mathcal{N}_\alpha$ there exists a set Y of reverse paths from a goal transition to fact transitions such that with any transition t of a path of Y its incidenting places $p \in {}^\bullet t \cup t^\bullet$ are nodes of a path of Y, too.

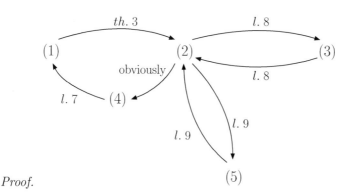

Proof.

References

[EzCoSi93] **J. Ezpeleta, J. M. Couvreur and M. Silva**. *A new technique for finding a generating family of siphons, traps and st-components. Application to colored Petri Nets.* Lecture Notes in Computer Science; Advances in Petri Nets 1993, 674:126–147, 1993.

[Gall87] **J.H. Gallier**. *Logic for Computer Science: Foundations of Automated Theorem Proving.* Wiley, 1987.

[GenThi76] **H. J. Genrich and G. Thieler-Mevissen**. *The Calculus of Facts.* Mazurkiewicz, *Mathematical Foundations of Computer Science 1976*, Lecture Notes in Computer Science 45, S. 588–595. Springer, 1976.

[Laut02] **K. Lautenbach**. *Reproducibility of the Empty Marking.* Javier Esparza and Charles Lakos, *Applications and Theory of Petri Nets 2002, 23rd International Conference, ICATPN 2002, Adelaide, Australia, June 24-30, 2002, Proceedings*, Band 2360 von *Lecture Notes in Computer Science*, S. 237–253. Springer, 2002.

[Murata89] **Tadao Murata**. *Petri Nets: Properties, Analysis and Applications.* Proceedings of the IEEE, S. 541–580, April 1989.

[PetMur89] **G. Peterka and Tadao Murata**. *Proof Procedure and Answer Extraction in Petri Net Model of Logic Programs.* IEEE Trans. Software Eng., 15(2):209–217, 1989.

[Reisig85] **W. Reisig**. *Petri Nets.*, Band 4. Springer-Verlag EATCS Monographs on Theoretical Computer Science, 1985.

Reactive Petri Nets for Workflow Modeling

Rik Eshuis[1][*] and Juliane Dehnert[2]

[1] LIASIT | CRP Henri Tudor
6 rue Coudenhove - Kalergi, L-1359 Luxembourg, Luxembourg
rik.eshuis@tudor.lu
[2] Institute for Computation and Information Structures (CIS),
Technical University Berlin,
Sekr.EN7, Einsteinufer 17, D-10587 Berlin, Germany
dehnert@cs.tu-berlin.de

Abstract. Petri nets are widely used for modeling and analyzing work-flows. Using the token-game semantics, Petri net-based workflow models can be analyzed before the model is actually used at run time. At run time, a workflow model prescribes behavior of a reactive system (the workflow engine). But the token-game semantics models behavior of closed, active systems. Thus, the token-game semantics behavior of a Petri net-based workflow model will differ considerably from its behavior at run time. In this paper we define a reactive semantics for Petri nets. This semantics can model behavior of a reactive system and its environment. We compare this semantics with the token-game semantics and prove that under some conditions the reactive semantics and the token-game semantics induce similar behavior. Next, we apply the reactive semantics to workflow modeling and show how a workflow net can be transformed into a reactive workflow net. We prove that under some conditions the soundness property of a workflow net is preserved when the workflow net is transformed into a reactive workflow net. This result shows that to analyze soundness, the token-game semantics can safely be used, even though that semantics is not reactive.

1 Introduction

Petri nets are a popular technique to formally model workflows [1,9,11,18]. They offer a formal counterpart for the bubbles and arrows that people draw when modeling workflows. Their formal token-game semantics enables analysis of Petri net based workflow models. Under the token-game semantics, the workflow model describes what behaviors are allowed. By computing behavior of a workflow model using the token-game semantics, errors in a workflow model can be spotted before the workflow model is actually put to use (cf. [1]).

A workflow model is put to use by feeding it to a workflow management system (WFMS). Heart of a WFMS is the workflow engine (WF engine), that does the actual management. WF engines are reactive systems. A reactive system

---

[*] Part of this work was done while the author was working at the University of Twente.

W.M.P. van der Aalst and E. Best (Eds.): ICATPN 2003, LNCS 2679, pp. 296–315, 2003.

runs in parallel with its environment and tries to enforce certain desirable effects in the environment [17]. It does so by reacting to changes, called events, in its environment. The response of the reactive system depends upon its current state. Giving a response may change the state of the reactive system.

The WF engine sees a workflow model as a prescription of what it has to do. The behavior of reactive systems is usually modeled using event-condition-action (ECA) rules [19], also known as production rules. The meaning of an ECA rule is that if the event in the environment occurs, and the condition is true, the reactive system does the action. Part of the condition can be a test of the state of the system. Part of the action can be a change of state. ECA rules can be easily incorporated in Petri nets by associating with every transition in the Petri net an ECA rule, that tells how the transition changes the state of the reactive system, in this case the state of the WF engine. This insight is already present in the pioneering work on workflow modeling done in the seventies [21]. Thus, in this case a Petri net workflow model models the behavior of the WF engine.

Unfortunately, the token-game semantics of Petri nets does not model behavior of reactive systems, and therefore does not model behavior of a WF engine [12,13]. The non-reactivity of the token-game semantics can be seen immediately from the definition of the firing rule. A transition in a Petri net is enabled once its input places are filled. The environment of the Petri net does not influence the firing of transitions. In contrast, in a reactive system a transition which is relevant, needs some additional input event to become enabled. So, the token-game semantics models closed systems, whereas a reactive system is open, otherwise it cannot interact with its environment.

Furthermore, in a reactive system an enabled transition *must* fire immediately, otherwise the system would fail to respond to a certain event. In the token-game semantics, an enabled transition *may* fire, but does not have to. In the worst case firing is postponed forever, or some transition that becomes enabled later fires before the enabled transition. Clearly, this contradicts reactivity.

To illustrate this point on a real-life example, consider the Petri net in Fig. 3. For an explanation of this example, we refer to Sect. 4. Now, suppose task check credit has just finished and that the new marking is [p1,p5]. Since task check credit has finished, presumably the WF engine now has to decide whether or not the credit was ok. (But the actual outcome of the decision depends upon the environment of the WF engine, which is modeled by the nondeterministic choice in place p5; see Sect. 4.) Then, under the token-game semantics, it is possible to fire transition check_order before firing transition ok or transition not ok. Suppose that the order is checked only after several days; then this firing sequence implies that the WF engine also takes days before it actually makes a decision. Clearly, this is inappropriate behavior: the decision should be made immediately when check credit has finished.

Since the token-game semantics is not reactive, a Petri net does not model the behavior of the WF engine. Thus, it is unclear how the behavior of the Petri net under the token-game semantics relates to the behavior of the WF engine when the workflow model is actually put to use. Consequently, it is also unclear whether the outcome of analysis of a workflow model using the token-game semantics carries over to the reactive setting. In other words, are analysis results in which the token-game semantics has been used still valid in a reactive setting?

The purpose of this paper is to define a reactive semantics for Petri nets and to relate this new semantics to the standard Petri net token-game semantics. In particular, we will study in what respect and under what conditions these two semantics induce similar behavior. As an application of this result, we will show that the soundness property [1] is preserved when transforming a workflow net into a reactive workflow net, i.e., a workflow net with a reactive semantics. Thus, we give a justification why the token-game semantics can safely be used to analyse workflow models for absence of deadlocks in a reactive setting, even though the token-game semantics is not reactive.

There are some commercially Petri net based workflow management systems available. We do not claim that our reactive semantics precisely describes the behavior of a WF engine in such a WFMS, but we do think that our semantics is closer to the behavior of such a WF engine than the token-game semantics.

The remainder of this paper is structured as follows. In Sect. 2 we will recapitulate some standard terminology and notions from Petri net theory. In Sect. 3 we define a reactive semantics for Petri nets and relate this semantics to the standard token-game semantics. In particular, we prove that under some conditions both semantics induce similar behavior. In Sect. 4 we recall the definition of a workflow net [1] and the soundness property. We present different interpretations for transitions in a workflow net. Using these different interpretations, in Sect. 5 we show how a workflow net can be transformed in a reactive workflow net. In Sect. 6 we prove that the soundness property is preserved when a workflow net is mapped into a reactive workflow net. In the proof we build upon the results obtained in Sect. 3. Related work is discussed in Sect. 7. We end with conclusions and further work.

2   Preliminaries

We recall the definition of a Petri net (P/T net).

A Petri net is a triple (P, T, F), where

- P is a finite set of places,
- T is a finite set of transitions, $(P \cap T = \emptyset)$
- $F \subseteq (P \times T) \cup (T \times P)$ is a finite set of arcs, the flow relation.

A transition t has input and output places. A place p is input (output) for transition t if there is a directed arc from p to t (from t to p). The input places of a transition t are denoted $\bullet t$. The output places of t are denoted $t\bullet$. A place can contain zero or more *tokens*. A token is represented by a black dot. The global *state* of a Petri net, also called a *marking*, is the distribution of tokens over places. Formally, a state or marking M is a function $M : P \to \mathbb{N}$ that assigns to every place p the number of tokens $M(p)$ that reside in p.

We now introduce some terminology.

- A transition t is *enabled* in marking M, written $M \xrightarrow{t}$, iff every input place of t contains at least one token.
- If a transition t is enabled in marking M, it *may fire*: from every input place one token is removed and to every output place one token is added. We write $M \xrightarrow{t} M'$ to denote that firing enabled transition t in marking M results in marking M'.

We write $M \to M'$ to indicate that by firing some transition t in M marking M' can be reached. We write $M \xrightarrow{\sigma} M'$ to denote that by firing sequence $\sigma = t_1 t_2 \ldots t_n$ from M marking M' can be reached, so $M_0 \xrightarrow{t_0} M_1 \xrightarrow{t_1} M_2 \ldots M_{n-1} \xrightarrow{t_{n-1}} M_n$, where $M_0 = M$ and $M_n = M'$. We write $M \xrightarrow{*} M'$ to denote that there is a sequence σ such that $M \xrightarrow{\sigma} M'$.

3   Reactive Nets

In this section we will adapt the definition of a Petri net to make it reactive. We call the new Petri net variant that we thus obtain a reactive net.

As explained in the introduction, the token-game semantics models closed systems, whereas a reactive system is open, otherwise it wouldn't be able to interact with its environment. This limitation of Petri nets can be circumvented by modeling the environment in the Petri net as well. We therefore distinguish external transitions of the environment from internal transitions of the reactive system. Thus, instead of a set T of transitions, we now have a set $T_{internal}$ of internal transitions and a set $T_{external}$ of external transitions.

We also explained in the introduction that the may firing rule used in Petri nets does not model reactivity. In a reactive system an enabled transition *must* fire immediately, otherwise the system would fail to respond to a certain event. In the token-game semantics, an enabled transition t *may* fire, but does not have to. In the worst case firing t is postponed forever, or some conflicting transition that becomes enabled later fires before t and disables t.

The most straightforward way to make the token-game semantics reactive is to change the firing rule from may firing into must firing, i.e., as many enabled transitions as possible should fire. This, however, is undesirable: since Petri nets model closed systems, the environment of the system is also included in the Petri net. The environment is active rather than reactive. For the environment, the may firing rule is more appropriate. Therefore, for internal transitions, done by the reactive system itself, we use a must firing rule, and for external transitions, done by the environment, we use a may firing rule.

However, if in a certain marking both an internal transition and an external transition are enabled, a conflict can arise. To avoid such a conflict, we require all internal transitions to fire with higher priority than external transitions. So, if both the environment and the reactive system can do a transition, the reactive system will fire first. This corresponds to the perfect synchrony hypothesis [5], an assumption frequently made in the design of reactive systems: The reactive system is faster than the environment it controls. Note that the perfect synchrony hypothesis is an assumption, not a guarantee.

We can informally describe the behavior of reactive Petri nets in the following way, borrowing some terminology from STATEMATE [16]. A state (marking) is *stable* if no internal transition is enabled, it is unstable otherwise. A stable state can become unstable if some external transition fires. In an unstable state, the reactive system must fire some enabled internal transitions. By firing these transitions, a new state is reached. If the new state is stable, the system has finished its reaction. Otherwise, the system again reacts by taking a transition and entering another new state. This sequence of taking a transition and entering a new state is repeated until finally a stable state is reached. It is possible that the system never reaches such a stable state: in that case the system diverges.

Definition. We now describe reactive nets and their behavior more formally. A *reactive net RN* is a tuple $(P, T_{internal}, T_{external}, F)$. Sets $T_{external}$ and $T_{internal}$ are transitions. A reactive semantics for a net $(P, T_{internal}, T_{external}, F)$ is defined as follows.

- An internal transition t is enabled in marking M iff all of t's input places are filled with a token, i.e. M is unstable.
- An external transition t is enabled in marking M iff all of t's input places are filled with a token, and there is no enabled internal transition in M, i.e. M is stable.

The firing of an enabled transition is as before: from every input place one token is removed and to every output place one token is added.

The must firing is encoded in the priority rule: enabled internal transitions have priority over external transitions. That unstable markings are instantaneous is an interpretation we attach to them. To model this explicitly in the semantics, we would have to switch to timed Petri nets.

Token-game semantics for reactive nets. It also possible to use a token-game semantics for a reactive net, by first transforming the reactive net into a Petri net using function *toPetri*, which takes the union of sets $T_{internal}$ and $T_{external}$. Function *toPetri* is defined as follows:

$$toPetri((P, T_{internal}, T_{external}, F)) = (P, T_{internal} \cup T_{external}, F)$$

To show the relation between the transition relation of RN under the token-game semantics, $\rightarrow^{tg}$, and the transition relation of RN under the reactive semantics, $\rightarrow^{r}$, we now define $\rightarrow^{r}$ in terms of $\rightarrow^{tg}$:

$$M \rightarrow^{r} M' \Leftrightarrow \exists t_i \in T_{internal} : M \xrightarrow{t_i}{}^{tg} M'$$
$$\vee \; (\; (\exists t_e \in T_{external} : M \xrightarrow{t_e}{}^{tg} M') \wedge (\nexists t_i \in T_{internal} : M \xrightarrow{t_i}{}^{tg}) \;)$$

Relation between reactive and token-game semantics. Indirectly, we have provided two different semantics for reactive nets: a reactive one and the traditional token-game semantics. In what respect do these semantics induce similar behavior?

Before we answer this question, let us look at the behavior of the reactive system in the reactive setting. In the reactive setting, the system moves typically from a stable state to another stable state. It is also possible that the system diverges: then there is a loop of internal transitions in the reactive net.

Of course, this reactive behavior can be simulated with the token-game semantics, since every transition enabled in marking M under the reactive semantics will also be enabled in M under the token-game semantics. The next theorem and corollary now follow immediately.

Theorem 1 *Given a reactive net RN that is in some state M. If under the reactive semantics t can fire in M and M' is reached, $M \xrightarrow{t} {}^r M'$, then t can also fire under the token-game semantics and M' is also reached, so $M \xrightarrow{t} {}^{tg} M'$.*

Corollary 2 *Given a reactive net RN. If M is a reachable state of RN under the reactive semantics, then M is a reachable state of RN under the token-game semantics.*

Clearly, the token-game semantics can be used to simulate any reaction, starting in some stable state, in the reactive semantics. The reverse, however, does not hold: not every behavior under the token-game semantics can be simulated using the reactive semantics. The reason for this is that in the reactive semantics external transitions are only enabled once all enabled internal transitions have fired. So, not every marking reachable under the token-game semantics will be reachable under the reactive semantics.

However, sometimes the *outcome* of the reaction of the system in the two semantics, i.e., the stable marking that is eventually reached, can be the same in both semantics. That is, under some constraints, if under the token-game semantics a particular stable marking is reached (where stable under the token-game semantics means that no internal transition is enabled), then this same marking can be reached under the reactive semantics. The constraints needed to enforces this, C1 and C2, are listed in Table 1.

Constraint C1 is necessary because under the reactive semantics internal transitions have priority over external transitions, whereas under the token-game semantics this is not the case. To see why constraint C2 is needed, consider the example reactive Petri net in Fig. 1. This reactive net does not satisfy C2, since t6 conflicts with both t2 and t4. Under the token-game semantics, in stable marking [p2, p3], there is a sequence t3, t5, t6 to stable marking [o]. All the intermediary markings in the sequence are unstable. Yet, it is impossible in the

Table 1. Constraints on reactive nets

C1 An external transition t_e does not conflict with an internal transition t_i:
$\bullet t_i \cap \bullet t_e \neq \emptyset$.

C2 For two internal transitions t and t', if $\bullet t \cap \bullet t' \neq \emptyset$, then either $\bullet t = \bullet t'$ (t and t' are free choice), or there is no reachable marking M, under the token-game semantics, such that $M \xrightarrow{t} {}^{tg}$ and $M \xrightarrow{t'} {}^{tg}$.

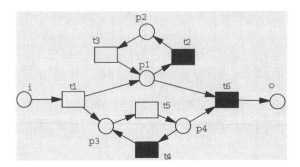

Fig. 1. Example reactive net to motivate constraint C2. Black transitions are internal, white transitions are external

reactive semantics to do such a sequence. Transition t6 is never taken; instead, t2 or t4 is taken. So marking [o] is unreachable under the reactive semantics.

Constraint C2 has been deliberately formulated on both the syntax and semantics of Petri nets, rather than only on the syntax. A syntactic constraint, e.g. "internal transitions are free choice", would have ruled out certain reactive Petri nets that constraint C2 allows. Figure 3 shows a Petri net, in which the two internal transitions cancel and pick satisfy constraint C2, but are not free choice.

We now proceed to prove one of the main theorems of this paper. The theorem states that under constraints C1 and C2, the token-game semantics and the reactive semantics have similar behavior. In the theorem, we use the terms 'stable' and 'unstable' states, introduced before.

Theorem 3 *Given a reactive net RN that satisfies constraints C1 and C2.*

Suppose, under the token-game semantics, there is a sequence σ of transitions from one stable state to another, such that all intermediary states are unstable,

$$M_0 \xrightarrow{t_0} {}^{tg} M_1 \xrightarrow{t_1} {}^{tg} M_2 \xrightarrow{t_2} {}^{tg} M_3 \xrightarrow{t_3} {}^{tg} \ldots \xrightarrow{t_{n-1}} {}^{tg} M_n$$

where M_0 and M_n are stable, and M_i, for $0 < i < n$, is unstable.

Then a permutation of σ (possibly σ itself) can be taken under the reactive semantics, and stable state M_n is reached.

Proof. If sequence σ is possible under the reactive semantics, we are done. So assume that σ is not possible under the reactive semantics. Then in some unstable state M_i an external transition $t_{external}$ is taken (so $t_i = t_{external}$). Under the reactive semantics, $t_{external}$ is disabled in M_i. Since M_i is unstable, there must be some enabled internal transitions in M_i. By C2 and since M_n is stable, one of these internal transitions, say t, is taken somewhere later in the sequence in some state M_j, where $j > i$. By C1, t and $t_{external}$ do not disable each other: the tokens in t's input places are not removed if $t_{external}$ is taken and vice versa. We modify σ by removing t_j from σ and inserting t just before $t_{external}$. Denote this

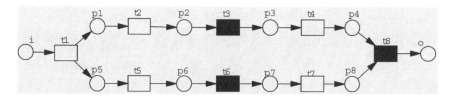

Fig. 2. Example reactive net to illustrate Theorem 3. Black transitions are internal, white transitions are external

modified sequence by σ'. Clearly, σ' can be taken under the token-game semantics; state M_n is then reached. Of course, sequence σ' may not be possible under the reactive semantics, because in some unstable state an external transition is taken. Then again the procedure sketched above has to be applied.

Thus, by repeatedly applying the procedure sketched above, finally a sequence σ_{final} is obtained. (The procedure terminates because the sequence is finite, and internal transitions are given a place earlier in the sequence.) In σ_{final}, in every unstable state an internal transition is taken. So σ_{final} can be taken in the reactive semantics.   □

Example. Consider the example reactive net in Fig. 2. It satisfies constraints C1 and C2. Under the token-game semantics, a possible sequence from stable state [p1,p5] to stable state [p3,p7] is t2,t5,t6,t3. All intermediate states in this sequence are unstable. This sequence cannot be taken under the reactive semantics, because in unstable state [p2,p5] external transition t5 is fired. By applying the procedure sketched in the proof, we obtain sequence t2,t3,t5,t6. This sequence can be taken under the reactive semantics. Note that this sequence has an intermediate stable state [p3,p5] not present in the original sequence.

4   Workflow Nets

In this section we recall the definition of Workflow nets [1] and give different interpretations for transitions in Workflow nets.

Definition. A *Workflow net* (WF net) is a Petri net with one input place i and one output place o such that:

– Place i does not have incoming arcs.
– Place o does not have outgoing arcs.
– Every node $n \in P \cup T$ is on a path from i to o.

WF nets use the standard Petri net token-game semantics.

Figure 3 shows a WF net for handling an incoming order for a mobile telephone. It is a reduced version of a real-life business process of a telephone company. The process involves two departments: the accountancy department, which handles the payment, and the sales department, which handles the distribution.

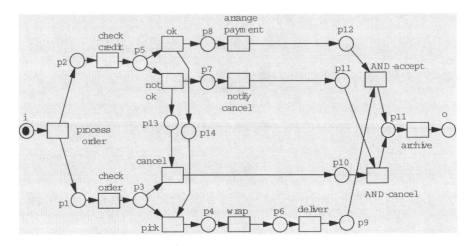

Fig. 3. WF net modeling the process "Handling of incoming order"

The process starts with an incoming order that is processed to the different departments. This is modeled by transition process_order which splits the execution into two parallel threads. The bottom part models the tasks on the sales side. Here the order is handled (executing tasks check_order, pick, wrap, and deliver). The top part models the tasks on the accounting side. Here the customer standing is checked first (check_credit). The result of this task is either ok or not_ok. In case the result is positive the payment is arranged (arrange_payment), in the latter case the order is refused (notify_cancel).

The cooperation of the two departments follows a pessimistic strategy. The sales department always waits for the outcome of the credit check performed by the accountancy. Depending on the outcome it either picks, wraps, and delivers the item or cancels it further processing.

Transitions in WF nets. In a WF net states are modeled via places, whereas transitions model active behavior. Transitions are used for different purpose. In the most common case they are used to model *activities* (or *tasks*). Examples of tasks in the process of Fig. 3 are check_order, pick, wrap, deliver, cancel, check_credit, arrange_payment, notify_cancel and archive.

Sometimes, transitions represent the making of *decisions*. Examples are the two transitions ok, not_ok representing the outcome of the task check_credit. Note that the outcome of the decision is determined by the environment of the WF engine; we come back to this issue below.

Transitions may also be employed to depict the occurrence of external *events*. Figure 4 gives an example. The WF net models part of the library process for returning books. Upon borrowing books, the system waits for an external event: This may either be the reader bringing the books back, or the reader asking for extension, or a timeout. Depending on the particular event occurring, a following task is executed.

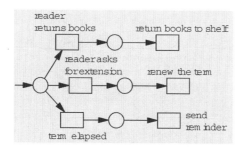

Fig. 4. Part of the library process "Return Books"

Finally, transitions can be there for the sole purpose of *routing* a case. This usually occurs when a case needs to be split into parallel parts (fork) or parallel parts of a case need to be merged (join). Examples for this case are process_order, AND_cancel, and AND_accept from Fig. 3.

Function *type* assigns to each transition the purpose of the transition.

$$type : T \rightarrow \{task, event, decision, routing\}$$

Van der Aalst [1] distinguishes task and decision transitions, but does not treat them differently. In WF nets, all transitions use the same firing rule. So there is no distinction between the different types of transitions. In Sect. 5, we will attach different semantics to these different types of transitions.

There exist some interesting dependencies between transitions of different type. Transition of type *decision* always follow a *task* transition. This task transition models the processing of some kind of test. For the evaluation it refers to external data. Transitions of type *decision* always occur within a choice. They are in conflict. Furthermore, it can be assumed that the corresponding choice is free choice. This means that only the evaluation of the external information (and nothing else) decides about the outcome of the choice.

If transitions of type *event* occur within a choice, we assume that choice to be free choice as well. This is a reasonable assumption as the occuring events should preclude each other. This means that the particular external event that occurs (and nothing else) decides the outcome of the choice. For instance, in the library example from Fig. 4 the timeout event occurs only if the reader neither returns the book (in person) nor asks for extension (e.g. via email or telephone). Depending on the kind of external event the books are either returned to the shelf, the term is renewed, or a reminder is send.

The *outcome* of the above mentioned choices depends either on the evaluation of external data or the event occurring. In [7] these choices have been clustered using the notation of a *non-controllable choice*. This notation suggests that the *outcome* of these choice depends on the environment. Choices whose outcome do not depend on the environment are called *controllable*; they can be controlled by the WF engine.

Table 2. Choice classification

	initiative	outcome
choice of *event* transitions (free choice)	environment	environment
choice of *decision* transitions (free choice)	WF engine	environment
choice of *task* and/or *routing transitions*	WF engine	WF engine

An orthogonal criterion to distinguish choices, is the moment of choice, i.e., the moment one of the alternative transitions is executed. This distinction was made by Van der Aalst and his coworkers [1,3]. They distinguish between implicit and explicit choices. An implicit choice (also deferred choice) is made the moment an external event occurs, hence it corresponds to a choice that consists of *event* transitions. An explicit choice is made the moment the previous task is completed. In our framework, explicit choices correspond to choices consisting of *decision* transitions.

In our framework, the moment of choice is determined by the one having the *initiative* to execute a transition. Events depict behavior of the environment, hence the *initiative* to execute an *event* transition is at the environment. In contrast, the initiative for transitions of type *decision* is at the side of the WF engine. The WF engine executes these transitions. In Sect. 5, we will map WF nets to reactive nets. *Event* transitions will be external, whereas *decision* transitions will be internal. Thus, in the reactive setting, an implicit choice behaves differently from an explicit choice.

Remaining choices are choices that consist of transitions of type task and of type routing. These are always controllable and explicit choices. These choices are furthermore not necessarily free choice. An example is the choice between the task transitions pick and cancel in Fig. 3.

Table 2 summarizes the possible influences of WF engine and environment on different choices.

Correctness criteria for WF nets. We only consider soundness as introduced by Van der Aalst [1]. Soundness requires that a WF net can *always* terminate with a single token in place o and that all the other places are then empty. In addition, it requires that there is no dead transition, i.e. each transition can be executed. We recall the definition of soundness as defined in [1]. To stress that soundness is defined on the token-game semantics, we label the transition relation with tg.

Definition 4. *(Soundness) A WF net is sound iff:*
(i) For every marking M reachable from marking i, there exists a firing sequence leading from marking M to marking o.

$$\forall_M([i] \xrightarrow{\;*\;}{}^{tg} M) \Rightarrow (M \xrightarrow{\;*\;}{}^{tg} [o]).$$

(ii) Marking $[o]$ is the only marking reachable from marking $[i]$ with at least one token in place o (proper termination).

$$\forall_M([i] \xrightarrow{\;*\;}{}^{tg} M \wedge M \geq [o]) \Rightarrow (M = [o])$$

(iii) The WF net does not have dead transitions.

$$\forall_{t\in T}\exists_{M,M'}([i] \xrightarrow{*} {}^{tg} M \xrightarrow{t} {}^{tg} M')$$

If a WF net is going to be used as input for a WF engine, soundness is essential. In a sound WF net every firing sequences terminates properly. Deadlocking executions, as well as executions where spare tokens remain in the net, are impossible. If the process description is used as base for operation at run time, soundness is a necessary requirement in order to guarantee a reliable execution.

However, the soundness criterion is defined using the token-game semantics. As we saw in the introduction, that semantics is not reactive. Therefore, in the next section we show how to transform a WF net into a reactive WF net. In Sect. 6 we show that the transformation preserves soundness. For reactive WF nets, soundness is defined by replacing in the definition above $\xrightarrow{}{}^{tg}$ with $\xrightarrow{}{}^{r}$.

5   From Workflow Nets to Reactive Workflow Nets

Workflow management is a current issue in many business (re-)engineering projects. It comprises support for the modeling, the analysis and the run-time execution of business processes. Many approaches aiming at providing support for workflow management are based on the use of Petri nets (e.g. [1,8,9]) or were mapped onto Petri nets (e.g. [2,8]). Even though these approaches cover the modeling and the analysis of business processes, they provide only limited support for the execution at run time. Reasons for that gap have been discussed in the introduction and concern the mismatch between the reactive behavior of the workflow (WF) engine and the active behavior of Petri nets (and thus WF nets) using the token-game semantics.

In this section we discuss how a WF net can be transformed into a reactive WF net. A reactive WF net is a Petri net with a reactive semantics. It can serve as input for a WF engine specifying what the WF engine should do.

The following three properties of WF nets makes that they are not entirely suitable as input for a WF engine.

1. Transitions fire instantaneously. This does not match with the requirement to model tasks as time consuming entities.
2. Usually, transitions in a WF net model tasks. The WF engine monitors tasks, but does not do them. Thus, it is hard to detect from the WF net the actual behavior of the WF engine.
3. Under the token-game semantics, a Petri net models an active system. But a WF engine is a reactive system. For such a system, the may-firing rule of the token-game semantics introduces unintended non-determinism, allowing either to execute an enabled task or to defer its execution.

We now show how we can overcome these three obstacles. We will change the perspective of a WF net from modeling a process to monitoring it. This way, we obtain a description of the desired behavior of a WF engine.

Task refinement. Recall from Sect. 4 that transitions in a WF net either model the occurrence of an event (type: *event*), the making of a decision (type: *decision*), the routing of tasks (type: *routing*) or the actual task that is executed by some external actor (type: *task*). The firing of transitions is considered to be instantaneous. This abstraction is adequate for transitions that model events, decisions or routing, but this generalization does not fit for tasks.

Changing the perspective from active task execution to only monitoring it, tasks performed by external actors should be modeled as time consuming. We therefore refine the modeling and depict a task as a sequence of transitions announce_task, begin_task, end_task, and record_task_completion. Figure 5 illustrates the described task refinement. The transition announce_task models the placing of the task to a possible actor. This may either mean that it is "pushed" into someones in-basket or that the task is put to a common list, from where it can be "pulled" by any actor. The precise implementation depends on the mode of the WFMS.

The actual processing of the task starts with transition begin_task and ends with end_task. This way the instantaneous firing of transitions can be retained now matching an acceptable abstraction. Note that the duration implicitly assigned to the execution of a task in a WF net is now assigned to a place in the refined WF net.

Division of powers between the WF engine and the environment. Changing the perspective of a WF net towards monitoring we have to distinguish precisely between the behavior of the WF engine and the environment. In WF nets, such a dinstinction is not made.

From a monitoring perspective, a transition is either executed by the WF engine, or executed by the environment of the WF engine. Taking the perspective of the WF engine, we call transitions executed by the WF engine *internal*, whereas transitions by the environment are *external*. We will therefore split the set of transitions T of a WF net $PN = (P, T, F)$ into disjoint sets of internal and external transitions: $T = T_{internal} \cup T_{external}$. Internal transitions are denoted by black whereas external transitions are represented by white boxes.

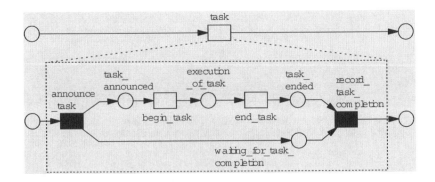

Fig. 5. Task refinement

Table 3. Mapping WF net to reactive WF net

transition type	internal	external
announce_task	x	
begin_task		x
end_task		x
record_task_completion	x	
event		x
decision	x	
routing	x	

Reviewing the four different transition types: task, event, decision, routing, we can classify transitions as internal or external as follows (see Table 3):

Tasks. The announce_task and record_task_completion transitions are internal and the begin_task and end_task transition are external. This denotes that the WF engine initiates the task but that an actor outside the WF engine does the actual task. The WF engine waits for completion of the task.

Events. These transitions are external. This is natural as such transitions model the occurrence of events coming from the environment.

Decisions. These transitions are internal. They are done by the WF engine, even though the outcome of the decision presumably depends upon the task that has been executed immediately before (cf. Sect. 4, in particular Table 2).

Routing. These transitions are internal, as routing is done by the WF engine.

Reconsidering Table 2, we can see that the party (WF engine or environment) having the initiative in taking a transition, also executes the transition.

Reactive semantics: Changing the firing rule. The last obstacle concerns the may firing rule of the token-game semantics. This rule states that if a transition t is enabled it may fire but does not have to. In the worst case, some conflicting transition that become enabled later fires before t, disabling t. This firing rule is not adequate to model behavior of the WF engine, which is modeled through internal transitions.

We therefore transform a refined WF net (P, T, F) into a reactive WF net $(P, T_{internal}, T_{external}, F)$, using the previously introduced distinction between internal and external transitions (Table 3). This reactive WF net can be mapped to the original refined WF net using function $toPetri$. Thus, we replace the token-game semantic of the refined WF net by a reactive semantics, i.e. replace relation $\rightarrow^{tg}$ by $\rightarrow^{r}$ (see Sect. 3).

To illustrate this mapping, Fig. 6 shows the reactive WF net corresponding to the example WF net of Fig. 3. Due to space limitations, we do not show how task transitions are refined, but just depict them with a shortcut: a transition subdivided into three sections: start and end black, middle white.

Note. We stated that reactive WF nets can be used by a WF engine to control and monitor processes. But not every transition of the reactive WF net

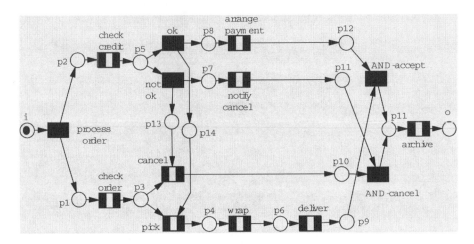

Fig. 6. A reactive WF net. Black transitions are internal, white transitions are external. Task transitions are decomposed as in Fig. 5

is executed by the WF engine; only internal transitions are. Thus, for a WF engine, a reactive WF net still contains too much information. By removing external transitions and external places (i.e. places filled by external transitions) from the WF net, a prescription for the WF engine is obtained. Note that by filling some places (e.g. place task_ended) in Fig. 5) with a token, the environment can trigger the WF engine to start doing something (e.g. do transition record_task_completion). Finally, we observe that reactive WF nets presupposes that the WF engine is faster than its environment, i.e., the WF engine must satisfy the perfect synchrony hypothesis (see Sect. 3).

6   Soundness of Reactive Workflow Nets

In the previous section we have defined a mapping from WF nets to reactive WF nets in two steps. First, we have refined task transitions to multiple transitions. Second, we have mapped a refined WF net to a reactive WF net. In this section, we show that the soundness property of a WF net is preserved when the WF net is mapped into a reactive WF net, provided the WF net meets the constraints defined in Table 1. (Although the constraints in Table 1 are not defined on WF nets, they can be lifted to WF nets by using the mapping defined in Table 3, provided function *type* is defined. Note that the task refinement (Fig. 5) satisfies the constraints C1 and C2.)

Given a WF net PN, we denote its refined variant by PN_{ref}. We denote the reactive variant of PN_{ref} by $PN_{reactive}$. For PN_{ref} we use the token-game semantics; for $PN_{reactive}$ the reactive semantics.

Theorem 5 *Sound PN $\Leftrightarrow$ sound PN_{ref}.*

Proof. Straightforward.   □

The following theorem shows that if a reactive WF net is sound, the WF net is sound as well. Note that the soundness property (see Sect. 4) is defined on the token-game semantics, not on the reactive semantics. To obtain the soundness property for reactive WF nets, replace in the definition $\to^{tg}$ by $\to^r$.

Theorem 6 $PN_{reactive}$ *is sound* $\Rightarrow$ PN_{ref} *is sound.*

Proof. Follows immediately from Theorem 1 and Corollary 2.   □

We now prove one of the main theorems of this paper. In the proof, we make use of the terminology of stable and unstable states, introduced in Sect. 3. State $[o]$ is stable by definition. We also assume state $[i]$ is stable. (It is possible to relax this constraint, but it will make the proofs more difficult.) To prove this theorem, we build on Theorem 3, so we need the constraints defined in Table 1.

Theorem 7 *Assume PN_{ref} (and thus $PN_{reactive}$) satisfies constraints C1 and C2. Then PN_{ref} is sound $\Rightarrow$ $PN_{reactive}$ is sound.*

Proof. We consider the three cases of the definition of soundness.

(i)   Let M be an arbitrary state in $PN_{reactive}$. By Corollary 2, M is also reachable in PN_{ref}. Since PN_{ref} is sound, there is at least one firing sequence from M to o under the token-game semantics. Denote this sequence by σ. There are two cases.
-    If M is stable, then σ can be split in subsequences $\sigma_0, \sigma_1, \ldots, \sigma_n$, such that $M \xrightarrow{\sigma_0}{}^{tg} M_1 \xrightarrow{\sigma_1}{}^{tg} M_2 \xrightarrow{\sigma_2}{}^{tg} \ldots \xrightarrow{\sigma_n}{}^{tg} [o]$, where M_1, M_2, $\ldots$, $M_n = [o]$ are all stable, and all other states visited in the sequence σ are unstable. By applying Theorem 3 on the sequences $\sigma_0, \sigma_1, \ldots, \sigma_n$, we have that there are permutations of these sequences $\sigma'_0, \ldots, \sigma'_n$ such that $M \xrightarrow{\sigma'_0}{}^r M_1 \xrightarrow{\sigma'_1}{}^r M_2 \xrightarrow{\sigma'_2}{}^r \ldots \xrightarrow{\sigma'_n}{}^r [o]$.
-    If M is unstable, then there is at least a sequence of transitions from some stable state M_0 to stable state M_n that leads through M. (There is at least one sequence from $[i]$ to $[o]$.) For $M_n \xrightarrow{*}{}^r [o]$, we can argue as in the previous case. The remainder then follows easily.
(ii)   Every reachable state in $PN_{reactive}$ is also a reachable state in PN_{ref} (Corollary 2). So if a state M with $M > [o]$ would be reachable in $PN_{reactive}$ it would also be reachable in PN_{ref}. But PN_{ref} is sound. So we have a contradiction.
(iii)   For every transition t, there is a state M in PN_{ref}, such that $M \xrightarrow{t}$. In PN_{ref}, there is at least one sequence σ of transitions, one of which is t, from some stable state M_0 to a stable state M_n, which passes M. (There is at least one sequence from $[i]$ to $[o]$.) By Theorem 3, a permutation of σ can be taken in $PN_{reactive}$. So t can be taken in $PN_{reactive}$. Therefore, there are no dead transitions in $PN_{reactive}$.

□

The desired result, that soundness is preserved when transforming a WF net into a reactive net, now follows immediately.

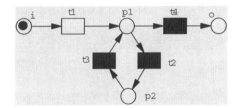

Fig. 7. Sound reactive WF net that can diverge. The net satisfies constraints C1 and C2. Black transitions are internal, white transitions are external

Corollary 8 *Assume* PN_{ref} *(and thus* $PN_{reactive}$*) satisfies constraints C1 and C2. Then* PN *is sound* $\Leftrightarrow PN_{reactive}$ *is sound.*

Proof. By Theorem 5, we have: PN is sound $\Leftrightarrow PN_{ref}$ is sound. By Theorems 6 and 7, we have: PN_{ref} is sound $\Leftrightarrow PN_{reactive}$ is sound.   □

Finally, we note that soundness of a reactive net does not guarantee that the net is divergence free. A reactive net diverges if there is a loop of internal transitions. It is easy to prove that in a sound reactive net satisfying constraints C1 and C2, a loop of internal transitions can always be exited by taking some internal transition t that leaves the loop. However, soundness only states that t can be taken, but does not guarantee that t *will* be taken if the system is in a loop. For example, in Fig. 7, the system can do t1,t2,t3,t2,t3,..., staying forever in loop t2, t3 without ever taking t4. So, to guarantee absence of divergence, an additional constraint is needed. A sufficient, but not necessary, constraint is to require that the input places of internal free choice transitions are only filled by external transitions. That constraint would rule out the reactive WF net in Fig. 7.

7   Related Work

Like the present work, Wikarski [20] argues that the may firing rule is not suited to prescribe behavior of a system. He therefore proposes to use the may firing rule when the Petri net describes behavior, and the must firing rule when the net prescribes behavior. Thus, a net either uses may or must firing but not both. In contrast, reactive nets are both descriptive (external) and prescriptive (internal transitions); reactive nets use a mixture of may and must firing.

Next, we look at other work that uses Petri nets for modeling reactive systems. Then we look at some related work done on priority nets. We also discuss the reactive semantics one of the authors has defined for UML activity diagrams.

Petri nets for modeling reactive systems. In the past, several other extensions for Petri nets have been proposed, that, like the present work, have been motivated by the desire to use Petri nets to control a process. We discuss two of those, Signal Event nets [15,14] and Grafcet [6].

Signal event nets [15,14] were introduced to model the combined behavior of some process and the controller that controls and monitors that process. Like in reactive nets, in Signal Event nets some transitions are by the environment (these are called spontaneous) whereas others are by the controller (these are called forced). Spontaneous transitions trigger forced transitions through transition synchronization. This way, it can be specified that a controller reacts to events in the environment. In reactive nets, an external transition triggers an internal transition t_i indirectly by filling all of t_i's input places.

Grafcet [6] is a standardized graphical language for specifying logic controllers. In Grafcet, Petri nets are extended with boolean variables, that are set by the environment of the system. These boolean variables represent the state of the environment. Every transition of a Grafcet model is a transition by the reactive system (i.e., the logic controller). Each transition has a corresponding event-condition-action (ECA) rule. If the event occurs and the condition is true and its input places are filled, the transition must fire immediately. Multiple transitions can fire simultaneously at the same time. In contrast, reactive nets use an interleaving firing rule.

There are some general differences between these approaches and reactive nets. First, reactive nets stay closer to the token-game semantics than these related approaches. Moreover, the definition of reactive nets is considerably simpler than the definition of signal-event nets and Grafcet models. Finally, none of these approaches attach a token-game semantics to a reactive model to relate it to the reactive semantics.

Priority nets. Reactive nets resemble Petri nets with priorities. The work on priority nets most close to ours is Bause [4]. Like the present work, Bause extends a net by defining a static priority ordering on its transitions. He shows that under a certain condition, properties like liveness of a net are preserved when the net is extended with static priorities. Bause's condition is similar to C1 and the constraint that internal transitions are free choice. However, our constraint C2 is more general, allowing internal transitions that are not free choice. For example, the Petri net in Fig. 3, does not satisfy Bause's condition, whereas it does satisfy our constraints C1 and C2. Motivated by the domain of stochastic Petri nets, Bause considers weighted Petri nets whose transitions are partitioned into multiple priority classes, whereas we, motivated by the domain of reactive systems, only consider unweighted Petri nets whose transitions are partitioned into two priority classes (internal and external transitions).

UML activity diagrams. Recently, one of the authors has defined two formal, reactive execution semantics for activity diagrams [12]. The design choices in both semantics are based on existing statechart semantics. The token-game semantics was not used, because that semantics is not reactive [12,13]. Even though reactive nets are more reactive than Petri nets having a token-game semantics, there still exist a lot of subtle differences between activity diagrams and reactive nets. For example, activity diagrams can refer to temporal events. These cannot be modeled in reactive nets; we would have to switch to timed Petri nets to model this. Moreover, activity diagrams can have data, whereas reactive nets can not. (Data can be modeled by switching to colored nets, but activity diagrams also differ from colored Petri nets [12,13].) Consequently, in reactive WF nets, conflicting decision transitions (for example ok and not ok in Fig. 3) are enabled at the same time: the net behaves non-deterministically, whereas in activity diagrams such decisions are deterministic.

8   Conclusion and Further Work

We have defined a reactive variant of Petri nets, called reactive nets. Reactive nets assume that the systems they model are perfectly synchronous. Reactive nets have a reactive semantics, which differs from the token-game semantics, but they also can use the traditional token-game semantics. We have shown that under some conditions, the reactive and token-game semantics induce similar behavior, i.e., the same stable states are eventually reached.

Reactive nets are motivated by the domain of workflow management. We have shown how a WF net can be transformed into a reactive WF net, and that under some conditions the soundness property is preserved. Thus, we have offered a justification why soundness can be analyzed on WF nets using the token-game semantics, even though that semantics does not model behavior of reactive systems, whereas a WF engine is reactive. However, our work shows that in addition to soundness some extra constraints are needed. Moreover, soundness does not rule out divergence. An interesting topic of future work is to investigate whether the extra constraints are not too restrictive, i.e., whether they cover a large class of workflow models.

Another topic of future work is extending the semantics of reactive nets with simple real-time constructs, or addition of data. Next, it might be interesting to see whether the definition of reactive nets can be changed such that the perfect synchrony hypothesis is no longer needed.

Acknowledgements. The work of the first author has been partially supported by the National Research Fund (FNR) of Luxembourg and has been partially performed within the scope of the LIASIT (Luxembourg International Advanced Studies in Information Technologies) Institute.

References

1. W.M.P. van der Aalst. The Application of Petri Nets to Workflow Management. *The Journal of Circuits, Systems and Computers*, 8(1):21–66, 1998.
2. W.M.P. van der Aalst, A. Hirnschall, and E. Verbeek. An alternative way to analyze workflow graphs. In *Proc. 13th Int. Conference on Advanced Information Systems Engineering (CAiSE 2002)*, volume 2348 of *Lecture Notes in Computer Science*. Springer Verlag, 2002.

3. W.M.P. van der Aalst, A. ter Hofstede, B. Kiepuszewski, and A. Barros. Advanced Workflow Patterns. In O. Etzion and P. Scheuremann, editors, *Proc. 7th IFCIS Int. Conference on Cooperative Information System (CoopIS 2000)*, volume 1901 of *Lecture Notes in Computer Science*, pages 18–29. Springer Verlag, 2000.
4. F. Bause. On the analysis of Petri nets with static priorities. *Acta Informatica*, 33(7):669–685, 1996.
5. G. Berry and G. Gonthier. The ESTEREL synchronous programming language: design, semantics, implementation. *Science of Computer Programming*, 19(2):87–152, 1992.
6. R. David. Grafcet: A powerful tool for specification of logic controllers. *IEEE Transactions on Control Systems Technology*, 3(3):253–267, 1995.
7. J. Dehnert. Non-controllable choice robustness: Expressing the controllability of workflow processes. In J. Esparza and C. Lakos, editors, *Proc. 23rd Int. Conference on Application and Theory of Petri Nets (ICATPN 2002)*, volume 2360 of *Lecture Notes in Computer Science*, pages 121–141. Springer Verlag, 2002.
8. J. Dehnert. Four steps towards sound business process models. In Ehrig et al. [10]. To appear.
9. J. Desel and T. Erwin. Modeling, simulation and analysis of business processes. In W. van der Aalst, J. Desel, and A. Oberweis, editors, *Business Process Management: Models, Techniques and Empirical Studies*, volume 1806 of *Lecture Notes in Computer Science*. Springer Verlag, 2000.
10. H. Ehrig, W. Reisig, G. Rozenberg, and H. Weber, editors. *Petri Net Technology for Communication Based Systems*, Lecture Notes in Computer Science. Springer Verlag, 2003. To appear.
11. C.A. Ellis and G.J. Nutt. Modelling and enactment of workflow systems. In M. Ajmone Marsan, editor, *Proc. 14th Int. Conference on Application and Theory of Petri Nets (ICATPN 1993)*, volume 691 of *Lecture Notes in Computer Science*, pages 1–16. Springer, 1993.
12. R. Eshuis. *Semantics and Verification of UML Activity Diagrams for Workflow Modelling*. PhD thesis, University of Twente, 2002.
13. R. Eshuis and R. Wieringa. Comparing Petri net and activity diagram variants for workflow modelling – a quest for reactive Petri nets. In Ehrig et al. [10]. To appear.
14. A. Foremniak and P.H. Starke. Analyzing and reducing simultaneous firing in signal-event nets. *Fundamenta Informaticae*, 43:81–104, 2000.
15. H.-M. Hanisch and A. Lüder. A signal extension for Petri nets and its use in controller design. *Fundamenta Informaticae*, 41(4):415–431, 2000.
16. D. Harel and A. Naamad. The STATEMATE Semantics of Statecharts. *ACM Transactions on Software Engineering and Methodology*, 5(4):293–333, 1996.
17. D. Harel and A. Pnueli. On the development of reactive systems. In K.R. Apt, editor, *Logics and Models of Concurrent Systems*, volume 13 of *NATO/ASI*, pages 447–498. Springer, 1985.
18. A. Oberweis. *Modellierung und Ausführung von Workflows mit Petri-Netzen (in German)*. Teubener-Reihe Wirtschaftsinformatik. B.G. Teubener Verlagsgesellschaft, Stuttgart, Leipzig, 1996.
19. R.J. Wieringa. *Design Methods for Reactive Systems: Yourdon, Statemate and the UML*. Morgan Kaufmann, 2003.
20. D. Wikarski. An introduction to modular process nets. Technical Report TR-96-019, International Computer Science Institute, 1996.
21. M.D. Zisman. *Representation, Specification and Automation of Office Procedures*. PhD thesis, University of Pennsylvania, Wharton School, 1977.

Distributed Diagnosis of Discrete-Event Systems Using Petri Nets*

Sahika Genc and Stéphane Lafortune

Department of Electrical Engineering and Computer Science,
University of Michigan,
1301 Beal Avenue, Ann Arbor, MI, 48109-2122 USA
{sgenc,stephane}@eecs.umich.edu; www.eecs.umich.edu/umdes

Abstract. The problem of detecting and isolating fault events in dynamic systems modeled as discrete-event systems is considered. The modeling formalism adopted is that of Petri nets with labeled transitions, where some of the transitions are labeled by different types of unobservable fault events. The Diagnoser Approach for discrete-event systems modeled by automata developed in earlier work is adapted and extended to on-line fault diagnosis of systems modeled by Petri nets, resulting in a centralized diagnosis algorithm based on the notion of "Petri net diagnosers". A distributed version of this centralized algorithm is also presented. This distributed version assumes that the Petri net model of the system can be decomposed into two place-bordered Petri nets satisfying certain conditions and that the two resulting Petri net diagnosers can exchange messages upon the occurrence of observable events. It is shown that this distributed algorithm is correct in the sense that it recovers the same diagnostic information as the centralized algorithm. The distributed algorithm provides an approach for tackling fault diagnosis of large complex systems.

1 Introduction

The problem of detecting and isolating faults in technological systems has received considerable attention due to its importance in terms of safety and efficiency of operation. A variety of complementary approaches have been proposed, based on the level of detail chosen for the model of the system and the kinds of faults that need to be diagnosed; see, e.g., [1]. In this paper, we consider technological systems that can be modeled at some level of abstraction as discrete-event dynamic systems [2]. This includes a wide variety of technological systems such as automated manufacturing systems, communication networks, heating, ventilation, and air-conditioning units, process control systems, and power systems. Faults are modeled as unobservable events, namely, events whose occurrence is not directly detected by the sensors. Rather, the occurrence of fault events must be inferred from the system model and future observations of the evolution of

* This research is supported in part by NSF grant ECS-0080406.

W.M.P. van der Aalst and E. Best (Eds.): ICATPN 2003, LNCS 2679, pp. 316–336, 2003.
© Springer-Verlag Berlin Heidelberg 2003

the system. This is often referred to as "model-based diagnostics." The faults of interest are those that cause a distinct change in the operation of the system but do not necessarily bring it to a halt. Examples of such faults include: equipment faults (e.g., stuck faults of valves, stalling of actuators, bias faults of sensors, controller faults, and degraded or worn-out components), as well as many types of process faults (e.g., overflow of buffers in manufacturing and communication networks, contamination in semiconductor manufacturing, and control software faults).

The discrete-event modeling formalism adopted in this paper is that of Petri nets with labeled transitions, where some of the transitions are labeled by different types of unobservable fault events. Our objective is to adapt and extend, in the context of Petri net models, a recently-proposed approach for fault diagnosis of discrete-event systems modeled by finite-state automata, termed the "Diagnoser Approach"; see [3] and the references therein, including [4,5]. That approach has been used successfully in a variety of application areas, including heating, ventilation, and air-conditioning units [6], intelligent transportation systems [7,8], document processing systems [9,10], and chemical process control [11]. In the Diagnoser Approach, a *diagnoser automaton*, or simply *diagnoser*, is contructed from (i) the finite-state automaton model of the discrete-event system, (ii) the set of unobservable events, (iii) the set of fault events, and (iv) the partition of the set of fault events into fault types. The states of the diagnoser contain information about the possible occurrence of faults, according to the system model. The diagnoser is then used for on-line fault diagnosis of the system as follows. Each observable event executed by the system triggers a state transition in the diagnoser. Examination of the current diagnoser state reveals the status of the different types of faults: fault(s) of Type $F1$ did not occur, fault(s) of Type $F1$ possibly occurred ("$F1$-uncertain state" in the terminology of [4]), fault(s) of Type $F1$ occurred for sure ("$F1$-certain state" in the terminology of [4]). It is this capability of diagnosers that we wish to extend to Petri net models of the system. Diagnosers can also be used to analyze the diagnosability properties of the system ("Can all fault types eventually be detected?"), but this aspect is not considered in this paper.

There are many reasons for extending the Diagnoser Approach to Petri nets. Our primary motivation is to take advantage of the modularity of Petri net models and thereby propose a modular/distributed version of the Diagnoser Approach that can help in mitigating the state space explosion problem that often occurs in discrete-event modeling of complex systems. Consequently, the contribution of this paper is two-fold. First, a centralized diagnosis algorithm based on the novel notion of "Petri net diagnosers" is presented in Section 2 for on-line diagnosis of systems modeled by Petri nets. The Petri net diagnoser associated with a Petri net has the same graphical structure as the Petri net but has a different state transition function. In addition, Petri net diagnosers include several markings of the net at any given time, corresponding to the notion in the Diagnoser Approach that the state of the diagnoser is a form of state estimate (of the system) together with fault type information. (Due to the simultaneous presence of different markings in Petri net diagnosers, we can think of them as a special kind of colored Petri nets.)

The second contribution of this paper is to present, in Section 3, a distributed version of the above-mentioned centralized algorithm. This distributed version assumes that the Petri net model of the system can be decomposed into two place-bordered Petri nets satisfying certain conditions. Moreover, it is assumed that the two resulting Petri net diagnosers can exchange messages upon the occurrence of observable events. A method to decompose the system into place-bordered nets is given, if such a decomposition is necessary. We refer the reader to ([12,13]) for modular modeling methodologies that result in place-bordered nets. We show that our distributed algorithm is correct in the sense that it recovers the same diagnostic information as the centralized algorithm. The distributed algorithm provides an approach for tackling fault diagnosis of large complex systems, in particular *networked systems* where the different system modules are connected by a communication network.

To the best of our knowledge, the present paper is the first to explore the extension of the Diagnoser Approach originally proposed in [4] to Petri net models. However, there has been prior work on the general problem of monitoring and fault diagnosis of dynamic systems using Petri net models. We mention in this regard the work done at IRISA/INRIA on alarm supervision in telecommunication networks [14,15] and the work done on detection of loss or creation of tokens in nets using matrix algebraic techniques in [16]. Our problem formulation and objective however differ from those in [14,15,16,17]. They also differ from the work on observability of Petri nets in [18].

The remainder of this paper is organized as follows. Section 2 starts by presenting our notation for labeled Petri nets and then presents the centralized algorithm for on-line diagnosis of dynamic systems using Petri net diagnosers. The distributed version of the centralized algorithm, termed Algorithm DDC for "distributed diagnosis with communication" is presented in Section 3. Algorithm DDC consists of two communicating Petri net diagnosers, whose respective states can be "merged" (in a technical sense made precise in Section 3) to recover the state of the corresponding centralized Petri net diagnoser. An illustrative example is used throughout the paper and conclusions are presented in Section 4.

2   Centralized Diagnosis Using Petri Net Diagnosers

In this section, we define the notion of a centralized Petri net diagnoser, or simply *diagnoser*, which is used as a tool to detect and isolate faults in the system. The system to be diagnosed is modeled by a labeled Petri net. The centralized diagnoser observes the system and determines the states the system can be in upon observation of an event. Note that upon observation of an event, the state of the system is not known exactly in general due to the presence of unobservable events in the set of transition labels. The Petri net diagnoser finds all the states the system can be in, namely, all the states that are consistent with the sequence of observable events seen thus far. Fault information is attached to these state estimates in the from of fault labels. The faults are explicitly modeled as events in the system. Figure 1 gives a block diagram of the system and its diagnoser interacting with each other (the notation in the figure is introduced below in Section 2.1 and 2.2).

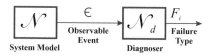

Fig. 1. Centralized diagnosis

This section first defines how the system and the diagnoser are modeled and gives their graphical representation. Then, we define the dynamics of the diagnoser. Although the diagnoser is modeled as a labeled Petri net graphically, its state transition function and states differ from regular Petri nets. We conclude the section by an example that builds the diagnoser and finds some of its states.

2.1 System Model

A Petri net graph[1] is a weighted bipartite graph

$$\mathcal{N} = \langle P, T, A, w \rangle \qquad (1)$$

where P is the finite set of places, T is the finite set of transitions, $A \subseteq (P \times T) \cup (T \times P)$ is the set of arcs from places to transitions and from transitions to places, and $w : A \to \mathbb{Z}_+$ is the weight function on the arcs. In a Petri net graph $\mathcal{N}$, given $t \in T$ we denote by $I(t) = \{p \in P : (p, t) \in A\}$ the set of input places to transition t, and similarly we denote by $O(t) = \{p \in P : (t, p) \in A\}$ the set of output places of t.

A marking of a Petri net graph is a mapping $x : P \to \mathbb{N}$. A state is represented by $x = [x(p_1), x(p_2), \ldots x(p_n)]$, where $p_1, p_2, \ldots, p_n$ is an arbitrary fixed enumeration of P and n is the number of elements of P. A Petri net is a pair $(\mathcal{N}, x_0)$, where $\mathcal{N}$ is Petri net graph and x_0 is the initial state. The state space of $(\mathcal{N}, x_0)$ is given by $X = \mathbb{N}^n$ and $x_0 \in X$. The state transition function $f : X \times T \to X$ of a Petri net $(\mathcal{N}, x_0)$ is defined for state $x \in X$ and transition $t \in T$ if $x(p) \geq w(p, t)$ for all $p \in I(t)$. That is, a transition t can fire from x if and only if t is feasible from x and when t fires, $f(x, t)$ gives the resulting state. If $f(x, t)$ is defined, then we set $x' = f(x, t)$ where

$$x'(p) = x(p) - w(p, t) + w(t, p), \text{ for all } p \in P. \qquad (2)$$

---

[1] The notation and terminology used in this paper mostly follow those in [2,19].

Not all the states in X are reachable in $(\mathcal{N}, x_0)$. In order to define the set of reachable states, denoted by $R(\mathcal{N}, x_0)$, of Petri net $(\mathcal{N}, x_0)$, we first extend the state transition function f from domain $X \times T$ to domain $X \times T^*$:

$$f(x, \varepsilon) := x \tag{3}$$
$$f(x, st) := f(f(x, s), t) \text{ for } s \in T^* \text{ and } t \in T \tag{4}$$

where ε is to be interpreted as the absence of transition firing and T^* denotes the Kleene-closure of T. The set of reachable states of Petri net $(\mathcal{N}, x_0)$ is

$$R(\mathcal{N}, x_0) := \{x' \in X : \exists s \in T^* \text{ such that } f(x_0, s) = x'\} \tag{5}$$

The system to be diagnosed is modeled by a labeled Petri net

$$(\mathcal{N}, \Sigma, l, x_0) \tag{6}$$

where Σ is the set of event labels for the transitions in T, $l : T \to \Sigma$ is the transition labeling function, and x_0 is the initial state. The event labeling function l is extended to $l : T^* \to \Sigma^*$ in the following manner: given $t, t' \in T$ and $a, a' \in \Sigma$,

$$l(t) = a \text{ and } l(t') = a' \Rightarrow l(tt') = l(t)l(t') = aa'. \tag{7}$$

The language generated by the labeled Petri net $(\mathcal{N}, \Sigma, l, x_0)$, denoted by $\mathcal{L}(\mathcal{N}, \Sigma, l, x_0)$, is the set of all traces of events that can be generated by $(\mathcal{N}, \Sigma, l, x_0)$ from its initial state x_0. $\mathcal{L}(\mathcal{N}, \Sigma, l, x_0)$ is formally defined as

$$\mathcal{L}(\mathcal{N}, \Sigma, l, x_0) = \{l(s) \in \Sigma^* : s \in T^* \text{ and } f(x_0, s) \text{ is defined}\}. \tag{8}$$

Some of the events in Σ are observable, i.e., their occurrence can be observed (detected by sensors), and while the other events are unobservable. Thus Σ is partitioned into observable and unobservable event sets: $\Sigma = \Sigma_o \cup \Sigma_{uo}$. The observable events in the system may be commands issued by the controller, sensor readings, and changes of sensor readings. On the other hand, unobservable events may be fault events and some events that cause changes in the system state that are not recorded by sensors.

We model faults as events. The set of fault events Σ_f is a subset of Σ. Since it is trivial to diagnose fault events that are observable, we assume $\Sigma_f \subseteq \Sigma_{uo}$. Our goal is to detect the occurrence of fault events, if any, from the observable of traces of events generated by the system.

We partition the set of fault events into disjoint sets where each disjoint set corresponds to a different fault type. The motivation for doing so is that it might not be necessary to detect uniquely every fault event, but only the occurrence of one among a subset (type) of fault events. We write

$$\Sigma_f = \Sigma_{F1} \dot{\cup} \cdots \dot{\cup} \Sigma_{Fk} \tag{9}$$

where Σ_{Fi} denotes the set of fault events corresponding to a type i fault, $1 \leq i \leq k$, where k is the number of fault types. When we write "a fault of type i has occurred", we mean that a fault event from the set Σ_{Fi} has occurred.

2.2   Petri Net Diagnoser

We now introduce the diagnoser. The diagnoser is a labeled Petri net built from the system model $(\mathcal{N}, \Sigma, l, x_0)$. This labeled Petri net performs diagnostics while observing on-line the behavior of $(\mathcal{N}, \Sigma, l, x_0)$.

The diagnoser for $(\mathcal{N}, \Sigma, l, x_0)$ is

$$\mathcal{N}_d = (\mathcal{N}, \Sigma, l, x_{d0}, \Delta_f) \tag{10}$$

where $\mathcal{N}, \Sigma, l$ are defined as before, x_{d0} is the initial diagnoser state and $\Delta_f = \{F_1, F_2, ..., F_k\}$ is the finite set of fault types. The diagnoser Petri net $\mathcal{N}_d$ keeps the graphical structure of the underlying system model. Up to this point $\mathcal{N}_d$ is not different from a labeled Petri net. However, its dynamics are different from those of a labeled Petri net since its state transition function is only defined for observable events.

The diagnoser gives the estimate of the current state of the system after the occurrence of an observable event. Hereafter when we say "state", we mean the state of the system model and when we say "diagnoser state", we mean the state of the diagnoser. The diagnoser state is a list of the set of states the system model can be in after observation of an event in Σ_o together with fault information. Fault information in a diagnoser state is coded by fault labels.

Every state in a diagnoser state has a fault label. A fault label is a vector of length k (the number of fault types) which has entries of "0" or "1". If we denote the fault label by l_f, then $l_f \in \Delta = \{0,1\}^k$. Thus, the number of possible fault labels is $|\Delta| = 2^k$. When the fault label is the zero vector, we say the fault label is "normal". The initial state has the "normal" fault label by definition.

We now define the fault label propagation function $LP : X \times \Delta \times T^* \to \Delta$. LP propagates the fault labels consistent with the traces of events. Let $x \in X$, $l_f \in \Delta$ and $s \in T^*$. Then $LP(x, l_f, s)$ is defined as

$$LP(x, l_f, s) = l_f + \sum_{i=1}^{k} b_i^s \tag{11}$$

where $b_i^s \in \Delta$ and

$$b_i^s = \begin{cases} [0, \cdots, 0, \underset{\uparrow i^{th}\ coloumn}{1}, 0, \cdots, 0], & \text{if } l(s) \text{ contains an event from } \Sigma_{Fi}, \\ [0, \cdots, 0, 0, 0, \cdots, 0], & \text{otherwise.} \end{cases} \tag{12}$$

Before we define the diagnoser state and the diagnoser state transition function, we need the notion of unobservable reach of a diagnoser state. To define the unobservable reach of a diagnoser state we first define the unobservable reach of a state.

Let $x_{d,i} = x_i l_i$ denote a state with fault label l_i in the diagnoser state x_d. The unobservable reach of $x_{d,i}$ is denoted by $UR(x_{d,i})$ and defined as follows:

$$UR(x_{d,i}) := \{x_i l_i\} \cup \{y l_y : \exists y \in R(\mathcal{N}, x_i), \exists s \in T^*, l(s) \in \Sigma_{uo}^*$$
$$(f(x_i, s) = y) \text{ and } (l_y = LP(x_i, l_i, s))\}. \tag{13}$$

The unobservable reach of the diagnoser state x_d is the listing of all distinct vectors $UR(x_{d,i})$ for all $x_{d,i}$ in x_d; it is denoted by $UR(x_d)$.

We can now define the initial diagnoser state x_{d0} of $\mathcal{N}_d$. x_{do} is the listing of all distinct elements of $UR(x_0 l_0)$, where x_0 is the initial state of the underlying labeled Petri net $(\mathcal{N}, \Sigma, l, x_0)$ and l_0 is the fault label of x_0 that is "normal" by definition.

We find the diagnoser states reachable from the initial diagnoser state by using the diagnoser state transition function. In order to define the diagnoser state transition function, we first define the feasible transitions and then the states reached by firing the feasible transitions.

We denote by $B(x_{d,i}, a)$ the feasible transitions from $x_{d,i} = x_i l_i$ where $x_i \in X$ is a state in diagnoser state x_d, l_i is the fault label of x_i, and a is an event in Σ_o. Formally, $B(x_{d,i}, a)$ is defined as

$$B(x_{d,i}, a) = \{t \in T : \ l(t) = a \text{ and for all } p \in I(t) \ (x_i(p) \geq w(p, t))\}. \quad (14)$$

We define $B(x_d, a)$ to be the set resulting from the union of $B(x_{d,i}, a)$ for all i, where $1 \leq i \leq r$, i.e.,

$$B(x_d, a) = \cup_{1 \leq i \leq r} B(x_{d,i}, a), \quad (15)$$

where r is the number of rows of x_d.

We denote by $S(x_d, a)$ the set of all distinct reachable states, together with their fault labels, when all transitions in $B(x_{d,i}, a)$ are fired from every $x_{d,i}$ in x_d. Namely, $S(x_d, a)$ is defined as

$$S(x_d, a) := \cup_{1 \leq i \leq r} \cup_{t \in B(x_{d,i}, a)} \{x_i' l_i' : \ x_i' = f(x_i, t), \ l_i' = l_i\} \quad (16)$$

where x_i is a state in x_d, l_i is the fault label of x_i, and r is the number of states in x_d. Since the fault events are unobservable, the label propagation function does not change the fault labels of the states in the process of building $S(x_d, a)$.

The diagnoser state transition function of $\mathcal{N}_d$ is $f_d : X_d \times \Sigma_o \rightarrow X_d$ where X_d is the state space of the diagnoser $\mathcal{N}_d$. Given the diagnoser state x_d and the event $a \in \Sigma_o$, $f_d(x_d, a)$ is defined if $B(x_{d,i}, a) \neq \emptyset$ for some $x_{d,i}$ in x_d. If $f_d(x_d, a)$ is defined, then $x_d' = f_d(x_d, a)$ and x_d' is the listing of the elements of the set

$$\cup_{s \in S(x_d, a)} UR(s). \quad (17)$$

The diagnostic information provided by a diagnoser state is given by examining the last k columns of that state: *(i)* if a column contains only 0's, then we know that no fault event of the corresponding type could have occurred; *(ii)* if a column contains only 1's, then we are sure that at least one fault event of that type has occurred; *(iii)* otherwise, if a column contains 0's and 1's, we are uncertain about the occurrence of a fault of that type. If the diagnoser is sure that a fault of type i has occurred, then it outputs "F_n" as indicated in Figure 1. This diagnostic infortmation is equivalent to that obtained from diagnoser automata in the Diagnoser Approach of [4].

Example 1. Consider the Petri net graph $\mathcal{N}$ given in Fig. 2. The set of places of $\mathcal{N}$ is $P = \{p_1, p_2, \ldots, p_{16}\}$. The set of transitions of $\mathcal{N}$ is $T = \{t_1, t_2, \ldots, t_{17}\}$. All arc weights are equal to 1. The initial marking x_0 is

$$x_0 = \begin{bmatrix} 1 & 1 & 1 & 0 & 0 & 0 & 0 & 0 & 0 & 0 & 0 & 0 & 0 & 0 & 0 & 0 \end{bmatrix} \tag{18}$$

The set of events is $\Sigma = \{a, e, g, h, \sigma_{uo}, f_1, f_2\}$. We do not explicitly write the event labeling function l, but the event label of every transition $t \in T$ is shown in Fig. 2. The set of unobservable events is $\Sigma_{uo} = \{\sigma_{uo}, f_1, f_2\}$ and all the remaining events in event set Σ are observable. There are two types of faults and the sets corresponding them are $\Sigma_{f1} = \{f_1\}$ and $\Sigma_{f2} = \{f_2\}$.

Let $\mathcal{N}_d = (\mathcal{N}, \Sigma, l, x^0, \Delta_f)$ denote the diagnoser. The initial diagnoser state x^0 is the listing of the elements of set $UR(x_0 l_0)$ where l_0 is "normal". Then, x^0 is found as

$$x^0 = \begin{bmatrix} 1 & 1 & 1 & 0 & 0 & 0 & 0 & 0 & 0 & 0 & 0 & 0 & 0 & 0 & 0 & 0 & |0\ 0 \\ 0 & 1 & 1 & 1 & 0 & 0 & 0 & 0 & 0 & 0 & 0 & 0 & 0 & 0 & 0 & 0 & |0\ 0 \\ 0 & 0 & 1 & 0 & 1 & 0 & 0 & 0 & 0 & 0 & 0 & 0 & 0 & 0 & 0 & 0 & |1\ 0 \\ 0 & 1 & 1 & 0 & 0 & 0 & 1 & 0 & 0 & 0 & 0 & 0 & 0 & 0 & 0 & 0 & |0\ 0 \end{bmatrix} \begin{matrix} \bullet \\ \blacksquare \\ \triangle \\ * \end{matrix} \tag{19}$$

These four states in the initial diagnoser state are shown in $\mathcal{N}_d$ in Fig. 2 by using four types of tokens.

We show in (20)-(22), the states of the diagnoser that are reached if the sequence of observable events is "*aeh*". An examination of the last two columns of x^1, x^2 and x^3 reveals that: *(i)* x^1 and x^2 are F_1-uncertain ($f1$ could have happened but we do not know for sure) and *(ii)* x^2 and x^3 are F_2-uncertain. The complete state space of $\mathcal{N}_d$ contains 28 diagnoser states. These states are not listed here due to space constraints. We note that we wrote a MATLAB program to generate the state space of Petri net diagnosers.

$$x^1 = f_d(x^0, a) = \begin{bmatrix} 1 & 0 & 0 & 0 & 0 & 1 & 0 & 0 & 0 & 0 & 0 & 0 & 0 & 0 & 0 & 0 & |0\ 0 \\ 0 & 0 & 0 & 1 & 0 & 1 & 0 & 0 & 0 & 0 & 0 & 0 & 0 & 0 & 0 & 0 & |0\ 0 \\ 0 & 1 & 1 & 0 & 0 & 0 & 0 & 1 & 0 & 0 & 0 & 0 & 0 & 0 & 0 & 0 & |0\ 0 \\ 0 & 1 & 1 & 0 & 0 & 0 & 0 & 0 & 1 & 0 & 0 & 0 & 0 & 0 & 0 & 0 & |1\ 0 \\ 0 & 0 & 0 & 0 & 0 & 1 & 1 & 0 & 0 & 0 & 0 & 0 & 0 & 0 & 0 & 0 & |0\ 0 \\ 0 & 1 & 1 & 0 & 0 & 0 & 0 & 0 & 0 & 1 & 0 & 0 & 0 & 0 & 0 & 0 & |0\ 0 \end{bmatrix} \tag{20}$$

$$x^2 = f_d(x^1, e) = \begin{bmatrix} 1 & 0 & 0 & 0 & 0 & 0 & 0 & 0 & 1 & 0 & 0 & 0 & 0 & 0 & 0 & 0 & |0\ 0 \\ 0 & 0 & 0 & 1 & 0 & 0 & 0 & 0 & 1 & 0 & 0 & 0 & 0 & 0 & 0 & 0 & |0\ 0 \\ 0 & 1 & 1 & 0 & 0 & 0 & 0 & 0 & 0 & 1 & 0 & 0 & 0 & 0 & 0 & 0 & |0\ 0 \\ 0 & 1 & 1 & 0 & 0 & 0 & 0 & 0 & 0 & 0 & 1 & 0 & 0 & 0 & 0 & 0 & |1\ 0 \\ 0 & 0 & 0 & 0 & 0 & 1 & 0 & 0 & 1 & 0 & 0 & 0 & 0 & 0 & 0 & 0 & |0\ 0 \\ 0 & 1 & 1 & 0 & 0 & 0 & 0 & 0 & 0 & 0 & 1 & 0 & 0 & 0 & 0 & 0 & |0\ 0 \\ 1 & 0 & 0 & 0 & 0 & 0 & 0 & 0 & 0 & 0 & 0 & 0 & 0 & 1 & 0 & 0 & |0\ 1 \\ 0 & 0 & 0 & 1 & 0 & 0 & 0 & 0 & 0 & 0 & 0 & 0 & 0 & 1 & 0 & 0 & |0\ 1 \\ 0 & 1 & 1 & 0 & 0 & 0 & 0 & 0 & 0 & 0 & 0 & 0 & 0 & 0 & 1 & 0 & |0\ 1 \\ 0 & 0 & 0 & 0 & 0 & 1 & 0 & 0 & 0 & 0 & 0 & 0 & 1 & 0 & 0 & 0 & |0\ 1 \end{bmatrix} \tag{21}$$

$$x^3 = f_d(x^2, h) = \begin{bmatrix} 0\ 1\ 1\ 0\ 0\ 0\ 0\ 0\ 0\ 0\ 0\ 0\ 0\ 0\ 1\ 0\ |0\ 0 \\ 0\ 1\ 1\ 0\ 0\ 0\ 0\ 0\ 0\ 0\ 0\ 0\ 0\ 0\ 0\ 1\ |0\ 1 \end{bmatrix} \qquad (22)$$

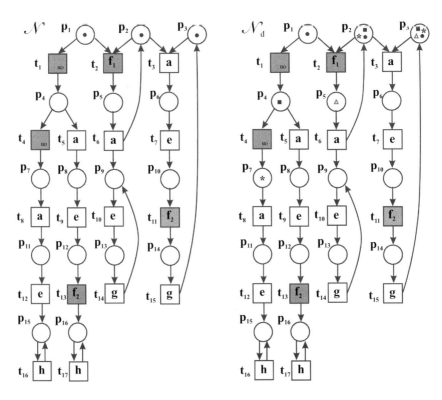

Fig. 2. The Petri net graph $\mathcal{N}$ with initial marking x_0 is given on the left, and its Petri net diagnoser $\mathcal{N}_d$ with the initial diagnoser state x^0 is given on the right

3   Distributed Diagnosis with Communication

In this section, we study the problem of distributed diagnosis with communication. In the case of centralized diagnosis, there is one diagnoser working on the entire system model and processing all the observations. We are interested in the situation depicted in Fig. 3 where there are two diagnosers, each containing only part of the model and each observing only a subset of the observed events. We allow these two diagnosers to communicate after the occurrence of an observable event.

We begin this section by defining the distributed system model and the distributed diagnosers. In the second part, we define the communication protocol

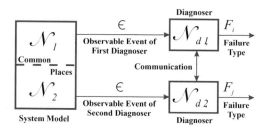

Fig. 3. Distributed diagnosis with communication

and give the algorithm of distributed diagnosis with communication at the end of the subsection. Then, we state the main result of this paper and give its proof.

3.1   System Model

The system to be diagnosed is given by the labeled Petri net (N, Σ, l, x_0). We wish to partition (N, Σ, l, x_0) into two "place-bordered" labeled Petri nets using a partition on the event set Σ. For this purpose, let Π_Σ be a partition on Σ such that $\Sigma = \Sigma_1 \cup \Sigma_2$ and $\Sigma_1 \cap \Sigma_2 = \emptyset$. Then, we define the place-bordered labeled Petri nets $(N_1, \Sigma_1, l_1, x_{01})$ and $(N_2, \Sigma_2, l_2, x_{02})$ where $N_1 = \langle P_1, T_1, A_1, w_1 \rangle$ and $N_2 = \langle P_2, T_2, A_2, w_2 \rangle$ by the following conditions

1. $\forall t \in T$ if $l(t) \in \Sigma_1$, then $t \in T_1$; $\forall t \in T$ if $l(t) \in \Sigma_2$, then $t \in T_2$.
2. $P_1 = \cup_{t \in T_1} (I(t) \cup O(t))$, $P_2 = \cup_{t \in T_2} (I(t) \cup O(t))$.

The corresponding Petri net graphs N_1 and N_2 have disjoint sets of transitions. However, the sets of places are not disjoint, i.e., there may exist t_1 and t_2 such that $O(t_1) \cap I(t_2) \neq \emptyset$ or $I(t_1) \cap O(t_2) \neq \emptyset$. A_1 and A_2 are the restrictions of A to $(P_1 \times T_1) \cup (T_1 \times P_1)$ and $(P_2 \times T_2) \cup (T_2 \times P_2)$, respectively. Similarly for w_1, w_2 and l_1, l_2. x_0 can be written as

$$x_0 = [x_{0,P_1-P_c}, x_{0,P_c}, x_{0,P_2-P_c}] \qquad (23)$$

where $P_c = P_1 \cap P_2$ denotes the set of common places, x_{0,P_1-P_c} denotes the columns of markings corresponding to the places in the set $(P_1 - P_c)$[we use $-$ to denote the set difference]; x_{0,P_c} and x_{0,P_2-P_c} are defined similarly. Then, x_{01} is defined as

$$x_{01} = [x_{0,P_1-P_c}, x_{0,P_c}], \qquad (24)$$

and x_{02} is defined as

$$x_{02} = [x_{0,P_c}, x_{0,P_2-P_c}]. \qquad (25)$$

Every labeled Petri net (N, Σ, l, x_0) can be partitioned into place-bordered nets $(N_1, \Sigma_1, l_1, x_{01})$ and $(N_2, \Sigma_2, l_2, x_{02})$ using the method described above.

But not all partitions allow us to do distributed diagnosis. Suppose $\mathcal{N}_{d,1} = (\mathcal{N}_1, \Sigma_1, l_1, x_{d,01}, \Delta_{f,1})$ and $\mathcal{N}_{d,2} = (\mathcal{N}_2, \Sigma_2, l_2, x_{d,02}, \Delta_{f,2})$ are diagnosers for $(\mathcal{N}_1, \Sigma_1, l_1, x_{01})$ and $(\mathcal{N}_2, \Sigma_2, l_2, x_{02})$, respectively. The sets $\Delta_{f,1}$ and $\Delta_{f,2}$ are the finite sets of fault types of $\mathcal{N}_{d,1}$ and $\mathcal{N}_{d,2}$, respectively, and $\Delta_{f,1} \cup \Delta_{f,2} = \Delta_f$. Then, $\mathcal{N}_{d,1}$ and $\mathcal{N}_{d,2}$ must satisfy the two following conditions to perform distributed diagnosis:

1. $\forall t \in T$ if $(I(t) \cup O(t)) \cap (P_1 \cap P_2) \neq \emptyset$, then $l(t) \in \Sigma_o$.
2. $\forall t_1 \in T_1$ and $\forall t_2 \in T_2$, if $l(t_1) \in \Sigma_{Fi}$ and $l(t_2) \in \Sigma_{Fj}$, then $i \neq j$.

The first condition says that the transitions putting tokens in or removing tokens from common places are labeled with observable events. The second condition ensures that if two transitions belong to different place-bordered nets then they belong to different types of faults, i.e, $\Delta_{f,1} \cap \Delta_{f,2} = \emptyset$. This assumption is made for the sake of simplicity; it could be relaxed at the price of extra communication between the diagnosers.

The initial diagnoser states $x_{d,01}$ and $x_{d,02}$ are listings of the elements of the sets $UR(x_{01} l_f^{01})$ and $UR(x_{02} l_f^{02})$, respectively. The fault label l_f^{01} is in Δ_1 where Δ_1 is the set of possible fault types in $\mathcal{N}_{d,1}$, and similarly the fault label l_f^{02} is in Δ_2 where Δ_2 is the set of possible fault types in $\mathcal{N}_{d,2}$.

3.2   Algorithm of Distributed Diagnosis with Communication

$\mathcal{N}_{d,1}$ and $\mathcal{N}_{d,2}$ diagnose $(\mathcal{N}_1, \Sigma_1, l_1, x_{01})$ and $(\mathcal{N}_2, \Sigma_2, l_2, x_{02})$, respectively. However, the individual estimates of $\mathcal{N}_{d,1}$ and $\mathcal{N}_{d,2}$ do not provide enough information for diagnosis, if $\mathcal{N}_{d,1}$ and $\mathcal{N}_{d,1}$ work in isolation. This is because $\mathcal{N}_1$ and $\mathcal{N}_2$ do not have disjoint sets of places and both nets can change the markings on the common places and then effect each other. If $\mathcal{N}_{d,1}$ and $\mathcal{N}_{d,2}$ are not informed of each others' changes of markings, then their state estimates are incomplete or otherwise wrong. We overcome this problem by defining a communication protocol between diagnosers. This protocol recovers the centralized diagnosis information by allowing the two diagnosers to send each other the change of markings on the common places.

We define the weighting vector $W(t)$ for a labeled Petri net graph as

$$W(t) = [w(t, p_1) - w(p_1, t), \; w(t, p_2) - w(p_2, t), \; \ldots, w(t, p_{|P|}) - w(p_{|P|}, t)] \tag{26}$$

where $t \in T$ and $p_i \in P$ for all $1 \leq i \leq |P|$. When we write $W_{P_c}(t)$, this means the columns of $W(t)$ corresponding to the common places of $\mathcal{N}_1$ and $\mathcal{N}_2$. We use the same notation in the states and the diagnoser states. That is, as was done for the initial state in the previous section if $x \in X$, then x_{P_1} denotes the columns of x corresponding to the places of $\mathcal{N}_1$ and if $x_d \in X_d$, then x_{d,P_1} denotes the columns of x_d corresponding to the places of $\mathcal{N}_1$.

In contrast with centralized diagnoser states, distributed diagnoser states carry message labels. Message labels record the actions on the common places. Let l_m be the message label of $x \in X$ and $t = t_1 t_2 \ldots t_{|t|} \in T^*$ be a string of transitions. If $x' = f(x,t)$, i.e., $x' = f(\cdots f(f(x,t_1),t_2),t_{|t|})$, then the message label propagation function MLP defines the message label l'_m of x' as

$$l'_m = MLP(x, l_m, t) = [l_m,\ W_{P_c}(t_1),\ W_{P_c}(t_2),\ \ldots, W_{P_c}(t_{|t|})]. \qquad (27)$$

The length of the message label l'_m is bounded by $(|l_m| + |t||P_c|)$.

The message label of a diagnoser state is the listing of the message labels for every state in the diagnoser state, i.e., every row of the diagnoser state. Given a diagnoser state x_d, we denote by $MLabel(x_d)$ the message label of x_d. Suppose that x_d is reached from x'_d by firing transition t labeled by observable event σ_o. If $W_{P_c}(t)$ is equal to the zero vector for every state reached in x_d, then the message label of x_d is equal to the message label of x'_d, i.e., $MLabel(x_d) = MLabel(x'_d)$.

As defined the length of the message labels will grow unboundedly as the number of observed events grows unboundedly. However, the message labels of diagnoser states can be truncated under the following conditions. Given the diagnoser state $x_d \in \mathbb{N}^{n \times k}$, let l^i_m denote the message label of the state in the ith row of x_d. Then for all i where $1 \le i \le n$ if the message labels are of the form $l^i_m = ab_i$, i.e., they have a common prefix a but different suffixes b_i), then these message labels are truncated to $l^i_m = b_i$.

Communication among the two diagnosers are triggered by the occurrence of observable events. When $\mathcal{N}_{d,1}$ observes event $\sigma_o \in \Sigma_{o,1}$, then $\mathcal{N}_{d,1}$ updates its diagnoser state and sends the message label of the resulting diagnoser state to $\mathcal{N}_{d,2}$. We assume that the message is correctly received by $\mathcal{N}_{d,2}$ without delay (or with delay that is less than the minimum interarrival time). Upon reception of the message, $\mathcal{N}_{d,2}$ uses the received message label to update its current diagnoser state. We will demonstrate that under this protocol, the centralized diagnoser state can be recovered from the diagnoser states of $\mathcal{N}_{d,1}$ and $\mathcal{N}_{d,2}$. We now formalize this protocol for distributed diagnosis with communication (DDC).

From now on when we denote a diagnoser state $x_{d,1}$ of the diagnoser $\mathcal{N}_{d,1}$, we will drop the subscript d and write x_1 instead of $x_{d,1}$. We use the same notation for the diagnoser states of diagnoser $\mathcal{N}_{d,2}$.

Algorithm DDC. *Given that the sequence $s = \sigma_{o0}\sigma_{o1} \ldots \sigma_{on}$ is observed where $|s| = n + 1$, initialize the algorithm $i := 0$.*

*Upon observation of σ_{oi} do { If $\sigma_{oi} \in \Sigma_1$, then go to **1**, else go to **2** }*

1 *{Master is $\mathcal{N}_{d,1}$ }*
  1.1 *Find the next diagnoser state of $\mathcal{N}_{d,1}$:*

$$x_1^{i+1} = f_{d,1}(x_1^i, \sigma_{oi}),$$

  where $f_{d,1}$ is the diagnoser state transition function of $\mathcal{N}_{d,1}$ and x_1^1 is the diagnoser state of $\mathcal{N}_{d,1}$ after the completion of the DDC algorithm for the event $\sigma_{o(i-1)}$.
  1.2 *If $W_{P_c}(t) = \bar{0}$ for all $t \in B(x_1^i, \sigma_{oi})$, then equate x_2^{i+1} to x_2^i and go to **1.4**.*

1.3 *Send a "message" to $\mathcal{N}_{d,2}$:*

$$message := MLabel(x_1^{i+1})$$

Upon reception of this message, $\mathcal{N}_{d,2}$ "updates" x_2^i to x_2^{i+1} as follows: Initialize diagnoser state x_2^{i+1} to empty matrix. For k from 1 to $\bar{r}$ where $\bar{r}$ denotes the number of rows of "message", do the following

1.3.1 *Given*

$$message_k = [message\_prefix_k, message\_present_k],$$

where $message\_present_k$ is the last $|P_c|$ columns of the $message_k$ and $message\_prefix_k$ is the rest of it, extract the set M of states of x_2^i with message labels that are equal to $message\_prefix_k$.

1.3.2 *Given M, construct the set $\overline{M}$ as follows*

$$\bar{s}_{P_c} = s_{P_c} + message\_present_k, \ \ \bar{s}_{P_2-P_c} = s_{P_2-P_c},$$

$$l_m^{\bar{s}} = [l_m^s, message\_present_k]$$

where $s \in M$, $\bar{s} \in \overline{M}$ and l_m^s and $l_m^{\bar{s}}$ are the message labels of s and $\bar{s}$, respectively.

1.3.3 *Append every element of $\overline{M}$ to x_2^{i+1} as a new row.*

1.4 *If possible, truncate message labels of both x_1^{i+1} and x_2^{i+1}.*

1.5 *Increment i.*

2 *{Master is $\mathcal{N}_{d,2}$ } Same as **1** but exchange 1 and 2 in every expression.*

End

Example 2. We use the same labeled Petri net in *Example 1*. We choose an arbitrary event partition Π_Σ such that $\Sigma_1 = \{a, \sigma_{uo}, f_1\}$ and $\Sigma_2 = \{e, g, h, f_2\}$. Given this event partition the set of transitions of $\mathcal{N}_1$ and $\mathcal{N}_2$ are $T_1 = \{t_1, t_2, t_3, t_4, t_5, t_6, t_8\}$ and $T_2 = \{t_7, t_9, t_{10}, t_{11}, t_{12}, t_{13}, t_{14}, t_{15}, t_{16}, t_{17}\}$, respectively. The set of common places is $P_c = \{p_3, p_6, p_8, p_9, p_{11}\}$. The resulting place-bordered Petri net graphs are shown in Fig. 4. In Fig. 4, common places are shown with dashed lines. Place-bordered Petri nets $(\mathcal{N}_1, \Sigma_1, l_1, x_{01})$ and $(\mathcal{N}_2, \Sigma_2, l_2, x_{02})$ satisfy the two conditions to be eligible for distributed diagnosis. That is, all the transitions putting tokens in or removing tokens from common places in both Petri net graphs are labeled with observable events, and the sets of fault types of these nets are disjoint.

Suppose event sequence "*aeh*" is observed. The diagnoser states of $\mathcal{N}_{d,1}$ and $\mathcal{N}_{d,2}$ at end of each iteration are

$$x_1^0 = \begin{bmatrix} 1\ 1\ 1\ 0\ 0\ 0\ 0\ 0\ 0\ 0\ |0 \\ 0\ 1\ 1\ 1\ 0\ 0\ 0\ 0\ 0\ 0\ |0 \\ 0\ 0\ 1\ 0\ 1\ 0\ 0\ 0\ 0\ 0\ |1 \\ 0\ 1\ 1\ 0\ 0\ 0\ 1\ 0\ 0\ 0\ |1 \end{bmatrix}, \ x_2^0 = \begin{bmatrix} 1\ 0\ 0\ 0\ 0\ 0\ 0\ 0\ 0\ 0\ 0\ |0 \end{bmatrix} \quad (28)$$

$$x_1^1 = \begin{bmatrix} 1\ 0\ 0\ 0\ 0\ 1\ 0\ 0\ 0\ 0\ |0|\ -1\ 1\ 0\ 0\ 0 \\ 0\ 0\ 0\ 1\ 0\ 1\ 0\ 0\ 0\ 0\ |0|\ -1\ 1\ 0\ 0\ 0 \\ 0\ 1\ 1\ 0\ 0\ 0\ 0\ 1\ 0\ 0\ |0|\ \ \ 0\ 0\ 1\ 0\ 0 \\ 0\ 1\ 1\ 0\ 0\ 0\ 0\ 1\ 0\ |1|\ \ \ 0\ 0\ 0\ 1\ 0 \\ 0\ 0\ 0\ 0\ 0\ 1\ 1\ 0\ 0\ 0\ |1|\ -1\ 1\ 0\ 0\ 0 \\ 0\ 1\ 1\ 0\ 0\ 0\ 0\ 0\ 0\ 1\ |1|\ \ \ 0\ 0\ 0\ 0\ 1 \end{bmatrix}, \quad x_2^1 = \begin{bmatrix} 0\ 1\ 0\ 0\ 0\ 0\ 0\ 0\ 0\ 0\ |0|\ -1\ 1\ 0\ 0\ 0 \\ 1\ 0\ 1\ 0\ 0\ 0\ 0\ 0\ 0\ 0\ |0|\ \ \ 0\ 0\ 1\ 0\ 0 \\ 1\ 0\ 0\ 1\ 0\ 0\ 0\ 0\ 0\ 0\ |0|\ \ \ 0\ 0\ 0\ 1\ 0 \\ 1\ 0\ 0\ 0\ 0\ 1\ 0\ 0\ 0\ 0\ |0|\ \ \ 0\ 0\ 0\ 0\ 1 \end{bmatrix}$$

$$(29)$$

$$x_1^2 = \begin{bmatrix} 1\ 0\ 0\ 0\ 0\ 0\ 0\ 0\ 0\ 0\ |0|\ -1\ 1\ 0\ 0\ 0\ 0\ -1\ \ \ 0\ \ \ 0\ \ \ 0 \\ 0\ 0\ 0\ 1\ 0\ 0\ 0\ 0\ 0\ 0\ |0|\ -1\ 1\ 0\ 0\ 0\ 0\ -1\ \ \ 0\ \ \ 0\ \ \ 0 \\ 0\ 0\ 0\ 0\ 0\ 0\ 1\ 0\ 0\ 0\ |1|\ -1\ 1\ 0\ 0\ 0\ 0\ -1\ \ \ 0\ \ \ 0\ \ \ 0 \\ 0\ 1\ 1\ 0\ 0\ 0\ 0\ 0\ 0\ 0\ |0|\ \ \ 0\ 0\ 1\ 0\ 0\ 0\ \ \ 0\ -1\ \ \ 0\ \ \ 0 \\ 0\ 1\ 1\ 0\ 0\ 0\ 0\ 0\ 0\ 0\ |1|\ \ \ 0\ 0\ 0\ 1\ 0\ 0\ \ \ 0\ \ \ 0\ -1\ \ \ 0 \\ 0\ 1\ 1\ 0\ 0\ 0\ 0\ 0\ 0\ 0\ |1|\ \ \ 0\ 0\ 0\ 0\ 1\ 0\ \ \ 0\ \ \ 0\ \ \ 0\ -1 \end{bmatrix},$$

$$(30)$$

$$x_2^2 = \begin{bmatrix} 0\ 0\ 0\ 0\ 1\ 0\ 0\ 0\ 0\ 0\ |0|\ -1\ 1\ 0\ 0\ 0\ 0\ -1\ \ \ 0\ \ \ 0\ \ \ 0 \\ 1\ 0\ 0\ 0\ 0\ 0\ 1\ 0\ 0\ 0\ |0|\ \ \ 0\ 0\ 1\ 0\ 0\ 0\ \ \ 0\ -1\ \ \ 0\ \ \ 0 \\ 1\ 0\ 0\ 0\ 0\ 0\ 1\ 0\ 0\ 0\ |0|\ \ \ 0\ 0\ 0\ 1\ 0\ 0\ \ \ 0\ \ \ 0\ -1\ \ \ 0 \\ 1\ 0\ 0\ 0\ 0\ 0\ 0\ 0\ 1\ 0\ |0|\ \ \ 0\ 0\ 0\ 0\ 1\ 0\ \ \ 0\ \ \ 0\ \ \ 0\ -1 \\ 0\ 0\ 0\ 0\ 0\ 0\ 0\ 1\ 0\ 0\ |1|\ -1\ 1\ 0\ 0\ 0\ 0\ -1\ \ \ 0\ \ \ 0\ \ \ 0 \\ 1\ 0\ 0\ 0\ 0\ 0\ 0\ 0\ 0\ 1\ |1|\ \ \ 0\ 0\ 1\ 0\ 0\ 0\ \ \ 0\ -1\ \ \ 0\ \ \ 0 \end{bmatrix}.$$

$$(31)$$

$$x_1^3 = x_1^2, \quad x_2^3 = \begin{bmatrix} 1\ 0\ 0\ 0\ 0\ 0\ 0\ 0\ 0\ 1\ 0\ |0|\ 0\ 0\ 0\ 0\ 1\ 0\ 0\ \ \ 0\ 0\ -1 \\ 1\ 0\ 0\ 0\ 0\ 0\ 0\ 0\ 0\ 1\ |1|\ 0\ 0\ 1\ 0\ 0\ 0\ 0\ -1\ 0\ \ \ 0 \end{bmatrix}.$$

$$(32)$$

The above diagnoser states of $\mathcal{N}_{d,1}$ and $\mathcal{N}_{d,2}$ were found by a MATLAB program that implements **Algorithm DDC**.

3.3 Recovering the Centralized Diagnoser State from the Distributed Diagnoser States

In this section we show how the centralized diagnoser state can be recovered under the communication protocol described in **Algorithm DDC** in the previous section. We verify the correctness of the recovery method by showing that it reconstructs the centralized diagnoser state after each observable event in the given observed sequence of events.

An iteration of **Algorithm DDC** is the completion of the algorithm for an observable event in the sequence. Let x and $\overline{x}$ be the diagnoser states of $\mathcal{N}_{d,1}$ and $\mathcal{N}_{d,2}$, respectively, at the end of an iteration. We denote by x_i the ith row of diagnoser state x. Similarly, we denote by $\overline{x}_j$ the jth row of diagnoser state $\overline{x}$. Let $l_m^{x_i}$ and $l_m^{\overline{x}_j}$ denote the message labels of x_i and $\overline{x}_j$, and $l_f^{x_i}$ and $l_f^{\overline{x}_j}$ denote the fault labels of x_i and $\overline{x}_j$, respectively. We denote by d_i and $\overline{d}_j$ the states of

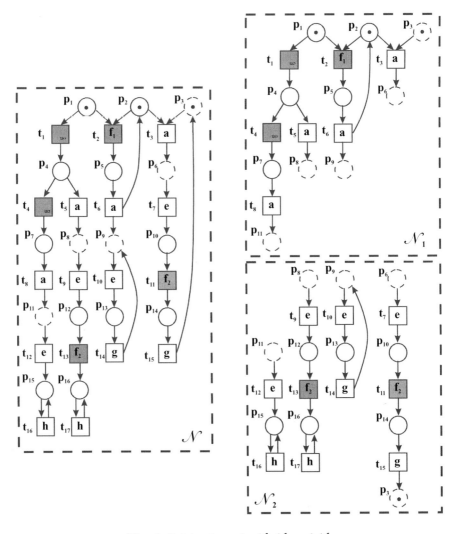

Fig. 4. Petri net graphs $\mathcal{N}$, $\mathcal{N}_1$ and $\mathcal{N}_2$

$(\mathcal{N}_1, \Sigma_1, l_1, x_{01})$ and $(\mathcal{N}_2, \Sigma_2, l_2, x_{02})$ in rows x_i and $\overline{x}_j$, respectively. Combining all these notations we write $x_i = d_i l_f^{x_i} l_m^{x_i}$ and $\overline{x}_j = \overline{d}_j l_f^{\overline{x}_j} l_m^{\overline{x}_j}$. If the number of rows of x and $\overline{x}$ are r and $\overline{r}$, respectively, then we define the set $Merge(x, \overline{x})$ as follows

$$Merge(x, \overline{x}) = \cup_{1 \le i \le r} \cup_{1 \le j \le \overline{r}} \{ [d_i \overline{d}_{j, P_2 - P_c} | l_f^{x_i} l_f^{\overline{x}_j}] : l_m^{x_i} = l_m^{\overline{x}_j} \}. \qquad (33)$$

Algorithm DDC results in $x_{i, P_c} = \overline{x}_{j, P_c}$ when $l_m^{x_i} = l_m^{\overline{x}_j}$. Therefore $Merge(x, \overline{x})$ can equivalently be defined as follows

$$Merge(x, \overline{x}) = \cup_{1 \le i \le r} \cup_{1 \le j \le \overline{r}} \{ [d_{i, P_1 - P_c} \overline{d}_j | l_f^{x_i} l_f^{\overline{x}_j}] : l_m^{x_i} = l_m^{\overline{x}_j} \}. \qquad (34)$$

We state the main result of this paper in the following theorem. The theorem claims that the centralized diagnoser state can be recovered from the diagnoser states of the distributed diagnosers by merging these diagnoser states as defined in (33) or (34).

Theorem 1. *Given the system $(\mathcal{N}, \Sigma, l, x_0)$, its diagnoser $\mathcal{N}_d$, and the event partition Π_Σ, let $(\mathcal{N}_1, \Sigma_1, l_1, x_{01})$ and $(\mathcal{N}_2, \Sigma_2, l_2, x_{02})$ be the resulting place-bordered Petri nets, and $\mathcal{N}_{d,1}$ and $\mathcal{N}_{d,2}$ be the corresponding diagnosers, respectively. Given a sequence of observable events $\sigma_{o0}\sigma_{o1}\ldots\sigma_{on}$ such that $x^{i+1} = f_d(x^i, \sigma_{oi})$ for all i where $0 \le i \le n$ and x^i is the diagnoser state of $\mathcal{N}_d$, if x^i_1 and x^i_2 are the diagnoser states of $\mathcal{N}_{d,1}$ and $\mathcal{N}_{d,2}$, respectively, at the end of the iteration of **Algorithm DDC** for σ_{oi}, then the set of rows of x^i is equal to $Merge(x^i_1, x^i_2)$.*

Proof (of Theorem). The proof is by induction. We first show that merging the initial diagnoser states of the distributed diagnosers results in the initial diagnoser state of the centralized diagnoser. The induction hypothesis states that merging the diagnoser states x^i_1 and x^i_2 of the distributed diagnosers results in the centralized diagnoser state x^i. In the induction step we prove that merging the diagnoser states x^{i+1}_1 and x^{i+1}_2 of the distributed diagnosers results in the centralized diagnoser state x^{i+1} using the induction hypothesis.
Induction Base. $Merge(x^0_1, x^0_2)$ is equal to the set of rows of x^0.

Proof (of Induction Base). The initial diagnoser state x^0 is the listing of the elements in the set $UR(x_0 l_0)$ where l_0 is "normal". If (s^s_f) is a row of x^0 then

$$s = x_0 + W^*(\bar{t}), \tag{35}$$
$$l^s_f = LP(x_0, l_0, \bar{t}), \tag{36}$$

where $\bar{t} = \bar{t}_1 \bar{t}_2 \ldots \bar{t}_{|\bar{t}|} \in \Sigma^*_{uo}$ and $W^*(\bar{t}) = \sum_{m=1}^{|\bar{t}|} W(\bar{t}_m)$.
Since addition is column-wise (36) can be separated into two equations as follows

$$s_{P_1} = x_{0,P_1} + W^*_{P_1}(\bar{t}) \tag{37}$$
$$s_{P_2} = x_{0,P_2} + W^*_{P_2}(\bar{t}) \tag{38}$$

From (24) and (25), $x_{0,P_1} = x_{01}$ and $x_{0,P_2} = x_{02}$. Thus $(s_{P_1} l_f s_{P_1})$ and $(s_{P_2} l^{s_{P_2}}_f)$ are rows in $UR(x_{01} l^{x_{01}}_f)$ and $UR(x_{02} l^{x_{02}}_f)$, respectively. By definition x^0_1 and x^0_2 are the listings of the elements in the sets $UR(x_{01} l^{x_{01}}_f)$ and $UR(x_{02} l^{x_{02}}_f)$, respectively. Thus, $(s_{P_1} l_f s_{P_1})$ and $(s_{P_2} l^{s_{P_2}}_f)$ are rows of x^0_1 and x^0_2, respectively. Conversely if $(s' l^{s'}_f)$ and $(s'' l^{s''}_f)$ are rows of x^0_1 and x^0_2, respectively, then $(s' s''_{P_2 - P_c} l^{s'}_f l^{s''}_f)$ is a row of x^0.

Induction Hypothesis. $Merge(x^i_1, x^i_2)$ is equal to the set of rows of x^i.
Induction Step. $Merge(x^{i+1}_1, x^{i+1}_2)$ is equal to the set of rows of x^{i+1}.

Proof (of Induction Step). We need to show that $Merge(x_1^{i+1}, x_2^{i+1})$ is equal to the set of rows of x^{i+1}. This is done by showing inclusion in both directions for these two sets of rows. Without loss of generality assume that $\sigma_{oi} \in \Sigma_1$.

($\Leftarrow$) If (sl_f^s) is a row of x^{i+1}, then there exist a row $(s'l_f^{s'})$ of x_1^{i+1} and a row $(s''l_f^{s''})$ of x_2^{i+1} such that $l_m^{s'} = l_m^{s''}$, $s = s's''_{P_2-P_c} = s'_{P_1-P_c}s''$ and $l_f^s = l_f^{s'}l_f^{s''}$, where l_f^s, $l_f^{s'}$ and $l_f s''$ are the fault labels of s, s' and s'', respectively, and $l_m^{s'}$ and $l_m^{s''}$ are the message labels of s' and s'', respectively.

If (sl_f^s) is a row of x^{i+1}, then $(sl_f^s) \in \cup_{z \in |S(x^i, \sigma_{oi})|} UR(z)$ [cf. (17)]. Thus, $(sl_f^s) \in S(x^i, \sigma_{oi})$ or there exist $(vl_f^v) \in S(x^i, \sigma_{oi})$ and $\bar{t} = \bar{t}_1 \bar{t}_2 \ldots \bar{t}_{|\bar{t}|} \in \Sigma_{uo}^*$ such that

$$s = v + W^*(\bar{t}) \quad \text{and} \quad l_f^s = LP(v, l_f^v, \bar{t}), \tag{39}$$

where $W^*(\bar{t}) = \sum_{m=1}^{|\bar{t}|} W(\bar{t}_m)$. The first case can be thought of as a special case of the second. When s is equal to v, $W^*(\bar{t}) = \bar{0}$. In this case, $l_f^s = l_f^v$. Then $(sl_f^s) \in S(x^i, \sigma_{oi})$ since $(sl_f^s) = (vl_f^v)$ and $(vl_f^v) \in S(x^i, \sigma_{oi})$. Thus, we will write the proof for the second case only.

If $(vl_f^v) \in S(x^i, \sigma_{oi})$, then there exists a row (dl_f^d) of x^i such that there exists a transition $t \in B(x^i, \sigma_{oi})$[cf. (15)] feasible from d, and

$$v = d + W(t) \quad \text{and} \quad l_f^v = l_f^d. \tag{40}$$

From the induction hypothesis, $d = d_{P_1} d_{P_2-P_c}$ and $l_f^d = l_f^{d_{P_1}} l_f^{d_{P_2}}$ where $(d_{P_1} l_f^{d_{P_1}})$ is a row of x_1^i and $(d_{P_2} l_f^{d_{P_2}})$ is a row of x_2^i, and $l_m^{d_{P_1}} = l_m^{d_{P_2}}$.

Since addition of vectors is column-wise, from (40) we get

$$v_{P_1} = d_{P_1} + W_{P_1}(t) \tag{41}$$

Note that since $l(t) \in \Sigma_1$, $t \in T_1$. Thus, t is feasible from d_{P_1} in $(\mathcal{N}_1, \Sigma_1, l_1, x_1^0)$. Then, $(v_{P_1} l_f^{v_{P_1}}) \in S(x_1^i, \sigma_{oi})$.

Consider (39); since addition of vectors is column-wise we get

$$s_{P_1} = v_{P_1} + W_{P_1}^*(\bar{t}) \tag{42}$$

$W_{P_c}(\bar{t}_m) = \bar{0}$ for all $\bar{t}_m$ in $\bar{t}$. Thus, $W_{P_c}^*(\bar{t}) = \bar{0}$. Then, $(s_{P_1} l_f^{s_{P_1}})$ is a row of x_1^{i+1} since $(s_{P_1} l_f^{s_{P_1}}) \in UR(v_{P_1} l_f^{v_{P_1}})$ and $(v_{P_1} l_f^{v_{P_1}}) \in S(x_1^i, \sigma_{oi})$. $(s_{P_1} l_f^{s_{P_1}}) \in UR(v_{P_1} l_f^{v_{P_1}})$ since if there exists some $\bar{t}_m$ in $\bar{t}$ where $1 \leq m \leq$ such that $l(\bar{t}) \in \Sigma_{uo,2}$, then $W_{P_1}(\bar{t}_m) = \bar{0}$. The message label $l_m^{s_{P_1}}$ of $(s_{P_1} l_f^{s_{P_1}})$ is given as

$$l_m^{s_{P_1}} = l_m^{v_{P_1}} = [l_m^{d_{P_1}}, \ W_{P_c}(t)] \tag{43}$$

if for some $t \in B(x_1^i, \sigma_{oi})$, $W_{P_c}(t) \neq \bar{0}$, i.e., a message is sent. Otherwise, $l_m^{s_{P_1}} = l_m^{d_{P_1}}$, i.e. no message is sent. Note that the transitions labeled with unobservable events do not change the markings of the common places.

We now show that $s_{P_2} l_f^{s_{P_2}}$ is a row of x_2^{i+1} and its message label $l_m^{s_{P_2}}$ is equal to the message label of s_{P_1}. Similar to (41) and (42) we can write

$$v_{P_2} = d_{P_2} + W_{P_2}(t) \tag{44}$$

and

$$s_{P_2} = v_{P_2} + W_{P_2}^*(\bar{t}) \tag{45}$$

If no message is sent, then $W_{P_c}(t) = \bar{0}$ for every $t \in B(x_1^i, \sigma_{oi})$. $W_{P_2 - P_c}(t) = \bar{0}$ since $t \in \Sigma_1$. Thus $W_{P_2}(t) = \bar{0}$ and from (44) we get $v_{P_2} = d_{P_2}$. If we substitute this result into (45) we see that $s_{P_2} l_f^{s_{P_2}} \in UR(d_{P_2} l_f^{d_{P_2}})$. Since $d_{P_2} l_f^{d_{P_2}}$ is a row of x_2^i and by definition (17) each element of $UR(d_{P_2} l_f^{d_{P_2}})$ is a row of x_2^i, $s_{P_2} l_f^{s_{P_2}}$ is a row of x_2^i. By **Algorithm DDC**, when no message is sent, $x_2^{i+1} = x_2^i$. Thus, $s_{P_2} l_f^{s_{P_2}}$ is a row of x_2^{i+1}. The message label of s_{P_2} is equal to $l_m^{d_{P_2}}$. Since $l_m^{d_{P_1}} = l_m^{d_{P_2}}$, the message label $l_m^{s_{P_2}} = l_m^{d_{P_1}}$. Since no message is sent, $l_m^{s_{P_1}} = l_m^{d_{P_1}}$ and $l_m^{s_{P_1}} = l_m^{s_{P_2}}$.

If a message is sent, then for some $t \in B(x_1^i, \sigma_{oi})$ $W_{P_c}(t) \neq \bar{0}$. If we substitute (44) into (45) we get

$$s_{P_2} = d_{P_2} + W_{P_2}(t) + W_{P_2}^*(\bar{t}) \tag{46}$$

Since $t \in T_1$, $W_{P_2 - P_c}(t) = \bar{0}$. It was also shown that $W_{P_c}^*(\bar{t}) = \bar{0}$. Then, (46) can be rewritten as

$$s_{P_2} = \underbrace{d_{P_2} + [\bar{0},\ W_{P_2 - P_c}^*(\bar{t})]}_{\bar{d}} + [W_{P_c}(t),\ \bar{0}] \tag{47}$$

Let $l_f^{\bar{d}}$ denote the fault label of $\bar{d}$. Then, $\bar{d} l_f^{\bar{d}} \in UR(d_{P_2})$ and the message label $l_m^{\bar{d}}$ of $\bar{d}$ is equal to the message label $l_m^{d_{P_2}}$ of d_{P_2}. Since $l_m^{d_{P_1}} = l_m^{d_{P_2}}$, then $l_m^{\bar{d}} = l_m^{d_{P_1}}$.

By **Algorithm DDC**, the message sent is $MLabel(x_1^{i+1})$. Since $s_{P_1} l_f^{s_{P_1}}$ is a row of x_1^{i+1}. There exists row k of $MLabel(x_1^{i+1})$ which is equal to $l_m^{s_{P_1}}$. Let $message_k$ denote row k of this message; then from (43) we extract that $message\_prefix_k = l_m^{d_{P_1}}$ and $message\_present_k = W_{P_c}(t)$. Thus, we get $message\_prefix_k = l_m^{\bar{d}}$. This results in $\bar{d} \in M$ in **Algorithm DDC**. If $\bar{d} \in M$, then there exists $\tilde{d} \in \overline{M}$ such that

$$\tilde{d}_{P_c} = \bar{d}_{P_c} + message\_present_k = \bar{d}_{P_c} + W_{P_c}(t),\ \tilde{d}_{P_2 - P_c} = \bar{d}_{P_2 - P_c} \tag{48}$$

and

$$l_m^{\tilde{d}} = [l_m^{\bar{d}},\ W_{P_c}(t)]. \tag{49}$$

From (47) and (48) we find that $\tilde{d} = s_{P_2}$ and $l_m^{\tilde{d}} = l_m^{s_{P_2}}$. From (43) and (49) we find that $l_m^{s_{P_1}} = l_m^{\tilde{d}}$. Thus, $l_m^{s_{P_1}} = l_m^{s_{P_2}}$.

As a result we showed that given that (sl_f^s) is a row of x^{i+1}, $(s_{P_1}l_f^{s_{P_1}})$ and $(s_{P_2}l_f^{s_{P_2}})$ are rows of x_1^{i+1} and x_2^{i+1}, respectively, and $l_m^{s_{P_1}} = l_m^{s_{P_2}}$. Thus, defining $s' = s_{P_1}$, $l_f^{s'} = l_f^{s_{P_1}}$, $s'' = s_{P_2}$ and $l_f^{s''} = l_f^{s_{P_2}}$ concludes the proof of one direction of inclusion.

($\Rightarrow$) If $(s'l_f^{s'})$ and $(s''l_f^{s''})$ are rows of x_1^{i+1} and x_2^{i+1}, respectively, and $l_m^{s'} = l_m^{s''}$, then there exists a row (sl_f^s) of x^{i+1} such that $s = s's''_{P_2 - P_c} = s'_{P_1 - P_c}s''$ and $l_f^s = l_f^{s'}l_fs''$, where l_f^s, $l_f^{s'}$ and $l_f^{s''}$ are the fault labels of s, s' and s'', respectively, and $l_m^{s'}$ and $l_m^{s''}$ are the message labels of s' and s'', respectively.

The proof of the above statement is similar to the proof of the converse statement proved in detail above, when the steps are followed in reverse order. First we find the rows of x_1^i and x_2^i from which $(s'l_f^{s'})$ and $(s''l_f^{s''})$ are reached, and using the induction hypothesis we show that the merging of these rows forms a row of x^i. Then we find the row in x^{i+1} when the transitions in $B(x_1^i, \sigma_{oi})$ or Σ_{uo} are fired, and show that it is the merging of $(s'l_f^{s'})$ and $(s''l_f^{s''})$. The details of this proof are omitted here.

This completes the proof of the induction step and the proof of the theorem.

$\square$

Example 3. We consider *Example 1* and *Example 2* since they give the centralized and distributed diagnoser states, respectively, for the event sequence "*aeh*".

Note that merging x_1^0 and x_2^0 in (28) results in x^0 given in (19). Similarly, merging x_1^1 and x_2^1 in (29) results in x^1 given in (20); merging x_1^2 and x_2^2 in (30) and (31), respectively, results in x^2 given in (21); merging x_1^3 and x_2^3 in (32) results in x^3 given in (22). Observe that x_1^3 in (32) contains state estimates for $\mathcal{N}_1$ that are not present in the centralized diagnoser state given in (22). This overestimation is due to the use of a partial system model, namely, $\mathcal{N}_1$ by $\mathcal{N}_{d,1}$. However, these overestimates disappear during the merge operation.

Remark: Each time the merge operation is invoked, we could send each diagnoser their part of the merged diagnoser state and they could use these parts as their new initial states. This resetting of the initial states allows the reset of the message labels as well, thus preventing their unbounded growth.

4   Conclusion

The Petri net diagnosers introduced in this work are different from the diagnoser automata in [4] in the sense that they perform on-line fault diagnosis on the same transition structure as the system model, namely the Petri net graph of the system. This feature can be exploited to allow for modular/distributed implementations of diagnosis algorithms, as was done in Section 3 based on a decomposition of the Petri net graph of the system into place-bordered subnets. This kind of modular decomposition often occurs naturally in the modelling of complex systems; it would be more difficult to achieve using automata models.

In the future work, it might be worthwhile to investigate other types of decomposition of Petri nets.

For the sake of generality, our presentation of Algorithm DDC focused on the main steps involved and on its correctness proof. Several improvements to it are possible in order to achieve more efficient implementations from the point of view of the communications required between the diagnosers. As was mentioned in Section 3, message labels can, and should, be truncated when all the rows in a message label share a common prefix. It may be possible to determine upper bounds on the size of message labels based on the structure of the Petri net. Another possible improvement is to attempt to reduce the frequency of communications. For instance, if the connectivity between the place-bordered subnets is "one-way", in the sense that one subnet only consumes tokens from common places but never puts tokens in them, and vice-versa for the other, then communication need only be one-way and in fact could possibly be delayed. In general, it may be possible to delay communications if the diagnosers use timestamps and their local clocks are synchronized. Detailed investigations of such improvements constitute interesting topics for future research.

References

1. Pouliezos, A.D., Stavrakakis, G.S.: Real time fault monitoring of industrial processes. Kluwer Academic Publishers (1994)
2. Cassandras, C.G., Lafortune, S.: Introduction to Discrete Event Systems. Kluwer Academic Publishers (1999)
3. Lafortune, S., Teneketzis, D., Sampath, M., Sengupta, R., Sinnamohideen, K.: Failure diagnosis of dynamic systems: An approach based on discrete event systems. In: Proc. 2001 American Control Conf. (2001) 2058–2071
4. Sampath, M., Sengupta, R., Lafortune, S., Sinnamohideen, K., Teneketzis, D.: Diagnosability of discrete event systems. IEEE Trans. Automatic Control 40 (1995) 1555–1575
5. Sampath, M., Sengupta, R., Lafortune, S., Sinnamohideen, K., Teneketzis, D.: Failure diagnosis using discrete event models. IEEE Trans. Control Systems Technology 4 (1996) 105–124
6. Sampath, M.: Discrete event systems based diagnostics for a variable air volume terminal box application. Technical report, Advanced Development Team, Johnson Controls, Inc. (1995)
7. Şimşek, H.T., Sengupta, R., Yovine, S., Eskafi, F.: Fault diagnosis for intra-platoon communication. In: Proc. 38th IEEE Conf. on Decision and Control. (1999)
8. Sengupta, R.: Discrete-event diagnostics of automated vehicles and highways. In: Proc. 2001 American Control Conf. (2001)
9. Sampath, M., Godambe, A., Jackson, E., Mallow, E.: Combining qualitative and quantitative reasoning - a hybrid approach to failure diagnosis of industrial systems. In: IFAC SafeProcess 2000. (2000) 494–501
10. Sampath, M.: A hybrid approach to failure diagnosis of industrial systems. In: Proc. 2001 American Control Conf. (2001)

11. García, E., Morant, F., Blasco-Giménez, R., Quiles, E.: Centralized modular diagnosis and the phenomenon of coupling. In Silva, M., Giua, A., Colom, J., eds.: Proceedings of the 6th International Workshop on Discrete Event Systems, IEEE Computer Society (2002) 161–168

12. Chehaibar, G.: Replacements of Open Interface Subnets and Stable State Transformation Equivalance, Springer-Verlag (1993) 1–25

13. Vogler, W.: Modular Construction and Partial Order Semantics of Petri Nets (Lecture Notes in Computer Science, vol. 625). Springer-Verlag (1998)

14. Aghasaryan, A., Fabre, E., Benveniste, A., Boubour, R., Jard, C.: Fault detection and diagnosis in distributed systems: An approach by partially stochastic petri nets. Journal of Discrete Event Dynamical Systems Vol. 8(2) (1998) 203–231

15. Benveniste, A., Fabre, E., Jard, C., Haar, S.: Diagnosis of asynchronous discrete event systems, a net unfolding approach. Technical Report Research Report 1456, Irisa (2002)

16. Hadjicostis, C.N., Verghese, G.C.: Monitoring Discrete Event Systems Using Petri Net Embeddings. Application and Theory of Petri Nets 1999 (Series Lecture Notes in Computer Science, vol. 1639) (1999) 188–207

17. Sifakis, J.: Realization of fault-tolerant systems by coding petri nets. Journal of Design Automation and Fault-Tolerant Computing Vol. 3 (1979) 93–107

18. Giua, A.: Petri net state estimators based on event observation. IEEE 36th Int. Conf. on Decision and Control (1997) 4086–4091

19. Desel, J., Esparza, J.: Free Choice Petri Nets. Cambridge University Press (1995)

Soundness and Separability of Workflow Nets in the Stepwise Refinement Approach

Kees van Hee, Natalia Sidorova, and Marc Voorhoeve

Eindhoven University of Technology
Department of Mathematics and Computer Science
P.O. Box 513, 5600 MB Eindhoven, The Netherlands
{k.m.v.hee,n.sidorova}@tue.nl, wsinmarc@win.tue.nl

Abstract. Workflow nets are recognized as a modelling paradigm for the business process modelling. We introduce and investigate several correctness notions for workflow nets, ranging from proper termination of cases to their mutual independence. We define refinement operators for nets and investigate preservation of correctness through these operators. This gives rise to a class of nets that are provably correct.

Keywords: Petri nets; workflow; modelling; verification; correctness; soundness; separability; serialisability.

1 Introduction

Petri nets are frequently used to model and analyse workflow processes in business process design (c.f. [1,3]). The nets used in this area are appropriately called workflow nets (WF-nets). In software engineering, the same WF-nets can be used for modelling the life cycles of objects. A case (transaction, object) starts as a token in the initial place of the WF-net and after a series of steps this token evolves into a marking consisting possibly of several tokens. An important property is *proper completion*: from such a marking, it must be possible to reach the final marking of one token in the final place. This property is called *soundness*, c.f. [1,4]. Soundness can be verified by more or less standard Petri net algorithms (e.g. coverability analysis).

In [4], it is argued that soundness alone is not compositional w.r.t. refinement; it is possible to refine a transition in a sound net with another sound net and obtain a non-sound result (see e.g. Figure 1 in Section 3). For this reason, soundness is considered in that work for free choice, safe and well-structured nets, and compositionality is proven for each of these classes.

In this paper we propose and investigate a generalization of the notion of soundness. We say that a workflow net is k-sound if any marking reached from k tokens in the initial place can reach the same k tokens in the final place. The original soundness becomes 1-soundness, and we propose to call workflow nets sound iff they are k-sound for each $k > 0$. A practical advantage of the new notion of soundness is introducing a possibility to avoid "earmarking" tokens

W.M.P. van der Aalst and E. Best (Eds.): ICATPN 2003, LNCS 2679, pp. 337–356, 2003.

to distinguish several cases processed in the net. Imagine processing n orders in the net. If the workflow net is 1-sound but not sound, every order has to be earmarked by adding a unique id-colour, thus guaranteeing a treatment of the order in isolation. If the net is sound, one can assure a proper completion of the task with k orders without introducing id's.

We show that the new notion of soundness is compositional w.r.t. refinement. Next, we prove several bisimilarity results, which allows to carry over temporal properties of nets to their refinements when the refinement is given by a sound (in the new sense) net. Unlike 1-soundness, no apparent verification algorithm for soundness exists, though we prove some classes of WF-nets to be sound.

Soundness is of course not the only correctness criterion. Analysis of a model can be done e.g. by proving temporal requirements specified in a temporal logic. It would be interesting to find a class of nets whose properties are the same for the WF-nets with removed earmarkings as for the original nets. The concept of serialisability in transaction processing [5] is based on the property that cases are independent of each other: the presence or absence of other cases does not influence the options for a specific case. This leads us to a similar concept of *serialisability* for WF-nets: the property that the set of traces of the WF-net with id-markings is equal to the set of traces of its abstraction. We show that state machines and cycle-free marked graphs are serialisable. On the negative side is the fact that serialisability is not a congruence w.r.t. refinement.

An attempt to soften the requirements results in a notion of *weak separability*: every marking reachable from the initial state with k tokens is representable as a sum of k markings each of which is reachable from a single initial token. Every serialisable net is clearly weakly separable. We show that weak separability together with 1-soundness imply soundness. Weak separability is a congruence w.r.t. place refinement, but not a congruence w.r.t. transition refinement. Looking for a compositional notion of separability, we come to a definition that is similar to serialisability, however, it does not require the trace equivalence between the net with id-tokens and its abstraction, but the equivalence of sets of Parikh vectors. One additional requirement turns this notion to the compositional one, that we call *split-separability*.

For business applications, weak separability is important because it formalizes the idea of independent cases: each marking is the sum of the markings of the individual cases and therefore all properties of the markings of a batch of cases are "cumulated properties" of the individual cases. The additional property of separability says that also the firings of a batch of cases is in fact the sum of the firings of the individual cases. If we associate to each firing the consumption of some resource, like money or energy, then separability implies that the consumption of the batch of cases equals the sum of the individual consumptions.

We prove state machines and cycle-free marked graphs to be sound, serialisable and split-separable. Combined with refinement, this fact gives rise to a class of WF-nets, that we call ST-nets, which are sound and split-separable by construction.

The rest of the paper is organized as follows. In Section 2, we sketch the basic definitions related to Petri nets and WF-nets. In Sections 3, we formulate the new notion of soundness and give weak bisimilarity results for sound refinements. In Section 4, we introduce and analyse the notions of soundness and separability. Afterwards, in Section 5 we define a class of separable by construction ST-nets and give a factorisation algorithm to invert refinement. We conclude in Section 6 with discussing the obtained results and directions for the future work.

2   Preliminaries

Let S be a set. A bag (multiset) m over S is a function $m : S \to \mathbb{N}$. We use $+$ and $-$ for the sum and the difference of two bags and $=, <, >, \leq, \geq$ for comparisons of bags, which are defined in a standard way, and overload the set notation, writing $\emptyset$ for the empty bag and $\in$ for the element inclusion. We list elements of bags between brackets, e.g. $m = [p^2, q]$ for a bag m with $m(p) = 2$, $m(q) = 1$, and $m(x) = 0$ for all $x \notin \{p, q\}$. The shorthand notation $k.m$ is used to denote the sum of k bags m.

For sequences of elements over a set T we use the following notation: The empty sequence is denoted with λ; a non-empty sequence can be given by listing its elements between angle brackets. The Parikh vector $\vec{\sigma} : T \longrightarrow \mathbb{N}$ of a sequence σ maps every element $t \in T$ to the number of occurrences of t in σ. $\vec{\sigma}(t)$ stands for the number of occurrences of t in σ. A concatenation of sequences σ_1, σ_2 is denoted with $\sigma_1\sigma_2$; $t\sigma$ and σt stand for the concatenation of t and sequence σ and vice versa. A projection of a sequence σ on elements of a set U (i.e. eliminating the elements from $T \setminus U$) is denoted as $\pi_U(\sigma)$. The shuffle $\sigma\|\gamma$ of two sequences is the set of sequences obtained by interleaving the elements of σ and γ; formally we have $\lambda\|\sigma = \sigma\|\lambda = \sigma$ and $a\sigma\|b\gamma = \{ax \mid x \in \sigma\|b\gamma\} \cup \{by \mid y \in a\sigma\|\gamma\}$.

Transition Systems. A *transition system* is a tuple $E = \langle S, Act, T \rangle$ where S is a set of *states*, Act is a finite set of *action names* and $T \subseteq S \times Act \times S$ is a *transition relation*. A *process* is a pair $\langle E, s_0 \rangle$ where E is a transition system and $s_0 \in S$ an initial state.

We denote (s_1, a, s_2) from T as $s_1 \xrightarrow{a} s_2$, and we say that a leads from s_1 to s_2. For a sequence of transitions $\sigma = \langle t_1, \dots, t_n \rangle$ we write $s_1 \xrightarrow{\sigma} s_2$ when $s_1 = s^0 \xrightarrow{t_1} s^1 \xrightarrow{t_2} \dots \xrightarrow{t_n} s^n = s_2$, and $s_1 \xrightarrow{\sigma}$ when $s_1 \xrightarrow{\sigma} s_2$ for some s_2. In this case we say that σ is a trace of E. Finally, $s_1 \xrightarrow{*} s_2$ means that there exists a sequence of transitions $\sigma \in T^*$ such that $s_1 \xrightarrow{\sigma} s_2$. We use action label τ to denote silent actions and write $s_1 \Longrightarrow s_2$ when $s_1 = s_2$ or $s_1 \xrightarrow{\tau} \dots \xrightarrow{\tau} s_2$. We write $s_1 \stackrel{a}{\Longrightarrow} s_2$ if $s_1 \Longrightarrow s_1' \xrightarrow{a} s_2' \Longrightarrow s_2$. To indicate that the step a is taken in the transition system E we write $s \xrightarrow{a}_E s'$, $s \stackrel{a}{\Longrightarrow}_E s'$ resp.

The *strong trace set* $ST(E, s_0)$ of a process $\langle E, s_0 \rangle$ is defined as $\{\sigma \in Act^* \mid s_0 \xrightarrow{\sigma}\}$. Two processes are strongly trace equivalent iff their strong trace sets are equal. $ST(E_1) = ST(E_2)$. The *weak trace set* $T(E, s_0)$ is defined as $\{\sigma \in$

$(Act \setminus \{\tau\})^* \mid s_0 \stackrel{\sigma}{\Longrightarrow}\}$. Two processes are weakly trace equivalent iff their weak trace sets are equal.

Bisimulation. Given two systems $N_1 = \langle S_1, Act, T_1 \rangle$ and $N_2 = \langle S_2, Act, T_2 \rangle$. A relation $R \subseteq S_1 \times S_2$ is a *simulation* iff for all $s_1 \in S_1, s_2 \in S_2$, $s_1 R s_2$ and $s_1 \stackrel{a}{\longrightarrow} s_1'$ implies that there exists a transition $s_2 \stackrel{a}{\longrightarrow} s_2'$ such that $s_1' R s_2'$. Relation R is a *bisimulation* [11] if R and R^{-1} are simulations.

Weak (bi)simulation is defined by copying the definitions for plain (bi)simulation and replacing $\stackrel{a}{\longrightarrow}$ by $\stackrel{a}{\Longrightarrow}$ throughout. Two processes $\langle E, s \rangle$ and $\langle F, r \rangle$ are called *(weakly) bisimilar* iff there exists a (weak) bisimulation R such that $s R r$. We often add the adjective "strong" to non-weak simulation relations. Strong and weak bisimilarity are equivalence relations. [1]

Petri nets. A *labelled Petri net* is a tuple $N = \langle S_N, T_N, F_N, l_N \rangle$, where:

- S_N and T_N are two disjoint non-empty finite sets of *places* and *transitions* respectively, the set $S_N \cup T_N$ are the *nodes* of N;
- F_N is a mapping $(S_N \times T_N) \cup (T_N \times S_N) \to \mathbb{N}$ which we call a *flow function*;
- $l_N : T_N \to Act$ labels each transition $t \in T_N$ with some action $l_N(t)$ from Act.

We assume that Act contains all transitions of all nets to be encountered. Unless stated otherwise, we assume that the labeling function maps a transition onto itself. If the identity function is not used, some transitions are labelled with the silent action τ.

We drop the N subscript whenever no ambiguity can arise and present nets with the usual graphical notation. A *path* of a net is a sequence $\langle x_1, \ldots, x_n \rangle$ of nodes such that $\forall i : 1 \leq i \leq n - 1 : F(x_i, x_{i+1}) > 0$.

Markings are states (configurations) of a net. We consider a *marking m* of N as a bag over S and denote the set of all markings reachable in net N from marking m as $\mathcal{M}(N, m)$.

Given a transition $t \in T$, the *preset* ${}^\bullet t$ and the *postset* $t^\bullet$ of t are the bags of places where every $p \in S$ occurs in ${}^\bullet t$ $F(p, t)$ times and in $t^\bullet$ $F(t, p)$ times. Analogously we write ${}^\bullet p, p^\bullet$ for pre- and postsets of places. To emphasize the fact that the preset/postset is considered within some net N, we write ${}^\bullet_N a, a^\bullet_N$. We overload this notation further allowing to apply preset and postset operations to a bag B of places/transitions, which is defined as the weighted sum of pre-/postsets of elements of B.

A transition $t \in T$ is *enabled* in marking m iff ${}^\bullet t \leq m$. An enabled transition t may fire, thus performing action $l(t)$. This results in a new marking m' defined by $m' \stackrel{\text{def}}{=} m - {}^\bullet t + t^\bullet$. For a firing sequence γ in a net N, we define ${}^\bullet_N \gamma$ and

---

[1] All systems proved to be weakly bisimilar in this paper are in fact branching bisimilar. This follows from Theorem 3.1 in [9]: a weak bisimulation where one of the related systems is τ-free is a branching bisimulation.

$\gamma_N^\bullet$ respectively as $\sum_{t \in \gamma} {}^\bullet_N t$ and $\sum_{t \in \gamma} t^\bullet_N$, which are the sums of all tokens consumed/produced during the firings of γ. So $m \xrightarrow{\gamma}_N (m + \gamma_N^\bullet - {}^\bullet_N\gamma)$.

We interpret a Petri net N as a transition system/process where markings play the role of states, firings of the enabled transitions define the transition relation and the initial marking corresponds to the initial state. The notions of reachability, traces, simulation and bisimulation, etc. for Petri nets are inherited from the transition systems. When $m_N \, R \, m_M$ for some markings m_N, m_M and bisimulation R we say that (N, m_N) and (M, m_M) are bisimilar, written $(N, m_N) \sim (M, m_M)$.

Workflow Petri nets. In this paper we primarily focus upon the *Workflow Petri nets (WF-nets)* [1]. As the name suggests, WF-nets are used to model the ordering of tasks in workflow processes. The initial and final nodes indicate respectively the initial and final states of cases flowing through the process.

Definition 1. *A Petri net N is a* Workflow net (WF-net) *iff:*

- *N has two special places (or transitions): i and f. Place (transition resp.) i is an initial place (transition): ${}^\bullet i = \emptyset$, and f is a final place (transition): $f^\bullet = \emptyset$.*
- *For any node $n \in (S \cup T)$ there exists a path from i to n and a path from n to f.*

We will call a WF-net sWF-net or tWF-net to indicate whether a WF-net has places or transitions as initial and final nodes. A tWF-net can be extended with an additional initial place and a terminal place up to an sWF-net.

3   Refinement and Soundness of Workflow Nets

When constructing models, the concept of refinement is very natural. A single task on a higher level can become a sequence of subtasks also involving choice and parallelism, i.e. it can be refined to a tWF-net. Similarly, being at some location (place of the net) resources (tokens) can undergo a number of operations, which can be reflected with a substitution of this place with an sWF-net. To build composed nets from WF-net components we will use two simple operations: Given two WF-nets L, M.

- *Place refinement* of a place $p \in S_L$ with sWF-net M yields a WF-net $N = L \otimes_p M$, built as follows: $p \in S_L$ is replaced in L by M; transitions from ${}^\bullet p$ become input transitions of the initial place of M and transitions from $p^\bullet$ become output transitions of the final place of M.
- *Transition refinement* of a transition $t \in T_L$ with tWF-net M yields a WF-net $N = L \otimes_t M$, built as follows: $t \in T_L$ is replaced by M; places from ${}^\bullet t$ become input places of the initial transition of M and places from $t^\bullet$ become output places of the final transition of M.

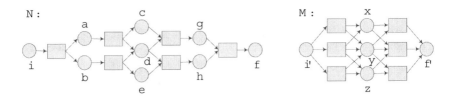

Fig. 1. Refining 1-sound nets

We consider transition and place refinements as basic techniques of our component-oriented design methodology. Note that the refinement of the initial and final places are legitimate operations, and are in fact sequential compositions of nets.

The refinement operators satisfy the following trivial equations.

Lemma 2. *Let A, B, C be WF-nets, $a, c \in S_A \cup T_A$, $c \neq a$ and $b \in S_B \cup T_B$. If $A \otimes_a B$, $B \otimes_b C$ and $A \otimes_c C$ are defined, then $(A \otimes_a B) \otimes_b C$ and $A \otimes_a (B \otimes_b C)$ are defined and equal as well as $(A \otimes_a B) \otimes_c C$ and $(A \otimes_c C) \otimes_a B$ are.*

Soundness of WF-nets. A natural requirement for WF-nets is that in any case the modelled process should be able to reach the end state, no matter what happens to it. This requirement has been called *soundness* by [1,3]. That formulation of soundness does not combine with refinement, though. In Figure 1, WF-nets N, M are depicted that are sound according to the standard definition of soundness. However, the net $L = N \otimes_d M$ is not sound: $[i] \xrightarrow{*}_L [c, e, i'^2] \xrightarrow{*}_L [c, e, y^2, f'] \xrightarrow{*}_L [c, y^2, h]$. From this last state, no successor state can be reached: it is a deadlock containing nonterminal nodes. The reason is that net M terminates properly when started from $[i]$ but not from $[i^2]$.

This example shows that we need a stronger notion of soundness that would require a correct outcome of the WF-net work for initial markings with an arbitrary number of tokens in the initial place. For this reason, we generalize the soundness notion; the original soundness becomes 1-soundness according to the new definition.

Definition 3. *An sWF-net $N = \langle S, T, F, l \rangle$ with initial and final places i and f resp. is k-sound for $k \in \mathbb{N}$ iff $[f^k]$ is reachable from all markings m from $\mathcal{M}(N, [i^k])$.*
A tWF-net N with initial and final transitions t_i, t_f respectively is k-sound iff the sWF-net formed by adding to S_N places p_i, p_f with $\bullet p_i = \emptyset, p_i^\bullet = [t_i], \bullet p_f = [t_f], p_f^\bullet = \emptyset$ is k-sound.
A WF-net is sound iff it is k-sound for every natural k.

The following simple property holds for sound WF-nets. This property will be used several times in the proofs that follow.

Fig. 2. Bisimilar Petri nets which are not WF-bisimilar nets

Lemma 4. *Let N be a sound WF-net and m its marking reached from some initial marking $[i^k]$. Then m can be represented as $m = [i^l] + m_1 + [f^n]$ where l, n are naturals and m_1 is a marking over places from $S_N \setminus \{i, f\}$, and there exists one and only one natural number $j \equiv k - l - n$ such that $[i^j] \xrightarrow{*} m_1$. Moreover, $m_1 \xrightarrow{*} [f^j]$ for this and only this j.*

Bisimulation of WF-nets. The notion of bisimulation for WF-nets must include the requirement of proper initialisation/termination. Consider e.g. nets N and M given in Figure 2. They are bisimilar Petri nets, however, N is sound while M has a deadlock and is not sound. We want to be able to transfer the conclusion about the soundness of a WF-net to all WF-bisimilar nets, therefore we do not consider nets N and M as WF-bisimilar.

Definition 5. *Given relation $R \subseteq (\mathbb{N} \to S_N) \times (\mathbb{N} \to S_N)$ on markings of sWF-nets N and M. R is a WF-simulation iff R is a simulation and*

$$(\forall k, x : [i_N^k] \, R \, x : x = [i_M^k]) \text{ and } (\forall k, x : [f_N^k] \, R \, x : x = [f_M^k]).$$

R is a weak WF-simulation iff R is a weak simulation and

$$(\forall k, x : [i_N^k] \, R \, x : [i_M^k] \Longrightarrow x) \text{ and } (\forall k, x : [f_N^k] \, R \, x : x \Longrightarrow [f_M^k]).$$

A strong/weak WF-simulation between tWF-nets N, M is a relation that can be extended (by adding pairs of markings) to become a strong/weak WF-simulation between the sWF-nets $\overline{N}, \overline{M}$ obtained by adding initial and terminal places to N, M respectively.

R is a strong/weak WF-bisimulation iff R and R^{-1} are strong/weak WF-simulations. We will say that the WF-nets N and M are strongly/weakly WF-bisimilar iff there exists a strong/weak WF-bisimulation R between N and M such that $\forall k :: [i_N^k] \, R \, [i_M^k] \wedge [f_N^k] \, R \, [f_M^k]$.

It is easy to show that WF-bisimilarity is an equivalence relation. Moreover, the following property holds:

Lemma 6. *Let N, M be WF-bisimilar WF-nets and N is sound. Then M is sound as well.*

Soundness and bisimulation of refinements. We prove that refinement with sound nets yields weakly WF-bisimilar nets.

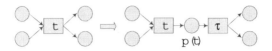

Fig. 3. Split refinement of transition t

Theorem 7. *Let M be a sound sWF-net with all transitions τ-labelled, L be a net with a place $p \in S_L$ and $N = L \otimes_p M$. Then L and N are weakly WF-bisimilar.*

Proof. Let $R = \{(m + [p^\ell], m + x) \mid m \in (S_L \setminus \{p\}) \to \mathbb{N} \wedge x \in \mathcal{M}(M, [i_M^\ell])$. We prove that this relation is a weak WF-bisimulation. Note that $[i_L^k] \, R \, [i_N^k]$ and $[f_L^k] \, R \, [f_N^k]$. Also, the "WF" requirement is satisfied, so it is sufficient to prove that (1) R is a weak simulation and (2) R^{-1} is a weak simulation.

(1): Suppose $(m + [p^\ell]) \, R \, (m + x)$ and $(m + [p^\ell]) \xrightarrow{a}_L (m' + [p^r])$ with $m' \in (S_L \setminus \{p\}) \to \mathbb{N}$. If the transition a does not affect p-tokens, we have $r = \ell$ and $(m + x) \xrightarrow{a}_N (m' + x)$, with $(m' + [p^r]) \, R \, (m' + x)$. Now suppose the transition consumes s and produces t p-tokens, so $s \le \ell$ and $r = \ell - s + t$. Since $x \in \mathcal{M}(M, [i_M^\ell])$, the soundness of M implies $x \xrightarrow{*}_M [f_M^\ell]$, and thus $x \xrightarrow{*}_N [f_M^\ell]$. Since the transitions of M are τ-labelled, we have $(m + x) \Rightarrow_N (m + [f_M^\ell]) \xrightarrow{a}_N (m' + [f_M^{\ell-s}] + [i_M^t])$, so $(m + x) \xRightarrow{a}_N (m' + y)$ with $y \in \mathcal{M}(M, [i_M^r])$, so $(m' + [p^r]) \, R \, (m' + y)$.

(2): Suppose $(m + [p^\ell]) \, R \, (m + x)$ and $(m + x) \xrightarrow{a}_N (m' + y)$. In case that $a \in T_M$, we have $m' = m$, $(m + x) \Rightarrow_N (m + y)$ and $y \in \mathcal{M}(M, [i_M^\ell])$, so $(m + [p^\ell]) \, R \, (m' + y)$. If $a \in T_N$, we set $F_L(a, p) = r$, $F_L(p, a) = s$. Thus by the construction of N, $x \ge [f_L^s]$ and $y = x - [f_L^s] + [i_L^r]$. Also, $(m + [p^\ell]) \xrightarrow{a}_L (m' + [p^{\ell-s+r}])$. Since M is sound and $x \in \mathcal{M}(M, [i_M^\ell])$, we have $y \in \mathcal{M}(M, [i_M^{\ell-s+r}])$, so $(m' + [p^{\ell-s+r}]) \, R \, (m' + y)$. □

We prove another bisimulation result for transition refinement with sound nets. First, we introduce a simple transition refinement —*split refinement* (see Fig. 3): transition t is replaced with the tWF-net Σ_t with places $\{p_t\}$ and transitions $\{i_t, f_t\}$ such that $^\bullet i_t = f_t^\bullet = \emptyset$, $i_t^\bullet = {}^\bullet f_t = [p_t]$. In this section, we suppose that i_t has label t and f_t has label τ.

Lemma 8. *Let N be a WF-net with $t \in T_N$ and $M = N \otimes_t \Sigma_t$. Then*

$$R \stackrel{def}{=} \{(m, m + [p_t^k] - k.t_N^\bullet) \mid m \in (S_N \to \mathbb{N}) \wedge m \ge k.t_N^\bullet\}$$

is a weak WF-bisimulation for N and M.

Proof. Consider some markings m, μ such that $m \, R \, \mu$, say $\mu = m + [p_t^k] - k.t_N^\bullet$. Suppose $m \xrightarrow{u}_N m'$. Then $\mu \Rightarrow_M m \xrightarrow{u}_M m'$, so $m \xRightarrow{u}_M m'$. Clearly $m' \, R \, m'$. Now suppose $\mu \xrightarrow{u}_M \mu'$. If the transition that fired is i_t, then $\mu' = \mu - {}^\bullet_M t + [p_t]$, so $m \ge {}^\bullet_M t$, so $m \xrightarrow{u}_N m'$ with $m' = m - {}^\bullet_M t + t_M^\bullet$ and $m' \, R \, \mu'$. If that transition is f_t, then $u = \tau$ and $\mu' = \mu - [p_t] + t_M^\bullet$, so taking $m' = m$, we have $m \Longrightarrow_N m'$ and $m' R \mu'$. In all other cases, the transition that fired was u and $^\bullet_N u = ^\bullet_M u$ and $u^\bullet_N = u^\bullet_M$. Since $\mu \geq {}^\bullet_N u$ and $\mu' = \mu - {}^\bullet_N u + u^\bullet_N$, we can take $m' = m - {}^\bullet_M u + u^\bullet_M$ and have $m \Longrightarrow_N m'$. This covers all cases. Clearly, R satisfies the additional requirements of a WF-bisimulation.   □

Theorem 9. *Let L be a net with $t \in T_L$ and M a sound tWF-net with all transitions except i_M labelled with τ. Then L and $N = L \otimes_t M$ are weakly WF-bisimilar.*

Proof. Let $\overline{M}$ be the extension of M with the initial and final places and all transitions relabelled with τ. By Lemma 2, $(L \otimes_t \Sigma_t) \otimes_{p_t} \overline{M} = ((L \otimes_t M) \otimes_{i_M} \Sigma_{i_M}) \otimes_{f_M} \Sigma_{f_M}$. Thus, by Lemma 8, $L \otimes_t M$ is weakly WF-bisimilar to $(L \otimes_t \Sigma_t) \otimes_{p_t} \overline{M}$, which by Theorem 7 is weakly WF-bisimilar to $L \otimes_t \Sigma_t$ and by Lemma 8 is weakly WF-bisimilar to L.   □

The theorems on weakly bisimilar refinements can be applied to yield soundness preservation.

Theorem 10. *Let L, M be sound WF-nets with $n \in S_L \cup T_L$ such that $N = L \otimes_n M$ is defined. Then N is sound.*

Proof. If $n \in S_L$, we use Theorem 7 after relabelling transitions of M with τ. L and N are weakly WF-bisimilar. Note that the relabelling does not influence the soundness. Suppose $[i^k_N] \xrightarrow{*} m$ within N, then there exists a state m' of L with $m R m'$ such that $[i^k_L] \xrightarrow{*} m'$. Since $m' \xrightarrow{*} [f^k_L]$ within L, there exists a state μ of N such that $m \xrightarrow{*} \mu$ and $\mu R [f^k_L]$. By the definition of WF-bisimulation we have $\mu = [f^k_N]$. If $n \in T_L$, we use Theorem 9 similarly.   □

4   Separability

With introducing the new notion of soundness, we extended the applicability of WF-nets for the compositional design process. However, soundness is not the only criterium for the correctness of the behaviour. In general, one looks for a Petri net model that meets its specification given e.g. by a temporal logic formula; note that the behaviour should satisfy the specification whatever number of tokens is chosen to be placed into the initial place of the WF-net. The challenge is to reduce the number of cases to be considered, when possible.

In this section we introduce a notion of *separability*, a behavioural property stating that the behaviour of a WF-net with k initial tokens can be seen in some sense as a combination of the behaviours of k copies of the net each of which has one initial token.

4.1   Workflow Nets with Id-Tokens and Serialisability

In this subsection, we extend the semantics of labelled Petri nets by introducing *id-tokens*: we consider a token as a pair (p, a), where p is a place and $a \in Id$ is an identifier (a primitive sort of a colour). We assume Id to be a countable set. A

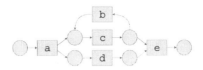

Fig. 4. Serialisable net with id-tokens that is not bisimilar to its abstraction

transition $t \in T$ is *enabled* in an id-marking m iff there exists $a \in Id$ such that m contains tokens $^{\bullet}t$ with identifier a. A firing of t results in consuming these tokens and producing tokens with identifier a to $t^{\bullet}$. To make it clear whether the firing happens in a classical Petri net or in a net with id-tokens, we write $\xrightarrow{t}$ for firings in nets with id-tokens. Later on, we will use the extended semantics when working with id-tokens, and the standard semantics for classical tokens.

Though being a very simple sort of coloured nets, WF-nets with id-tokens are often expressive enough to reflect the essence of a modelled business process taking care of separating different cases which are processed in the net concurrently.

A net with id-tokens can be abstracted into a classical labelled Petri net in a natural way by removing the token id's. We denote the abstraction function as α and a marking obtained from a coloured marking m as $\alpha(m)$ resp. It is easy to see that the obtained net is a sound abstraction of the original net, i.e. it shows more behaviour (see [6] for the definition of sound abstraction):

Lemma 11. *Let N be a Petri net and m its id-marking. Then there exists a simulation relation between (N, m) and $(N, \alpha(m))$.*

Proof. It is trivial to show that $R = \{(m, \alpha(m) \mid m \in \mathcal{M}(N)\}$ is a simulation relation.   □

There is no simulation between (N, m) and $(N, \alpha(m))$ in general (consider e.g. the trace *adcbace* for the WF-net in Figure 4).

Still, it would be interesting to see whether there exists a class of nets whose behaviour is *trace equivalent* to the behaviour of nets with id-tokens. For this purpose, we introduce a notion of *serialisability*.

Definition 12. *An sWF-net N is* serialisable *iff for any $k \in \mathbb{N}$, any firing sequence σ such that $[i^k] \xrightarrow{\sigma}$ there exist firing sequences $\sigma_1, \ldots, \sigma_k$ such that $[i] \xrightarrow{\sigma_1}, \ldots, [i] \xrightarrow{\sigma_k}$ and $\sigma \in (\sigma_1 \| \ldots \| \sigma_k)$.*

Theorem 13. *Let N be an sWF-net. Then N is serialisable iff for any id-marking M such that $\alpha(M) = [i^k]$ for some $k \geq 0$, we have $\{\sigma \mid [i^k] \xrightarrow{\sigma}_N\} = \{\sigma \mid M \xrightarrow{\sigma}_N\}$.*

Proof. ($\Rightarrow$): Let N be a serialisable net, $M = \sum_j [(i, c_j)]$ and k be given as specified. By Lemma 11 every trace of (N, M) is a trace of $(N, [i^k])$. We only have to prove that every trace of $(N, [i^k])$ is a trace of (N, M). Let σ be a trace of $(N, [i^k])$. Due to the serialisability of N there exist $\sigma_1, \ldots, \sigma_k$ such that

$[i] \xrightarrow{\sigma_1}, \dots, [i] \xrightarrow{\sigma_k}$ and $\sigma \in (\sigma_1 \| \dots \| \sigma_k)$. Then we have $[(i, c_j)] \xrightarrow{\sigma_j}$ (since all the tokens produced and consumed in the firings of σ_j have colour c_j). Hence, $\sum_j [(i, c_j)] \xrightarrow{\sigma}$ and σ is a trace of (N, M).

($\Leftarrow$): Now assume we have the described property for N and we have to prove that N is serialisable. Let $[i^k] \xrightarrow{\sigma}$ for some k. Consider a marking $M = \sum_j [(i, c_j)]$ with $j \neq l \Rightarrow c_j \neq c_l$ for all $j, l \in \{1, \dots, k\}$. Since σ is also a trace of (N, M) and all tokens of M have different colours, we can split σ according to the colours of firings into $\sigma_1, \dots, \sigma_k$ such that $\sigma \in (\sigma_1 \| \dots \| \sigma_k)$. We have $[(i, c_j)] \xrightarrow{\sigma_j}$ for all $j \in \{1, \dots, k\}$. Hence, $[i] \xrightarrow{\sigma_j}$ for all $j \in \{1, \dots, k\}$. So N is serialisable. □

Trace equivalence between the nets with id-tokens and their abstractions for serialisable nets allows to perform the verification of trace properties, e.g. LTL-properties, on the abstractions of the nets, thus simplifying the verification task. The same holds for some problems of the performance analysis. If one associate time or price to every transition of the net independent of token's id's, then the analysis results obtained with an abstracted net hold for the original net as well.

4.2 Serialisable Subclasses of WF-Nets

In this subsection we consider two subclasses of WF-nets which we prove to be serialisable.

Definition 14. *Let $N = \langle S, T, F \rangle$ be a Petri net. N is a state machine (SM) iff $\forall t \in T :| {}^\bullet t | \leq 1 \wedge | t^\bullet | \leq 1$.*

Definition 15. *Let $N = \langle S, T, F \rangle$ be a Petri net. N is a marked graph (MG) iff $\forall p \in S :| {}^\bullet p | \leq 1 \wedge | p^\bullet | \leq 1$.*

Marked graphs are dual to state machines in the graph-theoretic sense and from the modelling point of view. State machines can represent conflicts by a place with several output transitions, but they can not represent concurrency and synchronization. Marked graphs, on the other hand, can represent concurrency and synchronization, but cannot model conflicts or data-dependent decisions.

We will refer to the marked graph tWF-nets and state machine sWF-nets as MGWF-nets and SMWF-nets respectively.

Theorem 16. *All SMWF-nets are sound and serialisable.*

Proof. Let N be an SMWF-net and $[i^k] \xrightarrow{\sigma} m$. We shall prove by induction on the length of σ that σ can be serialized. The case $\sigma = \epsilon$ is trivial, so let the statement hold for σ', $[i^k] \xrightarrow{\sigma'} m'$ and we prove the statement for $\sigma = \sigma' t$. By the induction hypothesis, σ' can be serialized into $\sigma'_1, \dots, \sigma'_k$ such that $[i] \xrightarrow{\sigma'_j} m_j$. Since ${}^\bullet t \leq m'$ and $| {}^\bullet t | \leq 1$, there exists an m_l such that $m_l \geq {}^\bullet t$. Thus $[i] \xrightarrow{\sigma_l t}$ and σ can be serialized. So N is serialisable.

In an SMWF-net, $| {}^\bullet t | = | t^\bullet | = 1$ for any transition t, meaning that the number of tokens in the marking cannot change with any firing. So only markings of the form $\sum_{1 \leq j \leq k} [p_j]$ are reachable from $[i^k]$. By the definition of WF-nets, every place lies on a path from i to f. Since N is an SMWF-net, the existence of a path from place p_j to place f is equivalent to $[p_j] \xrightarrow{*}_N [f]$. Hence, $\sum_{1 \le j \le k} [p_j] \xrightarrow{*} [f^k]$ and N is sound.   □

Cycle-free MGWF-nets are sound and serialisable. The proof depends upon the following lemma.

Lemma 17. *Let N be a cycle-free MGWF-net with transitions $t, u \in T_N$ and a place $a \in S_N$ such that there exist paths in N from t to a and from a to u and $\emptyset \xrightarrow{\sigma}_N m$ for some σ, m. Then we have $m(a) \le \vec{\sigma}(t) - \vec{\sigma}(u)$.*

Proof. Note that N is a tWF-net, so traces σ with $\emptyset \xrightarrow{\sigma}_N$ must start with a firing of the initial transition i. Since N is cycle-free, the existence of a path between nodes implies that these nodes are different. We use induction on the length of the path from t to u. For the path of length 2, i.e. $a \in t^\bullet$ and $a \in {}^\bullet u$, the proof is immediate. If the path is longer, we can e.g. find $b \in S_N, v \in T_N$ such that there exists a path $tbv \dots a \dots u$. By the induction hypothesis, $m(a) \le \vec{\sigma}(v) - \vec{\sigma}(u)$ and $m(b) \le \vec{\sigma}(t) - \vec{\sigma}(v)$. Hence, $m(a) + m(b) \le \vec{\sigma}(t) - \vec{\sigma}(u)$, so $m(a) \le \vec{\sigma}(t) - \vec{\sigma}(u)$ □

As a corollary, if σ contains one firing of the initial transition i and a firing of a transition t, the marking m with $\emptyset \xrightarrow{\sigma} m$ satisfies $m(p) = 0$ for any place p on a path between i and t. We can now prove our theorem.

Theorem 18. *All cycle-free MGWF-nets are sound and serialisable.*

Proof. Let $\le$ be the following partial order on sequences: $\sigma \ge \rho$ iff $\forall t :: \vec{\sigma}(t) \ge \vec{\rho}(t)$. We use induction like in the proof of Theorem 16, strengthening the induction hypothesis: σ can be serialized into $\sigma_1, \dots, \sigma_k$ in such a way that $\sigma_1 \ge \dots \ge \sigma_k$. So let N be the extension of an MGWF-net with the initial an terminal places and suppose $[i^k] \xrightarrow{\sigma t}$. By the induction hypothesis on σ we have a decreasing serialization $\sigma_1, \dots, \sigma_k$ with $[i] \xrightarrow{\sigma_j}$ and $\sigma \in \sigma_1 \| \dots \| \sigma_k$. Let m be a marking such that $[i^k] \xrightarrow{\sigma} m$ and m_j such that $[i] \xrightarrow{\sigma_j} m_j$ for all $j \in \{1, \dots, k\}$. From Lemma 17, we know that σt has at most k occurrences of t, so σ has at most $k - 1$ occurrences of t. Due to the ordering of σ_j's, we can conclude that σ_k does not contain t. Let $n \le k$ be the smallest index such that σ_n does not contain t. We have $m_j \cap {}^\bullet t = \emptyset$ for $j < n$ by the corollary of Lemma 17. Moreover, for $j \ge n$ we have $m_j \cap {}^\bullet t \le m_n \cap {}^\bullet t$, since there exist traces γ_j not containing t such that $\sigma_j \gamma_j = \sigma_n$. Since N does not contain multiple edges, we deduce from ${}^\bullet t \le m = \sum_j m_j$ that ${}^\bullet t \le m_n$. So we can serialize σt into traces $\sigma_1, \dots, \sigma_{n-1}, \sigma_n t, \sigma_{n+1}, \dots, \sigma_k$. Since σ_{n-1} contained t, we have $\sigma_{n-1} \ge \sigma_n t$, completing the induction step. Again, 1-soundness follows from the WF-net property.   □

Serialisability is not compositional. As it normally happens with the notions based on the traces of the systems, the serialisability is not compositional. Figure 5 shows a net obtained as a place refinement of a marked graph with a state machine, which are both serialisable as we know from the theorems proven above. The trace *aecabf* of the refined net cannot be serialized.

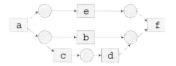

Fig. 5. Not serialisable net

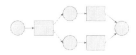

Fig. 6. A weakly separable net that is not 1-sound

4.3   Weak Separability

Our next approach is to look at the markings of the net only:

Definition 19. *An sWF-net N is* weakly separable *iff for any $k \in \mathbb{N}$ and any marking m, $[i^k] \overset{*}{\longrightarrow} m$ implies that there exist markings $m_1, \ldots, m_k$ such that $m = m_1 + \ldots + m_k$ and $[i] \overset{*}{\longrightarrow} m_j$ for $j = 1, \ldots, k$.*
We say that a tWF-net N is weakly separable iff the sWF-net obtained by adding a place with the outgoing arc to the initial transition of N and a place with ingoing arc from the final transition of N is weakly separable.

Property 20. Serialisability implies weak separability.

Proof. If $[i^k] \overset{*}{\longrightarrow} m$, there exists a σ such that $[i^k] \overset{\sigma}{\longrightarrow} m$, so there exist $\sigma_1, \ldots, \sigma_k, m_1, \ldots, m_k$ such that $[i] \overset{\sigma_1}{\longrightarrow} m_1, \ldots [i] \overset{\sigma_k}{\longrightarrow} m_k$ and $\sigma \in (\sigma_1 \| \ldots, \| \sigma_k)$. Clearly, $m = m_1 + \ldots + m_k$. □

Requirements that weak separability puts on a net are essentially weaker than the ones of serialisability, which also means that we loose some options for analysis on the class of weakly separable nets in comparison to the serialisable nets. However, weak separability is sufficient to reduce the problem of soundness to 1-soundness:

Theorem 21. *Let N be a weakly separable and 1-sound net. Then N is sound.*

Proof. Consider a marking m reachable from $[i^k]$ where k is an arbitrary positive natural number. Since N is weakly separable, there exist $m_1, \ldots, m_k$ such that $m = m_1 + \ldots + m_k$ and $[i] \overset{*}{\longrightarrow} m_1, \ldots, [i] \overset{*}{\longrightarrow} m_k$. Since N is 1-sound, $m_1 \overset{*}{\longrightarrow} [f], \ldots, m_k \overset{*}{\longrightarrow} [f]$, which means that $m \overset{*}{\longrightarrow} [f^k]$. So N is sound. □

A legitimate question would be whether weak separability implies soundness even without additional requirements. The answer to this question is negative: Figure 6 gives a weakly separable net which is not sound, and moreover not 1-sound.

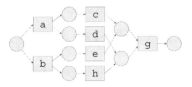

Fig. 7. A sound net that is not weakly separable

Corollary 22. *The class of all weakly separable nets is not a subclass of all sound nets.*

The reverse is also not true: Figure 7 shows a sound free-choice[2] net (see [8]) which is not separable.

Corollary 23. *The class of all sound free-choice nets is not a subclass of all weakly separable nets.*

Thus, the notion of separability is in some sense orthogonal to the notion of soundness.

Like soundness and unlike serialisability, weak separability is a congruence with respect to the place refinement:

Theorem 24. *Let L, M be weakly separable WF-nets, moreover M is a sound sWF-net and $p \in P_L$. Then the net $N = L \otimes_p M$ is weakly separable.*

Proof. We may assume L to be an sWF-net since a tWF-net could be transformed to the sWF-net just by adding initial and final places. Let i, f, i_M, f_M be respectively the initial and final places of L and M. We shall prove that N is weakly separable.

Let $[i^k] \xrightarrow{*}_N m$. Then there is a trace σ such that $[i^k] \xrightarrow{\sigma}_N m$. As the nodes of L and M are disjoint, m can be represented as $m_L + m_M$ for some m_L, m_M where m_L is a marking over $P_L \setminus \{p\}$ and m_M is a marking over P_M. Similarly, trace σ can be projected into two traces σ_L, σ_M such that $\sigma_L \in T_L^*$, $\sigma_M \in T_M^*$. Note that $\sigma \in \sigma_L \| \sigma_M$.

Since $(N, [i^k])$ and $(L, [i^k])$ are weakly bisimilar (Theorem 7), σ_L is a trace of $(L, [i^k])$: $[i^k] \xrightarrow{\sigma_L}_L \overline{m}_L$. Due to the weak separability of L, $\overline{m}_L$ can be split into a sum $\overline{m}_{L,1} + \ldots + \overline{m}_{L,k}$, such that $[i] \xrightarrow{*}_L \overline{m}_{L,1}, \ldots, [i] \xrightarrow{*}_L \overline{m}_{L,k}$. Due to the soundness of M, we can prove by induction on the length of σ that $\overline{m}_L = m_L + [p^n]$ and $[i_M^n] \xrightarrow{*} m_M$ for some n. Due to the weak separability of M, m_M can be split into a sum $m_{M,1} + \ldots + m_{M,n}$, such that $[i] \xrightarrow{*}_M m_{M,1}, \ldots, [i] \xrightarrow{*}_M m_{M,n}$. Now we choose an arbitrary bijective function that maps every occurrence of p in each of $\overline{m}_{L,i}$ to a $m_{M,j}$, replace every p by $m_{M,j}$ according to the chosen mapping and thereby get the splitting of m we are looking for. $\square$

---

[2] N is a *free-choice Petri net* iff $\forall\, t_1, t_2 \in T,\ {}^\bullet t_1 \cap {}^\bullet t_2 \neq \emptyset$ implies ${}^\bullet t_1 = {}^\bullet t_2$.

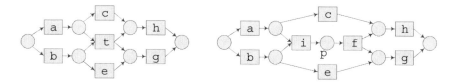

Fig. 8. Not weakly separable transition refinement of a weakly separable net

Applying transition refinement in the same way does not necessarily result in a weakly separable net. Figure 8 gives a weakly separable net and a refinement of this net where transition t is substituted with a sound weakly separable tWF-net. The resulting net is not weakly separable: marking $[p]$ is reachable from the initial marking $[i^2]$, however, it cannot be split into a sum of two markings reachable from $[i]$. Note that the net is nevertheless *sound*!

4.4   Separability

Finally, we shall try and introduce a notion stronger than weak separability but not as restrictive as serialisability. Moreover, we shall look for a subclass of nets where this notion is compositional w.r.t. refinements.

Definition 25. *An sWF-net N is separable iff for any $k \in \mathbb{N}$, any firing sequence σ such that $[i^k] \xrightarrow{\sigma}$, there exist firing sequences $\sigma_1, \ldots, \sigma_k$ such that $[i] \xrightarrow{\sigma_1}, \ldots, [i] \xrightarrow{\sigma_k}$ and $\vec{\sigma} = \vec{\sigma_1} + \ldots + \vec{\sigma_k}$.*

The following properties follow immediately from the corresponding definitions:

Property 26. (1) Serialisability implies separability. (2) Separability implies weak separability.

Note that the class of serialisable nets is strictly included in the class of separable nets: a not serialisable WF-net from Figure 5 *is* separable: e.g. the problematic trace $aecabf$ can now be separated into $aebf$ and ac.

For business applications of WF-nets, separability can be used to provide cost-effective management by simplifying the cost analysis. If costs are associated to every transition firing, the total cost of processing of k orders given by a trace in a WF-net is equal to the sum of costs of processing of k individual orders, each given by a trace with 1 initial token.

Unlike serialisability, separability turns out to be a congruence w.r.t. the place refinement operation:

Theorem 27. *Let L, M be separable WF-nets. If $p \in P_L$ and M is a sound sWF-net then $L \otimes_p M$ is separable.*

Proof. We may assume that L is an sWF-net; if not, we extend it. Let $N = L \otimes_p M$ and write $i_L = i_N$ as i. Assume $p \neq i$. Suppose $[i^k] \xrightarrow{\sigma}_N m$ for some m. We shall construct $\sigma_1, \ldots, \sigma_k$ such that $\vec{\sigma} = \sum_{1 \leq j \leq k} \vec{\sigma_j}$, where $[i] \xrightarrow{\sigma_j}_N$ for all j.

Let $\ell = \sigma_N^\bullet(p)$, $n = {}_N^\bullet\sigma(p)$. We define γ, ρ as the projections of σ on T_L, T_M respectively, so $\vec{\sigma} = \vec{\gamma} + \vec{\rho}$. Due to the existence of a weak WF-bisimulation between N and L, we have $[i^k] \xrightarrow{\gamma}_L$. The separability of L implies then the existence of $\gamma_1, \ldots, \gamma_k$ such that $\vec{\gamma} = \sum_j \vec{\gamma}_j$ with $[i] \xrightarrow{\gamma_j}_L$ for all j. Let $\ell_j = \gamma_j{}^\bullet_L(p)$, $n_j = {}^\bullet_L\gamma_j(p)$ for all j. Since $\gamma^\bullet_N(i_M) = \gamma^\bullet_L(p) = \ell$ and likewise ${}^\bullet_N\gamma(f_M) = n$, we have $\sum_{1 \le j \le k} \ell_j = \ell$ and $\sum_{1 \le j \le k} n_j = n$. We have $[i^\ell_M] \xrightarrow{\rho}_M m'$ with $m'(f_M) = n$. Due to the separability of M, we can find $\rho_1, \ldots, \rho_\ell$ such that $[i_M] \xrightarrow{\rho_j}_M m'_j$ with n of the traces ρ_j complete (i.e. $[i_M] \xrightarrow{\rho_j} [f_M]$ and $\vec{\rho} = \sum_{1 \le j \le \ell} \vec{\rho}_j$). Since $\sum_j \ell_j = \ell$, $\sum_j n_j = n$, we can partition the ρ_j's into disjoint sets $R_1, \ldots, R_k$ with respectively $\ell_1, \ldots, \ell_k$ elements such that m_j of the traces in R_j are complete for each j. We construct σ_j with $[i] \xrightarrow{\sigma_j}_N$ by adding transitions t from γ_j one by one preceded by ${}^\bullet_L t(p)$ completed traces from R_j. Since ${}^\bullet_L\gamma_j(p) = n_j$, all the completed traces in R_j are used in this process. We add the incomplete traces in any sequential order at the end. This we do for $1 \le j \le k$. These σ_j satisfy the requirement. If $p = i$ we can copy the above proof, setting $\ell = k$ and $\ell_j = 1$ for all j.   □

Transition refinement is still a problem: Figure 9 shows a refinement of a separable net which yields an inseparable net: the trace $abicg$ cannot be separated. Therefore, we constrict the class of separable nets in the following way:

Definition 28. *An sWF-net N is* split-separable *iff* $\mathcal{S}(N) = (\ldots(N \otimes_{t1} \Sigma_{t1}) \otimes_{t2} \ldots) \otimes_{tn} \Sigma_{tn}$, $T_N = \{t_1, \ldots, t_n\}$ (the net obtained by applying the split-refinement to every transition of N), is separable.

Due to Lemma 2, the order of split-refinements in the above definition is not important.

Property 29. Split-separability implies separability.

Proof. Let L be a split-separable net and $N = \mathcal{S}(L)$ its split-refinement, so N is separable. We shall prove that L is separable. We label i_t with t and f_t with τ for every transition $t \in T_L$. Then N and L are weakly bisimilar. Now let σ be a trace of L, $[i^k] \xrightarrow{\sigma}_L$ and σ' some corresponding trace of N, $[i^k] \xrightarrow{\sigma'}_N$. σ'

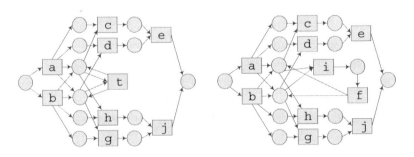

Fig. 9. Not separable transition refinement of a separable net

can be separated into $\sigma'_1, \ldots, \sigma'_k$, $[i] \xrightarrow{\sigma'_1}_N, \ldots, [i] \xrightarrow{\sigma'_k}_N$ and $\overrightarrow{\sigma'} = \overrightarrow{\sigma'_1} + \ldots + \overrightarrow{\sigma'_k}$. Due to bisimilarity, $\sigma_1, \ldots, \sigma_k$ obtained from $\sigma'_1, \ldots, \sigma'_k$ by replacing every i_t with t and removing all f_t's are the traces of L and $\overrightarrow{\sigma} = \overrightarrow{\sigma_1} + \ldots + \overrightarrow{\sigma_k}$. So L is separable. $\qquad\square$

Lemma 30. *Let L be a split-separable net and $N = L \otimes_t \Sigma_t$ for some $t \in T_L$. Then N is split-separable.*

Lemma 31. *Any split refinement of a split-separable net is split-separable.*

Proof. Notice that the net $\mathcal{S}(\mathcal{S}(N))$ can be obtained from $\mathcal{S}(N)$ by place refinement. Since $\mathcal{S}(N)$ is separable, $\mathcal{S}(\mathcal{S}(N))$ is also separable (Theorem 27). $\qquad\square$

Theorem 32. *Let L, M be split-separable WF-nets. (1) If $p \in P_L$ and M is a sound sWF-net then $L \otimes_p M$ is split-separable. (2) If $t \in T_L$ and M is a sound tWF-net then $L \otimes_t M$ is split-separable.*

Proof. (1) Let $N = L \otimes_p M$. We have to prove that N is separable, i.e. $N' = \mathcal{S}(N)$ is separable. Since L and M are split-separable, $L' = \mathcal{S}(L)$ and $M' = \mathcal{S}(M)$ are separable too. Due to Lemma 2, $\mathcal{S}(N) = \mathcal{S}(L) \otimes_p \mathcal{S}(M)$. Hence, by Theorem 27, $\mathcal{S}(N)$ is separable.

(2) By Lemma 30, $L' = L \otimes_t \Sigma_t$ is split-separable. Now construct M' by adding initial and final places to M (M' is split-separable as well) and consider $N' = L' \otimes_{p_t} M'$. N' is split-separable due to part (1) of this theorem. Label i_M and f_t in N' to τ, then N' is weakly bisimilar to N, where all labels are visible. So N is split-separable too. $\qquad\square$

Theorem 33. *SMWF-nets and acyclic MGWF-nets are split-separable.*

Proof. Since SMWF-nets and acyclic MGWF-nets are serialisable, they are also separable. Now notice that the classes of SMWF-nets and acyclic MGWF-nets are closed under the split-refinement operation, hence, these nets are split-separable. $\qquad\square$

5   ST-Nets

It is hard to find algorithms that check soundness and/or separability for an arbitrary WF-net, but we can define classes of nets that are sound and (split) separable by construction. One such class, called ST-nets, is treated in this section. These nets are constructed from state machines and marked graphs by means of refinement. In many cases, modelling problems can be solved by (provably correct) ST-nets.

Definition 34. *The set $\mathcal{N}$ of ST-nets is the smallest set of nets N defined as follows:*
- *if N is an acyclic MGWF-net, then $N \in \mathcal{N}$;*
- *if N is an SMWF-net, then $N \in \mathcal{N}$;*
- *if $N \in \mathcal{N}, s \in S_N$ and $M \in \mathcal{N}$ is an sWF-net, then $N \otimes_s M \in \mathcal{N}$;*
- *if $N \in \mathcal{N}, t \in T_N$ and $M \in \mathcal{N}$ is a tWF-net, then $N \otimes_t M \in \mathcal{N}$.*

Algorithm 37 (CheckST(N)).

$\Delta := CompDistEnd(N);$	compute distances to the end node
$X := S_N \cup T_N \setminus \{f_N\};$	initialise search for x
while $X \neq \emptyset$ do	search loop
pick $x \in X;$	pick a candidate
$M := FindFactor(N, x, \Delta);$	search for a factor
if $M \neq S_N \cup T_N \wedge CheckSMMG(M)$	SM/MG factor found
then return($CheckST(Quotient(N, M)))$	recursive call
else $X := X \setminus \{x\}$	continue search
od;	No smaller SM/MG factor found
return($CheckSMMG(N)$)	

Algorithm 38 (FindFactor(N, x, Δ)).

$X = x^\bullet;$	initialise possible internal nodes
$Y = \emptyset;$	initialise possible end nodes
while $X^\bullet \not\subseteq (X \cup Y) \vee {}^\bullet X \not\subseteq (X \cup \{x\})$ do	stop when augmentation stabilises
$X := X \cup X^\bullet \cup ({}^\bullet X \setminus \{x\}) \cup {}^\bullet Y;$	augmentation step
$Y := \{y \in X \mid \Delta(y) = Min_{z \in X}\Delta(z)\};$	compute candidates for y
if $\exists f : Y = \{y\} \wedge x \neq y \wedge type(x) = type(y)$	test for candidate y
then $X := X \setminus Y$ else $Y := \emptyset$ fi	adjust X, Y
od;	augmentation stable
return($X \cup Y \cup \{x\}$)	$S_N \cup T_N$ if unsuccessful

Fig. 10. Factorization algorithm

Property 35. Let N be an ST-net, $N = L \otimes_n M$ for some WF-net L, $n \in S_L \cup T_L$ and ST-net M. Then L is an ST-net as well.

Theorem 36. *All ST-nets are sound and split-separable.*

Proof. Follows immediately from Theorems 16, 18, 10, 33 and 32. $\square$

Algorithm 37 checks whether a given WF-net N is an ST net. It looks for a subnet of N, which is an STWF- or MGWF-net, i.e. $N = L \otimes_n M$ for some node n of a WF-net L. We call such a net M a *factor* of N and L the *quotient*. By Definition 34 and Property 35, N is an ST-net iff L is an ST-net. So the algorithm proceeds recursively with checking whether L is an ST-net. There exist various ways to speed up the algorithm but we choose the given presentation for the sake of simplicity.

The algorithm starts by computing the distance function $\Delta : (S_N \cup T_N) \to \mathbb{N}$ that gives the length of the shortest path of a node x to f_N. Then we pick up an arbitrary node $x \neq f_N$ and compute the smallest SM/MG factor (if any) with initial node x, with following Algorithm 38. Note that a factor of N with initial node x and terminal node y corresponds to a set S of nodes containing x, y and all successors of nodes $n \in S \setminus \{y\}$ and predecessors of nodes $n \in S \setminus \{x\}$, i.e. such that ${}^\bullet(S \setminus \{x\}) \subseteq S$ and $(S \setminus \{y\})^\bullet \subseteq S$. This observation allows us to compute the smallest such S by successive augmentation, starting with the set S containing all nodes from $x^\bullet$. The candidate for being the terminal node y in each augmentation step is the node that is nearest to the end node f_N, which is the reason for calculating Δ. The algorithm uses the function *type* on nodes that returns either "place" or "transition". The minimal distance computation (*CompDistEnd*), SM/MG check (*CheckSMMG*) and quotient computation (*Quotient*) are trivial and have not been elaborated further.

6   Conclusion

In this paper we studied workflow nets that allow "batched" cases. This perspective led to a strengthened notion of soundness. Advantages of this notion is that sound in the new sense WF-nets can be used freely as components without restricting their use to e.g. safe nets. Bisimilarity results speed up verification of temporal properties for composite nets.

Comparison of 1-soundness and (strengthened) soundness led to the notion of separability: independency of individual cases within a batch. Weakly separable and 1-sound nets are (strongly) sound. We introduced a notion of split-separability and proved its compositionality w.r.t. refinement, allowing a hierarchical approach to modelling and validation. A particular application of this approach are the processes that can be modelled by ST-nets, which are "sound by construction" and split-separable.

Future work. We investigated a strengthening of 1-soundness, though as argued in [7], 1-soundness is too strong a notion for some applications. It is interesting to investigate e.g. separable nets that are not fully sound.

Decidability and computability are an issue. Clearly, 1-soundness can be assessed by coverability analysis (c.f. [12]), but soundness and separability are a different matter. A decision algorithm for separability of 1-sound WF-nets can be found, but as yet not an efficient one. Soundness is probably undecidable in general, as well as separability, but it is still a question for further investigations.

Soundness and separability of communicating WF-subnets (c.f. [10]) will need extension of our class of operators that preserve soundness and separability. We intend to develop component-oriented strategies for connecting nets. Wider classes of nets than WF-nets can be considered as well. The use of net components with several entry and/or exit nodes enables a component-based modelling strategy that allows more freedom than refinement alone.

Acknowledgment. We are grateful to the referees for their constructive remarks and suggestions.

References

1. W.M.P. van der Aalst. *Verification of Workflow Nets.* In Azéma, P. and Balbo, G., editors, *Proceedings ATPN '97*, LNCS 1248, Springer 1997.

2. W.M.P. van der Aalst, J. Desel and A. Oberweis, editors *Business Process Management, Models, Techniques and Empirical Studies*. LNCS 1806, Springer 1998.
3. W.M.P. van der Aalst and K.M. van Hee. *Workflow Management: models, methods and systems*. The MIT Press, 2000.
4. W.M.P. van der Aalst. *Workflow Verification: Finding Control-Flow Errors using Petri-net-based techniques*. In [2], pages 161–183.
5. S. Ceri and G. Pelagatti. *Distributed Databases: Principles and Systems*. McGraw-Hill 1984.
6. P. Cousot and R. Cousot. Abstract interpretation: A unified lattice model for static analysis of programs by construction or approximaton of fixpoints. In *Fourth Annual Symposium on Principles of Programming Languages (POPL) (Los Angeles, Ca)*, pages 238–252. ACM, January 1977.
7. J. Dehnert and P. Rittgen. *Relaxed Soundness of Business Processes*. In K.R. Dittrich, A. Geppert and M.C. Norrie, editors, *Proceedings CAISE '01*, LNCS 2068, pages 157–170, Springer 2001.
8. J. Desel and J. Esparza. *Free Choice Petri Nets*. Cambridge University Press, 1995.
9. R.J. van Glabbeek and R.P. Weijland. *Branching Time and Abstraction in Bisimulation Semantics (extended abstract)*. In G.X. Ritter, editor, *Proceedings IFIP '89*, pages 613–618. North Holland 1989.
10. E. Kindler, A. Martens and W. Reisig. *Inter-operability of Workflow Applications: Local Criteria for Global Soundness*. In [2], pages 235–253.
11. R. Milner. Operational and algebraic semantics of concurrent processes. In J. van Leeuwen, editor, *Handbook of Theoretical Computer Science, vol. B*, chapter 19, pages 1201–1242. Elsevier Science, 1990.
12. H.M.W. Verbeek, T. Basten, and W.M.P. van der Aalst. Diagnosing workflow processes using Woflan. *The Computer Journal*, 44(4):246–279, 2001.

On Synchronicity and Concurrency in Petri Nets

Gabriel Juhás[1], Robert Lorenz[1], and Tomáš Šingliar[2]

[1] Lehrstuhl für Angewandte Informatik, Katholische Universität Eichstätt-Ingolstadt
{gabriel.juhas,robert.lorenz}@ku-eichstaett.de
[2] Department of Computer Science, University of Pittsburg
tomas@cs.pitt.edu

Abstract. In the paper we extend the algebraic description of Petri nets based on rewriting logic by introducing a partial synchronous operation in order to distinguish between synchronous and concurrent occurrences of transitions. In such an extension one first needs to generate steps of transitions using a partial operation of synchronous composition and then to use these steps to generate process terms using partial operations of concurrent and sequential composition. Further, we define which steps are true synchronous. In terms of causal relationships, such an extension corresponds to the approach described in [6,7,9], where two kinds of causalities are defined, first saying (as usual) which transitions cannot occur earlier than others, while the second indicating which transitions cannot occur later than others. We illustrate this claim by proving a one-to-one correspondence between such extended algebraic semantics of elementary nets with inhibitor arcs and causal semantics of elementary nets with inhibitor arcs presented in [7].

1 Introduction

There are many extensions of Petri Nets that improve their modelling, suitability and/or their expressive power. These are almost all based on the same original model, augmented by capacities, context arcs, data structures or even object-oriented features. Conceptually, structures like capacities impose restrictions on the set of legal markings, whereas context arcs introduce new arrow types, which have to be accounted for in the occurrence rule. The definition of sequential semantics for these extensions of Petri Nets can be directly obtained by "iterating" the occurrence rule. However, we are even more interested in obtaining the non-sequential semantics[1]. For the classes of systems mentioned, this is obtained at the time being in an ad-hoc way. Naturally there arises the question if these approaches can be unified by defining some more general framework. In [5] there was recently presented one such unifying concept for extensions of Petri Nets based on a restriction of the occurrence rule. The approach extends and generalizes the idea of Winkowski (see [16]) that non-sequential semantics of elementary nets can be expressed in terms of concurrent rewriting. The principles of the approach are briefly described in the following paragraphs. For more details see [4,5].

---

[1] For comparative treatment of sequential and non-sequential semantics see [1].

W.M.P. van der Aalst and E. Best (Eds.): ICATPN 2003, LNCS 2679, pp. 357–376, 2003.

A transition is understood to be an elementary rewrite term that allows replacing the marking $pre(t)$ by $post(t)$. A marking m is also considered to be a rewrite term that rewrites m by m itself. Assume that a suitable operation $+$ on the set of markings is given for each class of Petri Nets in interest such that for each transition $m \xrightarrow{t} m'$ there exist a marking x such that $x + pre(t) = m$ and $x + post(t) = m'$. Occurrence of t at m will be expressed by term $x \parallel t$. It may be the case that not all markings $x + pre(t)$ enable t. In that situations, x and t cannot be composed by $\parallel$. To describe such restriction, we introduce an abstract set of information I and the notion of independence of information elements. Each elementary term has an associated initial marking, final marking and an information set consisting of all information elements of elementary terms from which it is generated. A composition is allowed if and only if the associated information elements are independent. *The non-sequential behavior of a net is described by set of process terms, constructed from the elementary terms using operators of sequential and concurrent composition ; and $\parallel$, respectively.*

There are several works relating algebraic characterization and partial-order based description of non-sequential behavior of place/transitions Petri Nets, see [3,14]. These papers in common stem from the paper [10], which has inspired many to continue in this research direction. Thus, algebraic characteristics of non-sequential behavior based of sequential and concurrent composition of rewriting terms represents a suitable axiomatic semantics for the classes of nets which operational causal semantics can be based on partial order.

In a series of papers [6,7,9] authors illustrate that a simple partial-order is not enough expressive to characterize some kinds of causalities. They define more fine causal semantics, where two kinds of causalities are used, first saying (as usual by a partial-order based semantics) which transitions cannot occur earlier than others, while the second indicating which transitions have to occur later than others. Mathematically, this finer causal semantics is described using a relational structure with two relations, a partial order describing the "earlier than" causality and a relation representing the "not later than" causality. In [7,9] the principle is illustrated for a variant of nets with inhibitor arcs, where testing for absence of tokens precedes the execution of a transition (so called a-priori semantics). Thus, if a transition f tests a place for zero, which is in a post-set of another transition e (see Figure 1), this means that f cannot occur later than e and therefore they cannot occur concurrently or sequentially in order e, f - but still can occur synchronously or sequentially in order f, e. Moreover, there are cases where the pair of events e and f is executable neither concurrently nor sequentially, but still the a-priori semantics of transition firings allows them to fire synchronously at the same time. Such a situation is shown on figure 2.

Therefore, in this paper we extend our approach [5] to define an algebraic semantics which corresponds to the idea of finer causal semantics described in [6,7, 9]. Namely, we introduce a new partial operation of *synchronous composition* $\oplus$ which enable us to distinguish between synchronous and concurrent occurrences of transitions. In such an extension of our approach one first needs to generate synchronous steps from transitions using a partial operation of synchronous

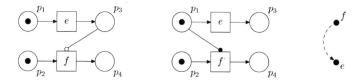

Fig. 1. A simple net with inhibitor arc (p_3, f) where partial ordered semantics does not describe veritably the behaviour of the net, a process of the net (where the inhibitor arc is modelled using an activator arc testing on presence of a token in place p_1, which is complementary to place p_3) and the associated relational structure. The "earlier than" partial order is empty, the "not later than" relation is represented by the dashed line.

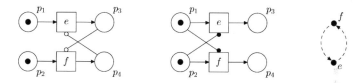

Fig. 2. A modification of the previous figure that shows a net with inhibitor arcs, a process of the net and and the associated relational structure. The "earlier than" partial order is again empty, the "not later than" relation is represented by the dashed lines.

composition and then to use these steps to generate process terms using partial operations of concurrent and sequential composition. Different process terms can still represent different partial or total sequentializations of the same run (the same process). Therefore, the process terms are further related modulo a set of axioms, which determine equivalence classes of process terms representing the same run. We will illustrate the approach on elementary nets with inhibitor arcs with a-priori semantics. As the main result, we prove a one-to-one correspondence between the new defined algebraic semantics of elementary nets with inhibitor arcs based on rewriting logic and causal semantics of elementary nets with inhibitor arcs presented in [7].

2   Mathematical Notation

We use the symbol id_A to denote the identity mapping on the set A. We write id to denote id_A whenever A is clear from the context.

We use partial algebra to define algebraic semantics of nets. A partial algebra is a set (called carrier) together with a couple of partial operations on this set (with possibly different arity). Given a partial algebra with carrier X, an equivalence $\sim$ on X satisfying the following conditions is a *congruence*: If op is an n-ary partial operation, $a_1 \sim b_1, \ldots, a_n \sim b_n$, $(a_1, \ldots, a_n) \in dom_{op}$ and $(b_1, \ldots, b_n) \in dom_{op}$, then $op(a_1, \ldots, a_n) \sim op(b_1, \ldots, b_n)$. If moreover $a_1 \sim b_1, \ldots, a_n \sim b_n$ and $(a_1, \ldots, a_n) \in dom_{op}$ imply $(b_1, \ldots, b_n) \in dom_{op}$

for each n-ary partial operation then the congruence $\sim$ is said to be *closed*. Thus, a congruence is an equivalence preserved by all operations of a partial algebra, while a closed congruence moreover preserves the domains of the operations. For a given partial algebra there always exists a unique greatest closed congruence. The intersection of two congruences is again a congruence. Given a binary relation on X, there always exists a unique least congruence containing this relation. In general, the same does not hold for closed congruences. Given a partial algebra $\mathcal{X}$ with carrier X and a congruence $\sim$ on $\mathcal{X}$, we write $[x]_\sim = \{y \in X \mid x \sim y\}$ and $X/_\sim = \bigcup_{x \in X}[x]_\sim$. A closed congruence $\sim$ defines the partial algebra $\mathcal{X}/_\sim$ with carrier $X/_\sim$, and with n-ary partial operation $op/_\sim$ defined for each n-ary partial operation $op : dom_{op} \to X$ of $\mathcal{X}$ as follows: $dom_{op/_\sim} = \{([a_1]_\sim, \ldots, [a_n]_\sim) \mid (a_1, \ldots, a_n) \in dom_{op}\}$ and, for each $(a_1, \ldots, a_n) \in dom_{op}$, $op/_\sim([a_1]_\sim, \ldots, [a_n]_\sim) = [op(a_1, \ldots, a_n)]_\sim$. The partial algebra $\mathcal{X}/_\sim$ is called factor algebra of $\mathcal{X}$ with respect to the congruence $\sim$.

Let $\mathcal{X}$ be a partial algebra with k operations $op_i^{\mathcal{X}}, i \in \{1, \ldots, k\}$, and let $\mathcal{Y}$ be a partial algebra with k operations $op_i^{\mathcal{Y}}, i \in \{1, \ldots, k\}$ such that the arity $n_i^{\mathcal{X}}$ of $op_i^{\mathcal{X}}$ equals the arity $n_i^{\mathcal{Y}}$ of $op_i^{\mathcal{Y}}$ for every $i \in \{1, \ldots, k\}$. Denote by X the carrier of $\mathcal{X}$ and by Y the carrier of $\mathcal{Y}$. Then a function $f : X \to Y$ is called homomorphism if for every $i \in \{1, \ldots, k\}$ and $x_1, \ldots x_{n_i^{\mathcal{X}}} \in X$ we have: if $op_i^{\mathcal{X}}(x_1, \ldots, x_{n_i^{\mathcal{X}}})$ is defined then $op_i^{\mathcal{Y}}(f(x_1), \ldots, f(x_{n_i^{\mathcal{X}}}))$ is also defined and $f(op_i^{\mathcal{X}}(x_1, \ldots, x_{n_i^{\mathcal{X}}})) = op_i^{\mathcal{Y}}(f(x_1), \ldots, f(x_{n_i^{\mathcal{X}}}))$. A homomorphism $f : X \to Y$ is called closed if for every $i \in \{1, \ldots, k\}$ and $x_1, \ldots x_{n_i^{\mathcal{X}}} \in X$ we have: if $op_i^{\mathcal{Y}}(f(x_1), \ldots, f(x_{n_i^{\mathcal{X}}}))$ is defined then $op_i^{\mathcal{X}}(x_1, \ldots, x_{n_i^{\mathcal{X}}})$ is also defined. If f is a bijection, then it is called an isomorphism, and the partial algebras $\mathcal{X}$ and $\mathcal{Y}$ are called isomorphic. In the paper we distinguish between partial algebras up to isomorphism.

There is a strong connection between the concepts of homomorphism and congruence in partial algebras: If f is a surjective (closed) homomorphism from $\mathcal{X}$ to $\mathcal{Y}$, then the relation $\sim \subseteq X \times X$ defined by $a \sim b \iff f(a) = f(b)$ is a (closed) congruence and $\mathcal{Y}$ is isomorphic to $\mathcal{X}/_\sim$. Conversely, given a (closed) congruence $\sim$ of $\mathcal{X}$, the mapping $h : X \to X/_\sim$ given by $h(x) = [x]_\sim$ is a surjective (closed) homomorphism. This homomorphism is called the *natural homomorphism w.r.t.* $\sim$. For more details on partial algebras see e.g. [2].

3   General Approach

An algebraic Petri Net according to [10] is a graph with vertices representing markings and edges labelled by transitions. Moreover, there is an operation $+ : M \to M$, which is the marking addition. Thus M, the set of markings, and $+$, together with neutral element e (the empty marking) form a commutative monoid $\mathcal{M} = (M, +)$.

To obtain a process term semantics, in [5] we assign to each marking and transition an information element, used for determining concurrent composability of processes. Now we use this information element also for synchronous composability of processes. The set of information elements is then equipped with operation

$\dot{\parallel}$ (as in [5]) and in addition the operation $\dot{\oplus}$, denoting the information of concurrent and synchronous composition, respectively. Since concurrent realization of events admits their synchronous realization, the domain of concurrent composition is a subset of that of synchronous composition and $\dot{\parallel}$ is the restriction of $\dot{\oplus}$ to this domain. The partial algebra $\mathcal{I}$ of information elements is formally defined as follows:

Definition 1. *Let* $\mathcal{I} = (I, dom_{\dot{\parallel}}, \dot{\parallel}, dom_{\dot{\oplus}}, \dot{\oplus})$, *where* I *is a set (of information elements),* $dom_{\dot{\parallel}} \subseteq dom_{\dot{\oplus}} \subseteq I \times I$, *and* $\dot{\parallel} : dom_{\dot{\parallel}} \to I$ *and* $\dot{\oplus} : dom_{\dot{\oplus}} \to I$ *satisfying* $\dot{\oplus}|_{dom_{\dot{\parallel}}} = \dot{\parallel}$ *and:*

- $\forall a, b \in I :$ *if* $a \dot{\oplus} b$ *is defined then* $b \dot{\oplus} a$ *is defined and* $a \dot{\oplus} b = b \dot{\oplus} a$. *Similarly, if* $a \dot{\parallel} b$ *is defined then* $b \dot{\parallel} a$ *is defined.*
- $\forall a, b, c \in I :$ *if* $(a \dot{\oplus} b) \dot{\oplus} c$ *is defined then* $a \dot{\oplus} (b \dot{\oplus} c)$ *is defined and* $(a \dot{\oplus} b) \dot{\oplus} c = a \dot{\oplus} (b \dot{\oplus} c)$. *Similarly, if* $(a \dot{\parallel} b) \dot{\parallel} c$ *is defined then* $a \dot{\parallel} (b \dot{\parallel} c)$ *is defined.*

In comparison with [5], we first need to generate steps from transitions using a partial operation of synchronous composition and then to use these steps to generate process terms using partial operations of concurrent and sequential composition.

The following explanations are now exactly the same as in [5]: A process term $\alpha: m_1 \to m_2$ represents a process transforming marking m_1 to marking m_2. Process terms $\alpha: m_1 \to m_2$ and $\beta: m_3 \to m_4$ can be sequentially composed, provided $m_2 = m_3$, resulting in $\alpha; \beta : m_1 \to m_4$. This notation illustrates the occurrence of β after the occurrence of α. The set of information elements of the sequentially composed process term is the union of the sets of information elements of the single process terms. The process terms can also be composed concurrently to $\alpha \parallel \beta : m_1 + m_3 \to m_2 + m_4$, provided the set of information elements of α is independent from (concurrent composable with) the set of information elements of β. The set of information elements of $\alpha \parallel \beta$ contains the concurrent composition of each element of the set of information elements of α with each element of the set of information elements of β. Since process terms have associated sets of information elements, we lift the partial algebra $(I, \dot{\parallel}, dom_{\dot{\parallel}})$ to the partial algebra $(2^I, \{\dot{\parallel}\}, dom_{\{\dot{\parallel}\}})$, where

- $dom_{\{\dot{\parallel}\}} = \{(X, Y) \in 2^I \times 2^I \mid X \times Y \subseteq dom_{\dot{\parallel}}\}$.
- $X\{\dot{\parallel}\}Y = \{x \dot{\parallel} y \mid x \in X \wedge y \in Y\}$.

For sequential composition of process terms we need information about the start and the end of a process term, which are both single markings. For concurrent composition, we require that the associated sets of information elements are independent. Two sets of information elements A and B do not have to be distinguished, if for each set of information elements C either both A and B are independent from C or both A and B are not independent from C. Therefore, we can use any equivalence $\cong \in 2^I \times 2^I$ that is a congruence with respect to the operations $\{\|\|\}$ (concurrent composition) and union $\cup$ (sequential composition) and satisfies $(A \cong B \wedge (A, C) \in dom_{\{\|\|\}}) \implies (B, C) \in dom_{\{\|\|\}}$, i.e. which is a *closed congruence* of the partial algebra $\mathcal{X} = (2^I, \{\|\|\}, dom_{\{\|\|\}}, \cup)$. The equivalence classes of the greatest (and hence coarsest) closed congruence represent the minimal information assigned to process terms necessary for concurrent composition. This congruence is unique ([2]). Now we can define an algebraic $(\mathcal{M}, \mathcal{I})$-net as given in [5].

Definition 2. *An algebraic $(\mathcal{M}, \mathcal{I})$-net is a quadruple $\mathcal{A} = (M, T, pre: T \to M, post: T \to M)$ together with a mapping $inf : M \cup T \to I$ satisfying*

(a) $\forall x, y \in M : \quad (inf(x), inf(y)) \in dom_{\dot{\oplus}} \implies inf(x + y) = inf(x) \dot{\oplus} inf(y)$.
(b) $\{inf(t)\} \cong \{inf(t), inf(pre(t)), inf(post(t))\}$.

Since $\dot{\|}$ is the restriction of $\dot{\oplus}$ to this domain, $\dot{\|}$ has the same property as $\dot{\oplus}$ in part (a). In the following definition we define steps of transitions, which represent their synchronous occurrences.

Definition 3. *Every step s has associated an initial marking $pre(s) \in M$, a final marking $post(s) \in M$, and an information element for concurrent and synchronous composition $inf(s) \in I$.*

The elementary step terms are transitions. If s, s' are step terms that satisfy $(inf(s), inf(s')) \in dom_{\dot{\oplus}}$, then their synchronous composition yields the step term $s \oplus s'$ with $pre(s \oplus s') = pre(s) + pre(s')$, $post(s \oplus s') = post(s) + post(s')$ and $inf(s \oplus s') = inf(s) \dot{\oplus} inf(s')$. The set of all step terms of $\mathcal{A}$ is denoted by $Step_{\mathcal{A}}$.

Finally, we are able to define inductively process terms for an algebraic $(\mathcal{M}, \mathcal{I})$-net. In comparison with [5] steps are used to be elementary process terms instead of single transitions.

Definition 4. *Let $\mathcal{A}$ be an algebraic $(\mathcal{M}, \mathcal{I})$-net. Every process term α has associated an initial marking $pre(\alpha) \in M$, a final marking $post(\alpha) \in M$, and an information for concurrent composition $Inf(\alpha) \in 2^I/\cong$. In the following, for a process term α we write $\alpha : a \longrightarrow b$ to denote that $a \in M$ is the initial marking of α and $b \in M$ is the final marking of α.*

For each $a \in M$, $id_a : a \longrightarrow a$ is a process terms with associated information $Inf(id_a) = [\{inf(a)\}]_\cong$. For each $s \in Step_{\mathcal{A}}$, $s : pre(s) \longrightarrow post(s)$ is a process term with associated information $Inf(s) = [\{inf(s)\}]_\cong$.

If $\alpha : a_1 \longrightarrow a_2$ and $\beta : b_1 \longrightarrow b_2$ are process terms satisfying $(Inf(\alpha), Inf(\beta)) \in dom_{\{\|\|\}}/\cong$, their concurrent composition yields the process term

$$\alpha \,\|\, \beta : a_1 + b_1 \longrightarrow a_2 + b_2$$

with associated information $Inf(\alpha \,\|\, \beta) = Inf(\alpha) \{\|\|\}/\cong Inf(\beta)$.

If $\alpha : a_1 \longrightarrow a_2$ and $\beta : b_1 \longrightarrow b_2$ are process terms satisfying $a_2 = b_1$, their sequential composition (concatenation) yields the process term

$$\alpha; \beta : a_1 \longrightarrow b_2$$

with associated information $Inf(\alpha; \beta) = Inf(\alpha) \cup/\cong Inf(\beta)$.

The partial algebra of all process terms with the partial operations concurrent composition and concatenation as defined above will be denoted by $\mathcal{P}(\mathcal{A})$.

Now, we define a congruence $\sim_t$ between process terms, saying when two terms are alternative descriptions of the same process.

Definition 5. *Let $\sim_t$ be defined on process terms as the smallest equivalence fulfilling the relations:*

1. $\alpha \parallel \beta \sim_t \beta \parallel \alpha$
2. $(\alpha \parallel \beta) \parallel \gamma \sim_t \alpha \parallel (\beta \parallel \gamma)$
3. $(\alpha; \beta); \gamma \sim_t \alpha; (\beta; \gamma)$
4. $\alpha = ((\alpha_1 \parallel \alpha_2); (\alpha_3 \parallel \alpha_4)) \sim_t$
   $\beta = ((\alpha_1; \alpha_3) \parallel (\alpha_2; \alpha_4))$
5. $\alpha \oplus \beta \sim_t \beta \oplus \alpha$

6. $(\alpha \oplus \beta) \oplus \gamma \sim_t \alpha \oplus (\beta \oplus \gamma)$
7. $(\alpha \oplus \beta) \sim_t (\alpha \parallel pre(\beta)); (post(\alpha) \parallel \beta)$
8. $(\alpha; post(\alpha)) \sim_t \alpha \sim_t (pre(\alpha); \alpha)$
9. $id_{(m+n)} \sim_t id_m \parallel id_n$
10. $(\alpha + id_\emptyset) \sim_t \alpha$

whenever the terms are defined (e.g. $\alpha \parallel \beta$ is defined iff $(A, B) \in dom_{\{\parallel\}}$, where $(A, B) = (Inf(\alpha), Inf(\beta))$. Axiom 4 holds whenever $Inf(\alpha) = Inf(\beta)$.

Henceforth, instead of writing id_m for any $m \in M$ we will usually only write just m. The only new axioms in comparison with [5] are axioms (5-7). Axioms (5) and (6) express commutativity and associativity of synchronous composition. The crucial axiom is the axiom (7), which enable to decompose (to sequentialize) synchronous steps. This axiom states that synchronous composition of two steps α and β and sequential composition of the step α (occurring concurrently with $pre(\beta)$) and the step β (occurring concurrently with $post(\alpha)$) are alternative decompositions of the same process whenever they are both defined. Surprisingly, as we prove later, this axiom is sufficient to identify all processes of nets with inhibitor arcs based on relational structures as defined in [7].

In the following the notion of a true synchronous step term is defined to be a set of transitions, which cannot be decomposed (cannot be sequentialized).

Definition 6. *Let s be a synchronous step term of an algebraic Petri net, such that for each pair of step terms s_1, s_2 satisfying $s_1 \oplus s_2 \sim_t s$ the term*

$$(s_1 \parallel pre(s_2)); (post(s_1) \parallel s_2)$$

is not a process term of the algebraic Petri net. Then s is called true synchronous step.

4   Elementary Nets with Inhibitor Arcs

In this section we will shortly describe the operationally defined a-priori step sequence and process semantics of elementary nets with inhibitor arcs, as defined in [7]. Although we restrict ourselves to elementary nets, the analogus results could be formulated for a-priori semantics of place/transition nets with inhibitor arcs defined in [9].

Definition 7. *A net is a triple* $N = (P, T, F)$, *where* P *and* T *are disjoint finite sets (of places and transitions, respectively) and* $F \subseteq (P \times T) \cup (T \times P)$. *An* elementary net with inhibitor arcs *is a quadruple* $ENI = (P, T, F, Inh)$, *where* (P, T, F) *is a net and* $Inh \subseteq P \times T$ *is an inhibitor relation satisfying* $(F \cup F^{-1}) \cap Inh = \emptyset$.

For a transition $t \in T$, ${}^\bullet t = \{p \in P \mid (p, t) \in F\}$ is the *pre-set* of t and $t^\bullet = \{p \in P \mid (t, p) \in F\}$ is the *post-set* of t, ${}^- t = \{p \in P \mid (p, t) \in Inh\}$ is the set of inhibiting places also called *negative context* of t. Elements of the inhibitor relation are graphically expressed by arcs ending with a circle (so called inhibitor arcs). Throughout the paper we assume that each transition has nonempty pre- and post-sets. A set $m \subseteq P$ is called *marking*. A transition t is enabled to occur in a marking m if no place from negative context ${}^- t$ belongs to m, every place from pre-set $\bullet t$ belongs to m, and no place from post-set $t\bullet$ belongs to m. The occurrence of t leads then to a new marking m', which is derived from m by removing the token from every place in $\bullet t$ and adding a token to every place in $t\bullet$. Thus, inhibiting places are tested on absence of tokens for the possible occurrence of a transition and this testing precedes the execution of the transition (so called a-priori semantics).

The following definitions introduce the basic notions of process semantics of elementary nets with inhibitor arcs introduced by [7].

Definition 8. *A* (labelled) occurrence net *is a labelled net* $ON = (B, E, R, l)$ *such that* $(\forall b \in B)(|{}^\bullet b| \leq 1 \geq |b^\bullet|)$, *the transitive closure* F^+ *of the relation* F *is irreflexive (i.e.* F^+ *is a partial order) and* l *is a labelling function for* $B \cup E$. *Elements of* B *are called* conditions, *elements of* E events.

Due to the fact that the absence of token in a place cannot be directly represented in an occurrence net, every inhibitor arc is replaced by an *activator* arc to a complement place. An activator arc (also called read arc, test arc, positive context arc) tests for the presence of a token in the place it is attached to. Moreover, the complement places remove possible contact situations, i.e. situations, when enabledness of a transition is violated by tokens in the post-set of the transition, i.e. by non-empty intersection of the actual marking and the post-set of the transition.

Definition 9. *Let* $ENI = (P, T, F, Inh)$ *be an elementary net with inhibitor arcs. Let* P' *be a set satisfying* $|P'| = |P|$ *and* $P' \cap (P \cup T) = \emptyset$, *let* $c : P \to P'$ *be a bijection. The complementation* $\overline{ENI} = (\overline{P}, T, \overline{F}, Act)$ *of* ENI *is defined by* $\overline{P} = P \cup \{c(p) \mid p \in P\}$, $\overline{F} = F \cup \{(t, c(p)) \mid (p, t) \in F \wedge (t, p) \notin F\} \cup \{(c(p), t) \mid (t, p) \in F \wedge (p, t) \notin F\}$ *and* $Act = \{(c(p), t) \mid (p, t) \in Inh\}$. *If initial marking* m_0 *of* ENI *is given, its complementation* $\overline{m_0}$ *is given by* $\overline{m_0} = m_0 \cup \{c(p) \mid p \in P \wedge p \notin m_0\}$.

Observe that this construction of $\overline{N}$ from N is unique up to isomorphism. In proofs we will take advantage of the fact that we define sets P and P' as disjoint though this is certainly very clumsy in graphical formalism. In most cases, only few places need to be equipped with complement places (co-places) for the system to become contact-free.

Definition 10. *A co-set of an occurrence net ON is a subset $S \subseteq B$ such that for no $a, b \in S: (a, b) \in R^+$. A slice is a maximal co-set. Let $Min(ON)$ denote the set of all minimal conditions of ON according to the partial order R^+. Similarly, let $Max(ON)$ denote the set of all maximal conditions of ON according to the partial order R^+.*

Definition 11. *Let $N = (P, T, F)$ be an elementary net and m_0 an initial marking. A process of N w.r.t. m_0 is a (labelled) occurrence net $ON = (B, E, R, l)$ such that these conditions are satisfied:*

1. *No isolated place of N is mapped by l to a co-place of N^2.*
2. *$l|_D$ is injective for every slice D of ON.*
3. *$l(Min(ON)) \cap P = m_0 \land l(Min(ON)) \subseteq \overline{m_0}$.*
4. *$\forall e \in E : l(^\bullet e) = {}^\bullet l(e) \land l(e^\bullet) = l(e)^\bullet$, where $^\bullet l(e)$ and $l(e)^\bullet$ refer to complementation $\overline{N}$ of N.*

We use $on(N, m_0)$ to denote the set of all processes of N w.r.t. m_0 and $on(N) = \bigcup_{m_0 \subseteq P} on(N, m_0)$ to denote the set of all processes of N.

Having defined the processes of "ordinary" elementary nets we may proceed to endow them with activator arcs. Let us first introduce the structure that is a generalization of the notion of partial order, suitable for the purpose of capturing both "earlier than" and "not later than" causality.

Definition 12. *A relational structure is a triple $\mathcal{S} = (X, \prec, \sqsubset)$. $\mathcal{S}$ is called a stratified order structure (so-structure) if the following conditions are satisfied:*

$$(C1) \; x \not\sqsubset x \qquad\qquad (C3) \; x \sqsubset y \sqsubset z \land x \neq z \Longrightarrow x \sqsubset z$$
$$(C2) \; x \prec y \Longrightarrow x \sqsubset y \qquad (C4) \; x \sqsubset y \prec z \lor x \prec y \sqsubset z \Longrightarrow x \prec z$$

Let $\mathcal{S} = (X, \prec, \sqsubset)$. The $\Diamond$-closure of $\mathcal{S}$ is the labelled relational structure

$$\mathcal{S}^\Diamond = (X, \prec_{\mathcal{S}^\Diamond}, \sqsubset_{\mathcal{S}^\Diamond}) = (X, (\prec \cup \sqsubset)^* \circ \prec \circ (\prec \cup \sqsubset)^*, (\prec \cup \sqsubset)^* \setminus id_X).$$

We say that a labelled relational structure $\mathcal{S}$ is $\Diamond$-acyclic if $\prec_{\mathcal{S}^\Diamond}$ is irreflexive.

It is easy to see that $(X, \prec)$ is a partially ordered set. Notice also that $\mathcal{S}^\Diamond$ is a labelled so-structure if and only if $\prec_{\mathcal{S}^\Diamond}$ is irreflexive. For these results see [7].

Definition 13. *A labelled activator occurrence net (ao-net) is a tuple $AON = (B, E, R, Act, l)$ such that: $ON = (B, E, R, l)$ is an occurrence net, $Act \subseteq B \times E$ are activator arcs, and the relational structure*

$$\mathcal{S}_{aux}(AON) = (E, \prec_{aux}, \sqsubset_{aux}) = (E, (R \circ R)|_{E \times E} \cup (R \circ Act), (Act^{-1} \circ R) \setminus id_E)$$

is $\Diamond$-acyclic.

---

[2] Isolated places of ON represent "unused" tokens of $\overline{m_0}$. We use only co-places which are necessary to get a contact-free system.

Definition 14. *An* activator process *of an elementary net with inhibitor arcs* $ENI = (P, T, F, Inh)$ *is an ao-net* $AON = (B, E, R, Act, l)$ *such that* $ON = (B, E, R, l) \in on(N)$ *(where* $N = (P, T, F)$*) and* $\forall b \in B \forall e \in E : (b, e) \in Act \iff (c^{-1}(l(b)), l(t)) \in Inh$*. The set of all activator processes of* ENI *is denoted by* $aon(ENI)$*.*

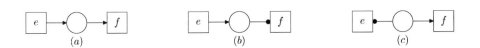

Fig. 3. Illustration of orders generation ([9]). Cases (a) and (b) generate $e \prec_{aux} f$, case (c) generates $e \sqsubset_{aux} f$.

In the diagrams, we draw activator arcs as arrows with black dots as heads and we write $^+t = \{b \mid (b, t) \in Act\}$.

Thus, $\prec_{aux}$ represent the causality "earlier than", while $\sqsubset_{aux}$ the causality "not later than". Because transitive closure of $\prec_{aux}$ is a partial order, there are no cycles formed by elements of $\prec_{aux}$. On the other hand, the transitive closure of $\sqsubset_{aux}$ is not necessary irreflexive, i.e. $\sqsubset_{aux}$ is not necessary acyclic. Then, cycles formed by elements of $\sqsubset_{aux}$ represent exactly the true synchronous step terms. Finally, the irreflexivity of $\prec_{s\diamond}$ expresses that there are no "combined" cycles, which obtain both elements of $\prec_{aux}$ and $\sqsubset_{aux}$.

5   Algebraic Representation of Elementary Nets with Inhibitor Arcs

In this section we represent elementary nets with inhibitor arcs with a-priori semantics as algebraic $(\mathcal{M}, \mathcal{I})$-nets. Let us consider the net $ENI = (P, T, F, Inh)$. We define $\mathcal{M} = (2^P, \cup)$. Further, let us define $pre(t) = {}^\bullet t$ and $post(t) = t^\bullet$ for each $t \in T$. To generate steps, we attach to transitions the information which consists of three disjoint components: the pre-set, the post-set, and the set of inhibiting places. So, we define (with $t \in T$ and $m \in 2^P$)

$$I = 2^P \times 2^P \times 2^P, \quad inf(t) = ({}^\bullet t, t^\bullet, {}^- t), \quad inf(m) = (m, m, \emptyset).$$

Synchronous composition of transitions t_1 and t_2 is possible only if places used by the transitions for token flow are disjoint, pre-set of t_1 is disjoint with the set of inhibiting places of t_2 and vice versa. In a-priori semantics, testing on absence of tokens precedes consuming/producing tokens. Therefore, we do not need to check whether the post-set places and the inhibiting places are disjoint in the case of the synchronous composition of transitions. Thus, we have $dom_{\dot\oplus} = \{((a, b, c), (d, e, f)) \in I \mid (a \cup b) \cap (d \cup e) = a \cap f = d \cap c = \emptyset\}$ and $(a, b, c) \dot\oplus (d, e, f) = (a \cup d, b \cup e, (c \cup f) \setminus (b \cup e))$.

For concurrent composition we want transitions to be independent and therefore we have to test whether both pre-sets and post-sets are disjoint with inhibiting places. Thus, we have $dom_{\parallel} = \{((a,b,c),(d,e,f)) \in I \mid (a \cup b) \cap (d \cup e) = (a \cup b) \cap f = c \cap (d \cup e) = \emptyset\}$.

Finally, we need to find the greatest closed congruence $\cong$ of $(2^I, \{\dot{\parallel}\}, dom_{\{\dot{\parallel}\}}, \cup)$. We define a mapping $supp$ which turns out to be the natural homomorphism of the congruence. The mapping yields the *support of the term*, i.e. the set of places that appear in the token flow, and the set of inhibiting places.

Definition 15. *Define two mappings $s_1, s_2 : 2^I \to 2^P$ by $s_1(A) = \bigcup_{(a,b,c) \in A}(a \cup b)$, $s_2(A) = \bigcup_{(a,b,c) \in A} c$ and $supp : 2^I \to 2^P \times 2^P$ by $supp(A) = (s_1(A), s_2(A) \setminus s_1(A))$.*

Lemma 1. *Denote $J = \{(x,y) \in 2^P \times 2^P \mid x \cap y = \emptyset\}$. Let $\circ$ be the binary operation on J defined by $(w,p) \circ (w',p') = (w \cup w', (p \cup p') \setminus (w \cup w'))$, $dom_{\overline{\parallel}} = \{((a,b),(c,d)) \in J \times J \mid a \cap c = b \cap c = a \cap d = \emptyset\}$ and $\overline{\parallel} = \circ|_{dom_{\overline{\parallel}}}$. Then the mapping $supp : (2^I, \{\dot{\parallel}\}, dom_{\{\dot{\parallel}\}}, \cup) \to (J, \overline{\parallel}, dom_{\overline{\parallel}}, \circ)$ is a surjective closed homomorphism.*

Lemma 2. *The closed congruence $\cong \subseteq 2^I \times 2^I$ defined by $A \cong B \iff supp(A) = supp(B)$ is the greatest closed congruence on $\mathcal{X} = (2^I, \{\dot{\parallel}\}, dom_{\{\dot{\parallel}\}}, \cup)$.*

The proofs of the previous two lemmas are similar to those in [5] for a-posteriori semantics of elementary nets with inhibitor arcs. Easy computation, using $(^\bullet t \cup t^\bullet) \cap {}^- t = \emptyset$ proves condition (b) of Definition 2, i.e. $supp(\{inf(t)\}) = supp(\{inf(t), inf(pre(t)), inf(post(t))\})$. The reader may observe that $\overline{\parallel}$ and J correspond to $\parallel$ and I in [5]; thus analogous results hold.

The partial algebra $(J, \overline{\parallel}, dom_{\overline{\parallel}}, \circ)$ is isomorphic with the greatest closed congruence on $\mathcal{X}$. Therefore, by construction of process terms (using concurrent and sequential composition) it is enough to save just the set of flow places and the set of inhibiting places which are not in the flow of the process as the information for deciding whether the processes are independent (concurrent composable). Now we are able to represent an elementary net with inhibitor arcs as an algebraic $(\mathcal{M}, \mathcal{I})$-net.

Theorem 1. *Let $ENI = (P, T, F, Inh)$ be an elementary net with inhibitor arcs, together with $\mathcal{M}, \mathcal{I}, pre, post, inf$ defined throughout this section. Then the quadruple $\mathcal{A}_{ENI} = (2^P, T, pre, post)$ together with the mapping inf is an algebraic $(\mathcal{M}, \mathcal{I})$-net.*

Definition 16. *The algebraic $(\mathcal{M}, \mathcal{I})$-net from the previous theorem is called the corresponding $(\mathcal{M}, \mathcal{I})$-net to the elementary net with inhibitor arcs ENI.*

Example 1. In the net from Figure 1 the expressions $(f \parallel \{p_1\}); (e \parallel \{p_4\})$, $e \oplus f$ and $f \oplus e$ are defined process terms, but $f \parallel \{p_3\}$ is not defined, because inhibiting place p_3 of f is also pre and post place of the elementary process term $\{p_3\}$

which means that the information elements of terms f and $\{p_3\}$ cannot be composed concurrently. By this the expression $(e \,\|\, \{p_2\}); (f \,\|\, \{p_3\})$ is not a defined process term, i.e. the sequence ef cannot be executed in the net from Figure 1. Similarly, in the net from Figure 2 neither expression $(e \,\|\, \{p_2\}); (f \,\|\, \{p_3\})$ nor expression $(f \,\|\, \{p_1\}); (e \,\|\, \{p_4\})$ are defined process terms, i.e. neither sequence ef nor sequence fe can occur in the net. The only expressions containing both e and f, which are defined process terms are $e \oplus f$ and $f \oplus e$, i.e. the only possibility to occur both e and f is to do it synchronously.

6   Activator Processes versus Process Terms

In this section we establish the main result on the relationship between the newly defined process terms and the activator processes of elementary nets with inhibitor arcs with a-priori semantics as introduced in [7]. We have the following pattern of the proof: First, we define a mapping τ that associates an activator occurrence net with each process term. Second, we prove that τ is surjective, i.e. every activator process can be represented by a process term. Then we show that $\alpha \sim_t \beta \Rightarrow \tau(\alpha) = \tau(\beta)$, i.e. equivalent process terms are mapped by τ on the same activator process. Finally, we prove that $\tau(\alpha) = \tau(\beta) \Rightarrow \alpha \sim_t \beta$. Hence, at the end of this section we have formulated the main theorem of the paper, which states that activator processes correspond bijectively to $\sim_t$-equivalence classes of process terms.

Because the process terms corresponding to markings are determined in a process term α by the value $pre(\alpha)$, in the following we will often omit them in process terms. In the sequel we consider an elementary net with inhibitor arcs $ENI = (P, T, F, Inh)$, its corresponding $(\mathcal{M}, \mathcal{I})$-net $\mathcal{A}_{ENI}$, and two actuator processes $K_1 = (B_1, E_1, R_1, Act_1, l_1)$ and $K_2 = (B_2, E_2, R_2, Act_2, l_2)$ of ENI. We will often state that $K_1 = K_2$ even if this is not exactly true and only the *graphical representations* of the nets are the same, i.e. there exists a bijective mapping that preserves labelling, flow and read arcs. If needed, we use $\approx$ to denote this isomorphism. Every time we use l without indices we mean the labelling function constructed as "union" of l_1 and l_2. Exactly, $l|_{B_1 \cup E_1} = l_1$ and $l|_{B_2 \cup E_2} = l_2$. From the definitions we can verify that it is always possible to construct such a function. We will commonly use the shorthand "places match". We precisely mean: Let $b_1 \in B_1, b_2 \in B_2$. We say that b_1 and b_2 match if $l_1(b_1) = l_2(b_2)$. In definitions of the compositional net operations we usually remove one of the matching places and attach the arcs adjacent with the removed place to the other. We then say these places were *glued*.

Definition 17 (Notation). *Let $K = (B, E, R, Act, l)$ be an ao-net. We define the set of* isolated places *by $^\emptyset K = \{b \in B \mid \forall e \in E : b \notin {}^\bullet e \cup e^\bullet \cup {}^+ e\}$, the set of* write places *(flow places) by $^\circ K = \bigcup_{e \in E}({}^\bullet e \cup e^\bullet) \cup {}^\emptyset K$ and the set of purely* read places *by $^+ K = (\bigcup_{e \in E} {}^+ e) \setminus {}^\circ K$.*

In the following definition we define the mapping τ associating activator processes to process terms representing markings and single transitions.

Definition 18.
Let $m \in M$ and $\alpha = m : m \to m$ be the related process term of $\mathcal{A}_{ENI}$. Define
$\tau(\alpha) = K_\alpha = (m, \emptyset, \emptyset, \emptyset, id_m)$. *Let* $t \in T$ *and* $\alpha = t : pre(t) \to post(t)$. *Define*
$\tau(\alpha) = K_\alpha = ({}^\bullet t \cup t^\bullet \cup {}^+t, \{t\}, ({}^\bullet t \times \{t\}) \cup (\{t\} \times t^\bullet), {}^+t \times \{t\}, id_{{}^\bullet t \cup t^\bullet \cup {}^+t \cup \{t\}})$,
where ${}^\bullet t$, $t^\bullet$ *and* ${}^+t$ *are with respect to* $\overline{ENI}$.

Observe, that K_m and K_t are activator processes. For a (step or process) term α we say $\tau(\alpha)$ to be the corresponding activator process. Beginning with processes representing markings and single transitions we define inductively the corresponding activator processes for synchronous composed step terms α_1 and α_2, and concurrent and sequential composed process terms α_1 and α_2, using activator processes K_1 and K_2 corresponding to the terms α_1 and α_2. In definitions we always assume that $B_1 \cap B_2 = E_1 \cap E_2 = \emptyset$. We can always achieve this by appropriate renaming. We also define $Int_\| = \{p \in P \mid (\exists b_1 \in {}^+K_1)(\exists b_2 \in {}^+K_2)(l_1(b_1) = l_2(b_2) = p)\}$ as the set of places that are used in both processes as purely read places.

To obtain the activator process corresponding to synchronous composed step terms we define the *synchronous interface* $Int_\oplus^i$ as the set of places that are used as read places in the process K_i and as write places in the other process. We delete places of $Int_\oplus^i$ and $Int_\|$ from K_i (places that are only used as read in one process remain), put the processes side-by-side, we glue read places of processes with their write counterparts in the other process. Then we add the "purely read" places in one copy and restore the *Act* relation.

Definition 19. *Define the* synchronous interfaces:
$$Int_\oplus^1 = \{p \in P \mid (\exists b_1 \in {}^+K_1)(\exists b_2 \in {}^\circ K_2)(l_1(b_1) = l_2(b_2) = p)\}$$
$$Int_\oplus^2 = \{p \in P \mid (\exists b_2 \in {}^+K_2)(\exists b_1 \in {}^\circ K_1)(l_1(b_1) = l_2(b_2) = p)\}$$
$$B_i' = (B_i \setminus (l_i^{-1}(Int_\oplus^i) \cap {}^+K_i)) \setminus l_i^{-1}(Int_\|)$$
$$Act_i' = Act_i \cap (B_i' \times E_i)$$
Define $\tau(\alpha_1 \oplus \alpha_2) = K_{\alpha_1 \oplus \alpha_2} = (B, E, R, Act, l) = (B, E_1 \cup E_2, R_1 \cup R_2, Act, l)$,
where $B = B_1' \cup B_2' \cup l_1^{-1}(Int_\|)$ *and*
$$Act = Act_1' \cup Act_2' \cup$$
$$\{(b_1, e_2) \mid b_1 \in B_1' \wedge (\exists b_2 \in l_2^{-1}(Int_\oplus^2))((b_2, e_2) \in Act_2 \wedge l_1(b_1) = l_2(b_2))\} \cup$$
$$\{(b_2, e_1) \mid b_2 \in B_2' \wedge (\exists b_1 \in l_1^{-1}(Int_\oplus^1))((b_1, e_1) \in Act_1 \wedge l_1(b_1) = l_2(b_2))\} \cup$$
$$\{(b, e_i) \mid b \in l^{-1}(Int_\|) \wedge e_i \in E_i \wedge (\exists b_i \in B_i)(l_i(b_i) = l(b) \wedge (b_i, e_i) \in Act_i),$$
for $i = 1, 2\}$.

Lemma 3. *If* $\alpha_1 \oplus \alpha_2$ *is a defined step term, then* $K_{\alpha_1 \oplus \alpha_2}$ *as defined in the previous definition is an activator process.*

Proof. (Sketch) Observe that the activator processes corresponding to step terms consist of three "layers". The first consists of pre-places, the second of transitions and the third of post-places. We show the following:

(i) (B, E, R, l) is a labelled occurrence net (that means R^+ is irreflexive). This follows immediately from $R = R_1 \cup R_2$.

(ii) (B, E, R, l) is a process (see definition 11). Only the injectivity of the labelling on all slices is not obvious. It follows from the facts, that every slice

D of the composed net is of the form $D = D_1 \cup D_2$ with slices D_1 of K_1 and D_2 of K_2, and that from the precondition ($\alpha_1 \oplus \alpha_2$ is a defined step term) the labelling images of the flow places of K_1 and K_2 are disjoint (see [5] for a detailed proof of a similar statement).

(iii) $K_{\alpha_1 \oplus \alpha_2}$ is an ao-net (that means $\mathcal{S}_{aux}(K_{\alpha_1 \oplus \alpha_2})$ is $\Diamond$-acyclic). This follows from the observation, that activator arcs are only connected with places from the pre-places layer.

(iv) $K_{\alpha_1 \oplus \alpha_2}$ is an activator process (that means the labelling respects the inhibiting relation of ENI). That is obvious.

To obtain the activator process corresponding to concurrent composed process terms, we remove the matching read places from K_2, put the two processes together and restore activator arcs incident with erased places, using same-labelled places of K_1.

Definition 20. *Set $B_2' = B_2 \setminus l_2^{-1}(Int_\|)$, $Act_2' = Act_2 \cap (B_2' \times E_2)$ and define $\tau(\alpha_1 \| \alpha_2) = K_{\alpha_1 \| \alpha_2} = (B, E, R, Act, l) = (B_1 \cup B_2', E_1 \cup E_2, R_1 \cup R_2, Act, l_1 \cup l_2)$, where $Act = Act_1 \cup Act_2' \cup \{(b_1, e_2) \in B_1 \times E_2 \mid (\exists b_2 \in {}^+K_2)(l_1(b_1) = l_2(b_2) \wedge (b_2, e_2) \in Act_2)\}$.*

Lemma 4. *If $\alpha_1 \| \alpha_2$ is a defined process term, then $K_{\alpha_1 \| \alpha_2}$ as defined in the previous definition is an activator process.*

Proof. (Sketch) The structure of the proof is the same as in the proof of the previous lemma. The statements (i), (ii) and (iv) can be proven analogously. Let $\mathcal{S}_{aux}(K_{\alpha_1 \| \alpha_2}) = (E, \prec_{aux}, \sqsubset_{aux})$, $\mathcal{S}_{aux}(K_1) = (E, \prec_{aux}^1, \sqsubset_{aux}^1)$ and $\mathcal{S}_{aux}(K_2) = (E, \prec_{aux}^2, \sqsubset_{aux}^2)$. Because no read place of the one process is matched and glued with a write place of the other process, we have $\prec_{aux} = \prec_{aux}^1 \cup \prec_{aux}^2$ and $\sqsubset_{aux} = \sqsubset_{aux}^1 \cup \sqsubset_{aux}^2$. Statement (iii) follows.

To obtain the activator process corresponding to sequential composed process terms we remove those minimal places of K_2 that match a maximal place in K_1 (the sequential interface $Int_;$), and attach the arcs originally attached to the minimal elements to these maximal elements.

Definition 21. *We define the* sequential interface *of the processes K_1, K_2 by $Int_;(K_1, K_2) := \{p \in P \mid (\exists b_1 \in Max(K_1))(\exists b_2 \in Min(K_2))(l(b_1) = l(b_2) = p)\}$.*

Set
$$B_2' = B_2 \setminus \{b_2 \in Min(K_2) \mid l(b_2) \in Int_;(K_1, K_2)\}$$
$$R_2' = R_2 \cap ((E_2 \times B_2') \cup (B_2' \times E_2))$$
$$Act_2' = Act_2 \cap (B_2' \times E_2) \text{ and}$$

define $\tau(\alpha_1; \alpha_2) = K_{\alpha_1; \alpha_2} = (B, E, R, Act, l) = (B_1 \cup B_2', E_1 \cup E_2, R, Act, l)$,

where
$$R = R_1 \cup R_2' \cup \{(b_1, e_2) \mid b_1 \in Max(K_1) \wedge$$
$$(\exists b_2 \in Min(K_2): l_1(b_1) = l_2(b_2) \wedge (b_2, e_2) \in R_2)\}$$
$$Act = Act_1 \cup Act_2' \cup \{(b_1, e_2) \mid b_1 \in Max(K_1) \wedge$$
$$(\exists b_2 \in B_2: l_1(b_1) = l_2(b_2) \wedge b_2, e_2) \in Act_2)\}.$$

Lemma 5. *If $\alpha_1; \alpha_2$ is a defined process term, then $K_{\alpha_1; \alpha_2}$ as defined in the previous definition is an activator process.*

Proof. (Sketch) The structure of the proof is the same as in the proof of the previous lemma. The statements (i), (ii) and (iv) follow from construction. Let $\mathcal{S}_{aux}(K_{\alpha_1;\alpha_2}) = (E, \prec_{aux}, \sqsubset_{aux})$. From construction we have $\forall e_1 \in E_1, e_2 \in E_2 :$ $e_2 \not\prec_{aux} e_1 \wedge e_2 \not\sqsubset_{aux} e_1$. Therefore (iii) is satisfied.

Remark 1. Let $\alpha_1; \alpha_2$ be a defined process term and let $\mathcal{S}_{aux}^{\diamond} = (E, \prec, \sqsubset)$ be the so-structure associated with $\tau(\alpha_1; \alpha_2)$. Let E_1, E_2 denote the set of events of $\tau(\alpha_1), \tau(\alpha_2)$, respectively. Then $\forall e_1 \in E_1, e_2 \in E_2 : e_2 \not\prec e_1 \wedge e_2 \not\sqsubset e_1$.

Naturally, the relationships between "earlier than" and "not later than" causalities on one side and definition domains of concurrent and synchronous composition play a crucial rôle in the proof of correspondence between the process term semantics and the activator process semantics.

If (copies of) transitions (in an activator process) are not ordered by "earlier than" causality, then they may be executed synchronously.

Lemma 6. *Let $K = (B, E, R, Act, l)$ be an activator process and let $e_1, e_2 \in E$. Let $\mathcal{S}_{aux}^{\diamond} = (E, \prec, \sqsubset)$ be the associated so-structure. If $e_1 \not\prec e_2$ and $e_2 \not\prec e_1$ then $l(e_1) \oplus l(e_2)$ is a defined process term.*

Proof. (Sketch) Assume the term $l(e_1) \oplus l(e_2)$ is not defined (although $e_1 \not\prec e_2$ and $e_2 \not\prec e_1$), i.e. the associated information elements of e_1 and e_2 are not composable by $\dot{\oplus}$. This means, one of the following cases must be fulfilled:

(i) $^{\bullet}l(e_1) \cap {}^{\bullet}l(e_2) \neq \emptyset$. Observe that $^{\bullet}e_1 \cap {}^{\bullet}e_2 = \emptyset$ and $e_1^{\bullet} \cap e_2^{\bullet} = \emptyset$ by the definition of process nets. So there are places $b_1, b_2 \in B$ with $b_1 \in {}^{\bullet}e_1$, $b_2 \in {}^{\bullet}e_2$, $l(b_1) = l(b_2)$ and $b_1 \neq b_2$. Because the labelling is injective on slices, b_1 and b_2 are in different slices, and therefore are ordered by the transitive closure R^+ of the flow relation. Since conditions are unbranched, e_1 and b_2 or e_2 and b_1 must be ordered by R^+. Since $R^+ \subseteq \prec = (\prec_{aux} \cup \sqsubset_{aux})^* \circ \prec_{aux} \circ (\prec_{aux} \cup \sqsubset_{aux})^*$, this contradicts $e_1 \not\prec e_2 \wedge e_2 \not\prec e_1$. The proof for $l(e_1)^{\bullet} \cap l(e_2)^{\bullet} \neq \emptyset$ is analogous.

(ii) $^{\bullet}l(e_1) \cap l(e_2)^{\bullet} \neq \emptyset$. If $^{\bullet}e_1 \cap e_2^{\bullet} \neq \emptyset$, we have directly a contradiction of $e_1 \not\prec e_2 \wedge e_2 \not\prec e_1$. If not, the proof is similar to (i). $^{\bullet}l(e_1) \cap {}^+l(e_2) \neq \emptyset$ and $^+l(e_1) \cap l(e_2)^{\bullet} \neq \emptyset$ are proven in the same way.

Moreover, if (copies of) transitions (in an activator process) are neither ordered by "earlier than" causality nor by "not later than" causality, then they may be executed concurrently.

Lemma 7. *Let $K = (B, E, R, Act, l)$ be an activator process and let $e_1, e_2 \in E$. Let $\mathcal{S}_{aux}^{\diamond} = (E, \prec, \sqsubset)$ be the associated so-structure. If $e_1 \not\prec e_2$, $e_2 \not\prec e_1$, $e_1 \not\sqsubset e_2$ and $e_2 \not\sqsubset e_1$, then $l(e_1) \| l(e_2)$ is a defined process term.*

Proof. (Sketch) Assuming the term $l(e_1) \| l(e_2)$ is not defined one can prove the lemma in a same way as the previous one. The only new inequality which has to be fulfilled is $^+l(e_1) \cap {}^{\bullet}l(e_2) \neq \emptyset$. Because $R^+ \subseteq \sqsubset = (\prec_{aux} \cup \sqsubset_{aux})^* \setminus id_E$, this can be proven analogously as (i) in the previous lemma.

Theorem 2 (τ is surjective). *For every activator process K of ENI there is a process term α, such that $\tau(\alpha) = K$.*

Proof. With lemma 6 we can prove the statement analogous to the correspond-ing theorem in [5] (replacing $\parallel$ by $\oplus$) using sequential composition of maximal synchronous step terms. The searched term α is of the form $\alpha = \alpha_0; \ldots; \alpha_n$, where α_k is a maximal synchronous step term.

If two process terms are alternative decompositions of the same process, then they should naturally have the same corresponding activator process. This is indeed so. The proof consists of verifying the claim for each of the ten axioms in definition 5 and we omit it. The interested reader may find the proof of the corresponding theorem in [5], proving the statement for the axioms (1)-(4) and (8)-(10). For the axioms (5) and (6) the statement is obvious. For axiom (7) the statement is proven in a similar way as for axiom (4).

Theorem 3. *Let α, β be process terms of $\mathcal{A}_N$. Then $\alpha \sim_t \beta \Longrightarrow \tau(\alpha) = \tau(\beta)$.*

The proof of the converse, the theorem similar to theorem 7 found in [5] (the crucial theorem of that paper) does not work anymore, since there can be transitions in a step term, which cannot be sequentialized, because they are true synchronous. So every process term can only be sequentialized down to true synchronous step terms. We want to identify subsets of events of an activator process, which correspond to true synchronous step terms via the mapping τ. Clearly a set of events, that can be cyclicly ordered by the "not later than" causality, cannot occur sequentially and therefore is a candidate for such a subset. We will need exactly such sets which are maximal w.r.t. the $\subseteq$-relation.

Definition 22. *We say that a process term α_{seq} is maximally sequentialized if and only if it is of the form $(a_1 \parallel s_1); \ldots; (a_k \parallel s_k)$, where s_i is a true synchronous step term and $a_i \in M$ for all $i \in \{1, \ldots, k\}$.*

Lemma 8. *Let α be a process term. Then there exists a term α_{seq} such that $\alpha \sim_t \alpha_{seq}$ and α_{seq} is a maximally sequentialized process term.*

Proof. (Sketch) Inductively, replace $\alpha \parallel \beta$ with $\alpha; \beta$. Replace $\alpha \oplus \beta$ with either $\alpha; \beta$ or $\beta; \alpha$, whichever is defined. If none is defined, then it can be proved that $\alpha \oplus \beta$ is a part of a true synchronous step term. The algorithm define functions denoted by $\cdot_{seq}$.

Definition 23. *Let $K = (B, E, R, Act, l)$ be an activator process, $\prec_{aux}$ be the associated "earlier than" causality on E and $\sqsubseteq_{aux}$ be the associated "not later than" causality on E. A set $\eta \subseteq E$ is called a cyclic event, if it either contains exactly one element or the following two conditions are fulfilled:*

(i)   The events from E are pairwise unordered w.r.t. $\prec_{aux}$.

(ii) There is a sequence $e_1 e_2 \ldots e_k$ of events from η, such that $e_i \sqsubseteq_{aux} e_{i+1}$ and $e_1 = e_k$ ($i \in \{1, \ldots, k-1\}$)[3]. In other words there is cycle w.r.t. $\sqsubseteq_{aux}$ through all events in η.

An synchronous event of K is a cyclic event, that is maximal w.r.t. the $\subseteq$-relation.

---

[3] Of course it is allowed, that some events of η appear more than once in the sequence

Clearly, synchronous events of an activator process are disjoint. Moreover, we can extend the lemma 6 as follows:

Lemma 9. *Let $K = (B, E, R, Act, l)$ be an activator process and let η be its synchronous event. Then $s = \bigoplus_{e \in \eta} l(e)$ is a defined process term.*

Applying this lemma, the property of $\mathcal{S}_{aux}^{\diamond}$, and the definition of $\|$ we can also extend the lemma 7.

Lemma 10. *Let $K = (B, E, R, Act, l)$ be an activator process, and η_1, η_2 its synchronous events. Let $\mathcal{S}_{aux}^{\diamond} = (E, \prec, \sqsubset)$ be the associated so-structure. , If $e_1 \not\prec e_2$, $e_2 \not\prec e_1$, $e_1 \not\sqsubset e_2$ and $e_2 \not\sqsubset e_1$ for some $e_1 \in \eta_1, e_2 \in \eta_2$, then $\bigoplus_{e \in \eta_1} l(e) \| \bigoplus_{e \in \eta_2} l(e)$ is a defined process term.*

We can characterize cyclic events in the following way.

Lemma 11. *Let $K = (B, E, R, Act, l)$ be an activator process. A set $\eta \subseteq E$ with at least two elements is a cyclic event, if and only if*

(i)   The events from η are pairwise unordered w.r.t. $\prec_{aux}$.
(ii) For every nonempty subset $\varphi \subset \eta$ we have: There are events $e_1, e_2 \in \varphi$, $f_1, f_2 \in \eta \setminus \varphi$, such that $e_1 \sqsubset_{aux} f_1$ and $f_2 \sqsubset_{aux} e_2$.

In this statement condition (ii) can be equivalently replaced by condition

(ii)' For every nonempty subset $\varphi \subset \eta$ we have: There are conditions $b \in \bigcup_{e \in \varphi} {}^{\bullet}e$, $b' \in \bigcup_{e \in \eta \setminus \varphi} {}^{\bullet}e$ and events $f \in \varphi$, $f' \in \eta \setminus \varphi$, such that $b \in {}^{+}f'$ and $b' \in {}^{+}f$.

Proof. The equivalence between conditions (ii) and (ii)' follows directly from the definition of $\sqsubset_{aux}$.

if: Assume η is not a cyclic event, that means there are two events $e, f \in \eta$ with $e \not\sqsubset_{aux}^{*} f$, although condition (ii) is fulfilled. Set $\varphi = \{e' \in \eta \mid e \sqsubset_{aux}^{*} e'\}$. Obviously $e' \not\sqsubset_{aux}^{*} f$ for all $e' \in \varphi$. This contradicts (ii).

only if: $e_i \sqsubset_{aux} e_{i+1}$ implies by the definition of $\sqsubset_{aux}$, that there is a condition $b_{i+1} \in {}^{\bullet}b_{i+1}$ with $b_{i+1} \in {}^{+}e_i$. Assume there is a subset $\varphi \subset \eta$, which does not fulfil condition (ii), although η is a cyclic event. Without loss of generality assume that there is no condition $b \in \bigcup_{e \in \varphi} {}^{\bullet}e$, such that there exists an event $f' \in \eta \setminus \varphi$ with $b \in {}^{+}f'$. In other words $(\bigcup_{e \in \varphi} {}^{\bullet}e) \cap (\bigcup_{e \in \eta \setminus \varphi} {}^{+}e) = \emptyset$, i.e. $\forall e \in \eta, \forall f \in \eta \setminus \varphi : f \not\sqsubset_{aux} e$. It follows, that the transitive closure of $\sqsubset_{aux}$ restricted to η cannot be symmetric, which is a contradiction to the fact, that η is a cyclic event.

Lemma 12. *Let s be a true synchronous step term and $\tau(s) = (B, E, R, Act, l)$ be the corresponding activator process. Then E is a synchronous event of $\tau(s)$.*

Proof. Assume E is not a synchronous event although s is a true synchronous step term. Then there exists $\varphi \subseteq E$ such that there are either no activator arcs from $\bigcup_{e \in \varphi} {}^{\bullet}e$ to $E \setminus \varphi$ or no activator arcs from $\bigcup_{e \in E \setminus \varphi} {}^{\bullet}e$ to φ. Without loss of generality assume that there are no activator arcs from $\bigcup_{e \in \varphi} {}^{\bullet}e$ to $E \setminus \varphi$. From lemma 6 and commutativity and associativity of $\oplus$, terms

$\bigoplus_{e \in \varphi} l(e)$ and $\bigoplus_{f \in E \setminus \varphi} l(f)$ are defined terms. From definition of $dom_{\parallel}$ the composition $\bigoplus_{e \in \varphi} l(e) \parallel post(\bigoplus_{f \in E \setminus \varphi} l(f))$ is defined and therefore also the term $(\bigoplus_{f \in E \setminus \varphi} l(f) \parallel pre(\bigoplus_{e \in \varphi} l(e)); (\bigoplus_{e \in \varphi} l(e) \parallel post(\bigoplus_{f \in E \setminus \varphi} l(f))$ is defined. This contradicts the fact that s is a true synchronous step term (i.e. that s cannot be sequentialized).

Lemma 13. *Let* $\alpha = s_1; \dots ; s_k$ *be a maximally sequentialized process term. Let* $SE = \{\eta_1, \dots, \eta_n\}$ *be the set of all synchronous events of the activator process* $\tau(\alpha)$. *Then* $n = k$ *and there exist a permutation* v *such that:* $\alpha = \bigoplus_{e \in \eta_{v_1}} l(e); \dots ; \bigoplus_{e \in \eta_{v_n}} l(e)$.

Proof. By induction: From previous lemma the statement is valid for $k = 1$. If it is valid for $i < k$, then from the proof of lemma 5 adding $\tau(s_{i+1})$ to $\tau(s_1; \dots, s_i)$ we cannot extend neither any existing synchronous event of $\tau(s_1; \dots, s_i)$ by an event from $\tau(s_{i+1})$ nor the added synchronous event of $\tau(s_{i+1})$ by an event from $\tau(s_1; \dots, s_i)$. Since the set of events E_{i+1} of $\tau(s_{i+1})$ forms itself a synchronous event satisfying $s_{i+1} = \bigoplus_{e \in E_{i+1}} l(e)$, the statement is valid for $i + 1$.

Corollary 1. *Let* $\alpha = s_1; \dots ; s_k$ *and* $\beta = r_1; \dots ; r_n$ *be maximally sequentialized process terms. If* $\tau(\alpha) = \tau(\beta)$ *then* $n = k$ *and there exist a permutation* v *such that* $\alpha = r_{v_1}; \dots ; r_{v_n}$.

Now we can prove the last part needed for the correspondence between process term semantics and activator process semantics.

Theorem 4. *Let* α, β *be process terms. Then* $\tau(\alpha) = \tau(\beta) \Longrightarrow \alpha \sim_t \beta$.

Proof. Lemma 8 provides us with maximally sequentialized process terms α_{seq} and β_{seq} such that $\alpha_{seq} \sim_t \alpha$ and $\beta_{seq} \sim_t \beta$. By theorem 3 we have $\tau(\alpha_{seq}) = \tau(\alpha) = \tau(\beta) = \tau(\beta_{seq})$. Thus, it suffices to show that $\alpha_{seq} \sim_t \beta_{seq}$. Denote $\alpha_{seq} = s_1; \dots ; s_k$ and $\beta_{seq} = r_1; \dots ; r_n$. By corollary 1 we have $n = k$ and there exists a permutation v such that $\alpha_{seq} = r_{v(1)}; \dots ; r_{v(n)}$. If $\alpha_{seq} \neq \beta_{seq}$, consider that i is the first index satisfying $v(i) \neq i$ (obviously $v(i) > i$). The idea is to "bubble-sort" step term $r_{v(i)}$ from the position $v(i)$ in β_{seq} backwards to the position i, which it has in α_{seq}, and repeat this procedure until there is no such i. Thus, it suffices to prove that $r_1; \dots ; r_{v(i)-1}; r_{v(i)}; \dots ; r_n \sim_t r_1; \dots ; r_{v(i)}; r_{v(i)-1}; \dots ; r_n$, i.e. that we can exchange $r_{v(i)-1}$ and $r_{v(i)}$ in β_{seq}. A sufficient condition for this is that $r_{v(i)-1} \parallel r_{v(i)}$ is a defined process term. Since i was the first index with the property $v(i) \neq i$, the position of $r_{v(i)-1}$ in α_{seq} is at least $i + 1$, i.e. $\alpha_{seq} = r_{v(1)}; \dots ; r_{v(i-1)}; r_{v(i)}; \dots ; r_{v(i)-1}; \dots r_{v(n)}$. Thus, from β_{seq} and Remark 1 we have $\forall e \in E_{v(i)-1}, f \in E_{v(i)} : f \not\lessdot e \wedge f \not\sqsubset e$, where $E_{v(i)-1}, E_{v(i)}$ denote the set of events of $\tau(r_{v(i)-1}), \tau(r_{v(i)})$, respectively. On the other hand, from α_{seq} and Remark 1 we have $\forall e \in E_{v(i)-1}, f \in E_{v(i)} : e \not\lessdot f \wedge e \not\sqsubset f$. From lemma 10 follows that $r_{v(i)-1} \parallel r_{v(i)}$ is a defined process term, what finishes the proof.

Now we are prepared to state the main result of the paper, which is an immediate consequence of theorems proved in this section.

Theorem 5. *Let* ENI *be an elementary net with inhibitor arcs. Let* $\mathcal{A}_{ENI}$ *be the corresponding algebraic* $(\mathcal{M}, \mathcal{I})$*-net. Then there is a one-to-one correspondence between* $\sim_t$*-equivalence classes of* $\mathcal{A}_N$ *and (isomorphism classes) of activator processes of* ENI.

7   Conclusion

In this paper we have presented an abstract axiomatic semantics for Petri nets, which enables to distinguish between synchronous and concurrent occurrence of steps. Within this framework, we have also defined notion of true synchronicity. The approach is based on rewriting logic with restricted definition domains of operations. We claim that our approach offers a dual description of processes based on "earlier than" and "not later than" causalities defined in the seminal papers [6,7]. We illustrate our claim proving a one-to one correspondence between our process semantics and those described in [7] for elementary nets with inhibitor arcs.

There are several works following the ideas of [7] in literature, e.g. [8] for nets with priorities and [9] for place/transition nets with inhibitor arcs. Many other papers also discuss weak causality of nets with inhibitor/read arcs [11,13, 12], or nets with read arcs [15]. However, they exclude the case where transitions are cyclically ordered by inhibitor arcs (or read arcs). In ([13], [12]) the weak causality is rather understood as an asymmetric conflict. In [15] the nets with read arcs are investigated. Duration is supposed and the causalities "e necessarily ends before f starts" and "e necessarily starts before f starts" are used. Thus, in the example from Figure 1, for e and f both to occur, f has to start before b. According [15], in the situation from Figure 2, e and f cannot both occur, because intuitively they block each other (to occur both, first the test on the *presence* of tokens has to be done (i.e. one event starts earlier) and after that the token is consumed (the second event starts), and after that occurrence is finished (tokens in post-sets are produced). A similar intuition (if a duration is assumed, then consuming a token take a time and during this time absence of tokens is post-sets can be tested) allow occurrence of both events in Figure 2.

An advantage of the algebraic approach we have presented is the fact, that it offers an abstract framework, where by "tuning" the underlying algebra of information elements and the definition domain of synchronous and concurrent composition, one can define non sequential semantics of different variants and dialects of nets in a unifying way. Moreover, true synchronous steps play a crucial role in Petri nets enriched by signals arcs, which are extensively used in modelling and control of engineering systems. Presently, we are working on non-sequential semantics for this class of nets. Another area of our present research consists in developing a suitable general mechanism which will allow to produce causal relations directly from the process terms.

References

1. E. Best and R. Devillers. Sequential and nonsequential behaviour in Petri nets. *Theoretical Computer Science*, 55:87–136, 1987.
2. P. Burmeister. *Lecture Notes on Universal Algebra – Many Sorted Partial Algebras.* TU Darmstadt, 2002.
3. P. Degano, J. Meseguer, and U. Montanari. Axiomatizing the algebra of net computations and processes. *Acta Informatica*, 33(7):641–667, 1996.

4. J. Desel, G. Juhás, and R. Lorenz. Process semantics of Petri nets over partial algebra. In Mogens Nielsen and Dan Simpson (Eds.): *Proc. of 21th International Conference on Applications and Theory of Petri Nets*, LNCS 1825, pp. 146–165, Springer-Verlag, 2000.
5. J. Desel, G. Juhás, and R. Lorenz. Petri nets over partial algebra. In H. Ehrig, G. Juhás, J. Padberg, G. Rozenberg (Eds.): *Unifying Petri Nets*, LNCS 2128, pp. 126–171, Springer-Verlag, 2001.
6. R. Janicki and M. Koutny. Structure of concurrency. *Theoretical Computer Science*, 112:5–52, 1993.
7. Ryszard Janicki and Maciej Koutny. Semantics of inhibitor nets. *Information and Computation*, 123(1):1–16, November 1995.
8. Ryszard Janicki and Maciej Koutny. On causality semantics of nets with priorities. *Fundamenta Informaticae*, 38:1–33, 1999.
9. H.C.M. Kleijn and M. Koutny. Process semantics of P/T-Nets with inhibitor arcs. In Mogens Nielsen and Dan Simpson (Eds.): *Proc. of 21th International Conference on Applications and Theory of Petri Nets*, LNCS 1825, pp. 261–281, Springer-Verlag, 2000.
10. J. Meseguer and U. Montanari. Petri nets are monoids. *Information and Computation*, 88(2):105–155, October 1990.
11. U. Montanari and F. Rossi. Contextual nets. *Acta Informatica*, 32(6):545–596, 1995.
12. A. Corradini P. Baldan and U. Montanari. Contextual petri nets, asymmetric event structures, and processes. *Information and Computation*, 171(1):1–49, 2001.
13. G. M. Pinna and A. Poigné. On the nature of events: another perspective in concurrency. *Theoretical Computer Science*, 138(2):425–454, February 1995.
14. V. Sassone. An axiomatization of category of Petri net computations. *Mathematical Structures in Computer Science*, 8:117–151, 1998.
15. W. Vogler. Partial order semantics and read arcs. *Theoretical Computer Science*, 286(1):33–63, 2002.
16. J. Winkowski. Behaviours of concurrent systems. *Theoretical Computer Science*, 12:39–60, 1980.

Analysing Properties of the Resource Reservation Protocol

María E. Villapol[1] and Jonathan Billington[2]

[1]School of Computer Sciences, Central University of Venezuela, Av. Los Ilustres, Los Chaguaramos, Caracas, Venezuela
mvillap@strix.ciens.ucv.ve
[2]Computer Systems Engineering Centre, University of South Australia, Mawson Lakes, Adelaide, SA, 5095, Australia
j.billington@unisa.edu.au

Abstract. The goal of the *Resource Reservation Protocol* (RSVP) is to establish *Quality of Service* information within routers and host computers of the Internet. This paper describes a model of RSVP and presents the analysis approach and results. A large part of RSVP is modelled using *Coloured Petri Nets*. The model provides a clear, unambiguous and precise definition of the considered features of RSVP, which is missing in the current protocol specification. The model is analysed for a set of general properties, such as correct termination, and a set of RSVP specific properties defined in this paper. The properties are checked by querying the state graph and its associated strongly connected component graph. As a first step, we analyse RSVP under the assumption of a perfect medium to ensure that protocol errors are not hidden by rare events of the medium. The results show that the RSVP model satisfies the defined properties.

1   Introduction

The *Resource Reservation Protocol* (RSVP) [2][5] is a signalling protocol developed to create and maintain resource reservations (e.g. buffer and data rate allocations) in Internet routers and host computers, to provide *Quality of Service (QOS)* guarantees for multimedia and real-time applications. For the desired QoS to be guaranteed it is essential that RSVP works correctly. The aim of our work is thus to provide a step towards the verification of the correctness of RSVP's mechanisms.

Formal methods provide techniques to support the design and maintenance of communication protocols [14]. It is uncommon for formal techniques to be applied to Internet protocols [6][12][16] and we have found little work related to a detailed study of RSVP except for our initial work [18][19]. In two similar papers, Creese and Reed [4] and Reed et al. [15] present an induction technique for proving properties of arbitrary configurations of nodes and illustrate their technique using one aspect of RSVP. This is that the reservation requests are acknowledged by the node where merging (of reservation requests) does not cause resulting state changes [2]. Our work is complementary, in that it does not address merging but instead covers a wide range of RSVP

W.M.P. van der Aalst and E. Best (Eds.): ICATPN 2003, LNCS 2679, pp. 377–396, 2003.
© Springer-Verlag Berlin Heidelberg 2003

features including the soft state mechanisms, which have not been considered in [4][15]. Also, we define an extensive set of properties against which RSVP is analysed.

Coloured Petri Nets (CPNs) [9] are a formal technique used for modelling many systems, particularly communication protocols [1]. In this paper, RSVP mechanisms are modelled using CPNs and analysed with a software tool called Design/CPN [11]. We define a set of desired properties for RSVP based on the specification [2]. The properties are checked by querying the state space using powerful ML [13] functions suited to the Design/CPN environment.

Our first attempt at modelling and analysis of RSVP was presented in [18]. Since then, the model has been significantly revised, restructured and refined to include more features (e.g. confirmation of reservations). The new model has been developed using an incremental modelling and validation methodology [20] to increase our confidence in its validity. The contribution of this paper is the presentation of the new model, the definition of RSVP's desired properties, and their verification.

The paper has been organised as follows. Sect. 2 presents an overview of RSVP, which includes its characteristics and operation. In order to define RSVP's properties, we summarise RSVP's application interface interactions, known as Service Primitives [19], in Sect. 3. Starting with a discussion of scope, Sect. 4 describes the RSVP CPN model. Desired RSVP properties are defined in Sect. 5 and verified in Sect. 6. Sect. 7 concludes and presents future research directions.

2   Resource Reservation Protocol Overview

RSVP is designed to be run on network routers and in end hosts to support a QoS application. It reserves resources for a data flow from the sender to one or more destinations (i.e. multicast destination). A *data flow* is a distinguishable packet stream, which results from using a single application (such as video conferencing) requiring a certain QoS. A packet stream includes all packets that travel from the same source to the same destination. Unlike other signalling protocols [5], RSVP destinations (receivers) request resource reservations. Those requests travel on the reverse path of the data flow by following the pre-established route setup by RSVP [2]. RSVP is also responsible for maintaining reservations on each node associated with the data flow. RSVP uses a *soft-state approach* where the reservation states must be refreshed periodically; otherwise they are automatically removed. The approach accommodates dynamic route changes, dynamic multicast group membership and dynamic QoS changes [2]. RSVP reserves resources for a *session*. A session includes all data flows from one or more senders to the same unicast (one receiver) or multicast destination (multiple receivers).

RSVP reservation requests are defined in terms of a *filter specification (filter spec)* and a *flow specification (flow spec)* [2][5]. A filter spec is used to identify the data flow that is to receive the QoS specified in a flow specification. A flow spec defines the desired QoS in terms of a service class, which comprises a *Reservation Specification (RSpec)*, and a *Traffic Specification (TSpec)*. A RSpec defines the reservation (i.e. desired QoS) characteristics of the flow, for example, the service rate (i.e. the data rate that a data flow can use). A *TSpec* defines the traffic characteristics of the flow, for example, the peak data rate (i.e. the maximum rate at which the sender is intended to send packets).

RSVP uses several messages in order to create, maintain, and release state information for a session between one or more senders and one or more receivers as shown in Fig. 1. We now describe the main RSVP features and their associated messages, structured to facilitate the definition of the desired properties of the protocol in Sect. 5.

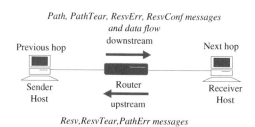

Fig. 1. RSVP messages

Path Setup. *Path messages* are used by the sender to set up a route to be followed by the reservation requests which uses the same routers as the corresponding data flow. These messages set up and maintain path state information (e.g. the Internet Protocol address of the previous router and the data flow's traffic characteristics).

Path Refresh. Path and reservation states have two timers associated with them: a *refresh timer* and a *cleanup timer*. A refresh timer determines when a path or reservation refresh message will be generated. The cleanup timer determines the maximum period of time that a node (i.e. router or host) can wait to receive a path or reservation refresh message, before it removes the associated state information. A path refresh is the result of either a path refresh timeout or a user request to modify the path state. Once a path is established, a node periodically (i.e. every refresh timeout period [2]) sends path refresh messages (i.e. *Path messages*).

Path Error. A node that detects an error in a Path message, generates and sends a *PathErr message* upstream towards the sender that created the error.

Path Release. RSVP tear down messages are intended to speed up the removal of path and reservation state information from the nodes. They may be triggered because a cleanup timeout occurs or an application wishes to finish a session. A *PathTear message* travels downstream from a sender to the receiver(s) and deletes any path state information and dependent resource reservation associated with the session and sender.

Reservation Setup. *Resv messages* carry reservation requests (e.g. for bandwidth and buffers) used to set up reservation state information along the route of the data flow. They travel upstream from the receiver(s) to the sender(s). Reservation requests, which arrive at a router, may be merged. The aim of merging is to control the overhead of reservation messages by making them carry more than one flow and filter specification [2]. Thus, the effective filter and flow specifications, which are carried in a reservation message, are the result of merging reservations from several requests.

Reservation Refresh. A reservation refresh is the result of either a reservation state refresh timeout or a receiver request to modify the reservation. Like path states, reservation states need to be refreshed. Thus, a receiver periodically sends reservation refresh messages (i.e. *Resv messages*) to the sender.

Reservation Release. *ResvTear messages* travel from the receiver(s) to the sender and remove any reservation state information associated with the receiver's data flow.

Reservation Error. If a node detects an error in a Resv message, it sends a *ResvErr message* downstream to the receiver that generated the failed Resv message.

Reservation Confirmation. A *ResvConf message* is used to notify the receiver that the reservation request was successful. In the simplest case, a ResvConf message is generated by the sender (Fig. 1).

3   RSVP Service Primitives

Service primitives [8] provide an abstract way to describe the interaction between the RSVP service user (i.e. QoS-aware application) and the RSVP service provider. A QoS-aware application interacts with RSVP to request reservation services. Since the RSVP specification [2] does not define the RSVP service, the authors [19] defined a set of service primitives for RSVP. They are used in the CPN model and in the definition of the desired RSVP properties.

Each primitive can be either a request or an indication. A *request (Req)* is used by the application to ask for a service from RSVP. An *indication (Ind)* is used by RSVP to notify the application of the invocation of a request primitive by its peer or to notify the user that the RSVP service provider detected an event. We have defined the following service primitives [19]. A sender application uses the *RSVP-Sender (Req/Ind)* primitive to establish or update the traffic characteristics of a data flow for a RSVP session. A receiver application uses the *RSVP-Reserve (Req/Ind)* primitive to establish or to modify a resource reservation during a session. A sender application uses the *RSVP-SenderRel (Req/Ind)* primitive to close a session. This means that the user data flow will eventually not have any QoS reserved. A receiver application uses the *RSVP-ReceiverRel (Req/Ind)* primitive to close a quality controlled session. The *RSVP-ResvConf (Ind)* primitive is used by the service provider to confirm that a reservation has been made. The *RSVP-SenderError (Ind)* primitive is used by the service provider to report an error in propagation or installation of the Sender's traffic characteristics. The *RSVP-ResvError (Ind)* primitive is used by the service provider to report a reservation failure inside the network.

4   CPN Model of RSVP

A model of RSVP is created with Coloured Petri Nets (CPNs) [9] using Design/ CPN [11].

4.1   Scope

Since RSVP is a complex protocol, the scope of the model is limited to make analysis tractable. The CPN model has been developed based on the protocol specification [2] and includes most of RSVP's features. The network topology comprises two hosts (sender and receiver) and a single router between them (Fig. 1). This allows us to consider router behaviour in RSVP's operation. Although multiple RSVP sessions can be open simultaneously, without loss of generality, just one session is modelled to study the functional behaviour of RSVP, since RSVP treats each session independently. Following an incremental methodology, it is important to firstly consider a unicast network. The multicast case will be considered in future work.

4.2   Assumptions

Although the Internet Protocol (IP) over which RSVP operates, does not guarantee that the order of sending messages is maintained at the receiver, message overtaking is not included in the model. This is because we want to be sure that RSVP will work under normal conditions (re-ordering is a rare event), before considering unusual events. Similarly IP may lose messages. As mentioned before, RSVP Path and Resv refresh messages deal with occasional loss of RSVP messages. Although, message loss is not considered in the model presented here, the mechanisms for dealing with loss are modelled.

4.3   Model Hierarchy

We deal with RSVP's complexity by using the hierarchical constructs of CPNs [9]. Hierarchies are built using the notion of a *substitution transition*, which may be considered a macro expansion. The model starts with a top-level CPN diagram, which provides an overview of the system being modelled and its environment. In hierarchical CPNs, this top-level diagram will contain a number of substitution transitions. Each of these substitution transitions is then refined by another CPN diagram, which may also contain substitution transitions. The top-level diagram and each of the substitution transitions is defined by a module, called a *page*. The relationships between the different pages are defined by a *hierarchy page*. The hierarchy page also includes the name of the page that defines the declarations required for the CPN inscriptions, the *Global Declaration node*.

The hierarchy page of the RSVP CPN model consists of eleven (11) pages as illustrated in Fig. 2. The top-level page is called RSVPNetwork (Fig. 4), which describes the network topology (Fig. 1) and interaction with the applications that use RSVP. This page includes substitution transitions for the *Sender*, *Router* and *Receiver*, which are defined by their own pages. These in turn also comprise substitution transitions for Path and Reservation management, which are defined at the lowest level of the hierarchy. These correspond to the major functions of RSVP described Sect. 2. The Path and Resv management pages include transitions that model the establishment, refreshment, release and error control of paths and reservations, respectively. Also included is the Global Declaration node (Page 11). Each page at the lower level of the hierarchy uses transitions to model service primitives and protocol actions, which include implementing RSVP functions (e.g. path refresh) or discarding messages that cannot be processed.

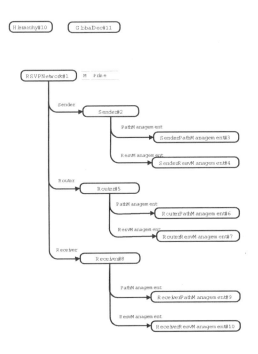

Fig. 2. Hierarchy page of the RSVP CPN model

4.4   Global Declaration

Fig. 3 shows the colour sets (types), variables, and functions from the global declaration node. The colour set ParameterValues is an enumeration type, which represents abstract values for both the *traffic specification* (*tspec*) and *flow specification* (*fspec*) parameters. The possible values are Ta, Tb, Fa, Fb and E (empty). These values allow us to test RSVP for changes to both the traffic specification and the reservations. The empty value (E) is used for initialisation. The colour sets, STSpec and SFSpec, are subsets of ParameterValues and represent the traffic specification stored as part of the path state information and the flow specification stored as part of the reservation state information, respectively. The colour set Status is an enumeration type, which defines the states of the RSVP entities. The values of this colour set have the following meanings:

```
color ParameterValues    = with E|Ta|Tb|Fa|Fb;
color STSpec             = subset ParameterValues with [E,Ta,Tb];
color SFSpec             = subset ParameterValues with [E,Fa,Fb];
color Status             = with SESSION|IDLE|WAITINGRESV|RESVREADY|
                                 RESVCONFIRMED|CLOSED;
color SenderStatus       = subset Status with [SESSION,WAITINGRESV,
                                 RESVREADY,CLOSED];
color ReceiverStatus     = subset Status with [SESSION,WAITINGRESV,RESVREADY,
                                 RESVCONFIRMED,CLOSED];
color RouterStatus       = subset Status with [IDLE,WAITINGRESV,RESVREADY];
color SenderState        = product SenderStatus * STSpec * SFSpec;
color ReceiverState      = product ReceiverStatus * STSpec* SFSpec;
color RouterState        = product RouterStatus * STSpec * SFSpec;
color TSpec              = subset ParameterValues with [Ta,Tb];
color FSpec              = subset ParameterValues with [Fa,Fb];
color TearMsgType        = with TEARDOWN|REL;
color ResvTear           =  product TearMsgType * FSpec;
color PathTear           =  product TearMsgType * TSpec;
color UpstreamMessages   = union
                             patherror: TSpec + resvtear: ResvTear +
                             resv: FSpec;
color DownstreamMessages = union
                             path: TSpec + resverror: FSpec + resvconf: FSpec +
                             pathtear: PathTear;
color UserInd            = with USR|NOUSR;
color Flag               = int with 0..1;
color FSpecXFlag         = product FSpec * Flag;
color TSpecXFlag         = product TSpec * Flag;
(* ==============    Application    ================= *)

color TSpecList = list TSpec;
color FSpecList = list FSpec;
 (* ================    Variables    ================= *)

var sta: Status;
var tspec,tspec1: STSpec;
var fspec,fspec1: SFSpec;
var flag: Flag;
var rmtusrind: UserInd;
var ttype: TearMsgType;
var tspeclist: TSpecList;
var fspeclist: FSpecList;

(* ================    Functions ==================== *)

fun pathexists (sn) = (sn= WAITINGRESV orelse sn=RESVREADY orelse
                          sn = RESVCONFIRMED);
fun resvexists (sn) = sn = RESVREADY orelse sn = RESVCONFIRMED;
```

Fig. 3. Global declarations

1. SESSION: indicates that the sender or receiver has opened a session, but no path information or reservation has yet been established.

2. IDLE: there exists neither path nor reservation information at the router.

3. WAITINGRESV: means that a request with the Sender's traffic information has been accepted by the entity (i.e. sender, router or receiver) and sent (if the entity is not the receiver) but as yet no reservation request has been received.

4. RESVREADY: means that a reservation request has been accepted and sent (if the entity is not the Sender).

5. RESVCONFIRMED: means that a reservation has been established and a confirmation has been received.

6. CLOSED: indicates that the sender or the receiver has left the session.

Three subsets of the colour set Status have been defined: the SenderStatus, RouterStatus and ReceiverStatus. The colour sets SenderState, RouterState and ReceiverState represent the states of the sender, router and receiver entities respectively. Each of them is the product of the status of the corresponding entity (i.e. SenderStatus, RouterStatus or ReceiverStatus), STSpec and SFSpec.

The colour sets, TSpec and Fspec, represent the parameters that may be carried in RSVP messages (see [2]) and are the traffic specification and flow specification, respectively. The colour set TearMsgType is intended to distinguish between a PathTear or ResvTear message generated as a result of the sender or receiver leaving the session (REL) or a path or reservation cleanup time-out (TEARDOWN). The other seven RSVP messages defined in Sect. 2 are represented by the colour sets UpstreamMessages and DownstreamMessages.

Colour set UserInd indicates whether the user at the other end has left the session (NOUSR) or not (USR). It is used to control the service primitive sequences allowed for each service user (see [20]).

Colour set FSpecXFlag indicates whether the RSVP-Reserve.Ind service primitive, which includes the requested FSpec, has occurred (flag is ON) or not (OFF). It is used to avoid multiple occurrences of this primitive with the same value of the flow specification. Similarly, colour set TSpecXFlag indicates whether the RSVPSender.Ind service primitive has occurred or not, avoiding multiple occurrences [20]. TSpecList and FSpecList are lists of traffic and flow specifications, respectively, that are requested by users.

The variables used in CPN inscriptions are typed in the declarations. As an example, the variable *rmtusrind* represents the state of a remote user. Two functions are used to simplify guard inscriptions. *Pathexists* returns true if the sender has path state information available, and similarly *resvexists* returns true if the receiver's state includes reservation information.

4.5   RSVP Network Page

The top-level CPN page, *RSVPNetwork*, is shown in Fig. 4. It shows the network topology including interaction with RSVP's users (applications). The three shaded rectangles (Sender, Router and Receiver) represent RSVP entities and are *substitution transitions*. A substitution transition is identified by a HS-tag in the lower left corner of the rectangle representing the transition. The ellipse, SenderUser, is a CPN *place* and is typed by TSpecList defined in Fig. 3. It represents a queue of possible traffic characteristics of the data flow requested by the sender user. Similarly the place ReceiverUser is typed by FSpecList and represents a queue of flow specifications requested by the receiver user. A *marking* of a place comprises a (multi) set of values (known as *tokens*) taken from the place's type. For example, SenderUser has the initial marking of a list of two traffic specifications [Ta,Tb]. The places SOutgoingMsgs and RIncomingMsgs have markings that represent RSVP messages travelling downstream, while ROutgoingMsgs and SIncomingMsgs have markings that represent RSVP messages travelling upstream. Places and substitution transitions are connected by *arcs*, which indicate the type of data required or produced by the substitution transitions.

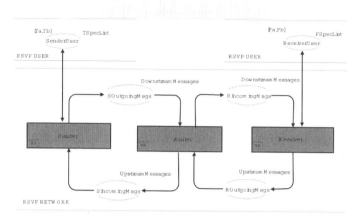

Fig. 4. RSVP Network page

The complete RSVP model [20] cannot be described is this paper due to space limitations. We therefore only describe the Sender and Sender-Path-Management pages (see Fig. 2), which are representative of the operation of the CPN model.

4.6   RSVP Sender Page

The *RSVP-Sender* page (Fig. 5) includes two substitution transitions, PathManagement and ResvManagement and a new place Sender. PathManagement is described below. ResvManagement models the establishment, refreshment, error control and release of reservation state information. The place Sender is typed by SenderState and models the status of the sender together with path and reservation information. The other places were described in the previous section (see Fig. 4).

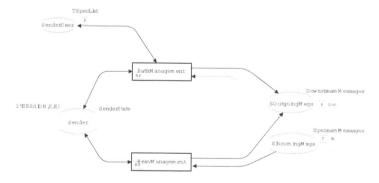

Fig. 5. Sender page

4.7   Path Management Page

The Sender-Path-Management page is shown in Fig. 6 and comprises five transitions and four places that have already been described. In Fig. 6, transition RSVPSender-Req models the establishment or updating of path state information and the occurrence of the service primitive RSVP-Sender.Req. RSVPSender.Req is enabled when the Sender is not CLOSED and a new *tspec* is waiting to be sent in SenderUser. Its occurrence updates the *tspec* to the one requested by the user, initially (i.e. if the sender status is SESSION) also updates the sender status to WAITINGRESV (i.e. ready to receive a reservation request), removes the *tspec* from the place SenderUser and sends it in a Path message by adding the corresponding token to SOutgoingMsgs.

The transition PathRfrTimeOut models periodic path refreshes required to maintain path state information. It is enabled after the initial sender request has occurred, but the session isn't closed. An occurrence of the transition does not change the state of the sender but adds a Path message carrying the corresponding *tspec,* to the place SOutgoingMsgs. The other three transitions model the sender processing errors or leaving the session.

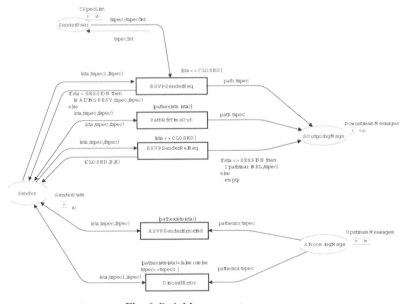

Fig. 6. Path Management page

5   Desired Properties of RSVP

We wish to verify RSVP against a set of desirable properties. They include standard protocol properties such as freedom from deadlock and absence of livelocks and RSVP specific properties concerning the setup, maintenance and release of both paths and reservations. They are formalised in the following subsections. In this description, the term *node* refers to a marking in the occurrence graph (OG); while *RSVP node* is a network device (e.g. a router), which supports RSVP.

Firstly, we summarise the notation that we need, some of which is standard CPN notation [9]:

- $B(t)$ denotes the set of all bindings of the variables associated with transition *t*.
- $M(p)$ denotes the marking of place *p*.
- $[M >$ denotes the set of markings that are reachable from marking M, and with $[M_0 >$ denoting the set of markings reachable from the initial marking M_0.
- DM denotes the set of all dead markings.
- SCC_T denotes the set of all terminal SCCs.
- $M \xrightarrow{(t,b)} M'$ denotes that marking M' is directly reachable from marking M by the occurrence of the binding element (t,b), $b \in B(t)$.
- $Proj_1 (x,y,z) = x$ is the first component projection function for a triple (x,y,z).
- $Proj_2 (x,y,z) = y$ is the second component projection function for a triple (x,y,z).
- $Proj_3 (x,y,z) = z$ is the third component projection function for a triple (x,y,z).
- $\emptyset$ is the empty (multi)set (depending on context).

5.1   Termination Property

RSVP must finish in a state where: all the message queues are empty; any path or reservation state information has been removed from the RSVP nodes; and the sender and receiver have left the session (i.e. are in the CLOSED state).

Definition 1.   *The RSVP CPN model terminates correctly iff*
$\forall M \in DM, Proj_1(M(Sender)) = Proj_1(M(Receiver)) = CLOSED \wedge Proj_1(M(Router)) =$
$IDLE \wedge (M(SOutgoingMsgs) = M(SIncomingMsgs) = M(ROutgoingMsgs) =$
$M(RIncomingMsgs) = \emptyset)$
where the marking of a state place is converted from a singleton multiset comprising one token to its token (e.g. we replace 1'(a,b,c) by (a,b,c)) before the projection function is applied.

5.2   Livelock Property

Livelock [7] occurs when protocol sequences are executed indefinitely without the possibility of making effective progress. Thus it is important to check that the RSVP model doesn't contain livelocks. Livelocks will be revealed as terminal strongly connected components of the OG that include cycles (and hence are not dead markings). We define a CPN to be livelock free if it has no terminal SCC (SCC_T) that contains cycles.

Definition 2.   *A CPN is livelock free if* $| SCC_T | = |DM|$

This is equivalent to checking that the terminal markings form a *home space*. A home space is a set of markings such that from each reachable marking, it is possible to reach at least one of these markings [9].

5.3   Path Setup Property

Once the RSVP-Sender.Req service primitive occurs, RSVP should be able to establish or update the path state in all RSVP nodes on the route of the data flow, unless the receiver has closed the session before the RSVP-Sender.Req service primitive occurs. Let *tsr* be the transition RSVPSenderReq. The property is defined as follows:

Definition 3.   *The RSVP CPN model satisfies the path set-up property iff*
For M,M' $\in$ *[M$_0$ >,*

$Proj_1(M(Receiver)) \neq CLOSED \wedge M \xrightarrow{(tsr,tspec=rtspec)} M' => \exists M'' \in [M' >:$
$Proj_2(M''(Sender))=Proj_2(M''(Receiver))=Proj_2(M''(Router))=$
rtspec

5.4   Path Maintenance Property

If the path state information, for a traffic specification that has been established along all RSVP nodes on the route of the data flow, is removed in any of those nodes as a result of the expiration of the path cleanup timer, it is possible for it to be established again. This cannot occur if either the sender or receiver user closes the session before the timer expires.

Let $\mathbb{T}pc$ = {RouterPathCleanup,ReceiverPathCleanup} be the set of path cleanup transitions and $tpc \in \mathbb{T}pc$. Let $\mathcal{M}_{PE}$ be the set of markings where a path has just been established in all RSVP nodes:

$\mathcal{M}_{PE} = \{M \in [M_0 > |$
$Proj_1(M(Sender))=Proj_1(M(Receiver))=Proj_1(M(Router))=WAITINGRESV\}$

Definition 4.   *The RSVP CPN model satisfies the path maintenance property iff*
For M $\in$ *[M$_{pe}$ >, where M$_{pe}$* $\in$ *$\mathcal{M}_{PE}$, and b* $\in$ *B(tpc)*

$Proj_1(M(Sender)) \neq CLOSED \wedge Proj_1(M(Receiver)) \neq CLOSED \wedge M \xrightarrow{(tpc,b)} M'$
$=> \exists M'' \in [M' >: Proj_1(M''(Sender))=Proj_1(M''(Receiver))=Proj_1(M''(Router))=$
WAITINGRESV

5.5   Path Release Property

Once the RSVP-SenderRel.Req service primitive occurs, the path and corresponding reservation state must be removed from all the nodes along the route of the data flow. Let *tpr* denote the transition RSVPSenderRelReq. The property is defined as follows:

Definition 5.   *The RSVP CPN model satisfies the path release property iff*
For M $\in$ *[M$_0$ >, and b* $\in$ *B(tpr)*

$M \xrightarrow{(tpr,b)} M' => \forall M'' \in [M' >, \exists M''' \in [M'' > | Proj_1(M'''(Sender))= CLOSED$
$\wedge Proj_1(M'''(Router))= IDLE \wedge (Proj_1(M'''(Receiver))=CLOSED \vee$
$Proj_1(M'''(Receiver))= SESSION)\}$

5.6   Resv Setup Property

If the RSVP-Reserve.Req service primitive occurs, RSVP should be able to establish or update the requested reservation along the route of the data flow, unless the sender has left the session before the reservation request is sent. Also, path state information must exist at any RSVP node before the reservation is established.

Let Trs={SenderRSVPReserveInd, SenderReserve, RouterResvEstablished, ReceiverReserveReq, ReceiverReserve} be the set of reservation setup transitions. $trs \in Trs$.

Definition 6.   *The RSVP CPN model satisfies the path set-up condition (i.e. path state information must exist before any trs transition occurs) iff*
For $b \in B(trs)$

$\forall M \in \{M' | M' \xrightarrow{(trs,b)} M'', M' \in [M_0>\} \ Proj_1(M(Sender)), Proj_1(M(Router)) \in \{WAITINGRESV, RESVREADY\} \wedge Proj_1(M(Receiver)) \in \{WAITINGRESV, RESVREADY, RESVCONFIRMED\}$*

Let trr denote the transition RSVPReserveReq. The property is defined as follows:

Definition 7.   *The RSVP CPN model satisfies the reservation set-up condition iff*
For $M,M' \in [M_0 >$,

$Proj_1(M(Sender)) \neq CLOSED \wedge M \xrightarrow{(trr, fspec=rfspec)} M' => \exists M'' \in [M' >:$
$Proj_3(M''(Sender))=Proj_3(M''(Receiver))=Proj_3(M''(Router))=$
$rfspec$

Definition 8.   The RSVP CPN model satisfies the reservation set-up property if Definition 6 and Definition 7 are satisfied.

5.7   Resv Maintenance Property

If the reservation state information for a traffic specification that has been established in all RSVP nodes on the route of the data flow is removed in any of those nodes as a result of expiration of the path or reservation cleanup timer, it can be established again. This cannot occur if either the sender or receiver user closes the session before the timer expires.

Let Trc = {RouterPathCleanup, ReceiverPathCleanup, RouterResvCleanup, SenderResvCleanup} be the set of cleanup transitions and $trc \in Trc$. Let M_{RS} be the set of markings where a reservation has just been established in all RSVP nodes and defined by:

$M_{RS} = \{ M \in [M_0 > | Proj_1(M(Sender))=Proj_1(M(Receiver))=Proj_1(M(Router))= RESVREADY\}$

Definition 9.   *The RSVP CPN model satisfies the resv maintenance property iff*
For $M \in [M_{rs} >, M_{rs} \in M_{RS}, b \in B(trc)$

$Proj_i(M(Sender)) \neq CLOSED \land Proj_i(M(Receiver)) \neq CLOSED \land M \xrightarrow{(trc,b)} M'$
$=> \exists M'' \in [M' >: Proj_i(M''(Sender))=Proj_i(M''(Receiver))=Proj_i(M''(Router))=$
$RESVREADY$

5.8   Resv Release Property

If the RSVP-ReceiverRel.Req service primitive occurs, the reservation must be removed from all RSVP nodes.

Let *trr* denote the transition RSVPReceiverRelReq. The property is defined as follows:

Definition 10.   *The RSVP CPN model satisfies the resv release property iff*
For $M \in [M_0 >,$ *and* $b \in B(trr)$

$M \xrightarrow{(trr,b)} M' => \forall M'' \in [M' >, \exists M''' \in [M''> \mid (Proj_i(M'''(Sender))=$
$WAITINGRESV \quad \lor \quad Proj_i(M'''(Sender))=CLOSED) \quad \land \quad (Proj_i(M'''(Router))=$
$WAITINGRESV \lor Proj_i(M'''(Router))= IDLE) \land$
$Proj_i (M'''(Receiver))= CLOSED \}$

6   Analysis and Verification

Design/CPN [11] is used to simulate and analyse the CPN model. The model is analysed by generating the OG [9] and its corresponding *Strongly Connected Component* (SCC) graph. Design/CPN was run on a 1.6 GHz Linux Pentium 4 PC with 1 GB RAM.

The RSVP model was developed and analysed incrementally by adding the features described in Sect. 2 (e.g. path refresh, path release) one at a time. This facilitates the validation task and increases confidence in the model. The model analysed in this paper includes all of these features.

The CPN model of Sect. 4 can generate an infinite state space due to periodic refreshes and unbounded communication places (e.g. see transition PathRfrTimeOut in Fig. 6). We thus modify the model so that the communication places have finite capacity using a standard approach [10].

The properties are checked by implementing the definitions in Sect. 5 as OG query functions in ML [13]. To illustrate the approach, this section describes the ML query for one of the Path Setup property and how livelocks are detected. The description of the queries for the others can be found in [20]. All properties, except for the termination and livelock properties, are verified by examining the nodes of the OG and checking reachability among multiple nodes by using the SCC graph (instead of the OG) because it is more efficient.

6.1   Initialisation

The RSVP model is initialised by distributing tokens to one or more places of the model to create the *initial marking*. Each of the places SenderUser and ReceiverUser contains a list of the requested traffic or reservation characteristics, respectively, as specified in Fig. 4. Each communication place contains a single a single empty slot token (*uslot(N)* or *dslot(N)*) indicating that there is space for one message. Each state place (Sender, Router and Receiver) contains a triple comprising null entries for path and reservation information and SESSION for the status of the sender and receiver and IDLE for the router status.

6.2   Occurrence Graph and SCC Graph Statistics

In the course of our investigations [20], a series of state spaces was generated for different models, that were incrementally developed to include more features, and for different initial markings. This was to gain further confidence in the model as it was developed. For example, in the simplest case, we just considered one sender traffic spec (Ta) and reservation user request (Fa) over a channel of capacity one. Here we only present the results for the fully featured model, for the initial marking given above.

The size of the OG and its SCC graph and the corresponding generation times using Design/CPN are shown in Table 1. The table also includes the number of dead markings and terminal SCC nodes. Design/CPN output shows that each terminal SCC node contains one marking, corresponding to a dead marking. Thus there are no livelocks. The size of the SCC graph is smaller than the OG indicating that there are cycles. This is expected due to the soft-state mechanisms.

Table 1. State space results

Statistical Information	OG	SCC graph
Number of Nodes	154986	56962
Number of Arcs	769880	428644
Number of Dead Markings	85	-
Number of Terminal SCC nodes	-	85
Calculation Time (hh:mm:ss)	3:27:20	00:05:57

A large number of dead markings (85) occurs because of the possible markings of places other than those concerned with communication and RSVP entity states, such as user places (e.g. SenderUser) and some control places that exist in the model.

6.3   General Properties

Boundedness and *liveness* [9] were investigated to validate and debug the model and to provide insight into RSVP's behaviour. This information is included in the standard report for the state space generated by Design/CPN [11].

Boundedness. Integer and multi-set bounds were analysed for the places of the model. For example, Table 2 lists the upper integer and multi-set bounds for the communication places. Upper Integer bounds describe the maximum number of tokens that can occur in a place, while multi-set bounds indicate which tokens can occur in a place [10].

Table 2. Upper bounds for communication places

Place	Integer Bounds	Multi-set Bounds
SOutgoingMsgs	1	1'path(Ta)++ 1'path(Tb)++ 1'resverror(Fa)++ 1'resverror(Fb)++ 1'resvconf(Fa)++ 1'resvconf(Fb)++ 1'pathtear((REL,Ta))++ 1'pathtear((REL,Tb))++ 1'dslot(N)
RIncomingMsgs	1	1'path(Ta)++ 1'path(Tb)++ 1'resverror(Fa)++ 1'resverror(Fb)++ 1'resvconf(Fa)++ 1'resvconf(Fb)++ 1'pathtear((TEARDOWN,Ta))++ 1'pathtear((TEARDOWN,Tb))++ 1'pathtear((REL,Ta))++ 1'pathtear((REL,Tb))++ 1'dslot(N)
SIncomingMsgs	1	1'patherror(Ta)++ 1'patherror(Tb)++ 1'resvtear((TEARDOWN,Fa))++ 1'resvtear((TEARDOWN,Fb))++ 1'resvtear((REL,Fa))++ 1'resvtear((REL,Fb))++ 1'resv(Fa)++ 1'resv(Fb)++ 1'uslot(N)
ROutgoingMsgs	1	1'patherror(Ta)++ 1'patherror(Tb)++ 1'resvtear((REL,Fa))++ 1'resvtear((REL,Fb))++ 1'resv(Fa)++ 1'resv(Fb)++ 1'uslot(N)

Each of the communication places may have a maximum number of one token. This corresponds with the maximum capacity of the message buffer (i.e. one). The multi-set bounds show that all RSVP messages can be generated by the protocol. For example, the place RIncomingMsgs can contain a token representing a Path message, which carries the traffic characteristics of the data flow (Ta). It may also contain a token representing a ResvErr message, which indicates an error. The message carries the reservation characteristics (Fa) for which the request failed. Similar information can be derived from the multi-set bounds for the other places.

The boundedness results are as expected and further confirm that the model is valid.

Dead Transitions. A *dead transition* is one that is not enabled in any reachable marking [9]. The Design/CPN standard report showed that there are no dead transitions. This is expected as there should be no 'dead code' in the specification.

6.4   ML Query for Path Set-up

The Path Setup property is checked using the following ML function:

```
1   fun PathSetup (des) = let
2   (* Returns the SCC nodes which include at least one
    path-setup node *)
3       val pathsetupasccs = PathSetupNodes (des);
4       val senderreqdasccs = SenderReqDestNodes(des);
5       fun PSAN x = ListMember (x,pathsetupasccs);
6       fun SRAN x = ListMember (x,  senderreqdasccs);
7   (* Checks if all SenderReqDest Nodes can reach any
    PathSetup Node *)
8       val v1,r1,m1) =Reachable ([~1],SRAN,PSAN,[],[],0);
9   in
10      r1
11  end;
```

The function, *PathSetup*, returns each of the destination markings of an occurrence of the RSVPSenderReq transition that cannot reach a marking where a path state for a requested traffic specification (des, line 1) (see Fig. 3) has been established. The function *PathSetupNodes* (line 3) returns the SCC nodes that include at least one OG node where a path has been established. This means that the status of the RSVP entities is equal to WAITINGRESV, RESVREADY or RESVCONFIRMED and the requested *tspec* corresponds to the *tspec* in the state of the entity (see Fig. 6). The function *SenderReqDestNodes* (line 4) returns the SCC nodes that include at least one destination OG node of the occurrence of the RSVPSenderReq transition. Each of the functions *PSAN* and *SRAN* (lines 5,6) is a predicate that checks if a SCC node is a member of the list returned by the function *PathSetupNodes* or *SenderReqDestNodes*, respectively.

We developed the algorithm *Reachable* to check if *multiple* nodes can reach at least one of the nodes in a list, since the in-built function provided by Design/CPN [11] only checks the reachability of one node from another. In order to check multiple nodes, the Design/CPN function must be invoked several times. This is inefficient when there are many nodes. The *Reachable* algorithm is described in detail in [20]. In line 8, the function *Reachable* traverses the SCC graph. It returns the SCC nodes (*r1*), which belong to the list returned by the function *SenderReqDestNodes*, that cannot reach any SCC node, which belongs to the list returned by the function *PathSetupNodes*. In order to say that the property is satisfied, the *PathSetup* function must return an empty list (line 10).

6.5 RSVP Specific Properties

The CPN model is verified against the desired RSVP properties by running all the corresponding queries together in batch mode. Table 3 lists the analysis results, which indicate that all the properties are satisfied (OK). It also includes the approximate time spent for running all the functions that implement the properties. The results show that the RSVP model works as expected under the assumptions we have made.

Table 3. Analysis results

Property	Result
Termination	OK
Livelock	OK
Path Setup	OK
Path Maintenance	OK
Path Release	OK
Resv Setup	OK
Resv Maintenance	OK
Resv Release	OK
Estimated Verification Time (hh:mm:ss)	8:54:10

7   Conclusions

Coloured Petri Nets have been used to model the main features of RSVP based on a number of assumptions. We use a simple representative network topology and a limited and reliable communication medium to verify that RSVP will operate correctly under ideal conditions. This is a necessary first step.

The main problem found during modelling was the lack of a well-defined specification of RSVP [2], where only a narrative description is provided. The resulting model provides a clear, unambiguous and precise definition of the considered features of RSVP. The model was developed incrementally and checked at each stage to reduce the possibility of modelling errors.

RSVP uses soft-state refresh and clean-up mechanisms, which are modelled in a non-deterministic way. Similar mechanisms are used in other Internet protocols (e.g. several routing protocols [5]). Therefore, the RSVP model can be used as a reference for modelling other protocols that use similar procedures.

The model is analysed based on general properties of the protocol and a set of eight desirable properties defined and formalised for the first time in this paper. The analysis of the protocol is carried out by querying the state space and SCC graph. For 6 of RSVP properties the use of the SCC graph greatly enhances the efficiency of the model checking process. The analysis of the model shows that RSVP works as expected under our simplifying assumptions.

Normally model checking uses a form of temporal logic (such as Computation Tree Logic (CTL)) to express properties. In our case, we have found the use of ML queries to be of immediate benefit with Design/CPN, without having to master another formalism. However, it is interesting to compare our current predicate logic/ML approach with using a temporal logic model checker. Design/CPN provides a temporal logic model checking library called ASK-CTL [3], which implements a CTL-like temporal logic. We translated our definitions of properties into ASK-CTL formulas (see [20]). Our current results indicate that the ML approach is significantly faster than the ASK-CTL approach. In the future, we intend to compare our model checking approach with the ASK-CTL approach in terms of expressiveness and elegance of property definition. It will also be beneficial to compare our use of the SCC graph, with the use of SCCs for optimisation of ASK-CTL verification as discussed in [3].

Further work will extend the model to include multicast operation and merging, and to relax the assumptions on the communication channels. However, analysing this model will be a challenge, due to its inherent complexity. Thus it will be necessary to explore ways of making the model tractable for analysis, including the use state space reduction techniques [9] [17].

This work extends the application of formal methods to a new protocol that uses a soft-state approach, proposed for providing QoS guarantees over the Internet. The results presented here are being extended to the verification of RSVP against its service specification [19][20].

Acknowledgements. We would like to acknowledge Dr Lars Kristensen for his advice in the analysis of the protocol and for suggesting the use of SCC graphs to improve the performance of the ML queries. We would also like to thank Dr Laure Petrucci for her comments on the formalisation of RSVP's properties.

References

[1] Billington J., Diaz M. and Rozenberg G. (eds), Application of Petri nets to Communication Networks, Advances in Petri Nets, LNCS, Vol. 1605, 1999.

[2] Braden R., et al. *Resource Reservation Protocol (RSVP) – Version 1: Functional Specification.* RFC 2205, IETF, September, 1997.

[3] Cheng A. and Christensen S. and Mortensen K. *Model Checking Coloured Petri Nets Exploiting Strongly Connected Components*, in Proc. Of the International Workshop on Discrete Event Systems, Institute of Electrical Engineers, University of Edinburgh, UK, August 1996, pp 169–177.

[4] Creese S and Reed J N, *Verifying End-to-End Protocols Using Induction with CSP/FDR*, Parallel and Distributed Processing, 13th International Parallel Processing Symposium and 10th International Symposium on Parallel and Distributed Processing, Puerto Rico, April 1999, LNCS 1586, Springer.

[5] Durham D. and Yavatkar R. *Inside the Internet's Resource Reservation Protocol.* Wiley, USA, 1999.

[6] Han B. and Billington J. *Validating TCP Connection Management.* Proceedings of the Workshops on Software Engineering and Formal Methods and Formal Methods Applied to Defense Systems, Adelaide, Australia, June 2002, pp 47–55.

[7] Holzmann G. *Design and Validation of Computer Protocols.* Prentice Hall. 1991.

[8] ITU-T *Convention for the Definition of OSI Services.* Recommendation X.210. 1994.

[9] Jensen K. Coloured Petri Nets: Basic Concepts, Analysis Methods and Practical Use. Vol. 1, 2 and 3. Springer-Verlag, 2nd edition, April, 1997.

[10] Kristensen L.M., Christensen S., and Jensen K. *The practitioner's guide to coloured Petri nets*. International Journal on Software Tools for Technology Transfer, Springer, 1998, Vol. 2, Number 2, pp 98–132.

[11] Meta Software Corporation. *Design/CPN Reference Manual for X-Windows*, Version 2, Meta Software Corporation, Cambridge, 1993.

[12] Ouyang C., Kristensen LM and Billington J. *A Formal and Executable Specification of Internet Open Trading Protocol*. Third International Conference, EC-Web 2002 Aix-en-Provence, France, September 2002, LNCS 2455, pp 377–387.

[13] Paulson L. *ML for the Working Programmer*. Cambridge University Press.1991.

[14] Proceedings of the (Joint) International Conferences on Formal Description Techniques for Distributed Systems and Communications Protocols (FORTE) and Protocol Specification, Testing & Verification (PSTV), 1997–2000.

[15] Reed JN, Jackson DM, Deianov B. and Reed GM. *Automated Formal Analysis of Networks: FDR Models of Arbitrary Topologies and Flow-Control Mechanisms*. Proceedings of Joint European Conferences on Theory and Practice of Software, ETAPS98, as part of Fundamental Approaches to Software Engineering, FASE98, Lisbon, LNCS 1382, Springer.

[16] Smith M.A. *Formal Verification of Communication Protocols*. In Reinhard Gotzhein and Jan Bredereke, editors Formal Description Techniques IX: Theory, Applications, and Tools FORTE/PSTV'96: Joint International Conference on Formal Description Techniques for Distributed Systems and Communication Protocols, and Protocol Specification, Testing, and Verification, Kaiserslautern, Germany, October 1996, pp 129–144. Chapman & Hall, 1996.

[17] Valmari A. *The State Explosion Problem*. Lectures on Petri Nets I: Basic Models, Vol. 1491, 1998, pp 429–528.

[18] Villapol M.E. and Billington J. *Modelling and Initial Analysis of the Resource Reservation Protocol using Coloured Petri Nets*, Proceedings of the Workshop on Practical Use of High-Level Petri Nets, Aarhus, Denmark, June 27, 2000, pp 91–110.

[19] Villapol M.E. and Billington J. *Generation of a Service Language for the Resource Reservation Protocol Using Formal Methods*, Proceedings of Eleventh Annual International Symposium of the International Council On Systems Engineering (INCOSE), 1–5 July 2001, on CD-ROM.

[20] Villapol M.E. *Modelling and Analysis of the Resource Reservation Protocol Using Coloured Petri Nets*. Doctoral Thesis, University of South Australia, March 2003.

Hierarchical Timed High Level Nets and Their Branching Processes*

Hans Fleischhack[1] and Elisabeth Pelz[2]

[1] Fachbereich Informatik
Carl-von-Ossietzky – Universität Oldenburg
D-26111 Oldenburg
fleischhack@informatik.uni-oldenburg.de
[2] LACL, Faculté de Sciences
Université Paris XII–Val de Marne
61 avenue du Général de Gaulle
F-94010 Créteil Cedex
pelz@univ-paris12.fr

Abstract. The paper aims at defining hierarchical time extensions of M-nets, a fully compositional class of high-level Petri nets. As a starting point, the class of classical timed M-nets are considered, where time intervals of duration are attached to each transition. This class is enriched by two new operations: timed refinement (which extends the class) and hierarchical scoping (which is shown to be a powerful feature for abstraction). It is argued that hierarchical timed M-nets permit the design of real-time systems in a top down manner. Moreover, a partial order semantics of hierarchical timed M-nets is defined based on branching processes. The definition is given directly for high level nets, without preliminary unfolding to low level nets. This semantics enables partial order model checking of hierarchical timed M-nets e.g. within the PEP-system.

Keywords: Timed and stochastic nets, partial order semantics

1 Introduction

In the last years time extensions of Petri nets have become more and more studied, see for instance the former volumes of this Petri Nets Conference. We can distinguish more classical approaches as time nets (cf. e.g. [25]) and timed nets (cf. e.g. [28,30]), where different kinds of timing constraints are explicitly added to certain net elements [1,2,3,10,22,23,27] from a quite different approach, considering causal time. This approach was first mentioned by [14,29], and has recently been applied to the M-net Calculus [19,20,21], a fully compositional class of high-level Petri nets [9,8]. In this paper, we will contribute to the classical approach. More specifically, we will consider timed nets, where the duration of

---

* This work has been partially supported by the Procope project PORTA (Partial Order Real Time Semantics).

W.M.P. van der Aalst and E. Best (Eds.): ICATPN 2003, LNCS 2679, pp. 397–416, 2003.

each action is given by an interval of [shortest firing duration, longest firing duration].

Some authors have already considered time(d) Petri nets in the context of PBC, the low level version of the M-net algebra [1,10,18,22]. We present them briefly in section 5.

What we would like to embody in this paper is a kind of global time constraint. Consider e.g. the requirement preparing a meal should take no more time than one hour. This could be implemented by a group of cooks, whose different manipulations of preparing dishes all have some particular duration. Only those implementations are acceptable, which respect the total limit of one hour. Or, as still a simpler problem consider the following sequence of two timed transitions in figure 1:

Fig. 1. Sequence of timed transitions

How would you model a constraint saying the consecutive firing of t_1 and t_2 consumes exactly 100 seconds in total? How would you express this constraint within a timed Petri net? To which parts of the net should constraints be imposed?

In PBC and M-net Calculus, one powerful operation is that of general transition refinement [5,12,13]. As a first solution to the above problem one could associate the time interval $[100, 100]$ to a transition t, which is then refined by the net in figure 1, i.e. by the sequence of t_1 and t_2. Unfortunately, the semantics of an M-net containing refinements yields a flat net, where all refined transitions have been replaced by appropriate sub-nets. So, for the above problem, a timed extension of general refinement should result in a flat net containing at least 101 alternative sequences of incarnations of t_1 and t_2, one for every allowed combination of firing times. Obviously, such an unfolding of time is not a desirable solution.

Let us illustrate this by a second, a bit more complicated, example:

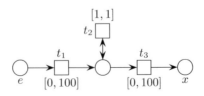

Fig. 2. Second sequence of timed transitions

Once more, the execution of the net of figure 2 should not consume more than 100 seconds in total. Now, the flat net with the same timed behavior, would have a branch for each combination of values c_1, c_2, c_3, c_4 such that $c_1 + c_2 + c_3 + c_4 \leq 100$ and c_1, c_3 stand for the time consumed by transitions t_1 and t_3, c_2 for the number of times transition t_2 fires (consuming each time exactly one second) and c_4 indicating the number of 'ticks' (or seconds) after the end of the firing of t_3. And, there are also some branches for executions with time-outs. Thus the equivalent flat net may look like the following net, where we only show some possible branches.

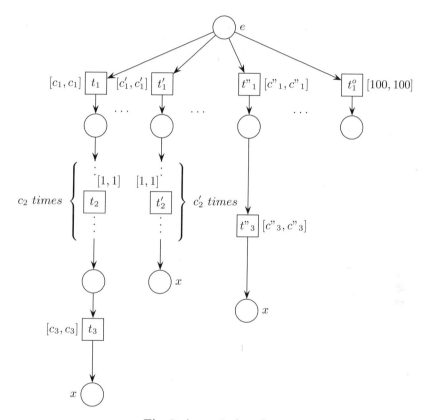

Fig. 3. An equivalent flat net

The time-constants in figure 3 are as follows: $c_1 + c_2 + c_3 \leq 100$, $c'_1 + c'_2 = 100$ and so the second branch ends by timeout, $c_1 + c_3 \leq 100$, and the right most branch ends by timeout too, after t^o_1 having consumed exactly 100 seconds.

Would it be possible to obtain such a flat net by using existing M-net operations, for instance by general refinement where figure 2 is the refining net? Surely not! The reason is that all existing operations only concern the interfaces of nets. Here, as shown in the last figure, the flat net depends on the possible executions: in fact each execution of the net in figure 2 respecting the global time constraint defines a branch in the flat net of figure 3. This cannot be the result of a composition in the existing M-net algebra.

Hence, in section 2 within the class of timed M-nets (also called duration M-nets) a totally new operation will be introduced, called *timed refinement*, $N[t \leftarrow N_t]$. Differing from the composition operators of the M-net calculus, application of timed refinement does not result in a (flat) timed M-net, but will yield a hierarchical net. The meaning of the timed refined transition is that of an additional clock which is visible within the refining net. In general, this operation leads to a tree-like nested structure representing the timely dependencies between the nets involved. The resulting class of nets will be called *hierarchical duration M-nets*, or *hd-M-nets*.

In section 3, the new concept of a timed maximal branching process is introduced and – based on this – a partial order semantics for hd-M-nets is defined. To our knowledge, it is the first approach to define directly a branching process semantics for high level nets with time. This concept will allow us to prove properties of hd-nets expressed in a suitable temporal logic with partial order model checking, e.g. within the PEP-system [10,16,17,24].

To facilitate even more a top down development, another original operation, called hierarchical scoping, will be added in section 4. It will allow to synchronize two transitions (for instance, a fax sender with a fax receiver) which will both be timed refined later on (here, by specific fax-machines) over abstract action symbols. Hierarchically scoped hd-nets will become usual (also called *concrete*) hd-nets after executing the scoping. So hierarchical scoping is a powerful device for abstraction, but no extension of the net class. The fax example will illustrate this later.

Some related work will be discussed in section 5, and in section 6, some conclusions are drawn.

2   Hierarchical Duration M-Nets

In this section, the notion of a hierarchical duration M-net is introduced and their transition rule is defined.[1] The section starts with a preliminary example illustrating the subsequent definitions.

2.1   An Introductory Example

The concepts introduced in this section are illustrated by means of an example. Consider a simple fax machine, which may either send or receive a fax[2]. Figure 4 shows an hd-Mnet N_0 modelling an abstract view of the fax. We assume that *fax* is an abstract action symbol, that $\alpha(t_{send}) = \{fax\}$, and that $\alpha(t_{receive}) = \{\overline{fax}\}$. Hence, t_{send} as well as $t_{receive}$ may be timed refined transitions.

---

[1] The M-net Calculus constitutes a fully compositional class of high-level Petri nets. Due to lack of space we are only able to present some notation here. For the full story, the reader is refered to [9].

[2] Note, that by no means we try to model a real fax machine. This is just to illustrate the following definitions.

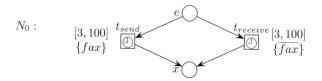

Fig. 4. A hd-Mnet N_0 modelling a simple fax machine

Figure 5 shows a timed refinement N_1 of transition t_{send} of the net N_0: After dialing (which consumes one time unit) the fax either gets connected (consuming 1 to 3 time units) and may send its data (consuming 1 to 5 time units), or it is not connected (which consumes one time unit) and has to wait for 20 time units and then has to dial again and so on. Note, that N_1 may unsuccessfully try to establish a connection for an unbounded amount of time. The timing restrictions of t_{send}, however, will stop N_1 after at most 100 time units.

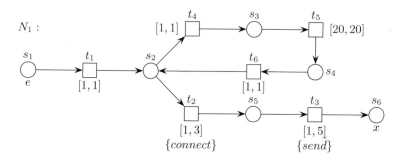

Fig. 5. Sending a Message

Figure 6 shows an (oversimplified) view of the receive action $t_{receive}$ by an hd-Mnet N_2. Hence altogether the fax is described by an hd-Mnet N such that

$$N = N_0[t_{send} \leftarrow N_1][t_{receive} \leftarrow N_2].$$

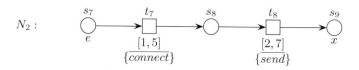

Fig. 6. Receiving a Message

2.2   A Short Introduction to the M-Net Calculus

In the sequel, we just like to introduce some notations necessary to understand the class of composable high-level nets, M-nets, which will be extended in this article. The reader interested in more details is invited to consult [9,12].

Let *Val* and *Var* be fixed but suitably large and disjoint sets of *values* and *variables*, respectively. The set of all well-formed *predicates* built from the sets *Val, Var* and a suitable set of operators is denoted by Pr.

We assume the existence of a set A of *synchronous action symbols*. Each $A \in \mathsf{A}$ has an arity $ar(A)$ and a *synchronous action* or just *action* is a construct $A(\nu_1, \ldots, \nu_{ar(A)})$, where $\forall j \in \{1, \ldots, ar(A)\} : \nu_j \in Var \cup Val$. The set of all actions is denoted by PA.

We assume the existence of a set X of *hierarchical actions*, which is the key to general refinements and recursions.

We denote by $\mathcal{M}(E)$ the set of all multi-sets over a set E and $\mathcal{M}_f(E)$ contains only the finite ones.

M-nets are equipped with double inscriptions: the second part of the inscriptions is the usual one for colored nets, while the first one is original for M-nets to define their interfaces. More formally:

An *M-net* is a triple (S, T, ι), where S and $T = T_c \uplus T_h$ are disjoint sets of *places* and *transitions*, and ι is an inscription function with domain $S \cup (S \times T) \cup (T \times S) \cup T$ such that:

- for every place $s \in S$, $\iota(s)$ is a pair $\lambda(s).\tau(s)$, where $\lambda(s) \in \{\mathsf{e}, \mathsf{i}, \mathsf{x}\}$ (for entry, inner and exit status) is the *label* of s, and $\tau(s) \subseteq Val$, is the *type* of s;
- for every communication transition $t \in T_c$, $\iota(t) = \alpha(t).\gamma(t)$, where $\alpha(t) \in \mathcal{M}_f(\mathsf{PA})$ is the *action label* and $\gamma(t)$, the *guard* of t, is a finite set of predicates from Pr;
- for every hierarchical transition $t \in T_h$, $\iota(t) = \lambda(t).\gamma(t)$, with $\lambda(t) \in \mathsf{X}$, and $\gamma(t)$ is as above;
- for every arc $(s, t) \in (S \times T) : \iota((s, t)) \in \mathcal{M}_f(Val \cup Var)$ is a multi-set of variables or values (analogously for arcs $(t, s) \in (T \times S)$).

In the following, we consider only *safe* M-nets, in the sense that, for every M-net N and for every reachable marking M, $M(s)$ is a set for each $s \in S$.

M-nets form an algebra. The general refinement $N[\mathcal{X}_i \leftarrow N_i \mid i \in I]$ intuitively means 'N where the $\mathcal{X}_i$-labelled transitions are refined into (i.e., replaced by a copy of) N_i, for each i in the indexing set I'. The operation of *synchronization* N **sy** A is defined as an unary operation consisting of an iteration of some basic synchronizations over a synchronous communication symbol A. The operation of *scoping* N **sc** A is defined as a *synchronization* followed by a *restriction* of the net N over symbol A. Please, see the quoted basic papers.

The control flow operations can be derived via general refinement of some special operator nets, illustrated in Figure 7. The arcs in the area of each transition in the given nets are inscribed by a single black token: $\{\bullet\}$.

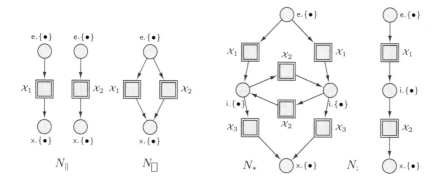

Fig. 7. Operator nets for derived composition operators.

For M-nets N_1, N_2 and N_3 the main operations are defined by

$$
\begin{aligned}
N_1 \parallel N_2 &= N_\parallel[\mathcal{X}_1 \leftarrow N_1, \mathcal{X}_2 \leftarrow N_2] & \text{parallel composition} \\
N_1 \,\square\, N_2 &= N_\square[\mathcal{X}_1 \leftarrow N_1, \mathcal{X}_2 \leftarrow N_2] & \text{choice} \\
[N_1 * N_2 * N_3] &= N_*[\mathcal{X}_1 \leftarrow N_1, \mathcal{X}_2 \leftarrow N_2, \mathcal{X}_3 \leftarrow N_3] & \text{iteration} \\
N_1 ; N_2 &= N_;[\mathcal{X}_1 \leftarrow N_1, \mathcal{X}_2 \leftarrow N_2] & \text{sequential composition.}
\end{aligned}
$$

2.3   Formal Definitions

A duration M-net is an M-net, where all standard definitions on M-nets can be found), in which the firing of events may consume time. The set of allowed firing times is given by an interval of nonnegative real numbers, the interval having rational bounds, but, following the discussion in [31], without loss of generality only integer firing times will be considered here.

Definition 1 (Duration M-net). *A duration M-net (d-Mnet for short) consists of an M-Net M together with an additional inscription*

$$\chi : T_c \to \mathbf{N} \times (\mathbf{N} \cup \{\omega\})$$

of transitions. For a transition $t \in T_c$, $\chi(t) = (sfd(t), lfd(t))$ denotes the possible range for the duration of an occurrence of t. Altogether, for each transition, the inscription of a transition t is of the form $\iota(t) = \alpha(t).\gamma(t).\chi(t)$. ◇

For a d-Mnet N,

$$Events(N) = \{t : \sigma | t \in T \text{ and } \sigma \text{ is an enabling binding for } t\}$$

denotes the set of possible firing events.

A firing event $t : \sigma$ of a d-Mnet N is enabled iff it is so in the underlying M-net, i.e. iff σ is an enabling binding of the variables occurring in the scope of t. After becoming enabled, $t : \sigma$ starts firing immediately, by removing the input tokens from its preplaces (unless it is prevented from doing so by a conflicting firing event). After that, $t : \sigma$ stays firing for some time delay δ s.t. $sfd(t) \leq \delta \leq lfd(t)$. After that, $t : \sigma$ ceases to fire by delivering the output tokens to its postplaces. Hence, for the definition of the transition rule for a d-Mnet, we have to consider three types of events, namely

1. *startfire events*, which remove tokens from the preplaces of a transition,
2. *endfire events*, which put tokens on the postplaces, and
3. *tick events*, which model passing of time.

So, in a d-Mnet, firing of an event does not only depend on markings but also has to take into account the timing constraints. Therefore a notion of state is introduced which consists of a marking part and a part keeping track of the relevant timing information: A *state* of a d-Mnet N is a pair (M, I), consisting of a marking M of N and a mapping I. For each firing event $t : \sigma \in Events(N)$, the clock vector I keeps track of the amount of time elapsed since $t : \sigma$ started to occur. Hence, I is a partial function, mapping each currently occurring firing event to a natural number. The *initial state* $S_0 = (M_0, I_0)$ of N is given by the standard initial marking M_0 of N and the totally undefined clock vector I_0.

The formal definition of the transition rule for d-Mnets is given by the case of not timed refined transitions in definition 3.

In a hierarchical d-Mnet N_1, a transition t may be equipped with another (hierarchical) d-Mnet N_t, in which case t is called *timed refined by N_t*. The intended meaning is that if an event $t : \sigma$ starts firing, a fresh copy $N_{t:\sigma}$ of N_t is created and put in parallel with N_1. Concurrently with N_1, $N_{t:\sigma}$ starts firing in its initial state S_0. If an endfire-event of $t : \sigma$ occurs, $N_{t:\sigma}$ is removed immediately. So, in some sense $t : \sigma$ acts as an additional global clock for N_t. This kind of semantics is called *timeout semantics*. Alternatively, $t : \sigma$ can be seen as specification of a time restriction which $N_{t:\sigma}$ has to meet, leading to a deadlock if it fails to do so. This type of semantics is called *deadlock semantics*. Of course, time refinements can be nested. These considerations lead to the following definition:

Definition 2 (Hierarchical duration M-net).
The set DN of hierarchical duration M-nets (hd-Mnets for short) is defined inductively as follows:

- *Every d-Mnet N is an hd-Mnet. The transitions of N are called not timed refined.*
- *If N_1, N_t are hd-nets and $t \in T_1$ such that t is not timed refined, then $N = N_1[t \leftarrow N_t]$ is an hd-Mnet. (N is called timed refinement of t in N_1 by N_t and t is called timed refined (by N_t).)*

$\diamond$

An inscription of a clock in the graphical representation of a transition t indicates that t is timed refined.

For an hd-Mnet N, $CON(N)$ denotes the set of constituent nets, defined by

$$CON(N) = \begin{cases} \{N\} & \text{if } N \text{ is a d-Mnet} \\ CON(N_1) \cup CON(N_2) & \text{if } N = N_1[t \leftarrow N_t] \end{cases}$$

Slightly abusing notation, for N,

$$S_N = \bigcup_{N' \in CON(N)} S_{N'}$$

will denote the set of all places of all constituent nets of N. Similar notation is used for the set of transitions of N etc.

Moreover,

$$Events(N) = \bigcup_{N' \in CON(N)} Events(N')$$

denotes the set of possible firing events of N and

$$Events_{ref}(N) = \{t : \sigma | t : \sigma \in Events(N) \text{ and } t \text{ is timed refined in } N\}$$

denotes the set of possible timed refined firing events.

For $t : \sigma \in Events_{ref}(N)$ such that t is timed refined by N_t, let $N_{t:\sigma}$ denote a fresh copy[3] of N_t, $S_{N_{t:\sigma}}$ its places, etc. . Then all places in the refining nets are defined as

$$S_{ref}(N) = \bigcup_{t:\sigma \in Events_{ref}(N)} S_{N_{t:\sigma}}.$$

A *marking* M of N is a set over S_N of the form

$$M = \bigcup_{s \in (S \cup S_{ref}(N))} M_s$$

such that, for all $s \in (S \cup S_{ref}(N))$, $M_s \in \mathcal{M}_f(\tau(s))$, i.e. a finite multi set over the type of s.

Let M be a marking of N. Then

$$Enabled(M) = \{t : \sigma | t : \sigma \in Events(N) \text{ and } M \text{ activates } t \text{ in mode } \sigma\}$$

denotes the set of firing events of N which can occur at a marking M.

$St = (M, I)$ is a state of N, iff

- M is a marking of N and
- $I : Events_{ref}(N) \to \mathbf{N} \cup \{\perp\}$, such that $I(t : \sigma) \neq \perp$ implies $I(t : \sigma) \leq lfd(t)$.

---

[3] We assume that different occurrences of a firing event are distinguished by different names of the bindings. Hence, a single copy of the timed refining net for each timed refined firing event is sufficient for the following definitions.

The set of all possible states of N is denoted by $States(N)$.

Assume that N is of the form

$$N = (...(N_0[t_1 \leftarrow N_1])[t_2 \leftarrow N_2])...[t_n \leftarrow N_n]),$$

where $N = N_0$ if $n = 0$. Then $St_0(N) = (M_0, I_0)$ such that

- $M_0 = M^e(N_0)$ is the standard initial marking of N_0, given by $M^e(s) = \tau(s)$ for every entry place s of N_0 and $M^e(s) = \emptyset$ for all other places and
- $I(t : \sigma) = \bot$ for all $t : \sigma \in Events(N)$

is called the *initial state* of N.

2.4   The Transition Rule

As for d-Mnets, for the transition rule of an hd-Mnet we distinguish three types of events. So, let N be an hd-Mnet. Then

1. *Startfire events:* A startfire event (denoted as $[t : \sigma)$ has to occur immediately if $t : \sigma$ becomes enabled in the underlying M-net. Upon occurrence of $[t : \sigma$, the input tokens are removed from its preplaces, the clock associated to $t : \sigma$ is started, and $t : \sigma$ is called *active*. Moreover, if t is timed refined, the net associated to $t : \sigma$ is set to its initial marking.
2. *Endfire events:* For the occurrence of an endfire event (denoted as $t : \sigma)$) we have to distinguish between transitions which are timed refined and those which are not.
   - If t is not timed refined, then $t : \sigma)$ may occur if the clock associated to $t : \sigma$ shows a time δ such that $sfd(t) \leq \delta \leq lfd(t)$. In this case, upon occurrence of $t : \sigma)$, the output tokens are delivered at the postplaces of t and the clock associated to $t : \sigma$ is reset.
   - Otherwise, i.e. if a net N_t is associated to t, $t : \sigma)$ may occur if either the copy $N_{t:\sigma}$ of N_t associated to $t : \sigma$ has reached its final marking $M^x(N_{t:\sigma})$ (which is defined by $M^x(s) = \tau(s)$ for every exit place s of $N_{t:\sigma}$ and $M^x(s) = \emptyset$ for all other places) and the clock associated to $t : \sigma$ shows a time δ such that $sfd(t) \leq \delta \leq lfd(t)$ or if the latest firing time is reached, i.e. if $\delta = lfd(t)$.

   For the effect of the occurrence of $t : \sigma)$ two different approaches are considered:
   - deadlock semantics: The net N (and all other nets of $CON(N)$) are stopped by removing all tokens from all places.
   - timeout semantics: The net $N_{t:\sigma}$ is stopped by removing all tokens from places in $S_{N_{t:\sigma}}$
3. *Tick events:* A tick event (denoted as $\checkmark$) is enabled iff there is no firing event which must either start firing or stop firing. Upon occurring, a tick event increments the clocks for all active firing events. Hence, tick events are global.

We write $S \xrightarrow{e}$ to denote that event e is enabled at state S and $S \xrightarrow{e} S'$ to denote that $S \xrightarrow{e}$ and that the occurrence of e at S leads to state S'.

Definition 3 (Transition rule).
  Let N be an hd-Mnet.

1. **startfire events:**
   - $(M, I) \xrightarrow{[t:\sigma}$ *iff*
     $t : \sigma \in Events(N)$ *and M activates t in mode σ.*
   - $(M, I) \xrightarrow{[t:\sigma} (M', I')$ *iff*
     $(M, I) \xrightarrow{[t:\sigma}$ *and*
     - **Case 1:** t *is not timed refined. Then*
       * $M' = M - (\bigcup_{s \in {}^\bullet t} \iota(s, t)[\sigma])$
       * $I'(t' : \sigma') = \begin{cases} 0 & \text{if } t = t' \text{ and } \sigma = \sigma' \\ I(t' : \sigma') & \text{otherwise} \end{cases}$
     - **Case 2:** $N = N_1[t \leftarrow N_2]$. *Then*
       * $M' = M - (\bigcup_{s \in {}^\bullet t} \iota(s, t)[\sigma]) \cup M^e_{N_{t:\sigma}}$
       * $I'(t' : \sigma') = \begin{cases} 0 & \text{if } t = t' \text{ and } \sigma = \sigma' \\ \perp & \text{if } t' : \sigma' \in Events(N_{t:\sigma}) \\ I(t' : \sigma') & \text{otherwise} \end{cases}$
2. **endfire events:**
   - **Case 1:** t *is not timed refined. Then*
     * $(M, I) \xrightarrow{t:\sigma\rangle}$ *iff*
       $sfd(t) \leq I(t : \sigma) \leq lfd(t)$.
     * $(M, I) \xrightarrow{t:\sigma\rangle} (M', I')$ *iff*
       $(M, I) \xrightarrow{t:\sigma\rangle}$ *and*
       * $M' = M + (\bigcup_{s \in t^\bullet} \iota(t, s)[\sigma])$

       * $I'(t' : \sigma') = \begin{cases} \perp & \text{if } t = t' \text{ and } \sigma = \sigma' \\ I'(t' : \sigma') & \text{otherwise} \end{cases}$
   - **Case 2:** $N = N_1[t \leftarrow N_2]$. *Then*
     * $(M, I) \xrightarrow{t:\sigma\rangle}$ *iff*
       either $[M \cap M_{t:\sigma} = M^x_{t:\sigma}$ and $sfd(t) \leq I(t : \sigma) \leq lfd(t)]$, or only
       $I(t : \sigma) = lfd(t)$
     * $(M, I) \xrightarrow{t:\sigma\rangle} (M', I')$ *iff*
       $(M, I) \xrightarrow{t:\sigma\rangle}$ *and*
       a) **deadlock semantics:**
          * $M' = \emptyset$
          * $I'(t' : \sigma') = \perp$ *for all $t' : \sigma' \in Events(N)$.*
       b) **time-out semantics:**
          * $M' = M + (\bigcup_{s \in t^\bullet} \iota(t, s)[\sigma] - M(N_{t:\sigma})$

          * $I'(t' : \sigma') = \begin{cases} \perp & \text{if } (t = t' \text{ and } \sigma = \sigma') \\ I'(t' : \sigma') & \text{otherwise} \end{cases}$

3. **tick events:**
- $(M, I) \xrightarrow{\checkmark}$ *iff*
  $\{t : \sigma \mid t : \sigma \in \textit{Events}(N) \textit{ and } M \textit{ activates } t \textit{ in mode } \sigma\} = \emptyset$ *and, for all* $t : \sigma \in \textit{Events}(N)$, $I(t : \sigma) \neq \bot$ *implies* $I(t : \sigma) < \textit{lfd}(t)$.
- $(M, I) \xrightarrow{\checkmark} (M', I')$ *iff* $I' = I \oplus 1$, $(M, I) \xrightarrow{\checkmark}$ *and* $M = M'$.
  where $(I \oplus 1)(t : \sigma) = I(t : \sigma) + 1$ *for all* $t : \sigma$. $\diamond$

All the usual dynamic concepts (occurrence sequences, step sequences, set of reachable markings etc.) follow from the transition rule in the standard way.

3   Partial Order Semantics of Hierarchical Duration M-Nets

In this section, the new notion of branching process of a hierarchical duration net is introduced. As for a P/T-net the maximal branching process β_m associates a partial order semantics to each safe hd-Mnet N [16]. Moreover, following the line of [16], a finite representation of β_m can be defined, which in turn can be used to analyse qualitative temporal properties of N, e.g. using the PEP-system.

A *causal process* describes a possible run of N, displaying the causal dependencies of the events that take place during the run. A *branching process* can represent several alternative runs of N in one structure and hence may be seen as the union of some causal processes. Therefore, the maximal branching process represents *all* causal processes within a single structure. A branching process consists of an occurrence net and a homomorphism.

An *occurrence net* ON $= (B, E, G)$ is an acyclic net such that $|{}^{\bullet}b| \leq 1$ for all $b \in B$, no $e \in E$ is in self conflict and, for all $x \in (B \cup E)$, the set of elements $y \in (B \cup E)$ such that $x \leq y$ is finite, where $\leq$ refers to the partial order induced by G on $B \cup E$. Nodes x_1, x_2 are in conflict $(x_1 \sharp x_2)$ if there are distinct events $e_1, e_2 \in E$ such that ${}^{\bullet}e_1 \cap {}^{\bullet}e_2 \neq \emptyset$ and $(e_1, x_1), (e_2, x_2) \in G^*$. A node $x \in (B \cup E)$ is in self conflict if $x \sharp x$. The elements of B and E are called *conditions* and *events*, respectively. *MIN(ON)* denotes the set of minimal elements of *ON* w.r.t. $\leq$.

The *homomorphism* is used to connect conditions and events of an occurrence net to places and transitions of the net whose behaviour is described. Here, in addition, we have to define some labels representing the time part. So let N be an hd-Mnet. To represent the clock vector part of a state, we introduce new place labels $(t : \sigma, i)$ with $i \in \mathbf{N}$ for any firing event $t : \sigma \in \textit{Events}(N)$. The intended meaning is that $(t : \sigma, i)$ indicates that event $t : \sigma$ is firing since i time units. The set of these clock labels is called $S_I(N)$. Thus a clock vector can be represented as part of a marking.

Next, we give a low level representation of a marking: Token v at place s is represented by a label s_v, and we let

$$S_D(N) = \{s_v | s \in (S_{ref}(N) \cup S(N)) \text{ and } v \in \alpha(s)\}$$

denote the set of data labels of N. The set of all low level labels is then given by

$$S^{ll}(N) = S_D(N) \cup S_I(N).$$

Moreover, define

$$\phi : States(N) \to \mathcal{M}_f(S^{ll}(N))$$

by

$$\phi(M, I)(x) = \begin{cases} M(s)(v) & \text{if } x = s_v \in S_D(N) \\ 1 & \text{if } x = (t : \sigma, i) \in S_I(N) \text{ and } I(t : \sigma) = i \\ 0 & \text{otherwise.} \end{cases}$$

So, for every state (M, I) of N, ϕ computes a multiset of labels in $S^{ll}(N)$ (which in fact is a set if N is safe). Note that ϕ is a partial bijection.

The set $\mathcal{FE}(N)$ of firing elements of N is defined by

- $Tick(N) = \{\checkmark_C | C \subseteq S_I(N) \text{ and } (t : \sigma, i) \in C \text{ implies } i < lfd(t)\})^4$
- $Startfire(N) = \{[t : \sigma \mid t : \sigma \in Events(N)\}$
- $Endfire(N) = \{t : \sigma\rangle_{(M,I)} \mid t : \sigma \in Events(N) \text{ and } (M, I) \in States(N)$
  and $sfd(t) \leq I(t : \sigma) \leq lfd(t)\}$
- $\mathcal{FE}(N) = Tick(N) \cup Startfire(N) \cup Endfire(N)$.

Let $fe \in \mathcal{FE}(N)$ and define

$$pre(fe) = \begin{cases} C & \text{if } fe = \checkmark_C \\ \phi(\bigcup_{s \in S} \iota(s, t)[\sigma]) & \text{if } fe = [t : \sigma \\ \{(t : \sigma, I(t : \sigma))\} & \text{if } fe = t : \sigma\rangle_{(M,I)} \text{ and } t \text{ not} \\ & \text{refined} \\ \{(t : \sigma, I(t : \sigma))\} \cup \phi(M, I)\lceil_{CON(N_{t:\sigma})}{}^5 & \text{if } fe = t : \sigma\rangle_{(M,I)} \text{ and } t \\ & \text{refined.} \end{cases}$$

and

$$post(fe) = \begin{cases} \{(t : \sigma, i + 1)|(t : \sigma, i) \in C\} & \text{if } fe = \checkmark_C \\ \{(t : \sigma, 0)\} & \text{if } fe = [t : \sigma \text{ and } t \text{ is not refined} \\ \{(t : \sigma, 0)\} \cup \phi(S_0(N_{t:\sigma})) & \text{if } fe = [t : \sigma \text{ and } t \text{ is refined} \\ \phi(\bigcup_{s \in S} \iota(t, s)[\sigma]) & \text{if } fe = t : \sigma\rangle_{(M,I)} \, . \end{cases}$$

Now a *homomorphism* from an occurrence net ON to an hd-Mnet N is defined to be a mapping

$$\pi : B \cup E \to (S^{ll}(N) \cup \mathcal{FE}(N))$$

such that

1. $\pi(B) \subseteq S^{ll}(N)$ and $\pi(E) \subseteq \mathcal{FE}(N)$;
2. for all $e \in E$, the restriction of π to ${}^\bullet e$ is a bijection from ${}^\bullet e$ to $pre(\pi(e))$;
3. for all $e \in E$, the restriction of π to $e^\bullet$ is a bijection from $e^\bullet$ to $post(\pi(e))$;

---

[4] Note, that if we are dealing with safe nets, there can at most be one entry $(t : \sigma, i)$ for any firing event $t : \sigma$.

[5] Denotes the restriction of $\phi(M, I)$ to the set of constituent nets of $N_{t:\sigma}$.

4. the restriction of π to $MIN(ON)$ is a bijection from $MIN(ON)$ to $\phi(M^e, I_0)$;
5. for all $e_1, e_2 \in E$ it holds that if $\pi e_1 = \pi e_2$ and $^\bullet e_1 = {}^\bullet e_2$ then $e_1 = e_2$.

The following algorithm nondeterministically constructs the maximal branching process $\beta_m = (B, E, G, \pi)$ of N in the following way: If any startfire event may occur, then the algorithm chooses among the enabled startfire and endfire events. Otherwise, if no endfire event is forced to occur, it will choose one of the tic events and the enabled endfire events. Otherwise, it chooses one arbitrary enabled endfire event.

Algorithm Maximal Branching Process:
Let $E = G = \emptyset$.
Construct B and π such that $\pi(B) = \phi(St_0(N))$.
LOOP
    IF *Possible(fe)* for some $fe \in Startfire$
        THEN
            Choose $fe \in Startfire(N) \cup Endfire(N)$ such that *Possible(fe)*
        ELSE
            IF (For all $fe \in Endfire$ it holds that (not *Necessary(fe)*)))
                THEN
                    Choose $fe \in Tic(N) \cup Endfire(N)$ such that *Possible(fe)*
                ELSE
                    Choose $fe \in Endfire(N)$ such that *Possible(fe)*
        FI
    FI
    Extend(β, fe, β').
POOL

where

- *possible(fe)* iff $fe \in \mathcal{FE}(N)$ and there exists a conflict free set $\{b_1, ..., b_k\} \subseteq B$ such that $\pi(\{b_1, ..., b_k\}) = pre(fe)$ and, for all $e \in E$, $((b_i, e) \in G$ for all $1 \le i \le k) \Rightarrow \pi(e) \ne fe$. Note, that $\{b_1, ..., b_k\}$ has to be a maximal conflict free set if $fe \in Tick(N)$.
- *Endfire(fe)* iff $fe \in \mathcal{FE}(N)$ and *Possible(fe)* and $(\pi(b_i) = (t : \sigma, lfd(t))$ for some $i \in \{1, ..., k\}$ and some firing event $t : \sigma)$.
- *Extend(β, fe, β')* iff *Possible(fe)* and β' is constructed from β by adding a new event e, new conditions $b'_1, ..., b'_l$, new arcs from $b_1, ..., b_k$ to e, new arcs from e to $b'_1, ..., b'_l$, and extending π such that $\pi(e) = fe$ and $\lambda(\{b'_1, ..., b'_l\}) = post(fe)$.

Figure 8 shows an initial part of the maximal branching process of the hd-Mnet N_1.

A *co-set* is a set $B' \subseteq B$ of conditions of β_m such that, for all $b \ne b' \in B'$, neither $b < b'$ nor $b' < b$ nor $b \# b'$. A *cut* is a maximal co-set C (w.r.t. set inclusion). The marking $Mark(C)$ associated to a cut C is given by $Mark(C) = \{\pi(b) | b \in C\}$.

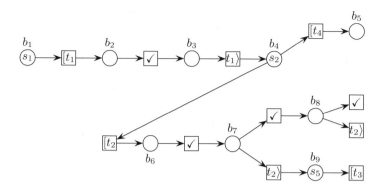

Fig. 8. Initial part of the maximal branching process of N_1 with the following clock labels: $\phi(b_2) = (t_1 : \sigma, 0)$, $\phi(b_3) = (t_1 : \sigma, 1), \phi(b_6) = (t_2 : \sigma, 0)$, $\phi(b_7) = (t_2 : \sigma, 1)$, $\phi(b_8) = (t_2 : \sigma, 3)$, where σ is always the trivial binding to the black token.

We have the following

Proposition 1. *Let N be an hd-Mnet.*

A state St is reachable in N iff the maximal branching process β_m contains some cut C such that $\phi(St) = Mark(C)$.

4   Operations on Hierarchical Duration M-Nets

In this section, we define composition operations on hd-Mnets which create a new hd-Mnet out of one, two, or three given hd-Mnets. We consider operators of two kinds: those concerning place interfaces and thus the control flow and those concerning transition interfaces and thus capabilities for communication.

The first class of operators consists of

- sequential composition ; ,
- parallel composition ——— ,
- choice composition ☐ and
- iteration composition [* *] .

These are (mutatis mutandis) defined as for M-nets.

On one hand, the second class of operators contains the classical communication operator **sc** for scoping based on synchronisation and restriction, which also is essentially the same as for M-nets. Let us just remark that for a new transition t_{12} arising out of t_1 and t_2 by basic synchronisation, the duration is calculated as follows:

$$\chi(t_{12}) = (max(sfd(t_1), sfd(t_2)), min(lfd(t_1), lfd(t_2))).$$

On the other hand, we define a new operation called *hierarchical timed scoping*. This operation can be used to synchronize over timed refined transitions and their respective refining hd-Mnets, thus allowing for a modular hierarchical specification of the timed behaviour of systems. The meaning of hierarchical timed scoping is the following: given an hd-Mnet N and transitions t_1, t_2 which are timed refined by hd-Mnets N_1 and N_2, respectively, such that t_1 and t_2 have to synchronize with respect to an action symbol A. Then, the hierarchical scoping of t_1 and t_2 (w.r.t. A) is given by the synchronized transition t_{12}, timed refined by the net which results from the parallel composition of N_1 and N_2, scoped with respect to the action symbols which N_1 and N_2 have in common. For the formal definition we have to bear in mind, that N_1 and N_2 in turn may contain timed refined transitions.

First, we introduce a special type of action symbols: Let $\mathcal{AA} \subseteq A$ such that $arity(a) = 0$ for all $a \in \mathcal{AA}$. The set $\mathcal{AA}$ is called the set of *abstract action symbols*. Abstract action symbols are used for the synchronisation of timed refined transitions. The remaining ones, $\mathcal{CA} = A \setminus \mathcal{AA}$ are called *concrete*. They are used for the (classical) synchronisation of not timed refined transitions. We fix that the action label of a transition t of a hd-Mnet, $\alpha(t)$, is **either** a (possibly empty) multi-set of concrete actions, if t is not timed refined, **or** a (possibly empty) multi-set of abstract actions, if t is timed refined.

Moreover, for an hd-Mnet N let $\mathcal{CA}(N)$ and $\mathcal{AA}(N)$ denote the set of concrete, respectively abstract action symbols occurring in N. A hd-Mnet is called concrete if all its action labels are in $\mathcal{CA}(N)$ (we get back the usual notion of hd-Mnets, as defined in
section 2).

Now, hierarchical scoping **hsc** is defined inductively in the following way:

Definition 4.   *1. Let N be an hd-Mnet such that N contains no timed refined transitions. Then the timed hierarchical scoping of N is just the net N:*

$$\text{hsc } (N) = N.$$

2. *Let N be an hd-Mnet and $t_1, t_2 \in T$ such that $var(t_1) \cap var(t_2) = \emptyset$ and t_1, t_2 are timed refined, i.e. $N = N_0[t_1 \leftarrow N_1, t_2 \leftarrow N_2]$ for hd-Mnets N_1 and N_2 and, for some $A \in \mathcal{AA}(N)$, $A \in \alpha(t_1)$ and $\overline{A} \in \alpha(t_2)$. Then the hierarchical basic scoping of t_1 and t_2 in N w.r.t. A is defined by*
   - $tsc_{basic}(N, t_1, t_2, A) =$
     $N'[t_{12} \leftarrow (N_1 || N_2 \text{ sc } (\mathcal{CA}(N_1) \cup \mathcal{CA}(N_2)) \text{ hsc } (\mathcal{AA}(N_1) \cup \mathcal{AA}(N_2))],$

     where N' is given by
     - $S_{N'} = S_N$.
     - $T_{N'} = T_{N_0} \cup \{t_{12}\} \setminus \{t_1, t_2\}$.
     - ${}^\bullet t_{12} = {}^\bullet t_1 \cup {}^\bullet t_2$ and $t_{12}{}^\bullet = t_1{}^\bullet \cup t_2{}^\bullet$.
     - $\alpha(t_{12}) = \alpha(t_1) \cup \alpha(t_2) - \{A, \overline{A}\}$.
     - $\gamma(t_{12}) = \gamma(t_1) \cup \gamma(t_2)$.
     - $\chi(t_{12}) = (max(sfd(t_1), sfd(t_2)), min(lfd(t_1), lfd(t_2)))$.

3. *Let N be an hd-Mnet. Then **hsc** $(N) = N$ **hsc** $(\mathcal{AA}(N))$, the timed hierarchical scoping of N w.r.t. all abstract action symbols of N, is constructed in two steps:*

step 1: *Construct N' such that N' is the smallest hd-Mnet satisfying*
a) $S = S'$.
b) Every transition of N (and its surrounding arcs) is also in N', with the same inscriptions as in N.
c) If t_1 is a transition of N and t_2 is a transition of N' such that one of them contains an abstract action symbol $A \in \mathcal{AA}(N)$ in its label and the other one the abstract action symbol $\overline{A}$, then any transition t arising through a basic hierarchical scoping over A out of t_1 and t_2 are also in N', as well as its surrounding arcs.

step 2: *Construct* **hsc** (N) *from N' by removing every timed refined transition t with $\alpha(t) \neq \emptyset$ (as well as its surrounding arcs) from N', i.e.* **hsc** (N) *is a concrete hd-Mnet.*

The following proposition is an immediate consequence of the definition of the hierarchical scoping operator **hsc** .

Proposition 2. *Let N be an hd-Mnet. Then* **hsc** (N) *is an hd-Mnet without any occurrence of abstract actions or hierarchical scoping operators. Thus* **hsc** (N) *is a concrete hd-Mnet.*

This means that **hsc** (N) will just denote an abstract view of a much more detailed net, which is an adequate approach for a top down design of concurrent systems with complicated timing requirements.

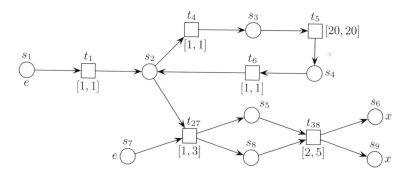

Fig. 9. Sending and Receiving a Message (Synchronized)

Hierarchical scoping may be applied to the hd-Mnet N of figure 4, which involves basic hierarchic synchronisation w.r.t. *fax* applied to t_{send} and to $t_{receive}$ (which of course is more sensible if t_{send} and $t_{receive}$ belong to different fax machines), which in turn leads to synchronisation of N_1 of FaxA (cf. figure 5) and N_2 of FaxB (cf. figure 6). The net of figure 9 shows

$$N_1 \| N_2 \ \textbf{sc} \ (\mathcal{CA}(N_1) \cup \mathcal{CA}(N_2)) \ \textbf{hsc} \ (\mathcal{AA}(N_1) \cup \mathcal{AA}(N_2)$$

$$= N_1 \| N_2 \ \textbf{sc} \ \{connect, send\} \ .$$

5   Related Work

Three time extensions of the low level Petri box calculus, but not on the M-net level have been presented at the last Petri Nets Conferences:

Koutny [22] defines a timed Petri box calculus by extending the PBC with waiting times for transitions and studies then its structural operational semantics. This approach corresponds to the time Petri net model and is fairly different from our one chosen in this paper.

In [1] an extension of the PBC with time (as duration of actions) is given and operational as well as denotational semantics are defined, thereby considering all operations of the PBC, including general refinement. The main difference, however, lies in the fact, that they do not consider inherently hierarchical nets and, hence, they also do not consider transitions whose timely behaviour may be constrained by the duration times of several transitions. So, in their approach only local time constraints can be modelled by finite timed Petri nets.

In [18], a method for proving qualitative temporal properties of time Petri nets was presented. As pointed out in section 3, this method can be lifted to hd-Mnets.

6   Conclusion

Starting from the notion of timed Petri net, we have introduced an extension: the model of hierarchical duration M-nets (hd-Mnets). This is in fact a double extension: on the one side a passage to compositional high level nets and on the other side the new possibility to treat global timing constraints for subnets. Resulting from the new operation of timed refinement, the duration of a transition occurrence in an hd-Mnet may be influenced by several hierarchically nested clocks (of the hierarchically superordinate transitions), where the hierarchical nesting is such that the time slots of two different clocks are either contained in one another or their intersection is empty. We have also added an abstract operation called hierarchical scoping which is a powerful device for modular top down development of distributed timed systems. This feature does not extend the expressive power of the class of hd-Mnets.

As an original contribution, we have adressed here for the first time a branching process semantics directly for high level nets with time: Partial order semantics of hd-Mnets have been introduced in terms of their timed maximal branching process to define their concurrent timed behavior. We propose this definition without preliminary unfolding to a low level hd-net: This involves a simultaneous treatment of bindings, choice, clock-counting, time respect (or time-outs).

The semantics so defined allows for proving qualitative temporal properties of systems described by hd-Mnets using already existing tools like the model checking component of the PEP system. Future work will be dedicated to extend the underlying temporal logics to cope also with quantitative temporal properties.

As already announced, our work has been partially led inside the Procope project PORTA, where several time extensions of M-nets have been developed.

Thus quite naturally, we have been interested in the comparison of hd-Mnets, presented here, with a quite different approach raised in PORTA, that of causal time M-nets, based on implicit treatment of time constraints (cf. [19,20,21]). We have been able to prove that the former can be simulated by the latter w.r.t. concurrent timed behaviour. Our results show other important differences: for instance, overlapping time requirements can not be modelled by an hd-Mnet, but it can with a causal time M-net. The full comparison of both approaches in the treatment of time can be found in [26].

As it is well known, timed Petri nets can be simulated by means of time Petri nets, but not vice versa. It is therefore interesting to define and study the notion of hierarchical time M-nets. The results will be presented within a forthcoming paper where the techniques needed for the translation will be totally different from those used here for timed nets.

References

1. O. M. Alonso and D. de Frutos Escrig. Extending the Petri Box Calculus with Time. In J. Colom, M. Koutny, (Eds.), *Advances in Petri Nets 2001*, Volume 2075 of *LNCS*, pages 303–322. Springer, (2001).

2. T. Aura, J. Lilius: Time Processes for Time Petri Nets. *Proc. ICATPN*, Toulouse (1997).

3. B. Berthomieu, M. Diaz: Modelling and Verification of Time Dependent Systems Using Time Petri Nets. *IEEE Transactions on Software Engineering*, Volume 17/3, 259–273 (1991).

4. E. Best: Partial Order Verification with PEP. Proc. *POMIV'96, Partial Order Methods in Verification*, G. Holzmann, D. Peled, and V. Pratt (eds.), American Mathematical Society (1996).

5. E. Best, R. Devillers, and J. Esparza. General refinement and recursion operators for the Petri box calculus. *LNCS 665:130–140*, (1993).

6. E. Best, R. Devillers, and J. G. Hall. The box calculus: a New Causal Algebra With Multi-Label Communication. In G. Rozenberg, (Ed.), *Advances in Petri Nets 92*, Volume 609 of *LNCS*, pages 21–69. Springer, 1992.

7. E. Best, R. Devillers, and M. Koutny. *Petri Net Algebra*. Springer-Verlag. EATCS Monographs on Theoretical Computer Science Series 2001.

8. E. Best, H. Fleischhack, W. Fraczak, R. P. Hopkins, H. Klaudel, and E. Pelz. A Class of Composable High Level Petri Nets. In G. De Michelis and M. Diaz, (Eds.), *Application and Theory of Petri Nets 1995*, Volume 935 of *LNCS*, pages 103–118. Springer, 1995.

9. E. Best, W. Fraczak, R. Hopkins, H. Klaudel, and E. Pelz. M-nets: An algebra of high-level Petri nets, with an application to the semantics of concurrent programming languages. *Acta Informatica*, 35, 1998.

10. B. Bieber and H. Fleischhack: Model Checking of Time Petri Nets Based on Partial Order Semantics. *Proceedings of ConCur'99*, Eindhoven (1999).

11. R. Devillers and H. Klaudel. Refinement and Recursion in a High Level Petri Box Calculus. *STRICT'95*. Springer, ViC, 144–159 (1995).

12. R. Devillers, H. Klaudel and R.-C. Riemann. General Refinement for High Level Petri Nets. *FST & TCS'97*, Springer, LNCS Vol. 1346, 297–311 (1997).

13. R. Devillers, H. Klaudel and R.-C. Riemann. General parameterised refinement for the M-net calculus. Theoretical Computer Science, to appear.

14. R. Durchholz. Causality, time, and deadlines. *Data & Knowledge Engineering*, 6:496–477, 1991.
15. J. Engelfriet: Branching Processes of Petri Nets, *Acta Informatica*, Volume 28, pages 575–591 (1991).
16. J. Esparza: Model Checking Using Net Unfoldings. *Science of Computer Programming*, Volume 23, 151–195, Elsevier (1994).
17. J. Esparza, S. Römer, and W. Vogler: An Improvement of McMillan's Unfolding Algorithm. *Proc. of TACAS'96* (1996).
18. H. Fleischhack and C. Stehno. Computing a Finite Prefix of a Time Petri Net. In *Advances in Petri Nets 2002*, Volume 2360 of *LNCS*, pages 163–181. Springer, 2002.
19. H. Klaudel and F. Pommereau. Asynchronous links in the PBC and M-nets. In *Proc. of ASIAN'99, LNCS 1742:190–200*, 1999.
20. H. Klaudel and F. Pommereau. A concurrent and compositional Petri net semantics of preemption. In W. Grieskamp, T. Santen, B. Stoddart (Eds.),*Proc. of IFM'00*, LNCS 1945, pages 318–337, Springer, 2000.
21. H. Klaudel and F. Pommereau. A Concurrent Semantics of Static Exceptions in a Parallel Programming Language. In J. Colom, M. Koutny, (Eds.), *Advances in Petri Nets 2001*, Volume 2075 of *LNCS*, pages 204–223. Springer, (2001).
22. M. Koutny. A Compositional Model of Time Petri Nets. In M. Nielsen, D. Simpson, (Eds.), *Advances in Petri Nets 2001*, Volume 1825 of *LNCS*, pages 303–322. Springer, 2000.
23. J. Lilius: Efficient state space search for time Petri nets. *MFCS Workshop on Concurrency* (1998).
24. K.L. McMillan: Using unfoldings to avoid the state explosion problem in the verification of asynchronous circuits. *Proc. CAV '92, Fourth Workshop on Computer-Aided Verification*, Vol. 663 of LNCS, 164–174 (1992).
25. P. Merlin, D. Farber: Recoverability of Communication Protocols – Implication of a Theoretical Study. *IEEE Transactions on Software Communications*, Vol. 24, 1036–1043 (1976).
26. E. Pelz and H. Fleischhack. High Level Petri Nets with Timing Constraints – a Comparison to appear in *Proc. of ASCD 03*, special volume of June 2003.
27. L. Popova: On Time Petri Nets. *Journal of Information Processing and Cybernetics*, Volume 1055 of *LNCS*, Springer (1991).
28. C. Ramchandani: *Analysis of Asynchronous Concurrent Systems by Timed Petri Nets*, MIT, Project MAC, Technical Report 120, (1974).
29. G. Richter. Counting interfaces for discrete time modeling. Technical Report rep-set-1998-26, GMD, 1998.
30. J. Sifakis: Performance Evaluation of Systems Using Nets. Volume 84 of *LNCS*, pages 307–319, Springer (1980).
31. P. Starke: Analyse von Petri-Netz-Modellen. Teubner Verlag, Stuttgart (1990) (in German).

A Heuristic Algorithm *FSDC* Based on Avoidance of Deadlock Components in Finding Legal Firing Sequences of Petri Nets

Satoshi Taoka, Shinji Furusato, and Toshimasa Watanabe

Graduate School of Engineering, Hiroshima University
1-4-1, Kagamiyama, Higashi-Hiroshima, 739-8527 Japan
Phone : +81-824-24-7666 (Taoka), -7662 (Watanabe)
Facsimile : +81-824-22-7028

{taoka, furusato, watanabe}@infonets.hiroshima-u.ac.jp

Abstract. The paper proposes a heuristic algorithm *FSDC* for solving the Maximum Legal Firing Sequence problem of Petri nets (**MAX LFS** for short) and evaluates it experimentally. *FSDC* is improved from the existing one *FSD* for **MAX LFS** by focusing on deadlock components, instead of D-siphons, and by incorporating efficient backtracking. As experimental evaluation, *FSDC* is applied to 3017 test problems to each of which existence of an exact solution is guaranteed, and it has produced an optimum solution to each of 2330 (77.2%) test problems, which is about 1.43 times more than that of *FSD*, while the average CPU time is about 1.82 times longer than that of *FSD*. For five related problems each of which contains **MAX LFS** as a subproblem, it is experimentally shown that incorporating *FSDC* for solving **MAX LFS** gives us five heuristic algorithms that are superior in capability to existing ones.

Topics: Analysis and synthesis, structure and behavior of nets

1 Introduction

1.1 The Subject

The subject of the paper is to propose a heuristic algorithm *FSDC* for solving the legal firing sequence problem of Petri nets (**LFS** for short). It is experimentally evaluated that *FSDC* has highest capability among existing ones.

Also proposed are *FSDB*, *FSDQ*, *GAD* and *GAB* for **LFS**. *FSDB*, *FSDC* and *FSDQ* are improved versions of *FSD* [21], and both *GAD* and *GAB* are genetic algorithms. These algorithms as well as existing ones are implemented and their capabilities are experimentally compared.

W.M.P. van der Aalst and E. Best (Eds.): ICATPN 2003, LNCS 2679, pp. 417–439, 2003.

1.2   Definitions of LFS and MAX LFS

The Legal Firing Sequence problem of Petri nets (**LFS**)[16] is defined by:

[**LFS**] "Given a Petri net N, an initial marking M and a firing count vector X, find a firing sequence (a sequence of transitions) that is legal on M with respect to X."

A component $X(t)$ of X denotes the prescribed total firing number of a given transition t. Without loss of generality we assume $X(t) > 0$ for any $t \in T$. We say that a firing sequence is legal on an initial marking M with respect to a given X if and only if the first transition of the sequence is firable on M, the rest can be fired one by one subsequently, and each transition t appears exactly $X(t)$ times in the sequence.

   LFS is very fundamental in the sense that it appears as a subproblem or a simpler form of various basic problems in Petri net theory, such as the well-known marking reachability problem (**MR** for short) [5,7], the minimum initial resource allocation problem [10,11,15,17,20], the liveness (of level 4) problem[9], the scheduling problem [17,18], and so on.

   For example, **MR** asks (i) computing a nonnegative integer solution X to a state equation $AX = M_F - M_I$, and (ii) detecting existence of (or possibly finding) a firing sequence δ that is legal on M_I with respect to X, where A is the place-transition incidence matrix of a given Petri net, and M_I or M_F is a given initial or final marking, respectively. Clearly (ii) is **RecLFS** (see 1.3) or **LFS** itself. The *minimum initial marking problem* of Petri nets (**MIM** for short), one of minimum resource allocation problems, is such another example. See its definition and other examples in Section 2.2.

   However, solving **LFS** generally is not easy: it is NP-hard even for Petri nets having very simple structures (see 1.3). This intractability of **LFS** may have been preventing us from producing efficient algorithms for those problems.

   In order to solve **LFS** heuristically, let us introduce the Maximum Legal Firing Sequence problem (**MAX LFS** for short) defined as follows.

[**MAX LFS**] "Given a Petri net N, an initial marking M and a firing count vector X, find a firing sequence δ such that δ is legal on M and satisfies the following (i) and (ii): (i) δ is legal on M with respect to some $X' \leq X$; (ii) the length $|\delta|$ of δ is maximum among those sequences satisfying (i), where $X' \leq X$ means that $X'(t) \leq X(t)$ for any $t \in T$."

Note that a solution δ with $|\delta| = |X|$ of **MAX LFS** is a solution to **LFS**.

1.3   Known Results

Known results on **LFS**, **RecLFS** and **MAX LFS** are briefly summarized.

[**LFS and RecLFS**] The recognition version of **LFS**, denoted as **RecLFS**, asks a "yes" or "no" answer on the existence of a solution δ to **LFS**. [8,16] proved that **RecLFS** is NP-hard even if N is a weighted free choice net or a weighted state machine. (See [4] for NP-hardness.) It is known that there is a pseudo-polynomial time algorithm for **LFS** if any one of the following (1)-(4) holds (see [8,16,3] for the definitions and the details):

(1) N is persistent;
(2) N is an unweighted state machine;
(3) N is a weighted directed tree;
(4) N is a weighted cactus and $X = \bar{1}$.

A polynomial time algorithm to solve **RecLFS** for each of (2), (3) and (4) is known.

[**MAX LFS**] Several heuristic algorithms for **MAX LFS** have already been proposed. Among others, *YWLFS* in [19] was first proposed, and *FSD* in [21] was known to have the highest capability among those existing ones at that time. *YWLFS* and *FSD* are explained in Section 5.

1.4   Organization of the Paper

After introduction in Section 1, we describe basic definitions on graphs and Petri nets in Section 2. Section 3 discusses the relationship between firability of transitions and existence of deadlocks. Section 4 briefly explains how to reduce state space of Petri nets. In Section 5, we consider solving **MAX LFS** heuristically. First we outline a heuristic algorithm *FSD* [21]. Then three improved algorithms *FSDB*, and *FSDC* and *FSDQ* are shown in 5.4, 5.5 and 5.6 respectively. Three genetic algorithms *GA*[12], *GAD* and *GAB* (to be proposed in this paper) are briefly explained in 6. In Section 7, experimental evaluation is given. Each of *FSD*, *FSDB* and *FSDC*(*GA*, *GAD* and *GAB*, respectively) is applied to 3017(2307) test problems, and *FSDQ* is applied to 800 test problems. Existence of an optimum solution to each test problem is guaranteed. Experimental evaluation is based on these results: *FSDC* shows the highest capability. Section 8 describes how using *FSDC* improves heuristic algorithms for the five Petri net problems **MCP**, **MIM**, **OFS**, **LUS** and **TPS**, each including **MAX LFS** as a subproblem (these definitions are going to be given in Section 2.2). It should be noted that **MR** is a subproblem of **MCP**.

2   Preliminaries

2.1   Basic Definitions

We assume that the readers are familiar with graph theory terminologies. (See [2,4] for definitions and notations not given here.) Let $G = (V, E)$ denote a directed graph or an undirected one with a vertex set V and an edge set E. A directed edge from u to v is denoted by (u, v). For disjoint sets V_1 and V_2, let $K(V_1, V_2) = \{(u, v) | u \in V_1, v \in V_2\}$. If $V_1 \cup V_2 \subseteq V$ then let $K(V_1, V_2; G) = \{(u, v) \in E \mid u \in V_1, v \in V_2\}$. Furthermore if $V_2 = V - V_1$ or $V_1 = V - V_2$ then we denote $K(V_1, \bullet; G)$ or $K(\bullet, V_2; G)$, respectively. For a fixed directed graph $G = (V, E)$, we denote $S^\bullet = K(S, \bullet; G)$ and $^\bullet S = K(\bullet, S; G)$ for $S \subseteq V$, and $^\bullet u = {}^\bullet S$ and $u^\bullet = S^\bullet$ if $S = \{u\}$ for some $u \in V$. A vertex u with $^\bullet u = \emptyset$ or $u^\bullet = \emptyset$ is called a *source* or a *sink* (of G), respectively. The outdegree $d_{out}(v)$ (indegree

$d_{in}(v)$, respectively) of a vertex $v \in V$ is the number of edges emanating from v (pointing to v).

Definitions and technical terms on Petri nets not given here can be found in [9]. A Petri net is an edge-weighted bipartite directed graph $N = (P, T, E, \alpha, \beta)$, where P is the set of places, T is that of transitions, $P \cap T = \emptyset$, $E = E_{in} \cup E_{out}$, $E_{in} \subseteq K(T, P)$, $E_{out} \subseteq K(P, T)$, α: $E_{out} \to Z^+$ (non-negative integers) and β: $E_{in} \to Z^+$ are weight functions. We denote $\alpha(p, t)$ or $\beta(t', p')$ for an edge $(p, t) \in K(P, T)$ or $(t', p') \in K(T, P)$, respectively, and we often omit edge weights if they are unity. We always consider N to be a simple directed digraph unless otherwise stated. A Petri net $N' = (P', T', E'\alpha', \beta')$ is called a *subnet* (with P' and T') of N if and only if $P' \subseteq P$, $T' \subseteq T$, $E' = \{(u, v) \in E \mid u, v \in P' \cup T'\} \subseteq E$, and both α' and β' are restrictions of α and β to E', respectively.

For a subset $P' \subseteq P$, the *restriction* of M to P' is a function $M \mid P'$: $P' \to Z^+$ such that $M \mid P'(p) = M(p)$ for any $p \in P'$. A *marking* M of N is a function $M: P \to Z^+$, and let $|M| = \sum_{p \in P} M(p)$. A transition t is *firable* on a marking M(of N) if $M(p) \geq \alpha(p, t)$ for any $p \in {}^\bullet t$. *Firing* such t on M is to define a marking M' such that, for any $p \in P$, we have $M'(p) = M(p) + \beta(t, p)$ if $p \in t^\bullet - {}^\bullet t$, $M'(p) = M(p) - \alpha(p, t)$ if $p \in {}^\bullet t - t^\bullet$, $M'(p) = M(p) - \alpha(p, t) + \beta(t, p)$ if $p \in {}^\bullet t \cap t^\bullet$ and $M'(p) = M(p)$ otherwise. For any transition t that is firable on M, let $M[t\rangle$ denote the marking after firing of t. Let $\delta = t_{i1} \cdots t_{is}$ be a sequence of transitions, called a *firing sequence*, and $\bar{\delta}(t)$ be the total number of occurrences of t in δ. $\bar{\delta} = [\bar{\delta}(t_1) \cdots \bar{\delta}(t_n)]^{tr}$ is the *firing count vector* of δ, where $n = |T|$. We say that δ is *legal* on M if and only if t_{ij} is firable on M_{j-1}, where $M_0 = M$ and $M_j = M_{j-1}[t_{ij}\rangle$, $j = 1, \cdots, s$. Let $M[\delta\rangle$ denote M_s, and it is said that $M[\delta\rangle$ is reachable from M. For M and an n-dimensional nonnegative integer vector X with component $X(t)$ being the total firing number of $t \in T$, we say that δ is *legal* on M with respect to X if and only if δ is legal on M and $\bar{\delta} = X$. Let us denote $|X| = \sum_{t \in T} X(t)$ and $\overline{X} = \max\{X(t) \mid t \in T\}$.

Let $\bar{0}$ ($\bar{1}$, respectively) denote a vector with every component equal to 0 (1). A nonempty set S of places of N is called a *siphon* (a *trap*, respectively) if and only if ${}^\bullet S \subseteq S^\bullet$ ($S^\bullet \subseteq {}^\bullet S$). A *minimal siphon* is a siphon such that no proper subset is a siphon. For a siphon S of N, let $N_S = (S, T_S, E_S, \alpha_S, \beta_S)$ be the subnet with S and $T_S = {}^\bullet S$ of N. N_S is called the *siphon net* of S (in N). N is called a siphon net if and only if P is a siphon and $N = N_P$.

N is a *state machine* if $|{}^\bullet t| \leq 1$ and $|t^\bullet| \leq 1$ for any $t \in T$.

2.2   Other Petri Net Problems Containing LFS as Subproblems

The definitions of five Petri net problems **MCP**, **MIM**, **OFS**, **LUS** and **TPS** are given in the following. Each of them contains **LFS** as a subproblem. Let N, M, X and δ be a Petri net, an initial marking, a firing count vector and a firing sequence.

[MIM (Minimum Initial Marking problem)] [15] Given N and X, find a minimum initial marking (an initial marking with the minimum total token number) M such that there is δ which is legal on M with respect to X.

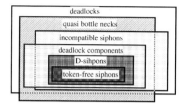

Fig. 1. Relationship among classes of D-siphons, of deadlock components and of quasi-bottlenecks.

[MCP (Marking Construction Problem)] [6] Given N, M and a target marking M', find a marking M'' such that M'' is reachable from M and $\Delta = \sum_{p \in P} |M'(p) - M''(p)|$ is minimum. (Note that **MR** asking for existence of M'' with $\Delta = 0$ is a subproblem of **MCP**.)

[OFS (Optimum Firing Sequence problem)] [16] Given N, M, X, two subsets of places P_1, $P_2 \subseteq P(P_1 \neq P_2)$, and two nonnegative integers K_1 and K_2, find δ that is legal on M with respect to X and such that $\sum_{p \in P_1} M'(p) \geq K_1$ and $\sum_{p \in P_2} M'(p) \geq K_2$ for any prefix δ' of δ, where $M' = M[\delta'\rangle$.

[LUS (Lower and Upper Bounded Submarking problem)] [16] Given N, M, X, a set of places $P_S \subseteq P$ and two nonnegative integers K_1 and K_2, find δ that is legal on M with respect to X and such that $K_1 \leq \sum_{p \in P_S} M'(p) \leq K_2$ for any prefix δ' of δ, where $M' = M[\delta'\rangle$.

[TPS (Timed Petri Net Scheduling problem)] (see [18] for example) Given a timed Petri net TN (in which time delay is associated with each transition firing), M and X, find δ which is legal on M with respect to X and which optimizes a given objective function.

Section 8 shows, through experimental results, how incorporating *FSDC* into heuristic algorithms for these problems improves their capabilities.

3   Deadlocks and MAX LFS

Given a Petri net N and a marking M, a *deadlock* is a situation such that no transition of N is firable on M, and then M is called a *dead marking*. Avoiding occurrence of deadlocks is one of main points in solving **MAX LFS**. Since its detection is intractable, we consider avoidance of *deficient siphons* (*D-siphons* for short) or of *deadlock components* instead: detection of their occurrence is easier. Also introduced is a *quasi-bottleneck* $T' \subseteq T$. Although it does not always result in a deadlock, avoiding its occurrence is preferable in solving **MAX LFS**. Fig. 1 shows the relationship among classes of D-siphons, of deadlock components, and of quasi-bottlenecks.

3.1   Deficient Siphons and Incompatible Siphons[21]

In this section, we explain three kinds of siphons: incompatible siphons, D-siphons and token-free siphons, where the first two siphons are introduced in [21]. Their definitions are as follows.

- A set of places $P_S \subseteq P$ is an *incompatible siphon* (of N with M and X) if and only if P_S is a siphon, $T_S(= {}^\bullet P_S) \neq \emptyset$ and the siphon net $N_S = (P_S, T_S, E_S, \alpha_S, \beta_S)$ of P_S in N has no firing sequence δ which is legal on M_S with respect to X_S, where $M_S : P_S \to Z^+$ satisfies $M_S(p) = M(p)$ for any $p \in P_S$ and $X_S : T_S \to Z^+$ does $X_S(t) = X(t)$ for any $t \in T_S$.
- A siphon $P_D \subseteq P$ is a *deficient siphon* (or a *D-siphon* for short) of N on M if and only if $T_D(= {}^\bullet P_D) \neq \emptyset$ and the siphon net $N_D = (P_D, T_D, E_D, \alpha_D, \beta_D)$ of P_D has no transition $t \in T_D$ such that any $p \in {}^\bullet t \cap P_D$ satisfies $M(p) \geq \alpha(p, t)$ (see Fig. 2).
- A set of places $P_F \subseteq P$ is a *token-free siphon* of N on M if and only if P_F is a siphon and $M(p) = 0$ for any $p \in P_F$.

Clearly, each token-free siphon S of N on M is a D-siphon of N on M if ${}^\bullet S \neq \emptyset$. We can easily prove the following lemmas.

Lemma 1. *If S is a D-siphon of N on M then so is S on any marking M' that is reachable from M.*   □

Lemma 2. *A D-siphon P_D of N on M is an incompatible siphon of N with M and X if $X(t) > 0$ for some $t \in T_D$.*   □

Given any siphon S of N and any marking M, detecting if S is a D-siphon on M can be done in $O(|E|)$ time, and the procedure *Forward_Deletion* proposed in [13] can extract a *maximal* D-siphon of any given Petri net in polynomial time.

Lemma 3. *If $P_S \subseteq P$ is an incompatible siphon of N with M and X then N has no firing sequence δ which is legal on M with respect to X.*   □

Lemma 3 shows that, once any incompatible siphon is created, we are unable to find a solution to **LFS**. Hence firing a transition t creating any incompatible siphon should be avoided. However, since detecting an incompatible siphon is intractable, it may be easier for us to consider prevention of creating D-siphons in designing heuristic algorithms.

3.2   Deadlock Components

A new concept, a *deadlock component*, is introduced. It is a set T' of transitions t_d such that, even if all transitions $t \in {}^\bullet({}^\bullet t_d) - T'$ are fired prescribed times, t_d is not firable if no transition $t' \in {}^\bullet({}^\bullet t_d) \cap T'$ is fired (see Fig. 3).

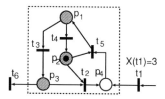

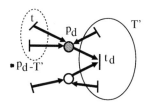

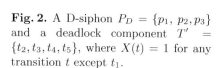

Fig. 2. A D-siphon $P_D = \{p_1,\ p_2, p_3\}$ and a deadlock component $T' = \{t_2, t_3, t_4, t_5\}$, where $X(t) = 1$ for any transition t except t_1.

Fig. 3. Schematic explanation of a deadlock component T' of N.

Definition 1. $T' \subseteq T$ *is a deadlock component of* N *(with* M *and* X*) if and only if each* $t_d \in T'$ *satisfies*

$$M(p_d) + \sum_{t \in {}^{\bullet}p_d - T'} \beta(t, p_d) \cdot X(t) < \alpha(p_d, t_d)$$

for some $p_d \in {}^{\bullet}t_d$. *The deadlock net of* T' *(in* N*) is a subnet* $N' = (P', T', E', \alpha', \beta')$ *of* N *with* $P' = {}^{\bullet}T'$. □

Note that $P'(= {}^{\bullet}T')$ is not necessarily a siphon (see the place set ${}^{\bullet}T'$ enclosed by broken lines in Fig. 2).

Lemma 4. *For* $N = (P, T, E, \alpha, \beta)$, M *and* X, *let* P_D *be any D-siphon on* M, *and* $N_D = (P_D, T_D, E_D, \alpha_D, \beta_D)$ *be the siphon net of* P_D *in* N. *Then* T_D *is a deadlock component of* N *with* M *and* X.

Proof. Assume the contrary that T_D is not a deadlock component of N with M and X. Because P_D is a siphon, ${}^{\bullet}p_d - T_D = \emptyset$ for any $p_d \in P_D$. By the assumption, N_D has a transition $t_d \in T_D$ such that $M(p_d) + \sum_{t \in {}^{\bullet}p_d - T_D} \beta(t, p_d) \cdot X(t) = M(p_d) \geq \alpha(p_d, t_d)$ for any $p_d \in {}^{\bullet}t_d \cap P_D$. This contradicts that P_D is a D-siphon. □

Lemma 5. *For* $N = (P, T, E, \alpha, \beta)$, M *and* X, *if* T' *is a deadlock component of* N *with* M *and* X, *then* N *has no firing sequence* δ *which is legal on* M *with respect to* X.

Proof. Let $N' = (P', T', E', \alpha', \beta')$ be the deadlock net of T' and $M' : P' \to Z^+$ be the marking of N' defined by

$$M'(p) = M(p) + \sum_{t \in {}^{\bullet}p - T'} \beta(t, p) \cdot X(t)$$

for any $p \in P'$. Then the definition of deadlock components implies that no $t' \in T'$ is firable on M'. And $M'(p)$ is the maximum number of tokens that can be brought into p by firing each $t \in T - T'$ $X(t)$ times.

This means that no $t' \in T'$ is firable on any marking M'' that is reachable from M. Thus N has no δ which is legal on M with respect to X. □

Lemma 5 shows that, once any deadlock component as in Lemma 5 is created before every transition is fired $X(t)$ times, we are unable to find a solution to **LFS**. Hence, firing any transition t creating such a deadlock component should be avoided.

We propose a procedure $DEADLOCK\_COMP$ to extract a maximal deadlock component of N with M and X, where $N_i = (P_i, T_i, E_i)$, $i \geq 0$, are subnets of N and M_i is a marking of each N_i. A marking M, a firing count vector X, a set of transitions F and a transition $t \in F$ are inputs, where $F = \{t' \in T \mid t'$ is firable on $M\}$. We are going to detect whether or not firing t creates any deadlock component. The procedure repeats both forcing any firable transition t to fire $X(t)$ times and then deleting t from the current Petri net until no firable one exists. If $P_i \neq \emptyset$ with $P_i \cap K(\bullet, \{t\}; N) \neq \emptyset$ is eventually left then T_i is a maximal deadlock component.

procedure $DEADLOCK\_COMP(M, X, F, t)$;
    begin
1. $i \leftarrow 0$; $M_i \leftarrow M$; $N_i \leftarrow N - \{t' \in T \mid X(t') = 0\}$;
2. $N_i \leftarrow N_i - \{p \in P_i \mid K(\bullet, \{p\}; N_i) \cup K(\{p\}, \bullet; N_i) = \emptyset\}$;
3. $F_{i+1} \leftarrow \{t' \in T_i \mid t'$ is firable on M_i of $N_i\}$; $R \leftarrow F_{i+1}$;
4.   **while** $R \neq \emptyset$ **do**
        begin
5. $i \leftarrow i + 1$; $M_i \leftarrow M_{i-1}$;
6.   **for** each $t' \in R$ **do**
            for each $p \in K(\{t'\}, \bullet; N_{i-1})$ **do**
                $M_i(p) \leftarrow M_i(p) + \beta(t', p) \cdot X(t')$;
7. $N_i \leftarrow N_i - R$;
8. $N_i \leftarrow N_i - \{p \in P_i \mid K(\bullet, \{p\}; N_i) \cup K(\{p\}, \bullet; N_i) = \emptyset\}$;
9. $M_i \leftarrow M_i \mid P_i$; $F_{i+1} \leftarrow \{t' \in T_i \mid t'$ is firable on M_i of $N_i\}$; $R \leftarrow F_{i+1}$;

        end;
10. **if** $(P_i \cap K(\bullet, \{t\}; N) \neq \emptyset)$ **then** $F \leftarrow F - \{t\}$;   /* If $P_i \neq \emptyset$ then T_i is a maximal */
    end;   /* deadlock component of N. */

Lemma 6. $DEADLOCK\_COMP(M, X, F, t)$ can extract, if any, a maximal deadlock component with M and X of N in $O(|T|(|P| + |T| + |E|))$ time.

Proof. We describe only the correctness since its time complexity is easily derived. We consider N_i, M_i and $F_{i+1}(= R)$ just before the execution of Step 5. If $t' \in F_{i+1}$, that is, t' is firable on M_i of N_i in Step 3 or 9 then any $p \in K(\bullet, \{t'\}; N_i)$ satisfies $M_i(p) \geq \alpha(p, t')$. That is, $M(p') + \sum_{t'' \in K(\bullet, \{p'\}; N) - T_i} \beta(t'', p') \cdot X(t'') = M_i(p') \geq \alpha(p', t')$ for any $p' \in K(\bullet, \{t'\}; N)$ since $p'' \notin K(\bullet, \{t'\}; N)$ for any $p'' \in P - P_i$. This means that any deadlock component of N is a subset of $T_i - F_{i+1}$. This is the reason why, in Step 7, $R(= F_i)$ is deleted from N_{i-1}(after setting $i \leftarrow i + 1$). Hence, at Step 10, if $P_i \neq \emptyset$ then T_i is a deadlock component with M and X of N and it is maximal. $\qquad\square$

3.3 Quasi-Bottlenecks

Another new concept, a *quasi-bottleneck*, is introduced. It is a set T' of transitions t_d having similar property to a deadlock component. The difference is relaxation of transition firability such that even if any one transition $t \in \bullet(\bullet t_d) - T'$ is fired once, $(\bullet t_d)\bullet$ includes some transition t' that is not firable if no transition $t'' \in \bullet(\bullet t_d) \cap T'$ is fired.

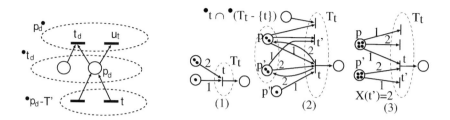

Fig. 4. Schematic explanation of a quasi bottleneck.

Fig. 5. Schematic explanation of the three conditions

Definition 2. T' *is a quasi-bottleneck if and only if* T' *consists of transitions* t_d *such that* $\bullet t_d$ *has a place* p_d *satisfying*

$$(\forall t \in {}^\bullet p_d - T')[M(p_d) + \beta(t, p_d) < \alpha(p_d, u_t) \text{ for some } u_t \in p_d{}^\bullet]$$

Fig. 4 schematically explains a quasi-bottleneck. Note that a quasi-bottleneck may not be a deadlock: some transition $t \in T'$ may be firable on a marking M' that is reachable from M. Then, even if a Petri net $N = (P, T, E)$ has a quasi-bottleneck $T' \subseteq T$, there may be a firing sequence δ which is legal on M with respect to X. Although a quasi-bottleneck implies only possibility of occurrence of a deadlock component, its detection is easy and it is experimentally observed that incorporating it into algorithms is worth trying.

We provide the following procedure to extract a maximal quasi-bottleneck.

procedure $QUASI\_BOTTLENECK(M, X, F, t)$;
   begin
1. $i \leftarrow 0$; $M_i \leftarrow M$; $N_i \leftarrow N - \{t' \in T \mid X(t') = 0\}$;
2. $N_i \leftarrow N_i - \{p \in P_i \mid K(\bullet, \{p\}; N_i) \cup K(\{p\}, \bullet; N_i) = \emptyset\}$;
3. $F_{i+1} \leftarrow \{t' \in T_i \mid t'$ is firable on M_i of $N_i\}$; $R \leftarrow F_{i+1}$;
4.   **while** $R \neq \emptyset$ **do**
      begin
5. $i \leftarrow i + 1$; $M_i \leftarrow M_{i-1}$;
6.   **for** each $t' \in R$ **do**
        $N_i \leftarrow N_i - \{p \in K(\{t'\}, \bullet; N_i) \mid M_i(p) + \beta(t', p) > \alpha(p, t'')$ for any
        $t'' \in K(\{p\}, \bullet; N_i)\}$;
7. $N_i \leftarrow N_{i-1} - F_i$;
8. $N_i \leftarrow N_i - \{p \in P_i \mid K(\{p\}, \bullet; N_i) \cup K(\{p\}, \bullet; N_i) = \emptyset\}$;
9. $M_i \leftarrow M_i \mid P_i$; $F_{i+1} \leftarrow \{t' \in T_i \mid t'$ is firable on M_i of $N_i\}$; $R \leftarrow F_{i+1}$;
     end;
10. **if** $(P_i \cap K(\bullet, \{t\}; N_i) \neq \emptyset)$ **then** $F \leftarrow F - \{t\}$;   /* If $T_i \neq \emptyset$ T_i is a maximal */
   end;   /* quasi-bottleneck of N. */

Lemma 7. $QUASI\_BOTTLENECK(M, X, F, t)$ *can extract, if any, a maximal quasi-bottleneck of* N *in* $O(|T|(|P| + |T| + |E|))$ *time.*

Proof. We briefly show only the correctness since the time complexity is similar to Lemma 6. If $t_d \in T_i$ at Step 10, then there is $p_d \in P_i \cap {}^\bullet t_d$. Let $t_m \in p_d{}^\bullet$ be such that $\alpha(p_d, t_m) = max\{\alpha(p_d, t') \mid t' \in p_d{}^\bullet\}$. Then

$$M(p_d) + \beta(t'', p_d) < \alpha(p_d, t_m) \text{ for any } t'' \in {}^\bullet p_d - T',$$

since otherwise $M(p_d) + \beta(t_1, p_d) \geq \alpha(p_d, t_m)$ for some $t_1 \in {}^\bullet p_d - T'$, meaning that $M(p_d) + \beta(t_1, p_d) \geq \alpha(p_d, t')$ for any $t' \in p_d{}^\bullet$. That is, p_d is deleted, a contradiction. Hence t_d is included in a quasi-bottleneck. The maximality of T_i is obvious. □

4   Preprocessing

In order to make algorithms efficient, we incorporate some preprocessing.

4.1   Changing Order of Transition Firing

Let us consider the following three conditions for any transition $t \in T$ in a given Petri net $N = (P, T, E, \alpha, \beta)$ with M and X. Let $T_t = ({}^\bullet t)^\bullet$ (see Fig. 5).

Condition 1 $T_t = \{t\}$. *(Fig. 5 (1))* □

Condition 2 *If $T_t - \{t\} \neq \emptyset$ then there is an edge (t, p) such that $\alpha(p, t) \leq \beta(t, p)$ for any $p \in {}^\bullet t \cap {}^\bullet(T_t - \{t\})$. (Fig. 5 (2))* □

Condition 3 $M(p) \geq \sum_{t' \in p^\bullet} X(t') \cdot \alpha(p, t')$ *for any $p \in {}^\bullet t$. (Fig. 5 (3))* □

Let
$$F = \{t \in T \mid t \text{ is firable on } M\},$$
and let $\#(t, M)$ denote the maximal possible firing number of t on M, that is,
$$\#(t, M) = min\{\lfloor M(p)/\alpha(p, t) \rfloor \mid p \in {}^\bullet t\}.$$
We obtain the following lemma.

Lemma 8. *If some transition $t \in F$ satisfies one of the three conditions, then $\#(t', M[t\rangle) \geq \#(t', M)$ for any transition $t' \in F - \{t\}$.*

Proof. If Condition 1 holds, then firing t removes no tokens from any $p \in {}^\bullet t'$ for any $t' \in F - \{t\}$. If Condition 2 holds, then firing t adds $\beta(t, p) - \alpha(p, t)$ (≥ 0) tokens into any $p \in {}^\bullet t \cap {}^\bullet t'$. If Condition 3 holds then firing t dose not disable any transition $t'' \in T_t - \{t\}$. Thus t' is firable on $M[t\rangle$ at least $\#(t', M)$ times if any one of the three conditions holds. □

Lemma 8 shows that if some $t \in F$ satisfies any one of the three conditions, then firing t does not affect subsequent searching of firing sequences.

We describe a procedure $CHECK\_CONFLICT(F)$ for checking these conditions in $O(|E|)$ time.

procedure $CHECK\_CONFLICT(F)$;
    begin
        $N_1 \leftarrow N - \{t \in T \mid X(t) = 0\}$;
        return $\{t \in F \mid t \text{ satisfies any one of the three conditions }\}$;
    end;

4.2   Reducing the Set of Candidate Markings

Next we consider reducing the set of candidate markings in backtracking operation. *FSDC* backtracks to a marking M on which firing some transition may have created an incompatible siphon, and then tries to find a transition whose firing creates no incompatible siphons. Note that, even if any incompatible siphon is created, some more transitions can be fired. However a dead marking eventually appears.

We provide a procedure *CHECK_SIPHON* that checks whether or not firing t on a marking M removes any token from some siphon S. The description is as follows.

procedure *CHECK_SIPHON*(t, F, X);
   begin
  1. $N_1 \leftarrow N - \{t \in T \mid X(t) = 0\}$;
  2. $N_1 \leftarrow Forward\_Deletion(N_1, t^\bullet)$;   /* see [13]*/
  3. $N_1 \leftarrow N_1 - \{p \in P_1 \mid {}^\bullet p \cup p^\bullet = \emptyset\}$;
  4. **if** $P_1 \cap {}^\bullet t \neq \emptyset$ **then return** $\{t\}$;   /* P_1 is a siphon with $P_1 \cap {}^\bullet t \neq \emptyset$. */
    else return F;   /* $F_M \leftarrow \{t\}$ or $F_M \leftarrow F$
    end;   in Step 16 of $FIRE^+$ in Section 5.*/

It should be noted that $Forward\_Deletion(N, t^\bullet)$[13] used in *CHECK_SIPHON* may fail to detect some siphon S with ${}^\bullet t \cap S \neq \emptyset$. Hence it is not guaranteed to detect every siphon S' with ${}^\bullet t \cap S' \neq \emptyset$. Its time complexity is $O(|P| + |T| + |E|)$.

Remark 1. Forward_Deletion$(N, t^\bullet)$ first deletes $t^\bullet \subseteq P$ from N and repeats deleting created source transitions and their output places. If any nonempty set $P' \subseteq P$ is left then it is a maximal siphon having no intersection with $t^\bullet$. See [13] for the formal description if necessary.

5   Heuristic Algorithms *FSDB*, *FSDC* and *FSDQ* for MAX LFS

We propose heuristic algorithms *FSDB*, *FSDC* and *FSDQ* in this section. They are improved versions of *FSD* in [21].

5.1   Summary of Procedures

We summarize main procedures appeared in these algorithms as well as $YWLFS$[19], and show how they are combined to make each algorithm (see Table 1).

SERCH_LFS$(M, X_{rest}, F, effect)$: computes two kind of values $max(p)$ for every place p and $visit(t_f)$ for a given transition t_f in $O(\bar{X} \cdot |E|)$ time. These values are used in *COMP_EFFECT*. (Notation: S_LFS)

COMP_EFFECT$(M, X_{rest}, F, effect)$: computes a value $effect(t_f)$ for each $t_f \in F$ (firable transitions on M) in $O(\bar{X} \cdot |T| \cdot |E|)$ time by using *SEARCH_LFS*. (Notation: C_EF)

Table 1. Summary of structure of algorithms, where $FIRE^+$ and BT^+ are improved versions of FIRE and BT, respectively, and $\bigcirc$ denotes the procedure is used

	S_LFS	C_EF	FIRE	FIRE+	D_SI	D_CO	QBN	BT	BT+
$YWLFS[19]$	$\bigcirc$	$\bigcirc$	$\bigcirc$						
$FSD[21]$	$\bigcirc$	$\bigcirc$	$\bigcirc$		$\bigcirc$				
$FSDB$	$\bigcirc$	$\bigcirc$	$\bigcirc$		$\bigcirc$			$\bigcirc$	
$FSDC$	$\bigcirc$	$\bigcirc$		$\bigcirc$		$\bigcirc$			$\bigcirc$
$FSDQ$	$\bigcirc$	$\bigcirc$		$\bigcirc$			$\bigcirc$		$\bigcirc$

FIRE(M,X_{rest},δ), $FIRE^+(M,X_{rest},\delta)$: construct a firing sequence as a solution to **MAX LFS**. It may include $BACKTRACK$ or $BACKTRACK^+$. (Notation: $FIER$, $FIRE^+$, respectively)

D_SIPHON(t_f,M_v,X_{rest},F) : repeats a depth-first search for finding a firable transition whose firing creates no D-siphons in $O(|P|+|T|+|E|)$ time: it detects if firing t_f creates any D-siphon by using $Forward\_Deletion$. (Notation: D_SI)

BACKTRACK$(next_t,F_M,M,X_{rest},\delta)$, $BACKTRACK^+(next_t,F_M,M,X_{rest},\delta)$: backtrack to a marking M on which firing some transition may have created any incompatible siphon (or a D-siphon), any deadlock components, or any quasi-bottleneck. (Notation: BT, BT^+, respectively)

DEADLOCK_COMP(M,X,F,t) : finds a maximal deadlock component in $O(|T|(|P|+|T|+|E|))$ time. (Notation: D_CO)

QUASI_BOTTLE_NECK(M,X,F,t) : finds a maximal quasi-bottleneck in $O(|T|(|P|+|T|+|E|))$ time. (Notation: QBN)

5.2   Outline of $YWLFS[19]$

The algorithm $YWLFS[19]$ has three procedures: $SEARCH\_LFS$, $COMP\_EFFECT$ and $FIRE$. $YWLFS$ computes a value $effect(t)$ for each transition t and chooses a transition $next\_t$ with the maximum value of $effect$ as the one to fire next.

5.3   Outline of $FSD[21]$

$FSD[21]$ is an improved version of $YWLFS$, in which a procedure D_SIPHON for detecting creation of D-siphons is added. FSD constructs a firing sequence by choosing an enabled transition t, one by one, based on the following two conditions: (a) firing t does not create any D-siphon; (b) the value $effect(t)$ is maximum among those transitions satisfying (a).

5.4 $FSDB$

$FSDB$ consists of FSD and $BACKTRACK$, which is incorporated in $FIRE$. In [11] the descriptions of the five procedures of $FSDB$ are corrected or improved from the original one, and these revised ones are going to be used in this paper.

Note that $BACKTRACK^+$ is improved from $BACKTRACK$ by incorporating $CHECK\_CONFLICT$ and $CHECK\_SIPHON$ mentioned in Section 4: statements in lines 7, 13 and 18 are inserted into $BACKTRACK^+$.

5.5   *FSDC*

FSDC is obtained by replacing:

(1) D_SIPHON of *FSDB* by $DEADLOCK\_COMP$ for detecting creation of a maximal deadlock component, and
(2) *FIRE* and $BACKTRACK$ by $FIRE^+$ and $BACKTRACK^+$ respectively, and they are going to be shown in the following.

FSDC is formally described as follows.

Algorithm *FSDC*;
/* **input:** N, M_0 and X; **output:** δ and X */
    begin
1. $\delta \leftarrow \emptyset$; $\mathcal{L} \leftarrow \emptyset$;
2. $FIRE^+(M_0, X, \delta)$;
3. Output δ and X;   /* if $X(t) \neq 0$ for some t then */
    end.   /* finding a solution is failed */

[**Description of** $FIRE^+(M, X_{rest}, \delta)$] It chooses a transition t_f, with $effect(t_f) = \max\{effect(t) \mid t \in F\}$, as the one to fire next, if firing t_f does not leave any deadlock component. Then t_f is fired and $\delta \leftarrow \delta t_f$ (concatenating t_f in the rear of δ). If F becomes empty during this procedure then *FSDC* begins backtracking operations. In order to reduce state space, $FIRE^+$ is devised by incorporating $CHECK\_CONFLICT$ and $CHECK\_SIPHON$ mentioned in Section 4 into the original *FIRE*[19].

procedure $FIRE^+(M, X_{rest}, \delta)$;
    begin
1. $F_M \leftarrow \{t \in T \mid X_{rest}(t) > 0 \text{ and } M(p) \geq \alpha(p, t) \text{ for any } p \in {}^\bullet t\}$;
2. **while** $F_M \neq \emptyset$ **do**
    begin
3. $F \leftarrow CHECK\_CONFLICT(F_M)$;
4.   **if** $F \neq \emptyset$ **then**
        begin
5.   choose t_f in F arbitrarily;   /* Firing t_f satisfying Conditions 1–3 */
6. $F_M \leftarrow \{t_f\}$; /* Removing M from the set of */
                          /* candidate markings in backtracking operations */
        goto Step 15;
        end;
7.   **for** each $t \in F_M$ **do**
        begin
8. $M_v \leftarrow M[t\rangle$; $X_{rest}(t) \leftarrow X_{rest}(t) - 1$;
9. $T' \leftarrow DEADLOCK\_COMP(M_v, X_{rest}, F_M, t)$;
        $X_{rest}(t) \leftarrow X_{rest}(t) + 1$;
        end;
    /* Now backtracking */
10. $BACKTRACK^+(t_f, F_M, M, X_{rest}, \delta)$;
11. **if** $F_M = \{t\}$ **then** $t_f \leftarrow t$;   /* Selecting t_f to */
12. **else**   /* fire next */
    begin
13. **for** each $t \in T$ **do** $effect(t) \leftarrow 0$;

14. $COMP\_EFFECT(M, X_{rest}, F_M, effect)$;   /* Computing $effect(t)$ for all $t \in F_M$ */
       choose t_f with $effect(t_f) = \max\{effect(t) \mid t \in F_M\}$;
       end; /* extending δ */
15. $X_{rest}(t_f) \leftarrow X_{rest}(t_f) - 1$; $M \leftarrow M[t_f\rangle$; $\delta \leftarrow \delta\, t_f$;   /* Concatenation of t_f
16. $F_M \leftarrow CHECK\_SIPHON(t_f, F_M, X_{rest})$;   at the end of δ */
17.   Push F_M to the top of $\mathcal{L}$;   /* for backtracking */

18. $F_M \leftarrow \{t \in T \mid X_{rest}(t) > 0 \text{ and } M(p) \geq \alpha(p,t) \text{ for any } p \in {}^{\bullet}t\}$;

    end
  end;

[Backtracking Operations] In the following we explain when, where and how backtracking operation is executed. We consider two kinds of backtracking, global one or local one, depending upon the marking to which we backtrack.

Let $N = (P, T, E, \alpha, \beta)$, M_0, and X be a given Petri net, a given initial marking and a given firing count vector, respectively. Let M denote the current marking such that $M = M_0[\delta\rangle$ for some firing sequence δ in which any transition t appears at most $X(t)$ times. Let $X_{rest}(t) = X(t) - \bar{\delta}(t)$ for any $t \in T$, and $F_M = \{t \in T \mid X_{rest}(t) > 0 \text{ and } t \text{ is firable on } M\}$. Suppose that we are given two nonnegative integers b_g and b_l. Let $\#g$ or $\#l$ denote, respectively, the number of global or local backtracking operations that have been executed. Consider any reachability tree that is represented as a directed tree such that every marking reachable from M_0 is represented as an individual vertex having at most one edge entering into it (see Fig. 6). Firing t on M is represented as a directed edge from the vertex v_M (corresponding to M) to the vertex $v_{M[t\rangle}$ (corresponding to $M[t\rangle$). If there is a directed path from $v_{M'}$ to v_M and at least two edges emanate from $v_{M'}$ then $v_{M'}$ (or M') is called a *branching ancestor* of v_M (or of M); if any other vertex, except v_M and $v_{M'}$, on the path from $v_{M'}$ to v_M has only one edge emanating from it, then $v_{M'}$ (or M') is called the *closest* branching ancestor of v_M (or of M). If $v_{M''}$ is a branching ancestor of v_M and any other vertex on the path from v_{M_0} to $v_{M''}$ has only one edge emanating from it, then $v_{M''}$ (or M'') is called the *highest* branching ancestor of v_M (or of M).

Now we are ready to explain how backtracking operation is done. This is schematically shown in Fig. 6. $BACKTRACK^+$ is called in Step 10 of $FIRE^+$. We prepare a list $\mathcal{L}$ to store the set $F_{M'}$ for each ancestor M' of the current marking M, and set $\#l \leftarrow 0$, $\#g \leftarrow 0$, $\delta_m \leftarrow nil$, $backtrack \leftarrow 1$ and $\mathcal{L} \leftarrow \emptyset$ initially. Let $T_\delta = \{t \in T \mid t \text{ appears in } \delta\}$. Let $\delta \backslash t$ denote removing t from the rear of δ if t is the last transition in δ. The description of $BACKTRACK^+$ is as follows.

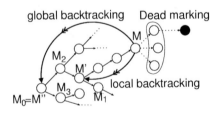

Fig. 6. Schematically explaining where to backtrack

procedure $BACKTRACK^+(next\_t, F_M, M, X_{rest}, \delta)$;
      begin
1. **if** $F_M \neq \emptyset$ **then return**   /* $F_M = \emptyset$ */
2. $M' \leftarrow M; X'_{rest} \leftarrow X_{rest}; \delta' \leftarrow \emptyset$;
3. Execute $FIRE(M', X'_{rest}, \delta')$ of FSD;
4. **if** $|\delta_m| < |\delta\delta'|$ **then** $\delta_m \leftarrow \delta\delta'$;
5. **if** $\#l < b_l$ **then**   /* Local backtracking */
       begin
       $\#l \leftarrow \#l + 1$;
6. Pop $(F_{M'})$ from the top of $\mathcal{L}$;
7. **while** $|F_{M'}| \leq 1$ **do**   /* Find the closest branching ancestor M' */
       begin   /* for backtracking (see M' in Fig. 6). */
8.   **if** $\mathcal{L} = \emptyset$ **then goto** Step 29;   /* Halt */
9.   Pop $(F_{M'})$ from the top of $\mathcal{L}$;
10.   Let $t \in F_{M'} \cap T_\delta$ be the transition that fired on M';
11. $X_{rest}(t) \leftarrow X_{rest}(t) + 1$;
12. $\delta \leftarrow \delta \backslash t$;   /* Deleting t from δ, where t is */
13.   **end**   /* the last one of δ */
14. $M \leftarrow M'; F_M \leftarrow F_{M'} - \{t\}$;   /* M' is the closest branching ancestor of M */
       end
15. **else if** $\#l = b_l$ and $\#g < b_g$ **then**   /* Global backtracking */
       begin
16. $\#g \leftarrow \#g + 1; \#l \leftarrow 0$;   /* Find the highest branching ancestor M'' */
17.   Let $F_{M''}$ be the set at the tail of $\mathcal{L}$;   /*M'' in Fig. 6 */
18.   **while** $|F_{M''}| \leq 1$ and $M'' \neq M$ **do**
19. $F_{M''} \leftarrow F'$;   /* F' is next to $F_{M''}$ in searching from the tail to the top of $\mathcal{L}$ */
       /* M'' is the highest branching ancestor of M */
20.   **if** $M'' = M$ **then goto** Step 29;   /* Halt */
21.   Let $F_{M'''}$ be the set existing on the top of $\mathcal{L}$;   /* Backtrack to $M'' \neq M$ */
22.   **while** $F_{M''} \neq F_{M'''}$ **do**
       begin
23.   Let $t \in F_{M'''} \cap T_\delta$ be the transition that fired on M''';
24. $X_{rest}(t) \leftarrow X_{rest}(t) + 1$;
25. $\delta \leftarrow \delta \backslash t$;   /* t is the last one of δ */
26.   Pop $F_{M'''}$ from the top of $\mathcal{L}$;
       end
27. $M \leftarrow M''; F_M \leftarrow F_{M''} - \{t\}$;
       end
28. **else** /* $\#l = b_l$ and $\#g = b_g$ */   /* Halt */
       begin
29.   Output the firing sequence $\delta \leftarrow \delta_m$;
30.   **halt**;
       end;
      end

5.6   *FSDQ*

FSDQ is obtained by replacing *DEADLOCK_COMP* of *FSDC* by
QUASI_BOTTLENECK for detecting creation of a maximal quasi-bottleneck.

5.7   Correctness and Time Complexity of *FSDB*, *FSDC* and *FSDQ*

It is clear that *FSDB* (*FSDC* or *FSDQ*, respectively) finds a firing sequence δ
that is legal on M, since procedure $FIRE(M, X_{rest})$ $(FIRE^+(M, X_{rest}))$ chooses
a transition $next\_t$, with $X_{rest}(next\_t) > 0$, such that it is firable on the current
marking M'. $FIRE^+$ and $FIRE$ run in $O(|T|(\bar{X}|X||E|+|T||P|+|T|^2))$, and this
is also the time complexity of *FSDC* and of *FSDQ*. It is $O(|X|^2(|X||E|+|P|+|T|)$
since $|T| \leq |X|$. On the other hand *FSDB* runs in $O(|X||T|(\bar{X}|E| + |P| + |T|))$
time, which is $O(|X|^2(|X||E| + |P| + |T|))$.

Remark 2. In local backtracking at Step 5 of $BACKTRACK^+$, if the transition
t_1 such that $M'[t_1\rangle = M_1$ in Fig. 6 has already been checked then we search for
the next closest branching ancestor M_2 in Fig. 6, and the iteration is continued
if necessary. Similarly searching for the next highest branching ancestor, if necessary, is done in global backtracking at Step 15 of $BACKTRACK^+$.   □

6   Genetic Algorithms *GAD* and *GAB* for MAX LFS

In [12], a genetic algorithm *GA* for solving **MAX LFS** is proposed. In this
paper we propose two genetic ones *GAD* and *GAB* by incorporating, into *GA*,
avoidance of D-siphons and of deadlock components, respectively.

   GA, *GAD* or *GAB* consists of the following procedures.
[Common structure of *GA*, *GAD* and *GAB*]
/* **Input**: a Petri net $N = (P, T, E, \alpha, \beta)$, a marking M, a firing count vector X,
two integers g (the upper bound on the number of generations) and h (the number of
strings) */
/* **Output**: a solution δ to **MAX LFS** */

    begin
  1. *Initialization*;
  2. $i \leftarrow 0$;
  3. **while** $(i \leq g)$ and $(p_\sigma < |X|$ for any string $\sigma)$ **do**
      begin
  4.   *Computing fitness*; /* fitness p_σ of each string σ is computed */
  5.   *Selection*;
  6.   *Crossover*;
  7. $i \leftarrow i + 1$;
      end;
  8. Output δ;
    end.

Each procedure is explained in the following.

[**Initialization**] We construct h sequences of transitions such that each $t \in T$ appears exactly $X(t)$ times, by allocating them randomly. Each sequence is of length $|X|$ and is called a *string* (see Fig. 7).

[**Computing fitness**] In Step 4, the algorithm GAD or GAB computes fitness p_σ for each string σ. We first explain how to compute fitness by using the following example.

Example 1. Suppose that a Petri net N shown in Fig. 8 is given as an input. Let us compute fitness p of the first string in Fig. 7 (see Fig. 9).

Step 1. We set $M = M_0$, $p = 0$ and $q = 0$, and let r denote the leftmost transition $t2$.

Step 2. Because $r(= t2)$ is not firable on M, we remove r from the current string, shift each transition to the left, and r to the rightmost position (the rear of the string).

Step 3. Since r is firable on M, we check whether or not firing r creates either a D-siphon or a deadlock component. Because firing r does not create any D-siphon or any deadlock component, we set $p = 1$ and $q = 0$. Similarly we proceed to Step 4, 5 and Step 6.

Step 7. We stop the procedure when $p + q = |X| = 5$, which means that there is no transition firable on the current marking M.

Hence the first string σ in Fig. 7 has fitness $p_\sigma = 3$.   □

We describe a procedure for computing fitness.

procedure *COMP_FITNESS*;
    begin
    for each string σ **do**
        begin
1. $M \leftarrow M_0$, $p_\sigma \leftarrow 0$ and $q \leftarrow 0$.
2.   Let r denote the transition in the $(p_\sigma + 1)$-th position from the left end of σ;
3.   **if** r is not firable on M **then goto** Step 6;
4.   **if** firing r creates any D-siphon (for GAD) or
        any deadlock component (for GAB) **then goto** Step 6;
5. $M \leftarrow M[r]$; $p_\sigma \leftarrow p_\sigma + 1$; $q = 0$; **goto** Step 7
6.   Remove r from σ;
        Shift each transition by one position to the left, and add r to the right end of the string;
        $q \leftarrow q + 1$;
7.   **if** either $p_\sigma = |X|$ or $p_\sigma + q = |X|$ **then** stop **else goto** Step 2;
        end;
    end

[**Selection**] We select two strings randomly as *parent1* and *parent2*.

[**Crossover**] We generate two strings *child1* and *child2* from *parent1* and *parent2*. Let $C(k)$ denote the k-th element from the left end of the string C. We explain **Crossover** by using the following example.

Example 2. First we prepare a *template* binary string that is randomly generated, as in Fig. 10. The length is $|X| = 5$. We construct a string *child1* (*child2*, respectively) by repeating the following for each i, $i = 1, \ldots, 5$:

    if $template(i) = 1(0)$ **then** $child1[i] \leftarrow parent1[i]$ $(child2[i] \leftarrow parent2[i])$;
    else $tmp1[i] \leftarrow parent1[i]$ $(tmp2[i] \leftarrow parent2[i])$;

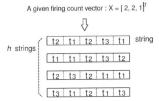

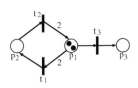

Fig. 7. An example of four strings of length $|X| = 5$ initially provided when $h = 4$

Fig. 8. An example of a Petri net N with an initial marking $M_0 = [2, 0, 0]^{tr}$ and a firing count vector $X = [2, 2, 1]^{tr}$.

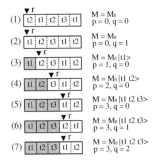

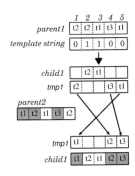

Fig. 9. An example of computing fitness of the first string of Fig. 7

Fig. 10. An example of the crossover operation

Next, we sort all nonempty elements $tmp1[i]$ $(tmp2[i])$ according to the order of their appearance in $parent2$ $(parent1)$ as shown in Fig. 10. Finally

$child1[i] \leftarrow tmp1[i]$ $(child2 \leftarrow tmp2[i])$ for each i, $i = 1, \ldots, 5$.

We compute fitness of $child1$ and $child2$. Among the four strings, we select the two strings having the largest fitness and the second largest one, and then set them as new $parent1$ and $parent2$.

GAD or GAB is obtained from GA by adding Step 4 into the procedure $COMP\_FITNESS$. For the details of GA, see [12].

7   Experimental Results and Evaluation

$FSD[21]$, $FSDB$, $FSDC$, $FSDQ$, $GA[12]$, GAD and GAB have been implemented on a personal computer (CPU: PentiumIII 800MHz, OS: FreeBSD 4.2) by means of the C programming code. We compare their capabilities through experiment, where FSD has been known to have the highest capability among existing algorithms so far.

[Construction of Test Problems] We outline how to construct Petri nets of general type.

(1) We first construct a number of test problems having Petri nets of small size such that existence of solutions to **LFS** is guaranteed: their solutions are found by means of a branch-and-bound method.

(2) We choose any two such Petri nets and select any pair of places, one from each, and then we combine the two Petri nets by identifying these selected places. Accordingly markings and firing count vectors are merged. By repeating the similar procedure, we obtain test problems of large size in which existence of solutions to **LFS** is guaranteed.

Petri nets of general type so generated are used as test problems. Since existence of a solution to each of these problems is guaranteed, capability of algorithms can be explicitly shown by using these problems as inputs.

[**Size of Test Problems**] Experiments are done for the following three groups of algorithms:

Group 1 : *FSD, FSDB, FSDC, GA, GAD, GAB*
Group 2 : *FSDC, GA, GAD, GAB*
Group 3 : *FSDB, FSDC, FSDQ*

We summarize the number and sizes of test problems and some statistical data.

- The number of test problems for **MAX LFS**: 3017 for Group 1; 2307 for Group 2; 800 for Group 3.
- Sizes of test problems :
  (for **Group 1** and **Group 2**)
  $7 \le |P| \le 500, 6 \le |T| \le 1642, 25 \le |E| \le 3532,$
  $6 \le |X| \le 1642, 1 \le X(t) \le 5$ for any $t \in T$,
  $1 \le \alpha(e), \beta(e') \le 10$ for any $e \in E_{out}$ and $e' \in E_{in}$.
  (for **Group 3**)
  $100 \le |P| \le 400, 238 \le |T| \le 1019, 476 \le |E| \le 2038,$
  $238 \le |X| \le 1019, X(t) = 1$ for any $t \in T$,
  $1 \le \alpha(e), \beta(e') \le 5$ for any $e \in E_{out}$ and $e' \in E_{in}$.

[**Experiment 1**] Each algorithm in **Group 1** is applied to total 3017 test problems, and the number of successful cases (meaning that solutions to **LFS** are obtained), CPU time in second, and the ratios $|\delta|/|X|$ are compared, where $|\delta|$ denotes the length of a firing sequence δ. The parameters are set as follows:

- $b_g = 10$ and $b_l = 10$ for *FSDB* and *FSDC*.
- $g = 10$ and $h = 10$ for *GA, GAD* and *GAB*.

Results obtained in Experiment 1 are shown in Tables 2 and 3 and in Figure 11.

[**Experiment 2**] For algorithms of **Group 2**, we compare *FSDC* with each of *GA, GAD* and *GAB*, by allowing them to spend only constant CPU time: we restrict *GA, GAD* and *GAB* to execute alternation of generations only for the time period spent by *FSDC*. The numbers of successful cases are compared.

Experimental results are shown in Table 4, where we have set $h = 10$.

Table 2. Successful cases (meaning that each algorithm finds solutions to **LFS**) and average CPU time (in second), where the number of test problems is 3017.

	FSD	$FSDB$	$FSDC$	GA	GAD	GAB
total	1683	1912	2330	1711	2078	2157
	(55.8%)	(63.4%)	(77.2%)	(56.7%)	(68.9%)	(71.5%)
time	45.7	309.7	83.4	1.36	5.23	4.88

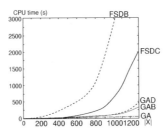

Fig. 11. $|X|$ versus CPU time

[**Experiment 3**] For algorithms of **Group 3**, we compare $FSDQ$ with each of $FSDB$ and $FSDC$, based on the results for 800 test problems as mentioned above. Table 5 shows $FSDQ$ has higher capability than $FSDB$ and $FSDC$ for these test problems.

[**Evaluation through Experimental Results**] Tables 2 and 3 and Figure 11 show that $FSDC$ has the highest capability and that avoiding creation of deadlock components and incorporating preprocessing improve capabilities of algorithms. Table 4 shows that GA, GAD and GAB are much faster than $FSDB$ and $FSDC$. Concerning solutions, however, $FSDC$ outperforms any one of GA, GAD and GAB. As total evaluation, $FSDC$ shows the highest capability among those algorithms we have tried in this experiment.

8   Using $FSDC$ in Solving Related Problems

MAX LFS is contained, as a subproblem, in the Optimum Firing Sequence Problem (**OFS**), the Lower and Upper Bounded Submarking Problem (**LUS**), the Minimum Initial Marking Problem (**MIM**), the Marking Construction Problem (**MCP**) and the Timed Petri Net Scheduling Problem (**TPS**). Heuristic algorithms $LBDC$, $LUDC$, $MKDB$ and $DBSC$ are proposed by adopting $FSDC$ for

Table 3. The ratio $|\delta|/|X|$

	FSD	$FSDB$	$FSDC$	GA	GAD	GAB
Max	1.000	1.000	1.000	1.000	1.000	1.000
Ave	0.926	0.941	0.984	0.708	0.811	0.827
Min	0.578	0.583	0.588	0.489	0.462	0.536

Table 4. Comparison of successful cases for total 2307 test problems

Total		GA	GAD	GAB	$FSDC$
2307		1380	1718	1777	1832
		(59.8%)	(74.5%)	(77.0%)	(79.4%)

Table 5. Comparison of successful cases and CPU time (in second) for *FSDB*,*FSDC* and *FSDQ*, where the notation "a/b/c" shows the worst (or shortest), the average and the best (or longest).

	FSDB	*FSDC*	*FSDQ*
#success	354	462	690
$\lvert\delta\rvert/\lvert X\rvert(\%)$	89.3/99.0/100	92.1/99.2/100	94.1/99.8/100
time (sec.)	0.9/158.5/712.1	1.2/172.8/887.7	1.0/329.2/1842.1

Table 6. Improvement by incorporating *FSDC*, where the notation " Results" shows the number of the successful cases (meaning that the algorithm finds optimum solutions among #inputs inputs) , and **MIM** is omitted: see [10,11].

Problems	**OFS**	**LUS**	**MCP**	**TPS**
Known	*LBD*[1]	*LUD*[1]	*WYMS*[6]	*SDS*[22]
Proposed	*LBDC*	*LUDC*	*MKDB*	*DBSC*
#inputs	1392	1005	490	995
Results	1094(78.6%)	813(80.8%)	–	682(68.5%)
Improved	+17.0%	+17.8%	−43.1%	+13.5%

· Existence of optimum solutions to all inputs (the total number is "#inputs") is guaranteed

· For **OFS**, **LUS** and **TPS**: "Improved" shows increase in the values of "Results", while for **MCP** it shows decrease in Δ (total sum of absolute values of token difference between target markings and produced ones)

solving the subproblem **MAX LFS**. It is experimentally shown that these algorithms outperform those existing ones *LBD*[1], *LUD*[1], *WYMS*[6] and *SDS*[22]. Because of space limitation we show only their experimental evaluation in Table 6. Note that, in [11,10] ([14], respectively), we reported new heuristic algorithms for **MIM** (**MCP**) and their experimental evaluation are reported.

9   Concluding Remarks

It is shown that the proposed algorithm *FSDC* has highest capability among existing ones for solving **MAX LFS**. Improving capabilities and/or shortening computation time of heuristic algorithms for **MAX LFS** is left for future research.

References

1. K. Awa, S. Taoka, and T. Watanabe, "The legal firing sequence problem of Petri nets with lower bounds on the number of tokens," Tech. Rep. CST99-3, IEICE of Japan, pp. 17–24, May 1999, (in Japanese).
2. S. Even, *Graph Algorithms*, Pitman, London, U.K., 1978.
3. T. Fujito, S. Taoka, and T. Watanabe, "On the legal firing sequence problem of Petri nets with cactus structure," *IEICE Trans. Fundamentals*, Vol. E83-A, No. 3, pp. 480–486, March 2000.

4. M. R. Garey and D. S. Johnson, *Computers and Intractability: A Guide to the Theory of NP-completeness*, Freeman, San Francisco, CA, 1978.
5. R. Kosaraju, "Decidability of reachability in vector addition systems," in *Proc. 14th Annual ACM Symposium on Theory of Computing*, pp. 267–280, 1982.
6. K. Maniwa, M. Yamauchi, and T. Watanabe, "The marking construction problem of Petri nets," in *Proc. of the 47th Joint Convention at the Chugoku Branch of Electrical 5-Societies of Japan*, pp. 290–291, October 1996, (in Japanese).
7. E. W. Mayr, "An algorithm for the general Petri net reachability problem," in *Proc. 13th Annual ACM Symposium on Theory of Computing*, pp. 238–246, 1981, See also SIAM J. Comput., Vol.13, pp.441-460 (1984). (with the same title by the author).
8. K. Morita and T. Watanabe, "The legal firing sequence problem of Petri nets with state machine structure," in *Proc. 1996 IEEE International Symposium on Circuits and Systems*, pp. 64–67, May 1996.
9. T. Murata, "Petri nets: Properties, analysis and applications," *Proc. IEEE*, Vol. 77, No. 4, pp. 541–580, April 1989.
10. S. Nishi, S. Taoka, and T. Watanabe, "An Improved Heuristic Algorithm *AAD* for Minimizing Initial Markings of Petri Nets," Tech. Rep. CAS2001-67, IEICE of Japan, pp. 23–30, November 2001.
11. S. Nishi, S. Taoka, and T. Watanabe, "A heuristic algorithm *FMDB* for the minimum initial marking problem of Petri nets," *IEICE Trans. Fundamentals*, Vol. E84-A, No. 3, pp. 771–780, March 2001.
12. K. Takahashi, M. Yamamura, and S. Kobayashi, "A GA approach to solving reachability problems for Petri net," *IEICE Trans. Fundamentals*, Vol. E79-A, No. 11, pp. 1774–1780, November 1996.
13. S. Tanimoto, M. Yamauchi, and T. Watanabe, "Finding minimal siphons in general Petri nets," *IEICE Trans. Fundamentals*, Vol. E79-A, No. 11, pp. 1817–1824, November 1996.
14. S. Taoka, T. Nojo, and T. Watanabe, "The marking construction problem of Petri nets MCP and its heuristic algorithm," in *Proc. 15th Karuizawa Workshop on Circuits and Systems*, pp. 447–452, April 2002.
15. T. Watanabe, Y. Mizobata, and K. Onaga, "Minimum initial marking problems of Petri nets," *Trans. IEICE of Japan*, Vol. E72, No. 12, pp. 1390–1399, December 1989.
16. T. Watanabe, Y. Mizobata, and K. Onaga, "Time complexity of legal firing sequence and related problems of Petri nets," *Trans. IEICE*, Vol. E72, No. 12, pp. 1400–1409, December 1989.
17. T. Watanabe, T. Tanida, M. Yamauchi, and K. Onaga, "The minimum initial marking problem for scheduling in timed Petri nets," *IEICE Trans. Fundamentals*, Vol. E75-A, No. 10, pp. 1407–1421, October 1992.
18. T. Watanabe and M. Yamauchi, "New priority-lists for scheduling in timed Petri nets," in *Application and Theory of Petri Nets 1993*, M. A. Marsan, Ed., Lecture Notes in Computer Science, No. 691, pp. 493-512. Springer-Verlag, Berlin, Germany, June 1993.
19. M. Yamauchi and T. Watanabe, "An approximation algorithm for the legal firing sequence problem of Petri nets," in *Proc. 1994 IEEE International Symposium on Circuits and Systems*, pp. 181–184, May 1994.
20. M. Yamauchi and T. Watanabe, "A heuristic algorithm for the minimum initial marking problem of Petri net," in *Proc. 1997 IEEE International Conference on Systems, Man and Cybernetics*, pp. 245–250, October 1997.

21. M. Yamauchi and T. Watanabe, "A heuristic algorithm *FSD* for the legal firing sequence problem of Petri nets," in *Proc. 1998 IEEE International Conference on Systems, Man and Cybernetics*, pp. 78–83, October 1998.
22. M. Yamauchi and T. Watanabe, "A heuristic algorithm *SDS* for scheduling with timed Petri nets," in *Proc. 1999 IEEE International Symposium on Circuits and Systems*, pp. VI–81–VI–84, May 1999.

PLC Programming with Signal Interpreted Petri Nets

Stéphane Klein[1,2], Georg Frey[1], and Mark Minas[3]

[1] Institute of Automatic Control, University of Kaiserslautern,
PO Box 3049, 67653 Kaiserslautern, Germany
{sklein, frey}@eit.uni-kl.de
[2] Univ. lab. of research in Automated Production (LURPA), ENS Cachan
61 av. du Pdt Wilson, 94235 CachanCedex, France
klein@lurpa.ens-cachan.fr
[3] Institute for Software Technology, Department of Computer Science,
Univ. of the Federal Armed Forces, Munich, 85577 Neubiberg, Germany
minas@acm.org

Abstract. In this paper a graphical editor to design Programmable Logic Controller (PLC) programs using Signal Interpreted Petri Nets (SIPN) is presented. SIPN are an extension of condition event Petri nets that allow the handling of input and output signals. The presented tool, SIPN Editor, has been developed using DiaGen which is an environment for rapidly developing diagram editors from a formal specification of the diagram language. The SIPN Editor supports the translation of SIPN into input code for the model checker SMV. Using SMV, the SIPN can be verified before it is automatically translated into Instruction List code according to the IEC 61131-3 standard. This code can be downloaded on nearly every PLC.

1 Introduction

The presented tool has been developed to support the design of Programmable Logic Controllers (PLC) using Signal Interpreted Petri Nets (SIPN). SIPN [1] are an extension of Condition Event Petri Nets enabling the handling of input and output signals. The SIPN Editor has been developed to edit and visualize SIPN and, since it is the purpose of this class of Petri nets, implement them on a PLC. Nevertheless, to cover the whole design process of a PLC control algorithm, it also supports the step of formal verification and validation that should be performed prior to the implementation of the algorithm. Therefore, a translation of SIPN into input code for the model checker SMV has been integrated in the tool. The SIPN Editor has been designed as a Java application so that it can be run on nearly every PC.

The paper is organized as follows: In section 1, the editor is presented, so as the concept of DiaGen on which it is based. In section 3, it will be shown how the SIPN Editor supports the analysis of the designed SIPN via model checking. Finally the generation of Instruction List (IL) code is presented in section 4.

W.M.P. van der Aalst and E. Best (Eds.): ICATPN 2003, LNCS 2679, pp. 440–449, 2003.
© Springer-Verlag Berlin Heidelberg 2003

2 Editing an SIPN

2.1 Signal Interpreted Petri Nets

Signal Interpreted Petri Nets [1] are an extension of Condition Event Petri Nets. Their goal is to provide the designer with a graphical language to easily design PLC control algorithms. Hence they have to be connected to the plant under control and that is achieved via the handling of signals: Transitions are associated with firing conditions and places with output functions.

Firing conditions are Boolean functions of the input variables of the SIPN. They are evaluated after the transition has been found enabled (pre places of the transition marked and post places unmarked) and produce the immediate firing of the transition when they are true.

An output function is a piece of PLC code that is executed when the considered place is marked. It affects the output signals of the PLC.

After the definition of the signal handling, the second important point in the description of an SIPN is the way it behaves. Opposite to ordinary Petri nets, an immediate and simultaneous firing of all the transitions that can fire simultaneously takes place. Moreover, the firing process is iterated until no more transition can fire under this input setting. The reached marking is said to be stable. After such a marking has been reached, the output functions are evaluated and given out.

To handle real problems, the SIPN has been extended with time and hierarchy concepts [2]. Hence, every arc going from a place to a transition can be associated with a time delay. This delay represents the minimal time a token has to spend in the place before the transition is enabled. Furthermore, as real controllers tend to be very large the concept of hierarchy has been introduced. It allows replacing an identified quite independent partition of an SIPN by a single hierarchical place. The extension of this place (subnet) contains the description of the identified process part.

2.2 Example

To illustrate the concept, let us present the example of the controller of a heating tank realized with a hierarchical timed SIPN (cf. Figure 1.). The informal specification of the tank under control is given by the following properties:

-   when the tank is empty and the *Start* button is pressed, the tank is filled,
-   if the temperature of the full tank is less than required, the liquid is heated until the required temperature is reached,
-   after a waiting of 20 seconds, the tank is emptied,
-   the tank remains stirred during the whole process.

Let us show in details how this SIPN works: In place P1, the control algorithm is waiting for the start of the process and all output signals are set to 0. After it has been checked that the tank is empty ($i1 = 1$) and the start button has been pressed ($i4 = 1$), transition T1 fires. Place P1 becomes unmarked whereas P2, P3 and P3A get marked. Hence the stirring motor is set ON ($o1 = 1$ in place P2) and the tank is being filled ($o2 = 1$ in place P3A). Once the tank is full ($i2 = 1$ in transition T3A and T3B), it is simultaneously checked whether the contents have the desired temperature or not. If they have not (i3 = 0 in transition T3B), they are heated (o4 = 1 in place P3B) until the desired temperature is reached (i3 = 1 in transition T3C). Once this temperature has been reached (T3A or T3C fires), the contents are stirred during 20 seconds (timer on place P3C). After this delay, transition T3D fires and the tank is being emptied (o3 = 1 in place P3D). The marking of P2, P3 and P3D enables transition T2 which fires as soon as the tank is empty (i1 = 1 in transition T2).

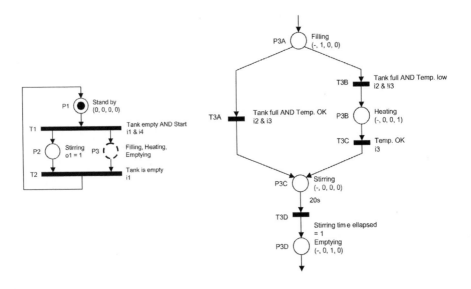

Fig. 1. SIPN for the heating tank.

2.3 SIPN Editor

The idea of the SIPN Editor is to have a tool to support the new way of graphical programming of PLC controllers offered by SIPN. Hence a Java application has been developed to achieve this goal. The SIPN Editor has been generated using DiaGen (Diagen Editor Generator), an environment for rapidly developing diagram editors from a formal specification of the diagram language based on hypergraph grammars and hypergraph transformation [3], [4], [5]. The main task of the editor is translating a "drawing" which is supposed to be a correct diagram (an SIPN) into a semantic representation (an equivalent IL code). During this translation process, the editor checks the drawing for correctness and has to provide feedback to the editor user if the drawing contains errors.

The editing tool (see Figure 2) consists of a drawing canvas which contains the SIPN diagram and the usual control widgets. The generated editor is a direct manipulation editor, i.e. any diagram component (place, transition, arrow or initial token) can be created and moved anywhere on the canvas by choosing the component on the toolbar and specifying its position with the mouse, or by selecting an existing component with the mouse and changing its position by means of drag handles which appear for selected components. A parser checks the syntactic correctness of the diagram and provides visual feedback by coloring of the erroneous component. A diagram layouter takes care of diagram beautification as of snapping to a gird and adjusts arrows when places and/or transitions are moved so that places, transitions and arrows remain connected. Using this SIPN Editor, it is possible to design and later implement hierarchical and timed SIPN.

Figure 2 shows a snapshot of the SIPN Editor during the design of the SIPN of the heating tank presented in the previous sub-section.

The snapshot shows on the left hand side the construction tree of the SIPN. In the main window, the upper level and the properties of place P2 are shown on the left side. On the right side, the definition of the hierarchical place P3 is given.

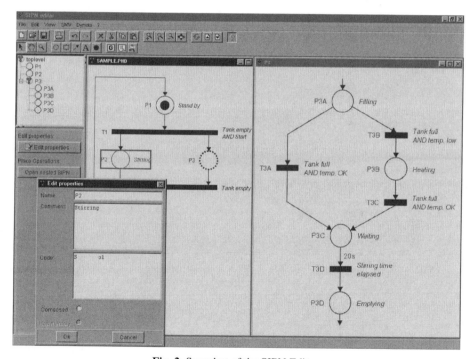

Fig. 2. Snapshot of the SIPN Editor

Let us now show how in detail the elements of an SIPN are handled in the SIPN Editor.

Places. The definition of a place consists of a name (mandatory and unique), a comment and a piece of code (optional). The name given to a place is a piece of text using the characters allowed by the IEC 61131-3 standard [6]. This name will later be used during the code generations (SMV model checker and IL) as a Boolean variable. The comment is a piece of text that will be displayed near the place on the editor screen. The associated code represents the output function of the place and must written using IL [6] statements. Since the editor does not check the syntax of this code, a formal analysis of the net should always be performed prior to the translation of the SIPN into IL code. Moreover, a place can be associated with a subnet. It is then drawn with a bold dashed line.

Transitions. As for places, a transition must be associated with a name and, optionally, with a comment and a piece of IL code. For the name and the comment, the same remarks apply as for the places. The IL code associated with a transition represent its firing condition. If no code is associated to a transition, it is implicitly supposed to be true. As for the output function of a place, this code will not be evaluated during the drawing of the net and it should be checked using formal methods before the translation into IL code.

3   SIPN Analysis

Since the graphical editor only checks the drawing, i.e. the structure, it must be made sure that the behavior of the designed SIPN is correct before it is implemented.

3.1 Verification & Validation

Let us remember the definition of verification and validation as it was given by Boehm [7] in 1979:
   Verification: Am I building the product right?
   Validation: Am I building the right product?

Verification. As given in the definition, it must be checked whether the designed SIPN is correct according to the criteria given in [1]. We only focus on the mandatory property which is: **Unambiguity**. Every control algorithm has to be defined unambiguously. This criterion can be subdivided into four sub-criteria:
   1. Determinism: A control algorithm has to be deterministic. If it were not, its behavior in a given situation would depend on implementation aspects. This of course could not be the aim of a correct design.
   2. Termination: In a cycle of a logic control algorithm, at least one marking must be stable. That means that the iterative firing process is broken in at least one transition. A cycle without stable marking leads to an algorithm that does not terminate. This in turn could lead to a hang-up of the implemented controller or to non-deterministic behavior, in the case when the implementation breaks up the endless loop at some unknown point.
   3. Defined output: If there were no specification for the value of an output signal at a reachable marking, again the behavior of the controller would depend on its implementation: In an implemented controller, there is no 'don't care'.
   4. Unambiguous output: If two places marked at the same time assign different values to an output signal, a contradictory output setting results. This is a clear design error.

Further Petri Net properties like deadlockfreeness, liveness or reversibility can also be checked on the SIPN. Nevertheless they are not mandatory for a correct SIPN. Actually the correctness of the SIPN guarantees that once the SIPN is implemented on a PLC, the behavior of the PLC is unambiguous. Hence an SIPN that has deadlocks or that is not reversible will let the PLC stuck in a given situation. This might not be a design failure but it might have been wanted by the designer (emergency stop, ...).

Validation. During the verification of the SIPN, it is checked that the designed SIPN is unambiguous, i.e. that its behavior does not depend on the way it is implemented in the PLC. However it has not yet been validated that the behavior of the SIPN fulfills the informal expectations formulated by the client. In other words, the problem specific functional properties must be checked. In the example used in section 2.2, such a property could be: The stirring motor is always on while the tank is being filled or heated.

3.2 Model Checking and Tool Support

To perform verification and validation, symbolic model checking is used. In this technique, a finite model of the system is built and the expected properties (specifications of the behavior) of the system are checked on this model. The system is modeled as a finite state transition system and the properties are expressed in a temporal logic [8]. A search procedure is then used to check whether the expected properties are verified on the finite state transition system or not.

Hence, to use a model checker, the "drawing" (SIPN) with its behavior has to be translated into a semantic representation (entrance code of the model checker).

An automatic code generator for the model-checker SMV [9] has been integrated in the SIPN Editor. Thus the translation of the SIPN into input code of SMV is transparent for the designer and the step of verification and validation made much easier. Actually performing verification and validation often implies the uses of three different formalisms. The first is used to design the control algorithm (SIPN in this case), the second is the model-checker formalism and the third is the one used to express the properties (Temporal logic). Producing automatically the description of the SIPN in the SMV formalism, the step of formal verification is simplified in the sense that the designer has one less formalism to know to use this helpful method.

Verification. During the translation of the SIPN, the editor generates automatically the CTL (Computation Tree Logic) formulae [8] corresponding to the mandatory properties every SIPN has to verify.

Validation. For the purpose of validation, a window allowing the input of CTL formulae has been integrated in the SMV menu of the SIPN Editor. In this window, properties and assumptions can be given. These will then be inserted in the SMV code during its generation.

After the SIPN has been verified and validated using the model checker SMV, it can be translated into IL code in order to be executed on a PLC.

4   Generation of IL Code

4.1 Principle – Goal

The generation of PLC code is the primary goal of the SIPN Editor. To use SIPN on a very large variety of PLC hardware, it is best to translate the resulting algorithm into one of the standardized PLC language as for example Instruction List (IL) [6], [11]. The direct implementation of an SIPN compiler would only work on a single PLC, whereas an IL according to the specifications of IEC 61131 standard can be executed on nearly any PLC.

For a structure-conserving conversion the token play of the SIPN has to be transferred to the PLC program. Therefore, for each place pi of the net, a Boolean variable Pi is defined that shows whether the corresponding place is marked (Pi = true) or unmarked (Pi = false).

Besides the differences in the specific Petri net type considered, the presented code generation differs from other approaches (see e.g. [13], [14]) in one main aspect. This is the one-to-one correspondence of net elements to code segments that is used. This correspondence allows to easily reinterpret produced code. This reinterpretation is of special importance if a user wants to understand and change the implemented code, which is commonplace in industrial applications.

In contrast to the SMV code generation where the SIPN is virtually flattened before the code is generated, the hierarchical structure remains unchanged during the generation of IL code. To achieve that, for each subnet a function block is defined. Once again, this allows an easy reinterpretation of the generated IL.

Based on this premise the net elements can be translated to PLC code step by step

4.2 Code Generation

Variable declaration. A Boolean variable is declared for each place of the SIPN. Its initial value depends on the presence of a token in the place.

Transitions. The compilation of a transition has to test whether the transition is enabled and whether the firing condition is fulfilled. If after the processing of this calculation the accumulator is set to 1, i.e. all conditions are fulfilled, then the transition fires. If not, a conditional jump to the next transition avoids firing. The firing unmarks all pre-places and marks all post-places. To optimize the code, the firing conditions are not evaluated if a transition is not enabled.

Since this firing process is iterative and only stops when no more transition can fire under the current setting of inputs, a stability variable has been introduced. This is set to 1 at the beginning of the process and set to 0 as soon as a transition fires. After all transitions have been computed, the stability variable is tested. If it is false, i.e. a transition has fired, the firing process is iterated. If it is true, the algorithm jumps to the computation of the outputs.

```
(* Set stability variable to TRUE *)
1_0:        S        stab
(* Transition T1 *)
1_1:        LD       P1      (* pre place Stand by *)
            ANDN     P2      (* post place Stirring *)
            ANDN     P3      (* post place P3 *)
            JMPCN    1_2
            AND      i1      (* Transition specific code *)
            AND      i2
            JMPCN    1_2
            R        P1
            S        P2
            S        P3
            R        stab    (* Stability Variable *)


(* Transition T2 *)
1_2:        ...

(* Stability check *)
1_8:        LD       stab
            JMPCN    1_0
```

Places. If a place is marked then the corresponding output function is executed (setting or resetting of variables). If it is not marked a conditional jump to the next place label is performed and the code segment of the output function is not executed. Note that the implementation of the output functions results in an output of zero or one (true or false) for all output variables according to the output of the SIPN. For an undefined output, the PLC code remains at the last defined value for this output. For a contradictory output it depends on the ordering of the involved place code segments if the variable is set to one or to zero. Both cases should be avoided using SIPN analysis prior to code generation. That's why verification and validation should always be performed before the translation of the SIPN into IL code.

```
1_9:        LD       P1      (* Place Stand by *)
            JMPCN    1_10
            R        o1      (* Place specific code *)
            R        o2
            R        o3
            R        o4

1_10:       LD       P2      (* Place Stirring *)
            ...
```

4.3 Download and Visualization

This code can then be downloaded on a PLC or a Soft-PLC running on an IPC and assure the command of an automated line. Furthermore, a visualization process has been implemented to visualize the machine state within the editor and its SIPN while the machine is running. Hence, the SIPN editor can also be used to find out failures or at least the state in which the machine is when it got stuck. During the generation of IL code, additional variables are generated to let the SIPN Editor known the current marking. In the visualization, these variables are used to shade the active places. In Figure 3, the current marking of the SIPN is (P2, P3, P3B). That means that the tank should being heated (Place P3B) while the stirring motor is ON (Place P2).

5   Conclusion

A graphical tool to support the design, verification, validation and implementation of Signal Interpreted Petri Nets has been presented. This tool, offers the designer a graphical editor to draw and visualize hierarchical and timed SIPN. After the design of the SIPN, it can be automatically translated into input code for the model checker SMV. Thus it can be made sure that the designed SIPN fulfills the expected properties before it is implemented on a PLC. This last step is also supported by the editor with the automatic generation of IL code according to the IEC 61131-3 standard. This code can then be downloaded on nearly every PLC since the IEC standard is not dependant on the PLC builder. Furthermore, with the appropriate communication between PLC and PC, the marking of the SIPN can be visualized while the machine is running.

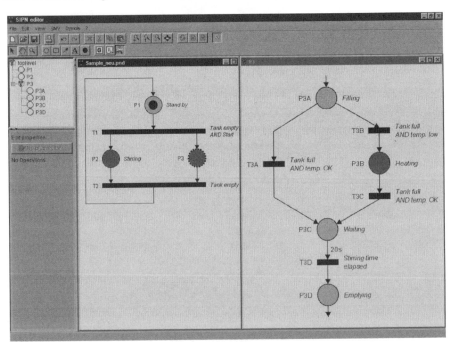

Fig. 3. Screenshot of the SIPN Editor during the visualization of the process.

The current version of the SIPN Editor does not allow to reuse a subnet (hierarchical extension of a place). One of the projected extensions of the SIPN Editor is to allow the reuse of subnets by creating a library of subnets and use them with different interfaces. This will also permit the verification of these independent parts before they are used in a larger SIPN.

Concerning the analysis via model checking, it is projected to integrate the V&V results in the SIPN Editor. That should work like the on-line visualization. Hence when a property is not satisfied, SMV gives a counter-example as a trace leaving from the initial marking to the faulty one. The goal is to show this trace directly by marking the concerned places of the SIPN.

The SIPN Editor is free software and available under
http://www.eit.uni-kl.de/litz/ENGLISH/software/SIPNEditor.htm or
http://www2.informatik.uni-erlangen.de/DiaGen/SIPN/

References

1. Frey, G.: Design and formal Analysis of Petri Net based Logic Control Algorithms – Entwurf und formale Analyse Petrinetz-basierter Steuerungsalgorithmen. (ISBN 3-8322-0043-6), Dissertation, University of Kaiserslautern, Shaker Verlag, Aachen, April 2002.
2. Frey, G.: SIPN, Hierarchical SIPN, and Extensions. In: Reports of the Institute of Automatic Control I19/2001, University of Kaiserslautern, Dec. 2001. Available at http://www.eit.uni-kl.de/litz/ENGLISH/members/frey/papers.htm
3. Minas, M.: Diagram editing with hypergraph parser support. In: Proceedings of 1997 Symposium on Visual Languages (VL'97), pages 230–237, 1997.
4. Minas, M.: Creating semantic representations of diagrams. In: M. Nagl and A. Schürr, (eds.), Int. Workshop on Applications of Graph Transformations with Industrial Relevance (AGTIVE'99), Selected Papers, LNCS 1779, pages 209–224. Springer, Mar. 2000.
5. Minas, M.: Concepts and realization of a diagram editor generator based on hypergraph transformation. In: Science of Computer Programming, 44(2):157–180, 2002.
6. IEC: International Standard 61131–3, 2nd Ed., Programmable Controllers – Programming Languages, 1999.
7. Boehm, B. W.: Guidelines for Verifying and Validating Software Requirements and Design Specifications. In: P.A. Samet (ed), Proceedings of the EURO IFIP 79, North-Holland Publishing Company, 1979.
8. Bérard, B., Bidiot, M., Finkel, A., Laroussinie, F., Petit, A., Petrucci, L., Schnoebelen, Ph.: Systems and Software Verification, Model-Checking Techniques and Tools. Springer, Berlin, New-York, 2001.
9. McMillan, K. L.: The SMV language. Cadence Berkeley Labs, Berkeley, Californy (USA), Mar. 23, 1999. www-cad.eecs.berkeley.edu/kenmc-mil
10. Canet, G., Couffin, S., Lesage, J.-J., Petit, A., Schnoebelen, P.: Towards the automatic verification of PLC programs written in Instruction List. In: Proceedings of the IEEE Conference on Systems Man and Cybernetics SMC 2000, pp. 2449–2454, Nashville, Tennessee (USA), Oct. 8–11, 2000.
11. John, K.-H., Tiegelkamp, M.: IEC 61131-3: Programming Industrial Automation Systems. Springer, Beriln, New-York, 2001.
13. Stanton, M. J., Arnold, W. F., Buck, A. A.: Modelling and control of manufacturing systems using Petri nets. In: Proceedings of the 13th IFAC World Congress, pp. 329–334, 1996.
14. Venkatesh, K., Zhou, M., Caudill, R. J.: Discrete event control design for manufacturing systems via ladder logic diagrams and Petri nets: A comparative study. In M. Zhou, editor, Petri Nets in Flexible and Agile Automation, pp. 265–304. Kluwer Academic Publish., 1995.

CPN Tools for Editing, Simulating, and Analysing Coloured Petri Nets

Anne Vinter Ratzer, Lisa Wells, Henry Michael Lassen, Mads Laursen,
Jacob Frank Qvortrup, Martin Stig Stissing, Michael Westergaard, Søren Christensen,
and Kurt Jensen

Department of Computer Science, University of Aarhus
IT-parken, Aabogade 34, DK-8200, Århus N, Denmark
cpn@daimi.au.dk

Abstract. CPN Tools is a tool for editing, simulating and analysing Coloured Petri
Nets. The GUI is based on advanced interaction techniques, such as toolglasses,
marking menus, and bi-manual interaction. Feedback facilities provide contextual
error messages and indicate dependency relationships between net elements. The
tool features incremental syntax checking and code generation which take place
while a net is being constructed. A fast simulator efficiently handles both untimed
and timed nets. Full and partial state spaces can be generated and analysed, and a
standard state space report contains information such as boundedness properties
and liveness properties. The functionality of the simulation engine and state space
facilities are similar to the corresponding components in Design/CPN, which is a
widespread tool for Coloured Petri Nets.

1 Introduction

CPN Tools is a tool for editing, simulating and analysing untimed and timed, hierar-
chical Coloured Petri nets (CPN or CP-nets) [1,2]. CPN Tools is intended to replace
Design/CPN [3], which is a widespread software package for CP-nets. In addition to
Design/CPN, CPN Tools can be compared to other Petri net tools such as ExSpect,
GreatSPN, and Renew which are all described in the Petri Nets Tool Database [4].

Design/CPN was first released in 1989 with support for editing and simulating CP-
nets. Since then a significant amount of time has been invested in developing efficient
and advanced support both for simulation and for generating and analysing full, partial,
and reduced state spaces. While the analysis components of Design/CPN have steadily
improved since 1989, the graphical user interface has remained virtually unchanged.

CPN Tools is the result of a research project, the CPN2000 project [5], at the University of Aarhus, sponsored by the Danish National Centre for IT Research (CIT), George
Mason University, Hewlett-Packard, Nokia, and Microsoft. The goal of the CPN2000
project was to take advantage of the developments in human-computer interaction, and
to experiment with these techniques in connection with a complete redesign of the GUI
for Design/CPN. The resulting CPN Tools combines powerful functionalities with a
flexible user interface, containing improved interaction techniques, as well as different
types of graphical feedback which keep the user informed of the status of syntax checks,

W.M.P. van der Aalst and E. Best (Eds.): ICATPN 2003, LNCS 2679, pp. 450–462, 2003.

simulations, etc. All models that are created in Design/CPN can be converted and then used in CPN Tools; the reverse, however, is not true.

This paper is organised as follows. Section 2 introduces the new interaction techniques and components of the GUI. Section 3 describes how to edit CP-nets in CPN Tools. Finally, Sect. 4 describes the simulation and state space facilities that are provided in CPN Tools.

2   The CPN Tools Interface

The CPN Tools interface requires a keyboard and at least one pointing device. Actually, the interface supports and encourages the use of two or more pointing devices. For a right-handed user we recommend using a mouse for the right hand and a trackball for the left hand. The mouse is used for tasks that may require precision, while the trackball is used for tasks that do not require much precision e.g. moving tools. For simplicity we assume a right-handed user in our description of interaction techniques. We describe how such a person would typically use the right or left hand, but it should be noticed that all operations can be done using either hand.

The interface has no menu bars or pull-down menus, and only few scrollbars and dialog boxes. Instead, it uses a combination of traditional, recent and novel interaction techniques, which are described below. Figure 1 shows the GUI for CPN Tools.

Workspace management makes it easy to manage the large number of pages that are typically found in industrial-sized CP-nets. The workspace occupies the whole screen and contains window-like objects called *binders*. Binders contain *sheets* where each sheet is equivalent to a window in a traditional environment. A sheet provides a view of either a page from a CP-net, or declarations, or a set of tools. Each sheet has a tab similar to those found in tabbed dialogs. Clicking the tab brings that sheet to the front of the binder. A sheet can be dragged to a different binder or to the background to create a new binder for it. Binders reduce the number of windows on the screen and the time spent organising them. Binders also help users organise their work by grouping related sheets together and reducing the time spent looking for hidden windows.

CPN Tools supports multiple views, allowing several sheets to contain different views of the same page. For example, one sheet can provide a close-up view of a small part of a page while another sheet can provide a view of the same page at a much smaller scale (see figure 1).

Direct manipulation (i.e. clicking or dragging objects) is used for frequent operations such as moving objects, panning the content of a view and editing text. When a tool is held in the right hand, e.g. after having selected it in a palette, direct manipulation actions are still available via a long click, i.e. pressing the mouse button, waiting for a short delay until the cursor changes, and then either dragging or releasing the mouse button.

Bi-manual manipulation is a variant of direct manipulation that involves using both hands for a single task. It is used to resize objects (binders, places, transitions, etc.) and to zoom the view of a page. The interaction is similar to holding an object with two hands and stretching or shrinking it. Unlike traditional window management techniques, using two hands makes it possible to simultaneously resize and move a binder, or pan and

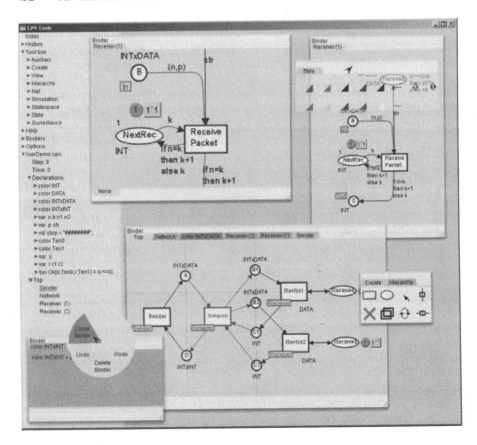

Fig. 1. The CPN Tools interface. The left column is called the index. The top-left and top-right binders contain sheets with different views of the same page. The bottom binder contains six sheets represented by tabs. The front sheet, containing the page named Top, shows a number of magnetic guidelines for easier alignment of objects. In the bottom-left binder, containing a declaration, a circular marking menu has been popped up. The small binder on the right-hand side contains two sheets with tool palettes. The palette in the front (Create) contains tools for creating CP-net objects and guidelines. A toolglass is positioned over the sheet in the top-right binder. This toolglass can be used to edit colours, line styles and line widths.

zoom the view of a page. This has been further generalised to allow an arbitrary number of hands, so two or more users can work together on the same computer.

Marking menus [6] are circular, contextual menus that appear when clicking the right button of the mouse. Marking menus offer faster selection than traditional linear menus for two reasons. First, it is easier for the human hand to move the cursor in a given direction than to reach for a target at a given distance, as in a traditional linear menu. Second, the menu does not appear when the selection gesture is executed quickly, which supports a smooth transition between novice and expert use. Kurtenbach and Buxton [6] have shown that selection times can be more than three times faster than with traditional menus.

Keyboard input is mainly to edit text. Some navigation commands are available at the keyboard to make it easier to edit several inscriptions in sequence without having to move the hands to the pointing devices. Keyboard modifiers and shortcuts are not necessary since most of the interaction is carried out with the two hands on the pointing devices.

Palettes contain tools represented by buttons. Clicking a tool with the mouse activates this tool, i.e. the user conceptually holds the tool in the hand. Clicking on an object with the tool in hand applies the tool to that object. Palettes can be moved with either hand, making it easy to bring the tools close to the objects being manipulated, and saving the time spent moving the cursor to a traditional menubar or toolbar. In many current interfaces, after a tool is used (especially a creation tool), the system automatically activates a "select" tool. This supports a frequent pattern of use in which the user wants to move an object immediately after it has been created but causes problems when the user wants to create additional objects of the same type. CPN Tools avoids this automatic changing of the current tool by ensuring that the user can always move an object, even when a tool is active, with a long click of the mouse. This mimics the situation in which one holds a physical pen in the hand while moving an object out of the way in order to write.

Toolglasses [7] like palettes, contain a set of tools represented by buttons, and are moved with the left hand, but unlike palettes, they are semi-transparent. A tool is applied to an object with a *click-through* action: The tool is positioned over the object of interest and the user clicks through the tool onto the object. The toolglass disappears when the tool requires a drag interaction, e.g. when creating an arc. This prevents the toolglass from getting in the way and makes it easier to pan the page with the left hand when the target position is not visible. This is a case where the two hands operate simultaneously but independently.

The index is positioned in the left side of the workspace and contains lists of all the available tools and net elements in CPN Tools (see figure 1). It is similar to, e.g., a tree view of files in Windows Explorer, and the entries can be opened and closed in the same way. in From the index, the user can drag tool palettes, CP-net pages, or declarations onto the workspace. It is also possible to edit declarations and file names for the loaded nets directly in the index. The index provides a feedback mechanism for locating CP-net objects connected to a particular declaration: if the cursor is held over a declaration, a blue halo or underline appears on all declarations, pages, and binders containing that declaration. This makes it easier to, e.g., make changes to a colour set and ensure that the changes are made on all objects using this colour set.

Magnetic guidelines are used to align objects and keep them aligned. Moving an object near a guideline causes the object to snap to it. Objects can be removed from a guideline by dragging them away from it. Moving a guideline moves all the objects attached to it, maintaining their alignment.

Preliminary results from our user studies make it clear that none of the above techniques is always better or worse. Rather, each emphasises a different, but common pattern of use. Marking menus work well when applying multiple commands to a single object. Palettes work well when applying the same command to different objects. Toolglasses work well when the work is driven by the structure of the diagram, such as working around a cycle in a CP-net.

3   Editing CP-Nets

Editing CP-nets in CPN Tools is easy, fast, and flexible since there is often more than one way to perform a particular task. For example, places can be created using marking menus, palettes and toolglasses. While a net is being edited, CPN Tools assists the user in a number of different ways, e.g. by providing a variety of graphical feedback regarding the syntax of the net and the status of the tool, or by automatically aligning objects in some situations. The syntax of a net is checked and simulation code for the net is automatically generated *while* the net is being constructed. This section describes how CP-nets can be created and edited in CPN Tools.

3.1   Tools for Editing CP-Nets

Most of the tools described here can be found both in the palettes and toolglasses that can be dragged out from the *Tool box* entry of the index (see figure 1) and in marking menus.

Create tools are used to create CP-net elements, i.e. places, transitions, and arcs. All net elements can be created using palettes, toolglasses and marking menus. Net elements can be positioned freely within a sheet, or they can be snapped to magnetic guidelines. CPN Tools assists users by automatically aligning objects in some situations, even if guidelines are not used. For example, if a place is connected to a transition, and the place is moved so that it is sufficiently close to being vertically aligned with the transition, then CPN Tools will snap the place to be perfectly vertically aligned with the transition.

Adding inscriptions to net elements is done by clicking on a net element. This will select a default inscription, e.g. the name of a place or the inscription for an arc, and the selected inscription can then be added, edited or removed through the keyboard. It is not necessary to use the mouse when editing the inscriptions for one particular object, since the TAB key can be used to move from one inscription to another for the object in question. Furthermore, CPN Tools assists in positioning inscriptions. All inscriptions have a default position, e.g. colour sets are positioned near the lower right-hand side of a place. A number of snap points around objects can be used to position inscriptions in alternative standard positions, and an inscription can also be positioned freely within a sheet. The alignment of an inscription is maintained when the text of the inscription is changed.

In CPN Tools it is possible to clone, i.e. copy, almost any type of object, and then to create new objects that are identical to the original object. Cloning an object clones all of the relevant information such as size, textual inscriptions, line colour and line width. This makes it very easy, for example, to create a number of places that have the same shape, line colour, and colour set inscription, or to add the same arc inscription to a number of different arcs. After an object has been cloned using a marking menu, the right hand holds a tool that can be used to create new objects. The cursor for the right hand indicates which object was cloned. Figure 2 shows a place (on the left) and the cursor (on the right) that was obtained after cloning the place .

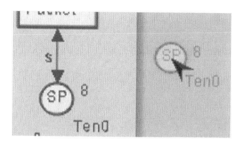

Fig. 2. Cursor indicating which object has just been cloned.

Style tools can be used to change the style of any net element. Each kind of net element has a default style which determines the size, line and fill colour, line width, and line style (solid, dashed, etc.) of newly created elements. Applying a style tool, e.g. a colour or a certain line width, to a guideline will apply it to all of the objects on the guideline.

View tools are used to define groups and to zoom in and out on a page. An arbitrary number of groups can be defined for each CP-net. Currently, each group may only contain objects from one page in a net. A group can, however, consist of different kinds of objects, such as places, arc inscriptions, and auxiliary nodes. Objects can be added and removed from groups via a marking menu, a tool palette or a toolglass. Creating a new object in a group adds the object to the group, as well as adding it to the appropriate page in the CP-net. If a tool, such as a style tool or a fusion set tool, is applied to a group member while in group-mode, then the tool is automatically applied to all (relevant) members in the group. The *View palette* also contains tools for zooming in and out on a page. These tools can be used as an alternative to the two-handed resizing technique that is described in Sect. 2.

Hierarchy tools are used to create hierarchical CP-nets. Tools exist for assigning an existing page as a subpage to a substitution transition, for turning a transition into a substitution transition and automatically creating a new page with interface places, for assigning port types to places, and for creating fusion sets. These tools support both top-down and bottom-up approaches to modelling. Marking menus can be used to navigate between superpages and subpages. When navigating from one page to another, the destination page is either brought to the front of a binder, if the page is already in a sheet, otherwise the page is opened in a sheet and is added to the current binder. Figure 3 shows an example of navigating from a superpage to a subpage.

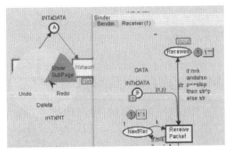
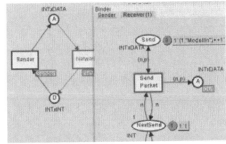

(a) Bring up marking menu on Sender subpage label. Receiver page is in front.

(b) Selecting *Show SubPage* brings the
Sender page to front.

Fig. 3. Navigating through marking menus.

3.2   Syntax Check and Code Generation

A common trait for many simulation tools is that the syntax of a model must be checked
and additional code must be generated before a simulation of the model can be executed.
In Design/CPN, users invoke syntax checks explicitly, either through a command in a
menu or through a switch to the simulation mode. In response to requests from users,
this explicit syntax check has been eliminated, and CPN Tools instead features a syntax
check that automatically runs in the background. Moreover, when changes are made in a
net, the syntax check will check only the parts of the net that are affected by the change.
For example, when a declaration is changed, the syntax checker does *not* recheck all
declarations, rather it will recheck only the declarations and the net inscriptions that
depend on the declaration that has been modified. This allows the user to do small cycles
of editing and simulation without having to wait for the syntax check to "catch up".
Immediately after a net has been loaded and while a net is being edited, CPN Tools
automatically checks to see if the net is syntactically correct, e.g. if all inscriptions are
of the right type and all ports and sockets are connected properly. The main drawback
to continually running the syntax check and code generation in the background is that
interaction with the GUI can occasionally be slowed down, particularly when large
portions of a CP-net are being checked.

   Syntax check feedback is updated while the syntax check runs, and the user can
follow the progress in the index as well as on the individual sheets. Coloured halos and
underlines indicate whether or not a net element has been checked and if it is syntactically correct. The colour-coded feedback not only lets the user know that something is
happening during the syntax check, but it also indicates the status and outcome of the
syntax check.

   When a net has just been loaded, all page entries in the index and all CP-net elements
on sheets are marked with orange to indicate that they have not yet been checked. Yellow
indicates that an object is currently being checked. Elements that are not marked with
halos or underlines have been successfully checked. When the user has finished a part
of a net to a certain degree, i.e. colour sets have been added to places, arcs have been

drawn between places and transitions, inscriptions have been added, etc., these objects are immediately syntax checked, and the halos disappear if the syntax check was successful. As the syntax check progresses, simulation information (enabled transitions, tokens, etc.) appears. Section 4.1 contains more details about simulation feedback.

Error feedback is provided for each object that has syntax errors. Objects with syntax errors are marked with red, and a speech bubble containing an error message appears with a description of the error, as shown in figure 4. Most of these error messages come directly from the simulation engine, which is implemented in Standard ML [8]. If the error is on an arc, the transition connected to it is also marked with red, since a transition is incorrect when at least one of its arcs is incorrect. The sheet tab and page entry in the index are also marked with red, making it easier for users to find all errors in a net. When the error is corrected, all red feedback and error messages disappear.

Code generation is connected to the syntax check. When portions of a net are found to be syntactically correct, the necessary simulation code is automatically generated incrementally. This saves time and eliminates the need for having two distinct modes for editing and simulating CP-nets. As a result, it is possible to simulate part of a CP-net even though other parts of the CP-net may have syntax errors or may be incomplete.

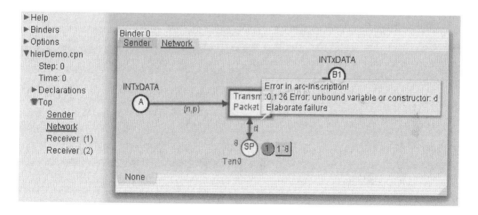

Fig. 4. Error feedback includes coloured halos and speech bubbles with error messages.

3.3 Additional Tools

Net tools are used to open, save and print CP-nets. In contrast to Design/CPN, multiple CP-nets can be opened, edited and simulated at the same time. Individual pages from CP-nets can be saved as Encapsulated Postscript (EPS) files using the *Print* tool. Figure 5 shows an example of a page that was saved as an EPS file. Pages can be saved in either black and white or colour, and either with or without current marking information.

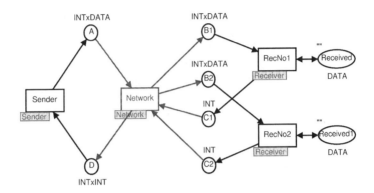

Fig. 5. CP-net that had been saved in EPS format.

History provides an overview of many of the operations that are executed during the construction of a CP-net. Typical operations that are shown in the History include: open and close operations for nets, sheets, and binders; create and delete operations for pages, net elements, and inscriptions; and style-change operations. Furthermore, *undo* and *redo* tools can be applied to all of the operations that are shown in the History. Some operation, such as movement of objects, are currently not saved within the History.

Help provides access to a number of web pages that are related to CPN Tools. This includes links to: the web-based user manual, the homepage for CPN Tools, and a web page for reporting bugs.

4   Analysing CP-Nets

CPN Tools currently supports two types of analysis for CP-nets: simulation and state space analysis. This section presents the *Simulation tools* and the *Statespace tools* that can be found under the *Tool box* entry in the index of CPN Tools.

4.1   Simulation

Simulations are controlled using the *Simulation tools*. As in many other simulation software packages, the icons for the simulation tools resemble buttons from a VCR (see figure 6). The *rewind* tool returns a CP-net to its initial marking. The *single-step* tool causes one enabled transition to occur. Applying this tool to different areas in the workspace has different results: on an enabled transition it causes that particular transition to occur, while on a page it will cause a random, enabled transition on that particular page to occur. The *play* tool will execute a user-defined number of steps, and the simulation graphics will be updated after each step. The *fast-forward* tool will also execute a user-defined number of steps, but the simulation graphics will not be updated until after the last step has been executed.

Simulation feedback is updated during the syntax check and during simulations. Figure 6 shows typical simulation feedback. Green circles indicate how many tokens are currently on each place, and current markings appear in green text boxes next to the places. Green halos are used to indicate enabled transitions. Pages containing enabled transitions are underlined with green in the index, and their page tabs are also underlined with green.

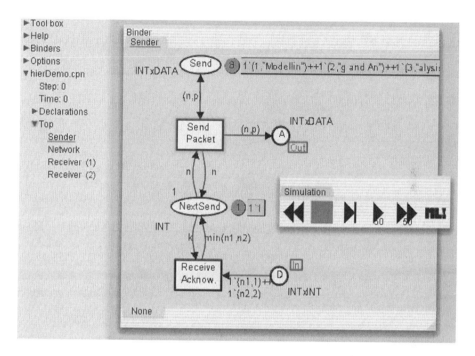

Fig. 6. Simulation tools have VCR-like icons. Simulation feedback includes current marking information and halos around enabled transitions. The Send Packet transition is enabled here.

As a simulation progresses, the simulation feedback changes with each step (if the single-step tool or the play tool are used), and the user can follow the simulation both in the index (through the green underlines) and on the individual pages. The green underlines in the tabs make it easy to see which pages currently have enabled transitions, without stealing the focus.

4.2   State Space Analysis

CPN Tools also contains facilities for generating and analysing full and partial state spaces for CP-nets. To facilitate the implementation of the state space facilities, we have added a few syntactical constraints which are important for state space generation and analysis but which are unimportant for simulation. For example, a state space cannot be generated unless all places and transitions in a page have unique names, and all arcs have inscriptions. The syntax checker will locate violations of these constraints, and graphical feedback will assist a user in locating potential problems. CP-nets that do not meet all of the constraints can still be simulated without problems.

<table>
</table>

```
                              Statistics
                              ---------------------
▼Statespace                   Occurrence Graph
    EnterStateSpace              Nodes:  54
  ►CalcSS                        Arcs:   1183
    CalcSCC                      Secs:   0
  ▼SStoSim                       Status: Partial
      state : 5                Scc Graph
    SimtoSS                      Nodes:  43
  ►SaveReport                    Arcs:   1128
                                 Secs:   0
```

(a) State space tools.   (b) Statistics from state space report.

Fig. 7. State space tools from the index and a state space report.

State space tools from the index are shown in figure 7(a). The *EnterStateSpace* tool is used first to generate net-specific code necessary for generating a state space, i.e. the state space code is not generated incrementally as the simulation code is. The *CalcSS* tool is the tool that generates the state space, while the *CalcSCC* tool calculates the strongly connected component graph of the state space. The user can set a number of options which will determine how much of a state space should be generated. For example, it is possible to stop generating a state space after a certain number of states have been generated or after a certain amount of time has passed. Options are changed by editing text in the index.

Two tools exist for switching between the simulator and a state space. The *SStoSim* tool will take a user-specified state (all states in the state space are numbered) from the state space and "move" it to the simulator. This makes it possible to inspect the marking of the CP-net and to see the enabled transitions. It is also possible to simulate the model starting at the state that was moved from the state space. Similarly, the *SimtoSS* tool will "move" the current state of the CP-net in the simulator to the state space. Once a (partial) state space has been generated, it is possible to seamlessly and instantaneously switch between the state space and the simulator. In figure 7(a), the SStoSim tool is configured to move state 5 to the simulator. A user can easily edit the text of the state number in order to select another state.

Standard state space reports can be generated automatically and saved using the *SaveReport* tool. Such reports contain information about one or more of the following: statistics about the generation of the state space, boundedness properties, home properties, liveness properties and fairness properties. Figure 7(b) shows a state space report containing only statistics regarding the generation of the state space.

Querying facilities are also available. The state space facilities of CPN Tools are very similar to the facilities in the Design/CPN Occurrence Graph Tool [9] (OG Tool). This means that the standard queries that are described in the user manual for the OG

Tool are also available in CPN Tools. However, currently there is no advanced interface that can be used to access these facilities. The method for accessing these queries is shown in figure 8. A query can be written using the *Auxiliary text* tool from the *Auxiliary tools*. The query is then evaluated by applying the *ML Evaluate* tool from the *Simulation tools* to the auxiliary text. The result of evaluating the query will be shown in a speech bubble. Currently, there is no support for drawing state spaces or parts of state spaces.

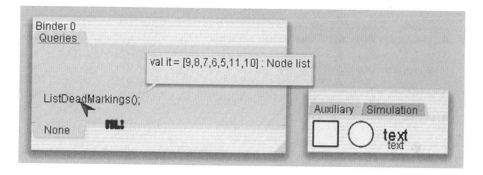

Fig. 8. Evaluating state space queries.

5   Conclusion and Future Work

CPN Tools combine advanced interaction techniques into a consistent interface for editing, simulating, and analysing Coloured Petri Nets. These interaction techniques have proven to be very efficient when working with Coloured Petri Nets. CPN Tools requires an OpenGL graphics accelerator and a PC running Windows 2000 or Windows XP. Furthermore, it is recommended that the CPU is at least a Pentium II, 400 MHz (or equivalent) and that there is at least 256 MB RAM. Future versions are expected to run on all major platforms including, Windows, Unix/Linux and MacOS. Additional information about CPN Tools, can be found at http://www.daimi.au.dk/CPNTools/.

CPN Tools does not currently provide all of the functionality that is available for Design/CPN. Future work will, however, extend the functionality of CPN Tools in several different ways. Facilities for collecting data, running multiple simulations, and calculating statistics are currently being integrated into CPN Tools, and these facilities are expected to be available by the end of 2003. Additional animation facilities, such as message sequence charts and domain-specific animation, are also being developed. However, these facilities will probably not be available before 2004. Additional plans include the design and implementation of a totally new generation of state space facilities.

Acknowledgements. We would like to thank Michel Beudouin-Lafon and Wendy E. Mackay who played a central role in designing the new interface for CPN Tools. We would also like to thank current and former members of the CPN Group at the University of Aarhus for their participation in the design and implementation of the tool.

References

1. Jensen, K.: Coloured Petri Nets: Basic Concepts, Analysis Methods and Practical Use, Volumes 1-3. Monographs in Theoretical Computer Science. Springer-Verlag (1992-1997)
2. Kristensen, L.M., Christensen, S., Jensen, K.: The practitioner's guide to coloured Petri nets. International Journal on Software Tools for Technology Transfer **2** (1998) 98–132
3. Design/CPN. Online: http://www.daimi.au.dk/designCPN/.
4. Petri Nets Tool Database. Online: http://www.daimi.au.dk/PetriNets/tools/db.html.
5. CPN2000 Project. Online: http://www.daimi.au.dk/CPnets/CPN2000/.
6. Kurtenbach, G., Buxton, W.: User learning and performance with marking menus. In: Proceedings of Human Factors in Computing Systems, ACM (1994) 258–264 CHI'94.
7. Bier, E., Stone, M., Pier, K., Buxton, W., Rose, T.D.: Toolglass and magic lenses: the see-through interface. In: Proceedings of ACM SIGGRAPH, ACM Press (1993) 73–80
8. Standard ML of New Jersey. Online: http://cm.bell-labs.com/cm/cs/what/smlnj/.
9. Jensen, K., Christensen, S., Kristensen, L.M.: Design/CPN Occurrence Graph Manual. Department of Computer Science, University of Aarhus, Denmark. (1996) Online: http://www.daimi.au.dk/designCPN/man/.

The Model-Checking Kit*

Claus Schröter, Stefan Schwoon, and Javier Esparza

Institutsverbund Informatik, Universität Stuttgart
{schroecs,schwoosn,esparza}@informatik.uni-stuttgart.de

Abstract. The Model-Checking Kit [8] is a collection of programs which allow to model finite state systems using a variety of modelling languages, and verify them using a variety of checkers, including deadlock-checkers, reachability-checkers, and model-checkers for the temporal logics CTL and LTL [7].

1 Introduction

Research on automatic verification has shown that no single model-checking technique has the edge over all others in any application area. Moreover, it is very difficult to determine a priori which technique is the most suitable for a given model. It is thus sensible to apply different techniques to the same model. However, this is a very tedious and time-consuming task, since each algorithm uses its own description language. The Model-Checking Kit [8] has been designed to provide a solution to this problem in an academic setting, with potential applications to industrial settings.

Out of the many different models for concurrent systems, we chose 1-safe Place/Transition nets as the basic model for the Kit for the following reasons: (i) They are a very simple model with nearly no variants, in contrast to most other models. For instance, communicating automata can be synchronous or asynchronous, and communication can be formalised in different ways. Process algebras have a wealth of different operators and semantics, and there exist many different high-level net models. (ii) Many different verification techniques which can deal with 1-safe P/T nets are available. Since P/T nets have a well-defined partial order semantics, partial order techniques like stubborn sets [21] and net unfoldings [18] can be applied (as a matter of fact, these techniques were originally introduced for Petri nets). Since a marking of a 1-safe P/T net is just a vector of booleans, symbolic techniques based on BDDs [5], like those implemented in SMV [17], can also be used. And, of course, the standard interleaving semantics of Petri nets allows to apply explicit state (finite-state) exploration algorithms, like those of SPIN [13].

For systems modelled in a language with a 1-safe net semantics, all the techniques listed above are in principle applicable. Since each of these techniques has both strengths and weaknesses, it would be highly desirable to apply them all

---

* http://www7.in.tum.de/gruppen/theorie/KIT/

W.M.P. van der Aalst and E. Best (Eds.): ICATPN 2003, LNCS 2679, pp. 463–472, 2003.

and to compare the results. However, employing two or more verification packages is not easy. In particular, the packages have different input formats, forcing the user to enter input data multiple times, a tedious and error-prone task.

To amend this situation the Kit provides a shell which allows the user to specify input data (i.e. systems and properties) in a variety of input languages. Once a system and a property have been specified, the user can choose any of the model checkers available in the shell to verify the property. The user is not required to be familiar with the different ways in which 1-safe P/T nets are represented to the different checkers. The Kit has been implemented as a command-line oriented tool without any GUI. While such a design may lessen the appeal of the Kit (especially to novices), it also has advantages: The Kit is easy to set up, can be run on many different platforms, and the incremental effort for adding further input languages and model checkers is small.

The paper is structured as follows. In Section 2 we show how the Kit works with a small example. In Section 3 we present the modelling languages and verification techniques supported by the Kit. Section 4 gives a brief overview of the available options in the Kit. In Section 5 we present some experimental results showing how the verification techniques differ in performance, before concluding in Section 6.

2   An Example

Figure 1 (a) shows Peterson's mutual exclusion algorithm [19] modelled in the language B(PN)^2, which is a parallel programming language developed as part of the PEP tool [10] and one of the Kit's input languages. The notation is mostly self-explanatory, but for the sake of clarity we would like to point out two things: (i) ⟨t'=1⟩ means value assignment, whereas ⟨t=1⟩ means test of equality; (ii) incs1: denotes a label which can be used in formulae to mark a program point between two actions.

Suppose we want to check a mutual exclusion property, i.e. whether the system can reach a global system state in which both processes enter their critical sections simultaneously. The Kit allows this property to be expressed as "incs1" & "incs2". The Kit translates the B(PN)^2 model and the formula into a 1-safe P/T net and a corresponding formula (e.g. P9 & P14, where P9, P14 are places in the net). Then the Kit invokes the model checker chosen by the user; in this case, the user would typically choose one of the reachability checkers. The model checker examines whether there exists a reachable marking with tokens on both P9 and P14 simultaneously and returns the result to the Kit. In Figure 1 (a) the answer is 'no', which means that the mutex property holds. What, then, happens in case of an error? Let us suppose that the user made a mistake and wrote ⟨i1'=0 or t=1⟩ in Process 2 instead of ⟨i1=0 or t=1⟩. This causes the mutex property to become violated, and the model checker returns a transition sequence to a state with tokens in both of the places representing the critical regions. As shown in Figure 1 (b), the Kit interprets this transition sequence at the level of B(PN)^2 and outputs the result to the user.

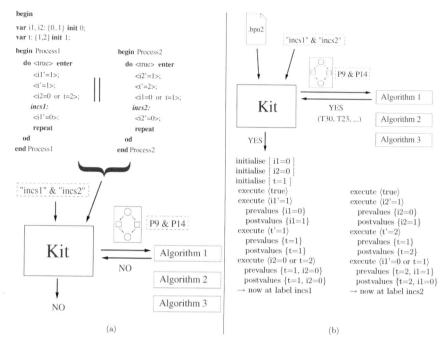

Fig. 1. Peterson's mutex algorithm in $B(PN)^2$

3   Modelling Languages and Verification Techniques

In this section we briefly introduce the different modelling languages and verification techniques supported by the Kit, and how properties are described.

3.1   Modelling a System

The languages offered by the Kit for modelling a system can be divided into so-called net languages and high-level languages that abstract from net details.

- **High-Level Languages**
  Loosely speaking, these description languages abstract from structural net concepts like places, transitions, and arcs. The Kit currently offers three such languages: **B(PN)²**, **CFA**, and **IF**. B(PN)² [3] (Basic Petri Net Programming Notation) is a structured parallel programming language with features such as loops, blocks, and procedures used in the PEP tool [10]. CFA [9, 8] (Communicating Finite Automata) is a language for describing finite automata that communicate via shared variables or channels of finite length. It offers very flexible communication mechanisms and is also one of the modelling languages of the PEP tool [10]. Finally, IF [4] (Interchange Format) is a language proposed in order to model asynchronous communicating real-time systems. It is the common model description language of the European ADVANCE [1] project. All three languages have a 1-safe Petri net semantics.

- **Net languages**

  The Kit supports two net languages, **PEP** and **SENIL**. In these languages one has to define places, transitions, and arcs explicitly. PEP [2] is the low level net language of the PEP tool [10]. It is supported by the Kit mostly because some tools can automatically export models into this format. SENIL [8] (Simple Extensible Net Input Language) is designed to make it easy to specify small P/T nets by hand; it is suitable for small nets with at most a few dozens of nodes, but not for larger projects.

  Users can use all available model checkers provided their nets are 1-safe. Whether a certain model checker can be used on non-1-safe nets depends on the tool in question (e.g., the translation into SMV fails in this case).

An interesting feature of the Kit is that, independently of the description language chosen by the user, all checkers can be applied to the same model. The counterexamples produced by the checker are presented to the user in terms of the description language used to model the system.

3.2 Describing Properties

The Kit can be used to check several types of properties, e.g. deadlock freeness, reachability, safety and liveness properties. Except for deadlock checkers (which always check the property of deadlock freeness) the model checkers expect a formula describing the property of interest. Formulae consist of atomic propositions and various logical or temporal connectives. The precise syntax and semantics of atomic propositions depends on the modelling language, e.g. net languages allow properties expressing that certain places are marked, whereas high-level languages allow to speak of properties of variables, channels and so forth. The syntax for logical connectives has been unified, so that, e.g., the same CTL formula can be used on all available CTL checkers.

Reachability properties. In our framework a reachability property is the question whether the system can reach a certain set of global states (the mutual exclusion property is an instance of a reachability property). Such a property can be expressed with a so-called state formula, a propositional logic formula consisting of atomic propositions and logical operators.

Temporal logics. The Kit supports the two temporal logics LTL and CTL. These two are arguably the most popular temporal logics, and together they can express all common safety and liveness properties. Here we give just a brief introduction to LTL and CTL according to [7].

- **LTL** (Linear-Time Temporal Logic) assumes that the underlying structure of time is a totally ordered set such that the time is discrete, has an initial moment with no predecessors and is infinite into the future. LTL formulae consist of atomic propositions, and boolean and temporal connectives. Temporal operators are $\mathbf{G}p$ ("always p"), $\mathbf{F}p$ ("eventually p"), $\mathbf{X}p$ ("nexttime p") and $p\,\mathbf{U}\,q$ ("p until q"). The Kit supports only the *next-free* fragment of LTL.

- **CTL** (Computation Tree Logic) is a logic whose underlying structure of time is assumed to have a branching tree-like nature. It corresponds to an infinite tree where each node may have finitely many successors and must have at least one successor. These trees have a natural correspondence with the computations of concurrent systems or nondeterministic programs. A CTL formula consists of a path quantifier [**A** (all paths), **E** (there exists a path)] followed by an arbitrary linear-time formula, allowing boolean combinations and nestings of linear-time operators (**G**, **F**, **X**, **U**).

3.3   Verification Techniques

As mentioned in the introduction many different verification techniques for 1-safe Petri nets are available. These techniques include among others the explicit construction of the state space, stubborn sets [21], BDDs [5], and net unfoldings [18]. The explicit construction of the state space is the classical approach, and still adequate in cases where the state space explosion is not very acute. Stubborn sets are used to avoid constructing part of the state space. They exploit information about the concurrency of actions. Using symbolic techniques (e.g. BDDs) one can succinctly represent large sets of states. They can reach spectacular compactification ratios for regularly structured state spaces. Approaches based on unfolding techniques make use of an explicitly constructed partial-order semantics of the system. It contains information not only on the reachability relation, but also on causality and concurrency. This technique is adequate for systems exhibiting a high degree of concurrency.

When planning the Kit we intended to integrate various checkers such that each of the verification approaches mentioned above is represented by at least one checker. This has led to the following selection:

- The **PEP** tool [10] (Programming Environment based on Petri nets) is a programming and verification environment for parallel programs written in $B(PN)^2$ or CFA. Programs can be formally analysed using methods which are based on the unfolding technique [18]. The PEP tool is distributed by the Theory group (subgroup Parallel Systems) of the University of Oldenburg. PEP contributes to the Kit a deadlock checker, a reachability checker, and a model checker for LTL.
- **PROD** [22] implements different advanced techniques for reachability analysis to mitigate the state explosion problem, including partial-order techniques like stubborn sets [21], and techniques that exploit symmetries. PROD is distributed by the Formal Methods Group of the Laboratory for Theoretical Computer Science at the Helsinki University of Technology. PROD contributes to the Kit a deadlock checker, a reachability checker, and model checkers for CTL and LTL.
- The **SMV** system [17] is a tool for checking finite state systems against specifications in the temporal logic CTL. The input language of SMV is designed to allow the description of finite-state systems that range from completely synchronous to completely asynchronous, and from the detailed

to the abstract. SMV is distributed by the Carnegie Mellon University. Its verification algorithms are based on BDDs [5], and it contributes to the Kit a deadlock checker and a CTL checker.

- **SPIN** [13] is a widely distributed software package that supports the formal verification of distributed systems. It can be used as a full LTL model-checking system, but it can also be used as an efficient on-the-fly verifier for more basic safety and liveness properties. Many of the latter properties can be expressed and verified without the use of LTL. SPIN uses explicit construction of the state space. It is distributed by the Formal Methods and Verification Group of Bell Labs. SPIN contributes to the Kit an LTL checker.
- The tool **MCSMODELS** [12] is a model checker for finite complete prefixes, i.e. net unfoldings [18]. It currently uses PEP [10] to generate the prefixes. These prefixes are then translated into logic programs with stable-model semantics, and the integrated Smodels solver is used to solve the generated problems. MCSMODELS is distributed by the Formal Methods and Logic Groups of the Laboratory for Theoretical Computer Science at the Helsinki University of Technology. MCSMODELS contributes to the Kit a deadlock checker and a reachability checker.
- **CLP** [14] is a linear-programming model checker. It uses net unfoldings [18] and can check among others absence of deadlocks, reachability, and coverability of a marking. CLP is distributed by the Parallelism Research Group of the University of Newcastle upon Tyne and contributes a deadlock checker.

4   How to Use the Kit

In this section we show how to use the Kit for verification tasks. The Kit is a command-line oriented tool (called `check`) without any graphical user interface. The available options are listed by calling `check` without any arguments. Figure 2 shows an overview of its current version. The Kit is called as follows:

```
check [options] <input>:<checker> <modelfile> <formulafile>
```

- `<input>` is a place holder for one of the available modelling languages, i.e. `cfa, bpn2, if, pep, senil`.
  Note: `<input>:` may be omitted; in this case the Kit guesses the input language by looking at the extension of `<modelfile>` (which should be `.cfa`, `.bpn2`, `.if`, `.ll_net`, or `.senil`, in the order of the languages mentioned above). The user is free to use arbitrary extensions, but then the language has to be specified explicitly.
- `<checker>` should be replaced by one of the available algorithms, see the list at the bottom of Figure 2.
- `<modelfile>` is the name of the file containing the system specification.
- `<formulafile>` is the name of the file containing the formula to be checked.
  Note: For deadlock-checking no formula file is needed.
- `[options]` are as follows (`-t` and `-v` should be self-explanatory):

```
Usage: check [options] <input>:<checker> <modelfile> <formulafile>
   or: check -r <name> [options] <checker> <formulafile>


   Options:
    -s <name>      save intermediate results under <name>
    -r <name>      resume from intermediate results saved under <name>
    -t <dir>       place temporary files in <dir> (default is '.')
    -o <opt>       specify options to pass to checker
    -v             run in verbose mode


   Available input formats:
   cfa            concurrent finite automata
   bpn2           B(PN) 2 language
   if             IF language
   pep            PEP low level net format
   senil          SENIL net format


   Available algorithms:
   CTL          : prod-ctl, smv-ctl
   LTL          : prod-ltl, pep-ltl, spin-ltl
   Deadlock     : prod-dl, smv-dl, pep-dl, mcs-dl, clp-dl
   Reachability: prod-reach, pep-reach, mcs-reach
```

Fig. 2. The Kit's available options

- -s <name>
  Temporary files representing intermediate results will be saved in a tar file by the name mckit_save_<name>.tar. Some algorithms profit from the reuse of intermediate results. For example, if one uses a method based on the unfolding technique for verifying many properties on the same system, it is sensible to compute the unfolding only once per system and not once for every property. This option allows the unfolding to be saved for reuse (see option -r).

- -r <name>
  With this option one can reuse intermediate results saved earlier with -s. This is sensible if one wants to check many properties on the same system. Then the translation from the modelling language into the correct input format for the checker should be done only once and not for every single property. When using this option one should omit the modelling language and the modelfile, i.e. the calling syntax becomes:

      check -r <name> [options] <checker> <formulafile>

  The selected checker can then take advantage of the files saved in the file mckit_save_<name>.tar.

- -o <opt>
  Many model checkers allow to influence their behaviour with a variety of options. The Kit uses every tool with a general-purpose set of options

	prod-dl	smv-dl	pep-dl	mcs-dl	clp-dl
peterson	7.04 (0.09)	0.24	0.04	0.05	0.03
plate(5)	46.68 (1.38)	*mem*	4.80	0.53	0.54
client/server	61.06 (0.79)	111.80	0.76	0.54	0.55
key(4)	37.63 (0.20)	*mem*	*mem*	*mem*	*mem*
fifo(30)	36.74 (0.72)	*mem*	*mem*	*mem*	*mem*

Fig. 3. Results for Deadlock-Checking

recommended by the authors, which is usually fine for educational purposes. Expert users of the Kit, however, may want to use the full power of the tools. The Kit passes the value of <opt> in the call to the chosen model-checker, and its meaning depends on that tool.

5   Experimental Results

In this section we compare the performances of the algorithms by means of experimental results on several systems. The results demonstrate the point we made in the beginning, namely that no single method has the edge over all others. We present results for checking deadlock freeness and some safety properties.

All experiments were performed on a Linux PC with 64 MByte of RAM and a 230 MHz Intel Pentium II CPU. The times are measured in seconds. The systems we used were as follows:

– peterson: Mutual exclusion algorithm [19].
– plate(5): Production cell handling 5 plates [11,15].
– client/server: Client/server system with 2 clients and 1 server [1].
– key(4): Manages keyboard/screen interaction in a window manager for 4 customer tasks [6].
– fifo(30): 1-bit-FIFO with depth 30 [16,20].

The systems are modelled in different languages. Peterson's mutual exclusion algorithm is modelled in $B(PN)^2$, and the client/server system in IF. All other examples are modelled in PEP's low-level net format.

Figure 3 shows the results for deadlock-checking. We split PROD's verification times because PROD generates a C program from the net description. The actual verification task is carried out by executing the resulting program. Since PROD spends most of its time generating the executable file, the pure verification times are quoted in parentheses.

Peterson's algorithm is a small example, and all techniques behave well. The systems plate(5) and client/server, however, exhibit big differences in the performances. The unfolding-based techniques outperform PROD and SMV here. Actually, SMV runs out of memory during the verification of plate(5) (shown as 'mem'). In contrast, PROD beats the other tools on key(4) and fifo(30).

	prod-ltl	pep-ltl	spin-ltl
plate(5)	67.22 (2.52)	1.20	*mem*
client/server	257.57 (30.83)	2.51	20.74
key(4)	4206.00 (4152.66)	*mem*	11.53

Fig. 4. Results for LTL-Checking

Figure 4 shows results for safety properties. These properties were expressed as LTL formulae and checked using LTL checkers. For the production cell we checked a mutual-exclusion property: exactly one of three places of the net carries a token. For the client/server system we checked that a buffer overflow can not occur. For the key(4) system we checked a mutual-exclusion property for two places. A look at the results confirms that the checkers behave quite differently here as well. We were able to verify the property for the production cell with PROD and PEP, but not with SPIN. On the other hand, SPIN handles the client/server example much faster than PROD. The key(4) system is an example that can be verified quickly with SPIN, but not with PEP.

Notice that the tools, when used as part of the Kit, may not perform as well as when used alone. It is possible that using SPIN to model-check a PROMELA system takes virtually no time, whereas, if the same system is modelled as a CFA and then checked again with SPIN, the verification can run out of memory. The reason is that the Kit transforms the initial CFA model into a 1-safe Petri net, and this net into PROMELA. If the model is data-intensive, the net (and hence the PROMELA model) can easily blow up.

6   Conclusions

The Model-Checking Kit is a collection of programs that allow to model a finite-state system using a variety of modelling languages, and (attempt to) verify it using a variety of checkers. It has been successfully applied in a lab course on automatic verification. Special care has been taken to design it in a modular way and to make it easy to use and install. In our experience, a moderately skilled user can install the tool and verify the first property of a small system within half an hour. Moreover, the Kit is highly portable since all programs are written in C, and we do not offer any graphical user interface. The Kit can be used for comparing the performances of different verification methods. However, it must be emphasised that, since each of the Kit's checkers has been optimised for its own modelling language, the Kit's internal language conversions can lead to performance losses. Finally, the Kit is an open library. Due to its modular design it is easy to add new input languages or checkers, and to replace old versions of checkers by new ones. Anyone interested in adding new languages and/or tools is very welcome to contact the authors.

References

1. ADVANCE – Advanced Validation Techniques for Telecommunication Protocols. http://verif.liafa.jussieu.fr/~haberm/ADVANCE/main.html.
2. E. Best and B. Grahlmann. PEP Documentation and User Guide 1.8. Universität Oldenburg, 1998.
3. E. Best and R. P. Hopkins. B(PN)^2 – a Basic Petri Net Programming Notation. In *PARLE'93*, LNCS 694, pages 379–390. Springer, 1993.
4. M. Bozga, J.-C. Fernandez, L. Ghirvu, S. Graf, L. Mounier, J. P. Krimm, and J. Sifakis. The Intermediate Representation IF. Technical Report. Vérimag, 1998.
5. R. E. Bryant. Graph-Based Algorithms for Boolean Function Manipulation. *IEEE Transactions on Computers*, C-35(8):677–691, Aug. 1986.
6. J. C. Corbett. Evaluating Deadlock Detection Methods, 1994.
7. E. A. Emerson. Temporal and Modal Logic. In *Handbook of Theoretical Computer Science*, volume B, pages 997–1067. Elsevier Science Publishers B. V., 1990.
8. J. Esparza, C. Schröter, and S. Schwoon. The Model-Checking Kit. http://www7.in.tum.de/gruppen/theorie/KIT/.
9. B. Grahlmann, M. Möller, and U. Anhalt. A new Interface for the PEP-tool – Parallel Finite Automata. 2nd Workshop of Algorithms and Tools for Petri nets. Oldenburg, 1995.
10. B. Grahlmann, S. Römer, T. Thielke, B. Graves, M. Damm, R. Riemann, L. Jenner, S. Melzer, and A. Gronewold. PEP: Programming Environment Based on Petri Nets. Technical Report 14, Universität Hildesheim, May 1995.
11. M. Heiner and P. Deusen. Petri net based qualitative analysis – A case study. Technical report I-08/1995. Brandenburg Technische Universität Cottbus, 1995.
12. K. Heljanko. *Combining Symbolic and Partial Order Methods for Model Checking 1-Safe Petri Nets*. PhD thesis, Helsinki University of Technology, 2002.
13. G. J. Holzmann. *Design and Validation of Computer Protocols*. Prentice Hall, 1991.
14. V. Khomenko. CLP. http://www.cs.ncl.ac.uk/people/victor.khomenko/home.formal/tools/tools.html.
15. C. Lewerentz and T. Lindner. Case Study Production Cell. In *Formal Development of Reactive Systems*, LNCS 891. Springer, 1995.
16. A. J. Martin. Self-timed FIFO: An exercise in compiling programs into VLSI circuits. In *From HDL Descriptions to Guruanteed Correct Circuit Designs*, pages 133–153. Elsevier Science Publishers, 1986.
17. K. L. McMillan. *Symbolic Model Checking – An approach to the state explosion problem*. PhD thesis, Carnegie Mellon University, 1992.
18. K. L. McMillan. Using Unfoldings to Avoid the State Explosion Problem in the Verification of Asynchronous Circuits. In *CAV'92*, LNCS 663, pages 164–174. Springer, 1992.
19. M. Raynal. Algorithms For Mutual Exclusion, 1986.
20. O. Roig, J. Cortadella, and E. Pastor. Verification of Asynchronous Circuits by BDD-based Model Checking of Petri Nets. In *ATPN'95*, LNCS 935, pages 374–391. Springer, 1995.
21. A. Valmari. On-the-Fly Verification with Stubborn Sets. In *CAV'93*, LNCS 697, pages 397–408. Springer, 1993.
22. K. Varpaaniemi, J. Halme, K. Hiekkanen, and T. Pyssysalo. PROD Reference Manual. *Technical Reports*, B(13):1–56, Aug. 1995.

Prototyping Object Oriented Specifications

Ali Al-Shabibi, Didier Buchs, Mathieu Buffo, Stanislav Chachkov,
Ang Chen, and David Hurzeler

Software Engineering Laboratory, Swiss Federal Institute of Technology Lausanne,
1015 Lausanne, SWITZERLAND
{didier.buchs,stanislav.chachkov,david.hurzeler}@epfl.ch

Abstract. CoopnBuilder is an integrated development environment (IDE) for Concurrent Object Oriented Petri Nets (COOPN). It comes with a complete set of tools enabling the user to view, edit, check, simulate and generate code from CO-OPN specifications. The Code Generation tool allows the user to develop applications in an open way: the produced code can be integrated in larger projects or use existing libraries. The code generation will be emphasized in this paper, and we will focus on ease-of-use and extensibility. CoopnBuilder is an open-source Java program and can be downloaded from http://cui.unige.ch/smv

1   Introduction

CO-OPN is an object-oriented modelling language, based on Algebraic Data Types (ADT), Petri nets, and IWIM (Idealized Worker, Idealized Manager) coordination models [1]. CO-OPN is designed for modelling systems where data, concurrency and problems related to effective distribution of system is of uttermost importance. In this kind of systems, it is necessary to provide techniques and tools to help developers deal with the high expressivity of the CO-OPN language. These tools should mainly guide users in finding the right model and reduce the time necessary for reaching an efficient implementation of a given model. We will make a survey of the basic tools of the Coopn-Builder environment and focus on the code generation techniques and their potential use in the development of real systems.

2   CO-OPN Specification Language

CO-OPN specifications are collections of algebraic abstract data types, and class and context (i.e. coordination) *modules*. Syntactically, each module has the same overall structure; it includes an *interface section* defining all accessible elements from the outside of the module, and a *body section* including the local aspects, private to the module. Moreover, class and context modules have convenient graphical representations, showing their underlying Petri net model. A more detailed description of CO-OPN can be found in [1].

From a semantic point of view, CO-OPN has a true concurrent semantics for objects, based on the classical model-based semantics of algebraic nets[2]. The dynamicity and mobility of objects (as for instance the dynamically evolving CO-OPN context structures) is captured by localization semantic rules [6].

W.M.P. van der Aalst and E. Best (Eds.): ICATPN 2003, LNCS 2679, pp. 473–482, 2003.

2.1   ADT Modules

CO-OPN ADT modules define data types by means of algebraic specifications. Each module introduces one or more sorts (i.e. names of data types), along with generators and operations on those sorts. The properties of the operations are given in the body of the module, by means of positive conditional equational axioms. Operations are partial deterministic functions.

An example of ADT is given by the CO-OPN standard library module Booleans (Figure 1).

```
ADT Booleans;
Interface
    Sort        booleans;
    Generator   true,false : -> booleans;
    Operation
    _ and _ :   booleans, booleans -> booleans;
Body
    Axioms
        X and true = X;
        X and false = false;
        ...
    Where
        X: booleans;
```

Fig. 1. ADT Booleans (fragment)

Any kind of data structure could be modelled in this language with a high level of abstraction and a clear declarative style. Even if we have a declarative approach, most of the usual data structures can be animated using rewriting techniques (essentially, re-write systems are obtained by orienting the equations). This is the main principle used for the code generation supported by our tools.

2.2   Classes

CO-OPN classes are described by means of modular algebraic Petri Nets with particular parameterized external transitions that are called methods or gates. Methods represent services provided by the class and gates - services required. As usual, instances of class-es are called objects. Objects are instantiated by using either user-defined creation methods or a predefined method *Create*.

The behavior of transitions is defined by so-called behavioral axioms, similar to the axioms in an ADT. A method call results in an external transitions synchronization, similar to the Petri Net transition fusion technique. The axioms have the following syntax (see Figure 2 for an example of corresponding graphical representation of axioms):

Cond => *eventname* **With** *synchro* : *pre* -> *post*

In which the terms have the following meaning:

- *Cond* is a set of equational conditions - the guard;
- *eventname* is the name of a method with the algebraic term parameters.
- *synchro* is the synchronization expression defining the policy of transactional in-teraction of this event with other events. Synchronization expressions are the CO-OPN equivalent of method calls. The dot notation is used to access events of spe-

cific objects. Synchronization expressions can be combined using the synchronization operators: sequence, simultaneity and nondeterminism.

- *Pre* and *Post* are the usual Petri Net flow relations determining what is consumed from and what is produced in the object state places.

The state of a CO-OPN object is stored in places, which are really named multisets of algebraic terms, object references or tuples of them.

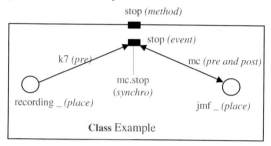

Fig. 2. Graphical Representation of an Example Class

The interface of a class is composed of methods and gates. Where methods are the services offered by the class, the gates can be seen as services requested by the class or outputs.

2.3   Contexts

Contexts are coordination entities, i.e. they coordinate activities among classes. Contexts, like classes, have an interface composed of methods and gates. They contain objects and/or sub-contexts. Coordination is done using behavorial axioms that connect methods and gates of embedded entities. A graphical representation of an example context is given in Figure 6.

Another important feature of CO-OPN contexts is locality. An object is always inside one (and only one) context. The CO-OPN language can be used to specify distributed systems and includes active object migration. This can be particularly useful for expressing complex localization problems. For instance, it seems natural to differentiate the behaviour of a phonecard (object of class PhoneCard in Figure 3) if this phonecard is in a wallet or if it is inserted into a telephone device. This can be expressed using a context representing the wallet and another context representing the telephone. Moving the card from the wallet to the telephone is modelled by a synchronization moving the object from the Wallet context to the Cabin context (Figure 3).

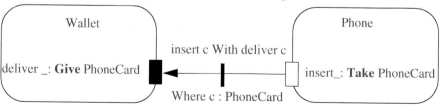

Fig. 3. PhoneCard specification with migration information

The migration of objects is done by transmitting the reference and the object itself between the contexts participating in the synchronization. The direction of the migration is specified by special keywords (**Give** and **Take**) decorating the parameters. It must be noted that if we omit these keywords, only references are copied. The CO-OPN semantics does not allow to access remote objects, even if their references are known. Any attempt to synchronize with a non-local object will result in failure.

3   CoopnBuilder Framework

CoopnBuilder is the successor of the CoopnTools environment[3]. The CoopnBuilder interface is inspired by the interfaces of most modern IDE. It is composed of three main parts: a package tree, an editor view and a message panel (Figure 4).

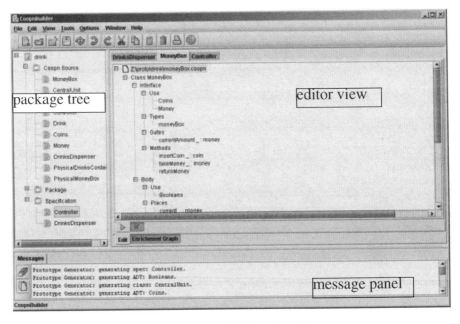

Fig. 4. CoopnBuilder main window

In CO-OPN, *modules* (source code) are grouped in *packages*. The package view enables the user to modify the package structure and launch tools associated to various kinds of modules. It is similar to "project views" in other IDE. The editor view is used to edit CO-OPN modules.

CoopnBuilder is an open framework. It is composed of a kernel part and a set of tools [Figure 5]. The tools are loaded at start-up, so it is easy to extend CoopnBuilder with new tools.

4   Tools

CoopnBuilder supports a very general software development model based on the classical waterfall model. It also supports various iterative development aspects at the specification level using constructs of the language (genericity and subtyping) and tuning of the generated code by means of generation strategy selection management (step 3 and 4 on Figure 5).

4.1   General Purpose Tools

As seen above, CO-OPN specifications can be edited (step 1 on Figure 5) using textual or graphical *editors*. A text editor also provides small tools that simplify the edition of a specification. The *checker* tool (2 on Figure 5) verifies that a CO-OPN specification is syntactically correct, and well-typed.

The textual and graphical representations of a specification can be printed or exported to various document formats. Graphical editors also support the integration of external layouters (dot [14]).

The modular nature of the CO-OPN specifications allows us to manage specifica-

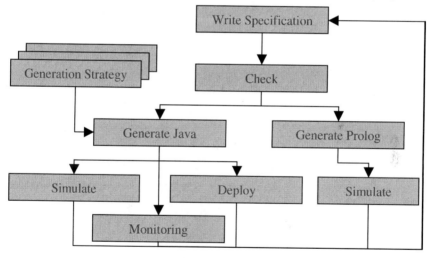

Fig. 5. Life-cycle of a specification with CoopnBuilder

tions as a set of interconnected components that are separately specified and composed by means of CO-OPN contexts. Special editors have been built to make this operation easier where it is possible to just interconnect components (Figure 6). The semantics of the interconnection follows the general idea of atomic synchronization (some kind of transaction) that we introduced in the CO-OPN model. Consequently, the workload of the specification can be considerably reduced if a well designed set of components already exists at the specification level; for some of the components, a dedicated implementation can make a link to already existing libraries. For instance, we have successfully integrated some of the Java media Framework librairies in some generated code.

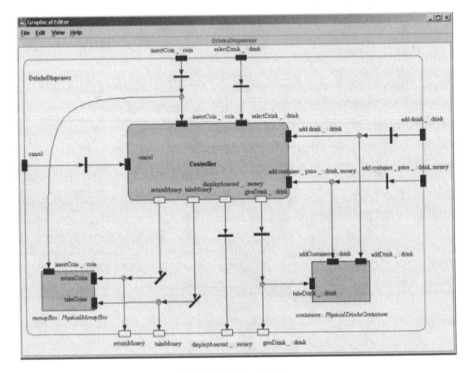

Fig. 6. Graphical Editor

4.2   The Abstract Data Type Evaluator and Simulator

After a CO-OPN specification has been "checked", we might want to execute, or simulate it. One of the approaches to do this is to transform it into a Prolog program[7]. The *evaluator* and *simulator* tools (5,9 on Figure 5) implement this technique, and are based on the resolution principle. While the *evaluator* tool enables the user to evaluate abstract data types terms, the *simulator* tool offers a similar service for classes and contexts. These tools accept term-based queries with logical variables.

4.3   The Java Code Generator and Interpreter

Another approach to execute CO-OPN specifications, is their translation into Java code [4] (step 4 on Figure 1). Basically, this corresponds to building rewrite systems for algebraic abstract data types, and implementing concurrency. Also, the particular semantics of the synchronizations between events in CO-OPN requires the implementation of transactional mechanisms [4]. The generated code can be used as a prototype implementation of a specification. CoopnBuilder has an "interpreter" tool, and this tool executes CO-OPN terms using the generated Java code (by using the java language's reflexive mechanisms). This tool also features a built-in debugger, which enables step-by-step execution of queries and/or exhibition of derivation trees at the end of a successful query (monitoring, step 7 on Figure 1). The evaluator, simulator and interpreter tools make the validation of CO-OPN specifications easier.

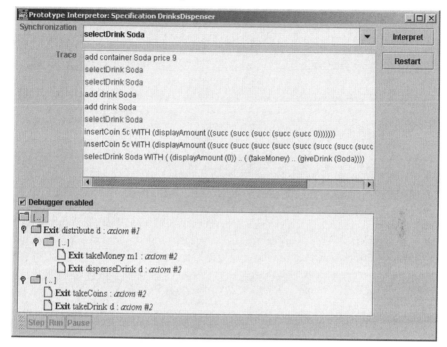

Fig. 7. Interpreter and Debugger

5   Code Generation

Simulation is only one of the many different ways the code generator can be used. In particular, it is designed to be embedded into an iterative development methodology where the implementation choices should evolve [8]. Moreover, the target systems may need drivers or user interfaces in order to manage human communication or hardware control. We will explain how the generated code may be integrated in larger systems, or use existing libraries.

5.1   Modularity and Configurability of the Code Generator

By default, the generated code uses rewriting techniques to evaluate terms. But this might be space/time-costly. Therefore, we have made the code generator flexible enough to allow users to improve the efficiency of evaluation by re-configuring the default code-generation algorithm.

For example, the evaluation of a specification often makes intensive use of numerical calculations. In this case, term representation of numbers with the zero and successor operations is not adapted to efficient evaluation. It is more interesting to represent numerical values occurring in specifications by numerical types of the target language - for example, the *int* type in Java. Our code generator can do that by allowing the choice of the appropriate specific code-generation strategy: the user may choose which representation of numbers he wants it to use.

Our generator allows the user to choose the strategy of code-generation for an individual module, a group of modules or an entire specification. Of course, the choice of code-generation strategies does not only address data representation alternatives. More generally, this technique allows various kinds of optimization and code instrumentations (for example, in order to allow debugging).

For ease-of-use purposes, a choice of pre-defined strategies for different kinds of modules (including standard library modules) is featured in the tool. Advanced users may write their own code-generation strategies. It will also be possible to compose a new code-generation strategy from pre-defined sub-strategies.

5.2   Integration of Generated Code

As stated above, another possible use for the generated code is integration into an application. Generated code can be integrated both as a server - you can call it -, or as a client - it will call your code. The specifics of the CO-OPN specification language imply that the generated code has to implement the non-determinism and transactional semantics of specifications. The user has the choice to either hide those aspects inside the generated code or use them for finer integration. For more details on how to handle non-determinism and transactional failures in non-reversible libraries, see [5].

5.3   Distributed Systems and Object Migration

5.3.1   Implementation of Migration

In order to implement the migration of objects, we have to carefully manage references, and differentiate references of local and non-local objects (see Figure 8). We use the classical Proxy mechanism to obtain an homogeneous access to objects. This indirect reference will present local and non-local objects to clients in a similar way. Usually, the Proxy just forwards synchronizations to the real object. In the case of an already moved object, the Proxy will always respond to inquiries by a failure.

The list of objects known by a context is managed in the KOT (Known Objects Table). In the case where an object returns back to a context, no new entries will be added to the KOT. Instead, the existing Proxy will be found and linked to the returned object. This general mechanism satisfies both centralized and distributed implementations.

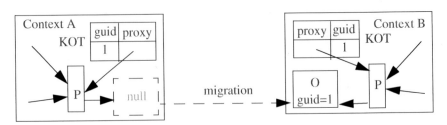

Fig. 8. Implementation of object migration. (P: proxy, O: object, KOT: known objects table of the context, GUID: globally unique id associated to object)

6   Related Work

We can compare the CO-OPN modelling language and its features in our example with various existing modelling languages. We should first note that CO-OPN is a formal modelling language, i.e. it has well defined formal semantics [1]. Algebraic Petri Nets are the basis of CO-OPN. The main differences between CO-OPN and mainstream Petri nets based languages (for example DesignCPN [9]) is object-orientation: in CO-OPN, Petri Nets are encapsulated into objects, which are instances of classes. Object-orientation enables the developer to obtain a clear decomposition of the modelled system into «modules». It also facilitates the transformation of a specification into a program in an object-oriented programming language.

We should also note that even if the structure of a CO-OPN specification is composed of classes and objects, we do not use any feature that could break the well defined semantics of CO-OPN. For example, we do not allow code fragments insertion in specifications, as some languages do (Cooperative Nets [10]).

Another important feature of CO-OPN useful for our example is object mobility. This has recently been added to our formalism, and the aim of this paper is to present an example where it is used. We are not aware of other object-oriented Petri net based languages in which mobility of objects or the notion of locality (expressed by Contexts in our formalism) is present. The notion of mobility can be modeled using formalisms such as Pi-calculus [11] or CSP [12]. Just as in the latters, communication between CO-OPN objects can be configured, using CO-OPN contexts.

Transactions are another important part of the CO-OPN semantics: interactions between CO-OPN objects obey transactional semantics. This is well illustrated in our example hereafter by the interactions between seller, buyer and bank account. Transactional semantics are not characteristic of Petri net based formalisms, or process calculi. From this point of view, CO-OPN is similar to Prolog or other logical programming languages [13].

Finally, interactions between CO-OPN objects are synchronous. From this point of view, CO-OPN can be compared to synchronous languages such as Esterel or Lustre [16].

7   Future Work and Conclusions

In this short description we have presented the CoopnBuilder environment and the main associated tools. The most usual tools - the checker, the structural editor, the viewers - are not detailed in this paper; other tools, such as the code generator, the evaluator or the simulator, have been shortly explained, and we have hopefully shown how they are evolutive, useful and powerful.

Our present effort evolves around the execution of CO-OPN specifications. Two aspects are investigated: how to use generated code for testing and how to use it for semantic analisys.

References

[1]   Didier Buchs and Nicolas Guelfi, "A Formal Specification Framework for Object-Oriented Distributed Systems," IEEE TSE, vol. 26, no. 7, July 2000, pp. 635-652.

[2]   Olivier Biberstein, Didier Buchs and Nicolas Guelfi, "Object-Oriented Nets with Algebraic Specifications: The CO-OPN/2 Formalism," Advances in Petri Nets on Object-Orientation, G. Agha and F. De Cindio and G. Rozenberg (Ed.), Lecture Notes in Computer Science, no. 2001, Springer-Verlag, May 2001, pp. 70-127.

[3]   Mathieu Buffo, Didier Buchs and Stanislav Chachkov, "CoopnTools a Toolkit for the support of CO-OPN," Proceedings of the Tools Demonstration of the 21th International Conference on Application and Theory of Petri Nets, Aahrus University, June 2000, pp. 2-6.

[4]   Stanislav Chachkov and Didier Buchs, "From Formal Specifications to Ready-to-Use Software Components: The Concurrent Object-Oriented Petri Net Approach," International Conference on Application of Concurrency to System Design, Newcastle, IEEE Computer Society Press, June 2001, pp. 99-110.

[5]   Stanislav Chachkov and Didier Buchs, "Interfacing Software Libraries from Non-deterministic Prototypes," International Workshop on Rapid System Prototyping, July 1-3, 2002, Darmstadt, Germany

[6]   Mathieu Buffo and Didier Buchs, "A Coordination Model for Distributed Object Systems," Proceedings of the Second International Conference on Coordination Models and Languages COORDINATION'97, September 1997, Lecture Notes in Computer Science, vol. 1282, Springer-Verlag, 1997.

[7]   D. Buchs and M. Buffo, "Rapid Prototyping of Formally Modelled Distributed Systems," Proceedings of the Tenth International Workshop on Rapid System Prototyping (RSP'99), Frances M. Titsworth (Ed.), IEEE, June 1999.

[8]   Stanislav Chachkov and Didier Buchs, "From an Abstract Object-Oriented Model to a Ready-to-Use Embedded System Controller," Rapid System Prototyping, Monterey, CA, IEEE Computer Society Press, June 2001, pp. 142-148.

[9]   K. Jensen, "Coloured Petri Nets. Basic Concepts, Analysis Methods and Practical Use," Volume 1, Basic Concepts. Monographs in Theoretical Computer Science, Springer-Verlag, 2nd corrected printing 1997. ISBN: 3-540-60943-1.

[10]   Sibertin-Blanc, C. "Cooperative Nets" In Valette, R.: Lecture Notes in Computer Science, Vol. 815; Application and Theory of Petri Nets 1994, Proceedings 15th International Conference, Zaragoza, Spain, pages 471-490. Springer-Verlag, 1994.

[11]   Robin Milner "Communicating and Mobile Systems: the -Calculus," Cambridge University Press, May 1999.

[12]   C.A.R. Hoare "Communicating Sequential Processes," Prentice Hall International Series in Computer Science, 1985.

[13]   Ulf Nilsson and Jan Maluszynski "Logic, Programming and Prolog (2ed)"

[14]   http://www.research.att.com/sw/tools/graphviz/

The Petri Net Markup Language:
Concepts, Technology, and Tools

Jonathan Billington[1], Søren Christensen[2], Kees van Hee[3],
Ekkart Kindler[4], Olaf Kummer[5], Laure Petrucci[6],
Reinier Post[3], Christian Stehno[7], and Michael Weber[8]

[1] University of South Australia, Computer Systems Engineering Centre,
j.billington@unisa.edu.au
[2] University of Aarhus, Department of Computer Science,
schristensen@daimi.au.dk
[3] Technische Universiteit Eindhoven, Department of Math. and Computer Science
{k.m.v.hee,r.d.j.post}@tue.nl
[4] University of Paderborn, Department of Computer Science,
kindler@upb.de
[5] CoreMedia AG, Germany,
olaf.kummer@coremedia.com
[6] Laboratoire Spécification et Vérification, CNRS UMR 8643, ENS de Cachan,
petrucci@lsv.ens-cachan.fr
[7] Carl von Ossietzky University Oldenburg, Department of Computing Science,
stehno@informatik.uni-oldenburg.de
[8] Humboldt-Universität zu Berlin, Computer Science Department,
mweber@informatik.hu-berlin.de

Abstract. The *Petri Net Markup Language (PNML)* is an XML-based interchange format for Petri nets. In order to support different versions of Petri nets and, in particular, future versions of Petri nets, PNML allows the definition of *Petri net types*. Due to this flexibility, PNML is a starting point for a standard interchange format for Petri nets. This paper discusses the design principles, the basic concepts, and the underlying XML technology of PNML. The main purpose of this paper is to disseminate the ideas of PNML and to stimulate discussion on and contributions to a standard Petri net interchange format.

1 Introduction

It has been recognised in the Petri net community for over a decade [29,18,2, 19,15] that it is useful to be able to transfer Petri net models between tools that may exist in different countries throughout the world. This would allow Petri Net tool users in geographically distributed locations to take advantage of newly developed facilities on other tools, for example, for analysis, simulation or implementation. The Petri net community would be able to exchange Petri net models that are of mutual interest, perhaps for teaching a course, or in a global development project where teams in different countries exchange design information. It would allow a library of Petri net models to be created that could

W.M.P. van der Aalst and E. Best (Eds.): ICATPN 2003, LNCS 2679, pp. 483–505, 2003.

be accessed worldwide via the Internet and edited, simulated and analysed on different tools. This idea can be extended to the transfer of analysis results. For example, one may wish to develop Petri net models with a tool, obtain the occurrence graph with another Petri net tool, and model check it on a third one.

To facilitate the transfer of Petri nets it is useful to develop a standard transfer format. This was recognised in the initial proposal to establish an International standard for High-level Petri nets in 1995 [12]. Three years ago, an initiative was taken to discuss the development of a standard interchange format for Petri nets by holding a workshop [1] as part of the Petri net conference in Aarhus, Denmark. The principles and objectives of such an interchange format were discussed and different proposals for XML-based interchange formats were presented. Since then, there have been several other meetings and discussions on such a format including standards meetings at the last two Petri net conferences, resulting in a mailing list being established to promote further discussion[1]. Still, there is no well-accepted interchange format for Petri nets to date.

The *Petri Net Markup Language* (PNML) [15] was one of the proposals for an interchange format at the first workshop. Though not generally accepted yet, it is currently supported by a couple of tools. Moreover, it is flexible enough to integrate different types of Petri nets and is open for future extensions. This makes it a good starting point for a future standard interchange format.

In this paper, we present the concepts of PNML, its realization in XML technology as well as some tools supporting its use. The purpose of this paper is to focus and promote the development of a standard interchange format. It will serve as a starting point for international standardization within the joint technical committee of the International Organization for Standardization and the International Electrotechnical Commission (ISO/IEC) and should stimulate the discussion of its concepts and its realization. A recent new work item proposal [14] on Petri Net techniques was accepted in 2002. It proposes the development of a 3 part standard for High-level Petri nets. Part 2 of this standard (ISO/IEC 15909-2) will develop the transfer format. (Part 1 [13] is concerned with basic concepts, definitions and graphical notation and Part 3 is reserved for extensions, such as the inclusion of time and modularity constructs.)

Though the basic concepts of PNML have been stable from its very beginning, there have been different extensions and minor changes. The current version is PNML 1.3, which will be discussed in this paper. However, due to its size, we cannot give here a complete description of PNML 1.3 and its realization in XML. For further technical details and examples, we refer the reader to the PNML homepage [24]. Here, we concentrate on the principles and concepts of PNML (Sect. 2) and its realization (Sect. 3).

One of the main guiding principles of PNML is extensibility to allow for the incorporation of future versions of Petri nets. To this end, PNML includes the definition of different *Petri Net Types*. In order to guarantee some compatibility between different Petri net types, PNML suggests an evolving *Conventions Document*, which maintains the syntax (and to some extent the semantics) of all

---

[1] See http://www.informatik.hu-berlin.de/top/PNX/ for details.

available features for defining Petri net types. In Sect. 4, we discuss the structure
of the Conventions Document and some of the features covered in its present
version. The graphical features of PNML are considered in Sect. 5, where we give
a transformation to Scalable Vector Graphics (SVG [11]) in order to provide a
reference for the graphical appearance of a PNML file. In Sect. 6, we present
some of the tools available for PNML and, in particular, tools that provide some
support in reading and writing PNML files. Section 7 briefly discusses the issues of modularity, and the transfer of analysis results. Finally we provide some
conclusions and suggestions for future work.

2   Concepts

In this section, the idea, the design principles and the concepts of PNML are
explained. PNML is designed to be a Petri net interchange format that is independent of specific tools and platforms. Moreover, the interchange format needs
to support different dialects of Petri nets and must be extensible. More recently,
the ideas to support the exchange of analysis information was added to this list.

An early proposal [3] for a standard Petri net transfer syntax appeared in
1988. The *Abstract Petri Net Notation* [2] is a more recent (1995) proposal for
such an interchange format. It tries to capture different versions of Petri nets
within a single format and provides limited features to extend it. PNML goes
one step further by providing an explicit concept for defining new features and
new Petri net types.

2.1   Design Principles

Starting from the above ideas, the design of PNML was governed by the principles
of *flexibility* and *compatibility* and the need for it to be *unambiguous*.

Flexibility means that PNML should be able to represent any kind of Petri net
with its specific extensions and features. PNML should not restrict the features
of some kinds of Petri nets, nor force us to ignore or to abstract from specific
information of a Petri net when converting it to PNML. In order to achieve this
flexibility, PNML considers a Petri net as a labelled graph, where all additional
information can be stored in labels that can be attached to the net itself, to the
nodes of the net or to its arcs.

Ambiguity is removed from the format by ensuring that the original Petri
net and its particular type can be uniquely determined from its PNML representation. To this end, PNML supports the definition of different *Petri net types*.
A *Petri net type definition* (PNTD) determines the legal labels for a particular Petri net type. By assigning a fixed type to each Petri net, the description
becomes unambiguous.

Compatibility means that as much information as possible can be exchanged
between different types of Petri nets. In order to achieve compatibility, PNML
comes with conventions on how to define a label with a particular meaning. In a
Conventions Document, the syntax as well as the intended meaning of all kinds

of extensions are predefined. When defining a new Petri net type, the labels can be chosen from this Conventions Document. When a new Petri net type complies with these conventions for defining its own labels, the meaning of its labels is well-defined. This allows other Petri net tools to interpret the net even if they do not know the new Petri net type itself.

2.2   PNML Structure

The different parts of PNML and their relationships are shown in Fig. 1. The *meta model* defines the basic structure of a PNML file; the *type definition interface* allows the definition of new Petri net types that restrict the legal files of the meta model; and the *feature definition interface* allows the definition of new features for Petri nets. These three parts are fixed once and for all. Another part of PNML, the *Conventions Document*, evolves. It contains the definition of a set of standard features of Petri nets, which are defined according to the feature definition interface. Moreover, there will be several *Standard Petri Net Types*, using some features from the Conventions Document and possibly others. New features can be added to the Conventions Document and new Petri net types to the standard types when they are of common interest. Due to their evolving nature, these documents are best published and maintained via a web site.

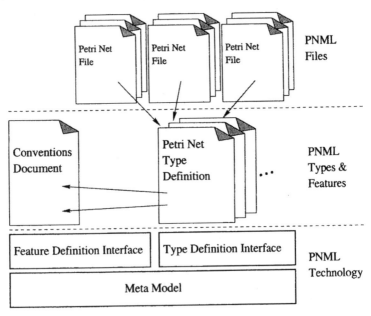

Fig. 1. Overview of the parts of PNML

2.3   Meta Model

Figure 2 shows the meta model of PNML in UML notation. We start with a discussion of the meta model of *basic PNML*, which consists of the classes with thick outlines. The other classes belong to *structured PNML* and will be explained later in this section.

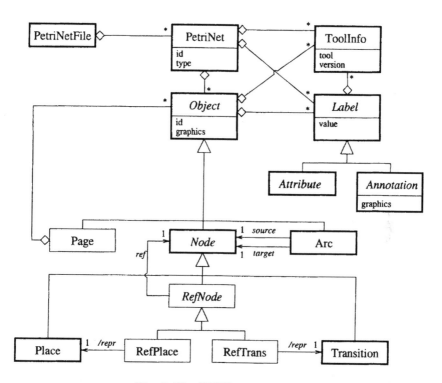

Fig. 2. The PNML meta model

Petri nets and objects. A file that meets the requirements of PNML is called a *Petri net file*; it may contain several *Petri nets*. Each Petri net consists of *objects*, which, basically, represent the graph structure of the Petri net[2]. Each object within a Petri net file has a unique *identifier*, which can be used to refer to this object. In basic PNML, an object is a *place*, a *transition* or an *arc*. For convenience, a place or a transition is called a *node*.

Labels. In order to assign further meaning to an object, each object may have *labels*. Typically, a label represents the name of a node, the initial marking of a place, the guard of a transition, or the inscription of an arc. In addition, the Petri net itself may have some labels. For example, the declarations of functions and variables that are used in the arc inscriptions could be labels of a high-level Petri net. The legal labels and the legal combinations of labels are defined by

---

[2] Note that the PNML meta model allows arcs between nodes of the same kind. The reason is that there are Petri net types with such arcs. Since Petri net types only restrict the meta model, the meta model should not forbid such arcs.

the Petri net type. The type of a Petri net is defined by a reference to a unique *Petri net type definition* (PNTD), which will be discussed in Sect. 3.2.

Two kinds of labels are distinguished: *annotations* and *attributes*. An annotation comprises information that is typically displayed as text near the corresponding object. Examples are names, initial markings, arc inscriptions, transition guards, and timing or stochastic information. In contrast, an attribute specifies a graphical property of an object such as colour, style, form or line thickness. For example, an attribute *arc type* could have domain normal, read, inhibitor, reset. Another example is an attribute for classifying the nodes of a net as proposed by Mailund and Mortensen [20]. PNML does not define how this is done, although the Conventions Document may provide directions.

Graphical information. Each object and each annotation is equipped with graphical information. For a node, this includes its position; for an arc, it includes a list of positions that define intermediate points of the arc; for an annotation, it includes its relative position with respect to the corresponding object[3]. There can be additional information concerning size, colour, and shape of nodes or arcs, or concerning colour, font, and font size of labels (see Sect. 3 and 5 for details).

Tool specific information. For some tools, it might be necessary to store tool specific information, which is not supposed to be used by other tools. In order to store this information, each object and each label may be equipped with such *tool specific information*. Its format depends on the tool and is not specified by PNML. PNML provides a mechanism for clearly marking tool specific information along with the name and the version of the tool adding this information. Therefore, other tools can easily ignore it, and adding tool specific information will never compromise a Petri net file.

Pages and reference nodes. Up to now, only basic PNML has been discussed. For structuring a Petri net, there is the more advanced *structured PNML*. Structured PNML allows us to separate different parts of a net into separate *pages* as known from several Petri net tools (e. g. Design/CPN [10]). A page is an object that may consist of other objects – it may even consist of other pages. An arc, however, may connect nodes on the same page only[4]. A *reference node*, which can be either a *reference transition* or a *reference place* represents an appearance of a node. It can refer to any node on any page of the Petri net as long as there are no cyclic references; this guarantees that, ultimately, each reference node refers to exactly one place or transition of the Petri net.

Reference nodes may have labels but these labels can only specify information about their appearance. They cannot specify any information about the referenced node itself, which already has its own labels for this purpose.

---

[3] For an annotation of the net itself, the position is absolute.

[4] The reason is that an arc cannot be drawn from one sheet of paper to another when printing the different pages.

Basic PNML is PNML without pages or reference nodes. A fixed transformation can "flatten" any PNML net to a semantically equivalent basic PNML net. To do so, the reference nodes are merged with the nodes they refer to and page borders are ignored (see [17] for details). This transformation can be achieved by a simple XSLT stylesheet [6]. By applying this stylesheet, a tool supporting only basic PNML can be used for arbitrary PNML nets[5]. A more powerful structuring mechanism, *modular PNML* [17], allows different PNML documents to reference each other. Modular PNML will be briefly discussed in Sect. 7.

2.4 Discussion of the Use of PNML

In this section, we briefly discuss the use of Petri net types and the definition of Petri net features to achieve our design goals. Every Petri net file has a unique type, which is a reference (a URI – Uniform Resource Identifier) to the corresponding Petri net type definition. This way, the syntax is unambiguously defined by the PNML meta model and the type definition. To be interpreted unambiguously, each Petri net type definition must have a formal semantics that is known by each tool that uses it.

The description of the semantics of Petri net features and Petri net types is not formalized yet. A formalism for defining the semantics of features and for combining the semantics of features to a semantics of a Petri net type is far from trivial. Working out such a formalism is a long-term project. The concepts of PNML provide a starting point for this work.

But what about compatibility? There are times when it would be very useful to be able to import a net, even when the semantics of the Petri net type is unknown. For example, when we have a hard copy of a net describing a complex system in a net dialect that our own tool does not support, but we wish to input the net. Manual input of the net would be very time consuming. A transferred file would go a long way to alleviating this problem. In this case, a tool might try to extract as much information as possible from the net type by guessing the meaning of some labels from their names, which may result in wrong interpretations. In order to allow the tool some more "educated guesses" on the meaning of labels, PNML provides the *features definition interface*. Each feature fixes the syntax for the corresponding labels along with a description of their semantics. Standard features defined according to this interface will be maintained in the Conventions Document. A new Petri net type definition may choose its features from these conventions by a reference to the Conventions Document (see Sect. 3.2 for details). If the Petri net type chooses all its features from the Conventions Document, a tool not knowing the new type, but knowing the feature from the Conventions Document, knows the meaning of the corresponding labels. Then it can try to extract all the relevant information and convert the net to a Petri net type it knows – possibly losing some information of the original net.

---

[5] It is also possible to use basic PNML with extra information to represent the page structure, and a pair of XSLT sheets for the conversion. This allows the page structure to be preserved even after processing a net with a tool that only supports basic PNML, if that tool supports the extra information.

3   PNML Technology

In this section, we will discuss how the concepts of PNML can be implemented in XML. Though the use of XML has some disadvantages[6], the advantages of XML clearly prevail. Aside from its popularity, it is platform independent, and there are many tools available for reading, writing and validating XML documents.

There are different XML technologies that could be used for implementing PNML. *RELAX NG* [7] was chosen for defining the structure of a PNML document, because it was one of the first technologies[7] with a module concept and a validator supporting it. This module concept was necessary for clearly separating the PNML meta model from the type definitions (PNTDs). Today, there is also tool support for XML Schema [26]. So PNML could be easily realized in XML Schema[8].

3.1   PNML Meta Model

The PNML meta model is translated into XML syntax in a straightforward manner. Technically, the syntax of PNML 1.3 is defined by a RELAX NG grammar, which can be found on the PNML web site [24].

PNML elements. Here, not the full grammar, but a more compact translation is presented: basically, each concrete class[9] of the PNML meta model is translated into an XML element. This translation along with the attributes and their data types is given in Tab. 1. These XML elements are the *keywords* of PNML and are called *PNML elements* for short. For each PNML element, the aggregations of the meta model define in which elements it may occur as a child element. We have omitted the Graphics class from the meta model shown in Fig. 2 so as not to clutter the diagram. The classes with associated graphical information are instead indicated by an attribute "graphics".

The data type ID in Tab. 1 describes a set of unique identifiers within the PNML document. The data type IDRef describes the same set, now as references to the elements of ID. The set of references is restricted to a denoted subset. For instance, a reference place transitively refers with its attribute ref to a place of the same net.

Labels. There are no PNML elements for labels because the meta model does not define any concrete ones. Concrete labels are defined by the Petri net types. An XML element that is not defined by the meta model (i. e. not occurring in

---

[6] One disadvantage is storage waste. But, this can be easily avoided by using a compressed XML format, which is supported by most XML APIs.

[7] RELAX NG replaces TREX (*Tree Regular Expressions for XML*) [5], which was used when first defining PNML.

[8] There are converters providing translation between several XML schema languages.

[9] A class in a UML diagram is concrete if its name is not displayed in italics.

Table 1. Translation of the PNML meta model into PNML elements

Class	XML element	XML Attributes
PetriNetFile	`<pnml>`	
PetriNet	`<net>`	`id`: ID
		`type`: anyURI
Place	`<place>`	`id`: ID
Transition	`<transition>`	`id`: ID
Arc	`<arc>`	`id`: ID
		`source`: IDRef (Node)
		`target`: IDRef (Node)
Page	`<page>`	`id`: ID
RefPlace	`<referencePlace>`	`id`: ID
		`ref`: IDRef (Place or RefPlace)
RefTrans	`<referenceTransition>`	`id`: ID
		`ref`: IDRef (Transition or RefTrans)
ToolInfo	`<toolspecific>`	`tool`: string
		`version`: string
Graphics	`<graphics>`	

Tab. 1) is considered as a label of the PNML element in which it occurs. For example, an `<initialMarking>` element could be a label for a place, which represents its initial marking. Likewise `<name>` could represent the name of an object, and `<inscription>` an arc inscription. A legal element for a label may consist of further elements. The value of a label appears as a string in a `<text>` element. Furthermore, the value may be represented as an XML tree in a `<structure>` element[10]. An optional PNML `<graphics>` element defines its graphical appearance, and further optional PNML `<toolspecific>` elements may add tool specific information to the label.

Graphics. PNML elements and labels include graphical information. The structure of the PNML `<graphics>` element depends on the element in which it appears. Table 2 shows the elements which may occur in the substructure of a `<graphics>` element.

The `<position>` element defines an absolute position and is required for each node, whereas the `<offset>` element defines a relative position and is required for each annotation. The other sub-elements of `<graphics>` are optional. For an arc, the (possibly empty) sequence of `<position>` elements defines its intermediate points. Each absolute or relative position refers to Cartesian coordinates (x, y). As for many graphical tools, the x-axis runs from left to right and the y-axis from top to bottom. More details on the effect of the graphical features can be found in Sect. 5.1.

---

[10] In order to be compatible with earlier versions of PNML, the text element `<value>` may occur alternatively to the `<text>` `<structure>` pair.

492   J. Billington et al.

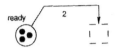

Fig. 3. A simple P/T-system

Listing 1. PNML code of the example net in Fig. 3

```
   <pnml xmlns="http://www.example.org/pnml">
     <net id="n1" type="http://www.example.org/pnml/PTNet">
       <name>
         <text>An example P/T-net</text>
5      </name>
       <place id="p1">
         <graphics>
           <position x="20" y="20"/>
         </graphics>
10       <name>
           <text>ready</text>
           <graphics>
             <offset x="-10" y="-8"/>
           </graphics>
15         </name>
         <initialMarking>
           <text>3</text>
         </initialMarking>
       </place>
20     <transition id="t1">
         <graphics>
           <position x="60" y="20"/>
         </graphics>
         <toolspecific tool="PN4all" version="0.1">
25         <hidden/>
         </toolspecific>
       </transition>
       <arc id="a1" source="p1" target="t1">
         <graphics>
30         <position x="30" y="5"/>
           <position x="60" y="5"/>
         </graphics>
         <inscription>
           <text>2</text>
35         <graphics>
             <offset x="15" y="-2"/>
           </graphics>
         </inscription>
       </arc>
40     </net>
   </pnml>
```

Table 2. Elements in the `<graphics>` element depending of the parent element

Parent element class	Sub-elements of `<graphics>`
Node, Page	`<position>` (required)
	`<dimension>`
	`<fill>`
	`<line>`
Arc	`<position>` (zero or more)
	`<line>`
Annotation	`<offset>` (required)
	`<fill>`
	`<line>`
	`<font>`

Listing 2. Label definition

```
 <define name="PTMarking"
   xmlns:pnml="http://www.informatik.hu-berlin.de/top/pnml">
   <element name="initialMarking">
     <interleave>
5      <element name="text">
         <data type="nonNegativeInteger"
           datatypeLibrary="http://www.w3.org/2001/XMLSchema-datatypes"/>
       </element>
       <ref name="pnml:StandardAnnotationContent"/>
10     </interleave>
   </element>
 </define>
```

Example. In order to illustrate the structure of a PNML file, we consider the simple example net shown in Fig. 3. Listing 1 shows the corresponding PNML code. It is a straightforward translation, where we have labels for the names of objects, for the initial markings, and for arc inscriptions. Note that we assume that the dashed outline of the transition results from the tool specific information `<hidden>` from an imaginary tool *PN4all*.

3.2   Petri Net Types

When defining a Petri net type, we firstly need to explain how labels are defined.

Label definition. Listing 2 shows the RELAX NG definition of the label `<initialMarking>`, which represents the initial marking of a place of a P/T-system (cf. List. 1). Its value (in a `<text>` element) should be a natural number, which is formalized by referring to the corresponding data type nonNegativeInteger of the data type system of XML Schema. Note that the optional graphical and tool specific information do not occur in this label definition; this is not necessary, because these standard elements for annotations are inherited from the standard annotation of PNML. Such label definitions can either be given explicitly for each Petri net type, or they can be included in the Conventions Document, such that Petri net type definitions can refer to these definitions. Some of the available labels in the Conventions Document will be discussed in Sect. 4.

Listing 3. PNTD for P/T-Systems

```
    <grammar ns="http://www.example.org/pnml"
             xmlns="http://relaxng.org/ns/structure/1.0"
             xmlns:conv="http://www.informatik.hu-berlin.de/top/pnml/conv">
      <include href="http://www.informatik.hu-berlin.de/top/pnml/pnml.rng"/>
5     <include href="http://www.informatik.hu-berlin.de/top/pnml/conv.rng"/>
      <define name="NetType" combine="replace">
        <text>http://www.example.org/pnml/PTNet</text>
      </define>
      <define name="Net" combine="interleave">
10       <optional><ref name="conv:Name"/></optional>
      </define>
      <define name="Place" combine="interleave">
        <interleave>
          <optional><ref name="conv:PTMarking"/></optional>
15         <optional><ref name="conv:Name"/></optional>
        </interleave>
      </define>
      <define name="Arc" combine="interleave">
        <optional><ref name="conv:PTArcInscription"/></optional>
20     </define>
    </grammar>
```

Petri Net Type Definitions (PNTDs). Listing 3 shows the Petri Net Type Definition (PNTD) for P/T-Systems as a RELAX NG grammar.

Firstly, it includes both the definitions for the meta model of PNML (pnml.rng) and the definitions of the Conventions Document (conv.rng) (Sect. 4), which, in particular, contains the definition from List. 2, a similar definition for arc inscriptions of P/T-systems, and a definition for names.

Secondly, the PNTD defines the legal labels for the whole net and the different objects of the net. In our example, the net and the places may have an annotation for names. Furthermore, the places are equipped with an initial marking and the arcs are equipped with an arc inscription. Note that all labels are optional in this type. The labels are associated with the net objects by giving a reference to the corresponding definitions in the Conventions Document. Technically, the definition extends the original definition of the net, places and arcs of the RELAX NG grammar for PNML.

3.3   High-Level Petri Nets

One of the driving forces for developing PNML was the standardisation process of high-level Petri nets [13]. This standard defines *High-Level Petri Net Graphs* (HLPNGs) as a syntactic representation of high-level Petri nets. Here, we will briefly sketch the idea of a PNTD for HLPNGs. Note that this definition is an example for having a textual as well as a structural representation of the value of a label.

The PNTD defines labels for the corresponding concepts of HLPNGs: *signatures, variables, initial markings, arc-inscriptions,* and *transition guards.* Arc-inscriptions and transition guards are *terms* over the signature and the variables of the HLPNG. No concrete syntax is defined, but terms are defined in the usual inductive manner, providing an abstract syntax.

In the PNTD for HLPNGs, the value of a label may be represented by using a concrete or abstract syntax. The value in a concrete syntax is represented as pure text within the label's `<text>` element, the value in an abstract syntax is represented within the label's `<structure>` element. An XML substructure within the `<structure>` element represents the inductive construction of a term from its subterms. This way, we can exchange high-level nets between two tools that use different concrete syntaxes, but have the same underlying structure (which is fixed for HLPNGs). Tools are not required to export or import abstract syntax, but if they do, interoperability with other tools is increased.

4   Conventions

The Conventions Document comprises the definitions of all standard labels, in order to guarantee compatibility among different Petri net types.

Technically, the Conventions Document consists of a sequence of more or less independent label definitions as discussed in Sect. 3.2. Each label definition in the Conventions Document must be assigned a unique *name.* In our example from List. 2, a label with name PTMarking has been defined. Note that it is not necessarily the name of the corresponding XML element. In the example from List. 2, the label PTMarking is defined as the XML element `<initialMarking>`. For each label, the Conventions Document gives a reference to its meaning and states in which PNML elements it may occur. The definitions in the Conventions Document can be used by referring to the name of a label in the Conventions Document as shown in List. 3.

Table 3 gives some examples of labels defined in the Conventions Document. The first column gives the name of the label. The second column gives the corresponding XML element. The third column gives the data type of the label. If it is a simple data type, the value of the label must be text in the XML element `<text>`. More complex data types (indicated by "structured" in Tab. 3), are represented both as strings in the element `<text>` and as an XML tree in the `<structure>` element of the annotation. The last column gives a short description of the label's meaning.

Table 3. Content of the Conventions Document

Label name	XML element	Domain	Meaning
Name	`<name>`	string	user given identifier of an element describing its meaning
PTMarking	`<initialMarking>`	nonNegativeInteger	initial marking of places in nets like P/T-nets
ENMarking	`<initialMarking>`	—	initial marking of places in nets like EN-systems
HLMarking	`<initialMarking>`	structured	term describing the initial marking of a place in a high-level net schema
PTCapacity	`<capacity>`	nonNegativeInteger	annotation describing the capacity of a place in nets similar to P/T-nets
HLSort	`<sort>`	structured	description of the sort of tokens on a place in high-level net schemas

Note that different labels in the Conventions Document can be represented by the same XML element. In the examples from Tab. 3, this applies to the labels PTMarking, ENMarking, and HLMarking. This is not a problem since these labels cannot be used within the same net anyway, and the name of the corresponding label definition in the Conventions Document can be retrieved[11] from the Petri Net Type Definition.

In contrast to PNML technology, the Conventions Document is a "living document", which requires some maintenance and continuous standardization. This requires both a maintenance policy and a team to maintain this Conventions Document. One policy should be that changes are always upward and downward compatible. This, basically, means that, once a label definition is in the Conventions Document, it cannot be changed anymore. Therefore, a definition should only be added to the Conventions Document when its definition is stable.

5 Layout

In this section, we discuss the graphical features of PNML and their effect on the graphical presentation of a PNML document. Section 5.1 gives an informal overview of all graphical features. In Sect. 5.2, we discuss a precise description of the graphical presentation of a PNML document, i.e. an XSLT transformation [6] from PNML to the *Scalable Vector Graphics* (SVG) [11]. SVG is a standard for

---

[11] An alternative to this solution would be to use different namespaces. But, this would result in quite complex definitions. Moreover, namespaces would be necessary even for simple Petri net types such as P/T-Systems. In order to keep PNML for simple Petri net types as simple as possible, the extensive use of namespaces is avoided.

two-dimensional vector graphics based on XML. In combination with a standard SVG viewer, this XSLT transformation provides us with a standard viewer for PNML documents.

5.1   Graphical Information in PNML

Table 4 lists the graphical elements that may occur in the PNML `<graphics>` element along with their attributes. The domain of the attributes refers to the data types of either XML Schema [26], or Cascading Stylesheets 2 (CSS2) [4], or is given by an explicit enumeration of the legal values.

Table 4. PNML graphical elements

XML element	Attribute	Domain
`<position>`	x	decimal
	y	decimal
`<offset>`	x	decimal
	y	decimal
`<dimension>`	x	nonNegativeDecimal
	y	nonNegativeDecimal
`<fill>`	color	CSS2-color
	image	anyURI
	gradient-color	CSS2-color
	gradient-rotation	{vertical, horizontal, diagonal}
`<line>`	shape	{line, curve}
	color	CSS2-color
	width	nonNegativeDecimal
	style	{solid, dash, dot}
`<font>`	family	CSS2-font-family
	style	CSS2-font-style
	weight	CSS2-font-weight
	size	CSS2-font-size
	decoration	{underline, overline, line-through}
	align	{left, center, right}
	rotation	decimal

The `<position>` element defines the absolute position of a net node or a net annotation, where the x-coordinate runs from left to right and the y-coordinate from top to bottom. The `<offset>` element defines the position of an annotation relative to the position of the object.

For an arc, there may be a (possibly empty) list of `<position>` elements. These elements define intermediate points of the arc. Altogether, the arc is a path from the source node of the arc to the destination node of the arc via the intermediate points. Depending on the value of attribute shape of element `<line>`, the path is displayed as a polygon or as a (quadratic) Bezier curve, where points act as line connectors or Bezier control points.

The <dimension> element defines the height and the width of a node. Depending on the ratio of height and width, a place is displayed as an ellipse rather than a circle.

The two elements <fill> and <line> define the interior and outline colours of the corresponding element. The value assigned to a colour attribute must be a RGB value or a predefined colour as defined by CSS2. When the attribute gradient-color is defined, the fill colour continuously varies from color to gradient-color. The additional attribute gradient-rotation defines the orientation of the gradient. If the attribute image is defined, the node is displayed as the image at the specified URI, which must be a graphics file in JPEG or PNG format. In this case, all other attributes of <fill> and <line> are ignored.

For a label, the <font> element defines the font used to display the text of the label. The complete description of possible values of the different attributes can be found in the CSS2 specification. Additionally, the align attribute defines the justification of the text with respect to the label coordinates, and the rotation attribute defines a clockwise rotation of the text.

Figure 4 shows an example of a PNML net, which uses most of the graphical features of PNML.

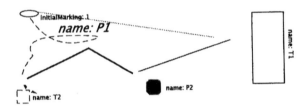

Fig. 4. Example of PNML graphical features

5.2   Portable Visualization Scheme

In order to give a precise description of the graphical presentation of a PNML document with all its graphical features, we define a translation to SVG. Petri net tools that support PNML can visualise Petri nets using other means than SVG, but the SVG translation can act as a reference model for such visualisations. Technically, this translation is done by means of an XSLT stylesheet. The basic idea of this transformation was already presented in [27]. A complete XSLT stylesheet can be found on the PNML web pages [24].

Transformations for basic PNML. The overall idea of the translation from PNML to SVG is to transform each PNML object to some SVG object, where the attributes of the PNML element and its child elements are used to give the SVG element the intended graphical appearance.

As expected, a place is transformed into an ellipse, while a transition is transformed into a rectangle. Their position and size are calculated from the `<position>` and `<dimension>` elements. Likewise, the other graphical attributes from `<fill>` and `<line>` can be easily transformed to the corresponding SVG attributes.

An annotation is transformed to SVG text such as `name: someplace`. The location of this text is automatically computed from the attributes in `<offset>` and the position of the corresponding object. For an arc, this reference position is the centre of the first line segment. If there is no `<offset>` element, the transformation uses some default value, while trying to avoid overlapping.

An arc is transformed into a SVG path from the source node to the target node – possibly via some intermediate points – with the corresponding attributes for its shape. The start and end points of a path may be decorated with some graphical object corresponding to the nature of the arc (e.g. inhibitor). The standard transformation supports arrow heads as decorations at the end, only. The arrow head (or another decoration) should be exactly aligned with the corresponding node. This requires computations using complex operations that are neither available in XSLT nor in SVG – the current transformation uses recursive approximation instead.

Transformations for structured PNML. Different pages of a net should be written to different SVG files since SVG does not support multi-image files. Unfortunately, XSLT does not support output to different files yet, but XSLT 2.0 will. Hence, a transformation of structured PNML to SVG will be supported once XSLT 2.0 is available.

The transformations for reference places and reference transitions are similar to those for places and transitions. In order to distinguish reference nodes from other nodes, reference nodes are drawn slightly translucent and an additional label gives the name of the referenced object.

Type specific transformations. Above, we have discussed a transformation that works for all PNML documents, where all annotations are displayed as text. For some Petri net types, one would like to have other graphical representations for some annotations. This can be achieved with customized transformations. The technical details of customized transformations are not yet fixed. Due to the rule-based transformations of XSLT, equipping the Type Definition Interface and the Feature Definition Interface of PNML with some information on their graphical appearance seems to be feasible. Basically, each new feature can be assigned its own transformation to SVG. Adding these transformations to the standard ones for PNML gives us a customized transformation for this Petri net type.

6   Tools and Reference Implementation

In this section, we describe how PNML can be used in Petri net tools, how they implement PNML and how XML techniques can help to validate and to parse PNML documents. Several Petri net tools inspired the development of PNML. The Petri Net Kernel (Sect. 6.2) implements an object model of Petri nets similar to PNML. Renew (Sect. 6.3) was the first tool that implemented a version of PNML. PEP (Sect. 6.4) features a Petri net based collection of tools. Design/CPN (Sect. 6.5) was one of the first Petri net tools that implemented an XML based file format. Currently, there are several Petri net tools implementing PNML as one (or as the only) file format (e. g. Renew [25], PNK [23], PEP [22], VIPtool [9]).

6.1   XML Techniques

Implementing a new file format such as PNML for an existing tool requires some extra work. Fortunately, there are different tools and Application Programming Interfaces (APIs) for implementing parsers for XML documents on different platforms and for different programming languages. Basically, there are two techniques supporting the parsing of XML documents, SAX and DOM. SAX is a lightweight, event-driven interface, which is well-suited for streaming large XML documents with minor modifications. SAX is not well-suited for implementing I/O-interfaces and, in particular, for implementing PNML. DOM (Document Object Model) provides a parser for XML documents. Then, a program has full access to the document and all its elements in a tree-like structure. Moreover, it provides a powerful reference mechanism for accessing the elements of the XML document.

The current version of PNML, the PNTDs, and the Conventions Document are defined in the XML schema language RELAX NG [7]. There are several tools for validating an XML document against a RELAX NG specification (e. g. Jing[12]). Some special features of PNML, however, cannot be expressed in RELAX NG yet. This concerns the correct use of a PNTD in a PNML document and the correctness of references. For example, a reference place must refer to a place or a reference place; it must not refer to a transition or to a reference transition. Moreover, references must not be cyclic. Currently, we are developing a Jing-based validator that takes the special features of PNML into account. See the PNML Web pages [24] for more details.

Another task is the validation of the syntactical correctness of the string values of labels. If the domain of a label is defined in a known data type system (e. g. the RELAX NG internal system or the XML Schema Datatype Library), Jing can validate these string values. Other more specific values must be validated by external tools.

---

[12] See URL http://www.thaiopensource.com/relaxng/jing.html for more details.

6.2   Petri Net Kernel

PNML has been strongly influenced by the development of the Petri Net Kernel (PNK) [28,23]. PNK is an infrastructure for building Petri net tools. Thus, it provides methods to manage Petri nets of different types. PNK implements a data model for Petri nets that is similar to that of PNML. Each place, transition, arc, or even the net may contain several labels according to the Petri net type. Standard operations on labels (e. g. setting a value, storing a string as an external representation of the label value, loading, etc.) are implemented in the API of PNK. Label specific operations (e. g. parsing the string representing the label value, operating on a label, etc.), however, are implemented by the label developer.

PNK stores one or more Petri nets in a PNML file. It uses the string representation of the current label values, which is stored in the <value> tag of the appropriate PNML label. PNK is able to read a PNML file even if it does not find a PNK implementation of the Petri net type. In this case, PNK assumes that all labels are simple string labels without special semantics. Therefore, PNK provides a universal editor for all Petri net types. Moreover, PNK provides an API for loading a net and for accessing its objects, and thus can be seen as a Document Object Model for PNML.

6.3   Renew

Renew [25] was one of the first Petri net tools to provide XML-based export and import. While its main focus lies on reference nets (object-based Petri nets), Renew was designed as a multi-formalism tool right from the start.

In order to keep the higher levels independent of the actual Petri net type, inscriptions are always edited textually and interpreted by a configurable net compiler later on. This approach was quite successful. As it is conceptually almost identical to the label concept of PNML, it gives credibility to the claim that labels are indeed expressive enough for practical purposes.

One special feature is that Renew distinguishes inscriptions syntactically, whenever possible. For instance, whether a label is a place type or an initial marking is not explicitly stored; that distinction is made by the compiler. If we want to conform to the PNML standard more closely, the actual type of a label must be computed at export time, so that the correct label element can be created.

It was evaluated whether a simple DTD suffices for the description of the format. A DTD that permitted all intended constructs was quickly given. But it turned out that certain important constraints were not easily expressible, so that they had to be checked by an external tool later on. This may be acceptable or not, but it justifies the use of more powerful grammars for the definition of PNML.

6.4 PEP

PEP [22] features a collection of tools and formalisms useful for the verification and development of concurrent systems, combined beneath a graphical user interface. PEP supports a number of different modelling languages, each of them having a Petri net semantics. Simulation and verification is based on its Petri net core. PEP uses stand-alone tools for most of the transformation and analysis tasks available. This allows easy extensions by new modelling languages or verification tools, but introduces a large number of interfaces for different file types.

PEP uses a common I/O library for accessing Petri net files in different formats. Thus, an extension to PNML files for all separate tools developed for PEP was easy to integrate. Although the original PEP file formats for Petri nets were not XML based, their structure was comparable to PNML. The implementation was therefore straightforward, once the Petri net types and the API support (by libxml2 library [30]) had been fixed.

Writing PNML files is supported by libxml2 with special output functions to adhere to XML encoding. For reading PNML, the syntax tree of the document is automatically generated. Further processing is based on user defined functions which perform the gathering of the current node's data and all of its subnodes. Parts of this processing may be delegated to further functions, allowing reuse of code for frequently used tags in PNML, e.g. `<graphics>`. Thus only one function `parseGraphics` is needed, which is called for any `<graphics>` element found in the input document. When a node is completely read, its data is stored in the internal data structure of the program. To resolve references, e.g. start and end coordinates of arcs, all possible reference targets are stored in a lookup table, allowing random access to any such element.

A converter was implemented to simplify access for tools, which use their own implementation of the original PEP file format. The converter is based on the functions from PEP's I/O library. The scripts which control these external tools now also take care of appropriate conversions, if necessary.

6.5 Design/CPN and CPN Tools

Design/CPN [10] and CPN Tools [8] both support Coloured Petri nets. Design/CPN has been available since 1989 and is being replaced by CPN Tools. Both tools support an XML based file format. Design/CPN exports and imports an external XML format, whereas CPN Tools has a native XML based file format. A DTD for both of these formats is publicly available. Petri net models are transferred between the two tools using an external file format converter.

Coloured Petri nets are hierarchical and tokens are complex data values. Arc inscriptions and guards comprise terms with variables, and the operations involved and the types of variables are defined as annotations of the net. This means that the XML format must be tailored to this information. CPN Tools will support PNML with a PNTD for Coloured Petri nets, most likely using an external converter.

7   Modularity and Analysis Issues

There are two other issues that are important to raise in this paper. They are *modular PNML*, and the integration of analysis results (such as net properties or occurrence graphs and related automata) with the net that is being analysed.

Modular PNML [17] allows Petri net modules to be defined where a module's interface definition is clearly separated from its implementation. This facilitates the modular construction of Petri nets from instances of the modules. Like PNML itself, this module concept for PNML is independent of the Petri net type. Moreover, a Petri net in modular PNML can be transformed into a Petri net in structured PNML by a simple XSLT stylesheet (similar to the transformation from structured PNML to basic PNML).

The second issue is *properties* and *analysis results*. For a net, we would like to represent its properties and analysis results in a uniform way. This would allow us to use different tools with different capabilities more efficiently, because one tool could use the results of others. For example, the analysis of a high-level Petri net might be too complex to perform. In this case, a way of obtaining partial results consists of analysing the net's skeleton (i.e. the Petri net obtained by removing the terms). Sufficient conditions (e.g. for deadlock analysis) can be checked by transferring the skeleton to a fast dedicated Petri net tool, and returning the results. PNML provides a technical hook for this purpose: an element `<properties>` that could contain this additional information. A uniform representation of Petri net properties and analysis results, however, is beyond the scope of this paper. This is an interesting and important research direction. The VeriTech project [16] and the Munich Model-Checking Kit [21] can serve as guidelines for this work.

8   Conclusion

In this paper, we have presented the principles and concepts of PNML and have sketched its realization with XML technology. We have discussed the need for the format to be extensible, to cater for evolving Petri net dialects and to include analysis results (such as occurrence graphs). This flexibility has been obtained by considering the objects that we wish to transfer as labelled directed graphs. We have introduced the notion of Petri net type definitions (PNTDs) to accommodate different Petri net dialects (and analysis results). PNTDs contain the set of legal labels for a particular net type. The concept of a Conventions Document that contains all the features for the various PNTDs and their semantics has been suggested as a mechanism for increasing compatibility between different type definitions and hence different tools. The paper does not address the difficult issue of providing a formal semantics for each feature and combining them to provide a semantics for each Petri net type. This is seen as important future work.

The work presented in this paper provides a starting point for experimenting with using PNML to transfer Petri nets between tools. We encourage the Petri net community to participate in these experiments and provide the full details of PNML [24] for this purpose. The experience gained in experimenting with the transfer format will lead to formulating a set of requirements and a relatively mature baseline document needed for the development of an International Standard within the work of project ISO/IEC 15909. Although not addressing the problem of combining features, the current Final Committee Draft of the Standard for High-level Petri Nets [13] does include an example of how the semantics of a Petri net type (in this case High-level Petri Net Graphs) may be provided (see clause 9). The work on semantics will need to be harmonised with ISO/IEC 15909.

Moreover, there are many other matters that will require significant future work. These include: user definable defaults for the graphical information of PNML elements; the realization of type specific graphical representations for PNML elements; and the policy and procedures required for maintaining the Conventions Document.

Acknowledgements. Many people from the Petri net community have contributed to the concepts of PNML during discussions of earlier versions, by making proposals, and by asking questions. Here is an incomplete list of those who contributed: Matthias Jüngel, Jörg Desel, Erik Fischer, Giuliana Franceschinis, Nisse Husberg, Albert Koelmans, Kjeld Høyer Mortensen, Wolfgang Reisig, Stephan Roch, Karsten Schmidt, as well as all former members of the DFG research group "Petri Net Technology". Thanks to all of them. Moreover, we would like to thank the DFG (German Research Council) for supporting the work on PNML.

References

1. R. Bastide, J. Billington, E. Kindler, F. Kordon, and K. H. Mortensen, editors. *Meeting on XML/SGML based Interchange Formats for Petri Nets*, Århus, Denmark, June 2000. 21st ICATPN.
2. F. Bause, P. Kemper, and P. Kritzinger. Abstract Petri net notation. *Petri Net Newsletter*, 49:9–27, October 1995.
3. G. Berthelot, J. Vautherin, and G. Vidal-Naquet. A syntax for the description of Petri nets. *Petri Net Newsletter*, 29:4–15, April 1988.
4. B. Bos, H. W. Lie, C. Lilley, and I. Jacobs (eds.). Cascading Style Sheets, level 2 – CSS2 Specification. URL http://www.w3.org/TR/CSS2, 1998.
5. J. Clark. TREX – tree regular expressions for XML. URL http://www.thaiopensource.com/trex/. 2001/01/20.
6. J. Clark (ed.). XSL Transformations (XSLT) Version 1.0. URL http://www.w3.org/TR/XSLT/xslt.html, 1999.
7. J. Clark and M. Murata (eds.). RELAX NG specification. URL http://www.oasis-open.org/committees/relax-ng/. 2001/12/03.
8. CPN Tools. URL http://www.daimi.au.dk/CPNtools. 2001/09/11.
9. J. Desel, G. Juhás, R. Lorenz, and C. Neumair. Modelling and validation with VipTool. In *Conference on Business Process Management, Tool Presentation*, 2003.
10. Design/CPN. URL http://www.daimi.au.dk/designCPN/. 2001/09/21.

11. J. Ferraiolo, F. Jun, and D. Jackson (eds.). Scalable Vector Graphics (SVG) 1.1 Specification. URL http://www.w3.org/TR/SVG11/, 2003.
12. ISO/IEC/JTC1/SC7. Subdivision of project 7.19 for a Petri net standard. ISO/IEC/JTC1/SC7 N1441, October 1995.
13. ISO/IEC/JTC1/SC7. Software Engineering - High-Level Petri Nets - Concepts, Definitions and Graphical Notation. ISO/IEC 15909-1, Final Committee Draft, May 2002.
14. ISO/IEC/JTC1/SC7 WG19. New proposal for a standard on Petri net techniques. ISO/IEC/JTC1/SC7 N2658, June 2002.
15. M. Jüngel, E. Kindler, and M. Weber. The Petri Net Markup Language. *Petri Net Newsletter*, 59:24–29, 2000.
16. S. Katz and O. Grumberg. VeriTech: Translating among specifications and verification tools. Technical report, The Technion, Haifa, Israel, March 1999.
17. E. Kindler and M. Weber. A universal module concept for Petri nets. An implementation-oriented approach. Informatik-Berichte 150, Humboldt-Universität zu Berlin, June 2001.
18. A. M. Koelmans. PNIF language definition. Technical report, Computing Science Department, University of Newcastle upon Tyne, UK, July 1995. version 2.2.
19. R. B. Lyngsø and T. Mailund. Textual interchange format for high-level Petri nets. In *Proc. Workshop on Practical use of Coloured Petri Nets and Design/CPN*, pages 47–63, Department of Computer Science, University ofÅrhus, Denmark, 1998. PB-532.
20. T. Mailund and K. H. Mortensen. Separation of style and content with XML in an interchange format for high-level Petri nets. In Bastide et al. [1], pages 7–11.
21. The Model-Checking Kit. URL http://wwwbrauer.in.tum.de/gruppen/theorie/KIT/. 2003/02/18.
22. The PEP Tool. URL http://parsys.informatik.uni-oldenburg.de/~pep/. 2002/07/29.
23. The Petri Net Kernel. URL http://www.informatik.hu-berlin.de/top/pnk/. 2001/11/09.
24. Petri Net Markup Language. URL http://www.informatik.hu-berlin.de/top/pnml/. 2001/07/19.
25. Renew: The Reference Net Workshop. URL http://www.renew.de. 2002/03/04.
26. M. Sperberg-McQueen and H. Thompson (eds.). XML Schema. URL http://www.w3.org/XML/Schema, April 2000. 2002-03-22.
27. C. Stehno. Petri Net Markup Language: Implementation and Application. In J. Desel and M. Weske, editors, *Promise 2002, Lecture Notes in Informatics* P-21, pages 18–30. Gesellschaft für Informatik, 2002.
28. M. Weber and E. Kindler. The Petri Net Kernel. In H. Ehrig, W. Reisig, G. Rozenberg, and H. Weber, editors, *Petri Net Technology for Communication Based Systems*, Lecture Notes in Computer Science 2472. Springer, Berlin Heidelberg, 2002. To appear.
29. G. Wheeler. A textual syntax for describing Petri nets. Foresee design document, Telecom Australia Research Laboratories, 1993. version 2.
30. The XML C library for Gnome. URL http://xmlsoft.org. 2003/01/23.

Author Index

Lecture Notes in Computer Science

For information about Vols. 1–2590

please contact your bookseller or Springer-Verlag

Vol. 2631: R. Falcone, S. Barber, L. Korba, M. Singh (Eds.), Trust, Reputation, and Security: Theories and Practice. Proceedings, 2002. X, 235 pages. 2003. (Subseries LNAI).

Vol. 2632: C.M. Fonseca, P.J. Fleming, E. Zitzler, K. Deb, L. Thiele (Eds.), Evolutionary Multi-Criterion Optimization. Proceedings, 2003. XV, 812 pages. 2003.

Vol. 2633: F. Sebastiani (Ed.), Advances in Information Retrieval. Proceedings, 2003. XIII, 546 pages. 2003.

Vol. 2634: F. Zhao, L. Guibas (Eds.), Information Processing in Sensor Networks. Proceedings, 2003. XII, 692 pages. 2003.

Vol. 2636: E. Alonso, D, Kudenko, D. Kazakov (Eds.), Adaptive Agents and Multi-Agent Systems. XIV, 323 pages. 2003. (Subseries LNAI).

Vol. 2637: K.-Y. Whang, J. Jeon, K. Shim, J. Srivastava (Eds.), Advances in Knowledge Discovery and Data Mining. Proceedings, 2003. XVIII, 610 pages. 2003. (Subseries LNAI).

Vol. 2638: J. Jeuring, S. Peyton Jones (Eds.), Advanced Functional Programming. Proceedings, 2002. VII, 213 pages. 2003.

Vol. 2639: G. Wang, Q. Liu, Y. Yao, A. Skowron (Eds.), Rough Sets, Fuzzy Sets, Data Mining, and Granular Computing. Proceedings, 2003. XVII, 741 pages. 2003. (Subseries LNAI).

Vol. 2641: P.J. Nürnberg (Ed.), Metainformatics. Proceedings, 2002. VIII, 187 pages. 2003.

Vol. 2642: X. Zhou, Y. Zhang, M.E. Orlowska (Eds.), Web Technologies and Applications. Proceedings, 2003. XIII, 608 pages. 2003.

Vol. 2643: M. Fossorier, T. Høholdt, A. Poli (Eds.), Applied Algebra, Algebraic Algorithms and Error-Correcting Codes. Proceedings, 2003. X, 256 pages. 2003.

Vol. 2644: D. Hogrefe, A. Wiles (Eds.), Testing of Communicating Systems. Proceedings, 2003. XII, 311 pages. 2003.

Vol. 2645: M.A. Wimmer (Ed.), Knowledge Management in Electronic Government. Proceedings, 2003. XI, 320 pages. 2003. (Subseries LNAI).

Vol. 2646: H. Geuvers, F, Wiedijk (Eds.), Types for Proofs and Programs. Proceedings, 2002. VIII, 331 pages. 2003.

Vol. 2647: K.Jansen, M. Margraf, M. Mastrolli, J.D.P. Rolim (Eds.), Experimental and Efficient Algorithms. Proceedings, 2003. VIII, 267 pages. 2003.

Vol. 2648: T. Ball, S.K. Rajamani (Eds.), Model Checking Software. Proceedings, 2003. VIII, 241 pages. 2003.

Vol. 2649: B. Westfechtel, A. van der Hoek (Eds.), Software Configuration Management. Proceedings, 2003. VIII, 241 pages. 2003.

Vol. 2651: D. Bert, J.P. Bowen, S. King, M, Waldén (Eds.), ZB 2003: Formal Specification and Development in Z and B. Proceedings, 2003. XIII, 547 pages. 2003.

Vol. 2652: F.J. Perales, A.J.C. Campilho, N. Pérez de la Blanca, A. Sanfeliu (Eds.), Pattern Recognition and Image Analysis. Proceedings, 2003. XIX, 1142 pages. 2003.

Vol. 2653: R. Petreschi, Giuseppe Persiano, R. Silvestri (Eds.), Algorithms and Complexity. Proceedings, 2003. XI, 289 pages. 2003.

Vol. 2656: E. Biham (Ed.), Advances in Cryptology – EUROCRPYT 2003. Proceedings, 2003. XIV, 649 pages. 2003.

Vol. 2657: P.M.A. Sloot, D. Abramson, A.V. Bogdanov, J.J. Dongarra, A.Y. Zomaya, Y.E. Gorbachev (Eds.), Computational Science – ICCS 2003. Proceedings, Part I. 2003. LV, 1095 pages. 2003.

Vol. 2658: P.M.A. Sloot, D. Abramson, A.V. Bogdanov, J.J. Dongarra, A.Y. Zomaya, Y.E. Gorbachev (Eds.), Computational Science – ICCS 2003. Proceedings, Part II. 2003. LV, 1129 pages. 2003.

Vol. 2659: P.M.A. Sloot, D. Abramson, A.V. Bogdanov, J.J. Dongarra, A.Y. Zomaya, Y.E. Gorbachev (Eds.), Computational Science – ICCS 2003. Proceedings, Part III. 2003. LV, 1165 pages. 2003.

Vol. 2660: P.M.A. Sloot, D. Abramson, A.V. Bogdanov, J.J. Dongarra, A.Y. Zomaya, Y.E. Gorbachev (Eds.), Computational Science – ICCS 2003. Proceedings, Part IV. 2003. LVI, 1161 pages. 2003.

Vol. 2663: E. Menasalvas, J. Segovia, P.S. Szczepaniak (Eds.), Advances in Web Intelligence. Proceedings, 2003. XII, 350 pages. 2003. (Subseries LNAI).

Vol. 2665: H. Chen, R. Miranda, D.D. Zeng, C. Demchak, J. Schroeder, T. Madhusudan (Eds.), Intelligence and Security Informatics. Proceedings, 2003. XIV, 392 pages. 2003.

Vol. 2667: V. Kumar, M.L. Gavrilova, C.J.K. Tan, P. L'Ecuyer (Eds.), Computational Science and Its Applications – ICCSA 2003. Proceedings, Part I. 2003. XXXIV, 1060 pages. 2003.

Vol. 2668: V. Kumar, M.L. Gavrilova, C.J.K. Tan, P. L'Ecuyer (Eds.), Computational Science and Its Applications – ICCSA 2003. Proceedings, Part II. 2003. XXXIV, 942 pages. 2003.

Vol. 2669: V. Kumar, M.L. Gavrilova, C.J.K. Tan, P. L'Ecuyer (Eds.), Computational Science and Its Applications – ICCSA 2003. Proceedings, Part III. 2003. XXXIV, 948 pages. 2003.

Vol. 2670: R. Peña, T. Arts (Eds.), Implementation of Functional Languages. Proceedings, 2002. X, 249 pages. 2003.

Vol. 2674: I.E. Magnin, J. Montagnat, P. Clarysse, J. Nenonen, T. Katila (Eds.), Functional Imaging and Modeling of the Heart. Proceedings, 2003. XI, 308 pages. 2003.

Vol. 2675: M. Marchesi, G. Succi (Eds.), Extreme Programming and Agile Processes in Software Engineering. Proceedings, 2003. XV, 464 pages. 2003.

Vol. 2676: R. Baeza-Yates, E. Chávez, M. Crochemore (Eds.), Combinatorial Pattern Matching. Proceedings, 2003. XI, 403 pages. 2003.

Vol. 2679: W. van der Aalst, E. Best (Eds.), Applications and Theory of Petri Nets 2003. Proceedings, 2003. XI, 508 pages. 2003.

Vol. 2686: J. Mira, J.R. Álvarez (Eds.), Computational Methods in Neural Modeling. Proceedings, Part I. 2003. XXVII, 764 pages. 2003.

Vol. 2687: J. Mira, J.R. Álvarez (Eds.), Artificial Neural Nets Problem Solving Methods. Proceedings, Part II. 2003. XXVII, 820 pages. 2003.

Vol. 2692: P. Nixon, S. Terzis (Eds.), Trust Management. Proceedings, 2003. X, 349 pages. 2003.

Vol. 2707: K. Jeffay, I. Stoica, K. Wehrle (Eds.), Quality of Service – IWQoS 2003. Proceedings, 2003. XI, 517 pages. 2003.